Light Scattering in
Solids

Proceedings of the Second Joint USA—USSR Symposium

Light Scattering in Solids

Edited by

Joseph L. Birman and Herman Z. Cummins

Department of Physics
The City College of the City University of New York
New York, New York

and

Karl K. Rebane

Institute of Physics
Academy of Sciences of the Estonian SSR
Tartu, Estonian SSR

PLENUM PRESS · NEW YORK AND LONDON

Library of Congress Cataloging in Publication Data

Joint USA-USSR Symposium on Light Scattering in Condensed Matter,
2d, New York, 1979.
  Light scattering in solids.

  "Proceedings of the second Joint USA-USSR Symposium on Light
Scattering in Condensed Matter, held in New York, New York, May 21—
25, 1979."
  Includes indexes.
  1. Solids—Optical properties—Congresses. 2. Light-Scattering—Congresses.
I. Birman, Joseph Leon, 1927-   II. Cummins, Herman Z., 1933-
III. Rebane, Karl Karlovich, 1926-   IV. Title.
QC176.8.06J64 1979   530.4'1   79-21683
ISBN 978-1-4615-7352-4   ISBN 978-1-4615-7350-0 (eBook)
DOI 10.1007/978-1-4615-7350-0

Proceedings of the Second Joint USA-USSR Symposium on Light Scattering
in Condensed Matter, held in New York, New York, May 21—25, 1979

© 1979 Plenum Press, New York
Softcover reprint of the hardcover 1st edition 1979
A Division of Plenum Publishing Corporation
227 West 17th Street, New York, N. Y. 10011

SECOND USA-USSR SYMPOSIUM ON
LIGHT SCATTERING IN CONDENSED MATTER
May 21-25, New York City

USA Executive Committee

J. L. Birman (Chairman)
H. Z. Cummins
M. Lax

USSR Organizing Committee

A. M. Prokhorov (Scientific Advisor)
K. K. Rebane (Chairman)
S. A. Akhmanov
L. A. Bureyeva (Secretary)

Organizing and Program Committee

J. L. Birman   P. Martin
E. Burstein   J. Ruvalds
R. K. Chang   P. Wolff
P. Fleury   J. M. Worlock
M. Lax   P. Yu
H. Z. Cummins (Chairman)

Sponsorship and Support

National Science Foundation (USA) Division of International Programs
National Academy of Sciences (USA)
Academy of Sciences (USSR) - General Physics and Astronomy Section
    and Commission of Spectroscopy
Science Division of The City College, City University of New York
Graduate Center of The City University of New York

PREFACE

The Second USA-USSR Symposium on Light Scattering in Condensed Matter was held in New York City 21-25 May 1979.

The present volume is the proceedings of that conference, and contains all manuscripts received prior to 1 August 1979, representing scientific contributions presented. A few manuscripts were not received, but for completeness the corresponding abstract is printed. No record was kept of the discussion, so that some of the flavor of the meeting is missing. This is particularly unfortunate in the case of some topics which were in a stage of rapid development and where the papers presented stimulated much discussion - such as the sessions on spatial dispersion and resonance inelastic (Brillouin or Raman) scattering in crystals, enhanced Raman scattering from molecules on metal surfaces, and the onset of turbulence in fluids.

The background and history of the US-USSR Seminar-Symposia on light scattering was given in the preface to the proceedings of the First Symposium held in Moscow May 1975, published as "Theory of Light Scattering in Condensed Matter" ed. B. Bendow, J. L. Birman, V. M. Agranovich (Plenum Press, N. Y. 1976). Strong scientific interest on both sides in continuing this series resulted in a plan for the second symposium to be held in New York in 1977. For a variety of reasons it was necessary to cancel the planned 1977 event, almost at the last minute. Despite this setback, the continued scientific enthusiasm for face-to-face interactions between American and Soviet scientists working in theory and experiment of "light scattering in a general sense" resulted in the second symposium coming to fruition in New York in 1979.

Now that two binational light scattering symposia have occurred, it is possible to take some stock of what has been achieved. The changing content of the two Symposia reflects changing emphasis of the field. Some topics continued to be emphasized such as: spatial dispersion (non-local effects) and phase transitions investigated by light scattering; others were

added such as: enhanced Raman scattering by molecules on metal
surfaces and the onset of turbulence as studied by light scattering; some topics in which interest has waned such as studies
on electron-hole drops do not appear.

Judging from reactions of participants, the exchange of
new results, ideas and points of view was most worthwhile.
The expansion of the circle of participants on both sides should
be noted. The composition of participants (about half theorists
and half experimentalists) reflects the ongoing vitality of this
field as an active branch of contemporary Condensed Matter
Physics.

Careful scrutiny of the topical contents of the two symposia
emphasizes and illustrates the meaning of the term "light scattering in a general sense". In actual fact a more apt description of the subject area encompassed by these symposia is
                "Optics of Matter - Light Scattering"
since the subjects continue to be: investigation of the physics
and processes of radiation-matter interactions, the use of
light as a weakly coupled probe of dynamical processes in matter,
and the regimes of strong light-matter coupling such as a polariton and non-linear optical effects. A basis exists for continued and strong interaction between American and Soviet Scientists working in these fields, some of which has already been
implemented as a result of the symposia in the form of joint
research projects, long-term visits to laboratories and the like.

The practical implementation of the Second Seminar-Symposium
was the result of the work and support of many individuals and
organizations. Essential financial support was provided by the
National Science Foundation - Division of International Programs;
this and other support is gratefully acknowledged. Dean Harry
Lustig, Science Division, City College, and President Harold
Proshansky, Graduate Center, City University graciously extended
scarce resources to assist the Symposium.

In this connection it is worthwhile to record the essence
of some remarks made by (then) President Robert E. Marshak of
City College, City University of New York, at the Symposium
dinner calling attention to the fact that these Binational
Symposia on Light Scattering have been outside the official
"umbrella" of US-USSR science exchanges. In this way additional
valuable channels of binational scientific cooperation have come
into being.

It is a very pleasant duty to record thanks to the following persons who helped in various ways during the Symposium and
also in bringing the Proceedings to fruition: Dr. L. Bureyeva,
Mr. M. Belic, Dr. T. K. Lee, Dr. T. Odagaki, Dr. H. R. Trebin,

Dr. D. N. Pattanayak, Dr. W. Yao, Mrs. F. Tritt, Mrs. E. de Crescenzo, Mrs. N. Odagaki. The American co-editors are grateful to their Soviet co-editor, Professor K. K. Rebane for his continued help and assistance.

New York, 24 September 1979   Joseph L. Birman
                                                Herman Z. Cummins

CONTENTS

SECTION III
NON–LOCAL AND TRANSIENT PHENOMENA
(CHAIRMEN: S. A. AKHMANOV AND M. CARDONA)

SECTION VII
MULTI-PHOTON SPECTROSCOPY
(CHAIRMEN: A. S. BOROVIK-ROMANOV AND J. M. FRIEDMAN)

SECTION VIII
RESONANCE SCATTERING AND SURFACE ENHANCED RAMAN SCATTERING
(CHAIRMEN: N. BLOEMBERGEN AND P. A. APANASEVICH)

OPENING REMARKS

JOSEPH L. BIRMAN
City College of the City University of New York

It is a high privilege and pleasure for me to open the Second Binational Light Scattering Symposium. This 1979 Symposium in New York City follows the First Symposium in Moscow in 1975.

On behalf of the Organizers I express our greetings and welcome to all participants. We welcome our scientific colleagues from the Soviet Union and from all over the United States. We will do our best to make your stay at the Symposium and in New York as fruitful and pleasant as possible.

The subject of our Symposium is Light Scattering. We interpret this as the fundamental physics of the interaction of matter —— especially condensed matter —— and light. We are concerned with Light "Scattering" in a "wider" or "general sense". We include the investigation of composite excitations like polaritons, excitons, and magnons, as well as linear optics, non-linear optics, local optics, and non-local optics. A glance at the program shows the diversity of topics included, such as: physics of fluids and turbulence, physical processes in atomic systems, in crystals, in liquid crystals, and in lower dimensional systems such as surfaces and two dimensional crystals. The unifying threads are the basic physical process of radiation-matter interaction, and the dynamics of scattering in these systems.

The Organizing Committee has worked hard to make an exciting and high level program —— from the first session on turbulence to the last one on "giant" or "enhanced" scattering from molecules on metal surfaces. Our Soviet colleagues also have an interesting program of laboratory and post-Symposium visits. Altogether a very exciting prospect for the next week or two.

Looking at the participants gathered here we may note the presence of practicing theorists and experimenters — or "experimenters well versed in theory". We note also some familiar American and Soviet scientists whom we already know from the First Symposium as well as others whom we hope to know during the days ahead.

Some of the several purposes of these Symposia were:
    To strengthen and deepen American-Soviet scientific
        cooperation in Light Scattering;
    To bring together experimenters and theorists, and
    To continue to widen the circle of participants.
To some extent we have succeeded in these goals.

While we are so happy to greet those of you who have come,
we must also remark on those absent. The absence of Professor
Rem Khokhlov who died following a mountain climbing accident is
particularly sad for us. He is greatly missed.

It is no great secret — as we note from the Band-Aid on the
official folder — that originally we planned the Second Symposium
in New York in 1977 — two years ago. But it was necessary to
cancel that plan because of various impediments and also lack of
vital communication between our two organizing committees. The
coming to pass of this Symposium today is a result of improvements
— both the removal of some impediments, and some better communication. Let us hope that this Symposium, and the Post-Symposium
program will contribute to the continued removal of impediments,
and the continued improvement in communications, — or as has been
said: to the "free flow of scientists and scientific information".
In this framework we look forward to further Symposia. We look
forward to establishing cooperative research activities, and working
groups, on this subject of Light Scattering, which is of great
interest to scientists in America and the Soviet Union.

In a few moments we turn to the scientific program. Those of
us fortunate enough to have participated in the First Symposium in
Moscow remember many "hot" discussions there and look forward to
hot discussions this week and next. For example, I myself recall
hot discussions in Moscow with Professor Vitaly Ginzburg and I
regret his inability to be present today. We hope indeed that
there will be future Symposia in the United States and that
Professor Ginzburg, and others will be present for them.

Now it is my happy duty to introduce Professor Karl K. Rebane,
President of the Academy of Sciences of the Estonian S. S. R.,
Tallin, and Corresponding Member of the Soviet Academy of Sciences
— who will make some remarks.

KARL K. REBANE
Academy of Sciences, Estonian S. S. R.

On behalf of the Soviet participants, I am pleased to say that
we are really happy to be in the USA; to be in the great city of
New York and to take part in our Second Joint Seminar-Symposium on
the Theory of Light Scattering in Condensed Matter.

We arrived yesterday after a long flight in the late hours of the night by Moscow time. We were warmly met at J. F. Kennedy airport by Professor Birman and Professor Lax. Everything in the program for the beginning of our stay here was carefully and well prepared by the Organizing Committee, and its Chairman, Professor Cummins. Everything worked smoothly. All of that provided us with so much new energy of high quality, that already we do not feel the time difference any more and are eager to start with the work of the Seminar-Symposium.

Among our delegation are people who have already visited the United States and New York City before — some of us several times. For a considerable part of our group — it is the first visit to your country. But for all of us, to be here again, or for the first time, is very useful as physicists, or simply as people living in our contemporary world.

One more point. As we know very well — it is much easier and pleasant to be guests than to be hosts of scientific meetings.

We are grateful to our hosts here, at the Second Seminar-Symposium. We wish our hosts every success in their work; and we are, naturally, ready to be as cooperative as possible with the matters of the Seminar-Symposium and our stay here!

Thank you very much.

JOSEPH L. BIRMAN

It is also no secret that among the participants of this Symposium are scientists of the City University of New York. Today, tomorrow and Wednesday we meet in the Graduate Center of City University — which has some analogy to one building of Lomonsov Moscow State University where some of our sessions were held in 1975. On Thursday and Friday we meet in the building of the New York Academy of Sciences — analogous let us say to the "Scientists Palace" in Moscow the site of some other 1975 sessions. The New York Academy of Sciences building was formerly a private home — which was given to the New York Academy.

I should now like to introduce some academic officials and colleagues of City University;

President Harold Proshansky, President of the Graduate Center of City University who is our host here at the City University Graduate Center.

I note that President Robert Marshak of my own City College will be present at the Symposium Banquet, Wednesday evening and I will introduce him then.

I have great pleasure and honor to introduce Dr. Robert Kibbee, the Chancellor of the City University of New York, the highest administrative official of the University.

ROBERT J. KIBBEE
The City University of New York

Distinguished Colleagues and Visitors:

It is a great honor and privilege for the City University to serve as host to the second US–USSR Light Scattering Symposium; and a particular pleasure for me to bring greetings on behalf of the University Community to our distinguished guests from the Soviet Union and from throughout the United States. Several summers ago I had the pleasure of visiting the Soviet Union in connection with the meetings of the International Association of Universities held at the University of Moscow. I hope that in the next few days my University can provide our guests from the Soviet Union the same kind of generous hospitality I received at Moscow.

Light Scattering is, of course, a technical term in the language of physics that encompasses various phenomena which will be discussed during your meetings. Yet this technical term is comprised of two common English words which, if reversed, speak to the very purpose of a University and to the heart of scholarly activity. The Scattering of Light about the nature and purpose of our universe, about the development and meaning of different cultures, about the human condition, and about how man reacts and adapts to his environment is what absorbs the energy of scholars and artists.

What absorbs us most as nations and as people is the elusive but cherished goal of international peace — a condition achieved when nations understand each other better, trust each other more and can work together for a common good. Meetings such as this contribute positively to this end and reflect the improved environment between our two great nations. For this reason it is of great significance to us to be your host.

Your presence further honors the University because of the value we place on the scientific endeavors in which you are engaged. The very fact that this Symposium is occurring under our aegis gives testimony to the commitment of the University, and especially of the City College, to the field of physics.

The president of City College, who sadly soon will be leaving us — Robert Marshak — is himself a world renowned Particle Physicist and he is also a moving force in the opening up of scientific exchanges between the United States and the Soviet Union, of both of which facts I am sure you are all well aware. Indeed, it is likely that we would all not be here this morning if it were not for the tireless efforts of President Marshak in the building of a superb science faculty at City College and in pressing for improved relations between the scientific communities of our two countries.

I would also like to take this opportunity to extend our gratitude to the two professors most responsible for bringing this Symposium to the University: Chairman of your Executive Committee, Professor Joseph L. Birman, and Chairman of your Organizing and Program Committee, Herman Z. Cummins.

You have come from great distances to meet with each other and share ideas and research. I do not wish to take any more of your time.

Let me again say welcome on behalf of the City University of New York, — it is a great privilege to have you amongst us.

JOSEPH L. BIRMAN

I now <u>end</u> the introductory Opening Session.

I open the scientific meeting. I ask the two co-chairmen to come forward. Professor Lev Pitaevskii of the Institute for Physical Problems; Professor Paul Martin of Harvard University.

21 May 1979

TO SCALE OR NOT TO SCALE? -- THE PUZZLE AT THE LAMBDA POINT OF

LIQUID ^{4}He

Richard A. Ferrell and Jayanta K. Bhattacharjee

Institute for Physical Science and Technology
        and Department of Physics and Astronomy
University of Maryland
College Park, Maryland 20782

ABSTRACT

    Light scattering measures both static and dynamic properties of
liquid ^{4}He. Recent work reconciles the dynamic scaling theory with
apparent experimental discrepancies. Some questions remain regarding
the statics.

I. INTRODUCTION

    The light scattering properties of liquid ^{4}He in the vicinity of
the λ-point are quite unique. The Landau-Placzek[1] central peak in the
spectrum of the scattered light is split upon entering the superfluid
He II phase into the second sound doublet. The detailed temperature
dependence of how this splitting sets in is the subject of the dynamic scaling theory.[2,3] Until recently it has been suggested[4] that
there were serious discrepancies between the predictions of dynamic
scaling and the experimental data. In this paper we want to review
how these discrepancies can be accounted for by certain natural and
necessary extensions of the theory.[5,6,7]

    Before beginning our discussion of the dynamics in Section III,
we first take up in Section II some important questions regarding the
statics. Section IV consists of a brief summary.

II.   STATICS

In this section we consider the total intensity of light scattering, which is related to the static, or equal-time correlation function. The determination of the spectrum of the fluctuations and how the total scattered intensity is distributed in frequency will be taken up in the next section. The total intensity is proportional to β_T, the isothermal compressibility. This has a critical behavior similar to that of c_p, the constant pressure specific heat. Because the critical exponent α is known to be very small, the critical temperature dependence of c_p is described quite accurately by

$$c_P = A_o \, \nu \, \ln(t_o/t), \tag{2.1}$$

where $t = (T - T_\lambda)/T_\lambda$ is the reduced temperature. t_o and A_o are constants. ν is the correlation length critical exponent. For the time being we limit our discussion to the He I phase $(T > T_\lambda)$. With the inverse correlation length given by

$$\kappa = \kappa_o \, t^\nu, \tag{2.2}$$

we can eliminate the temperature from Eq. (2.1) to obtain

$$c_P = -A_o \, \ln \kappa + \text{const.} \tag{2.3}$$

When $T \to T_\lambda$, Eq. (2.3) can be expected to fail. This will happen when κ becomes comparable to k, the scattering wave number, because thermodynamics is not valid at distances smaller than the correlation length, κ^{-1}. . Thus the total scattering intensity will not increase indefinitely as $\kappa \to 0$, as might be expected from Eq. (2.3). Instead it will assume some finite limiting value corresponding to an effective κ, κ_{eff}. From static scaling, κ_{eff} must be proportional to k, so that

$$\kappa_{eff} = k/\ell, \tag{2.4}$$

where ℓ is a dimensionless constant. It is convenient to rewrite Eq. (2.3) in the form

$$c_P = A_o \, C_{TH}(y) - A_o \, \ln k + \text{const.}, \tag{2.5}$$

where $y = \kappa/k$ and

$$C_{TH}(y) = -\ln y \tag{2.6}$$

is the thermodynamic limit (i.e., the $y \to \infty$ form) of the scaling function for the energy-energy correlation. The true scaling function $C(y)$ will have a finite $y = 0$ limit, which is related to κ_{eff} by

$$C_{TH}(\kappa_{eff}/k) = C(0). \tag{2.7}$$

Substitution of Eqs. (2.4) and (2.6) yields $\ln \ell = C(0)$ or

$$\ell = e^{C(0)} \tag{2.8}$$

The calculation of $C(y)$ for three-dimensional space remains an unsolved problem. The $D = 4$ form for $C(y)$ is, however, suggestive of the way in which $C(y)$ can be expected to behave for $D = 3$. The $D = 4$ limit is represented by a noninteracting polarization diagram, which yields[8]

$$C(y) = 1 - \ln y - (4y^2 + 1)^{\frac{1}{2}} \ln\left[\frac{1}{2y} + \left(1 + \frac{1}{4y^2}\right)^{\frac{1}{2}}\right] \tag{2.9}$$

Note that for $y \to \infty$ $C(y) \approx C_{TH}(y)$, as required, while for $y = 0$ we find $C(0) = 1$, giving $\ell = e = 2.718$. $C(y)$ is plotted as the upper of the two heavy curves in Fig. 1. In the language of spectral theory, as discussed further below, ℓ is an effective cutoff. This suggests the approximation

$$C(y) \approx -\frac{1}{2} \ln\left[y^2 + \ell^{-2}\right], \tag{2.10}$$

which is shown by the dotted curve in Fig. 1. It is evident that the deviation produced by replacing the true smooth threshold by a sharp

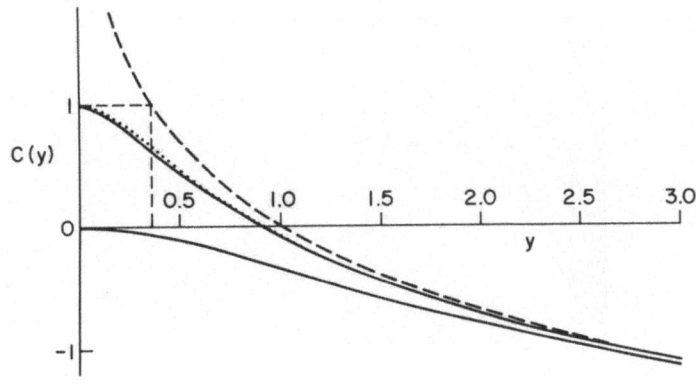

Fig. 1. Nonlocal specific heat scaling function for the four-dimensional model. $y = \kappa/k$ is the ratio of the inverse correlation length to the wave number. The intersection of the vertical dashed line with the horizontal axis determines the threshold parameter ℓ by $\kappa_{eff}/k = \ell^{-1} = e^{-1} = 0.37$.

threshold is not very great. For the order parameter correlation
function this approximation corresponds to the Fisher-Burford[9]
approximant. The dashed curve represents $C_{TH}(y)$. The construction
of horizontal and vertical dashed lines intersecting this curve
illustrates the determination of the threshold by $\kappa_{eff}/k = \ell^{-1} = e^{-1}$,
as described above in Eq. (2.4). The lower solid line shows for comparison Eq. (2.10) for $\ell = 1$, a much lower threshold, and obviously
too crude an approximation.

For D = 3 the single loop polarization diagram[10] gives $\ell = \pi$.
The screening approximation[11,12] (n^{-1} expansion[13,14]) takes interaction into account and again yields $\ell = \pi$. But the associated specific
heat exponent is $\alpha = -1$, making the calculation unsuitable for liquid
^{4}He where $\alpha \simeq 0$. We therefore adopt a phenomenological approach,
making use of the spectral function.[15] The application of this method
to the order parameter correlation function has been discussed at the
previous meeting of this symposium.[16] From physical considerations,
we require that the spectral function F vanish below the two-particle
threshold at $|k| = 2\kappa$. This is shown in Fig. 2 where F(u) for D = 4
is plotted vs. $u = |k|/2\kappa$ as the dot-dash curve. The vertical dot-
dash line shows the equivalent Fisher-Burford[9] type sharp cutoff at
$u = \ell/2 = e/2 = 1.36$. The phenomenological spectral function[15] based
on the Fisher-Langer terms, and linearized[17] in α, is

$$F(u) = 1 - u^{-3/2}(1 + B \ln u) \tag{2.11}$$

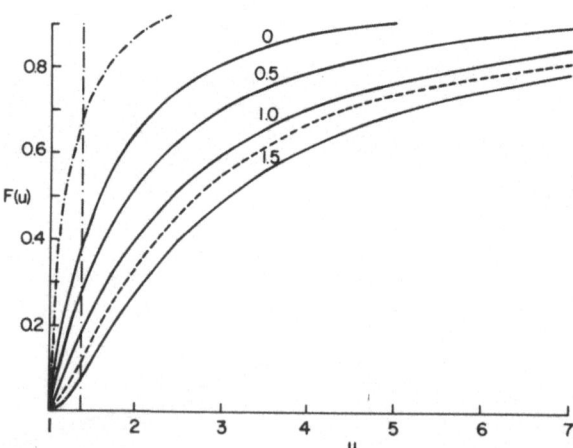

Fig. 2. Nonlocal specific heat spectral function vs. $u = |k|/2\kappa$
where k is the wave number and κ^{-1} the correlation length.
The dot-dash curve shows the D = 4 spectral function while
the other curves illustrate the D = 3 phenomenological
forms.

and is shown by the solid curves in Fig. 2 for B = 0, 0.5, 1.0, and 1.5. The dashed curve is Bray's[18] best estimate of F.

By means of Cauchy's theorem the scaling function for arbitrary y can be written as an integral over F. But we are primarily interested here in the y = 0 value, for which the integral takes on the simple form

$$C(0) = \ln 2 + \int_1^\infty \frac{du}{u} (1 - F) \qquad (2.12)$$

Substitution into Eq. (2.8) then yields

$$\ell = 2 \exp \int_1^\infty \frac{du}{u} (1 - F). \qquad (2.13)$$

$F(u) = (1 - u^{-2})^{\frac{1}{2}}$ for D = 4 gives $\ell = e$, in agreement with the previous result. The substitution of the phenomenological D = 3 form, Eq. (2.11), leads, however, to the much higher values

$$\ell = 2 \, e^{\frac{2}{3}(1 + \frac{2}{3} B)} \qquad (2.14)$$

ranging over the interval $3.9 \leqslant \ell \leqslant 7.6$. It is important to note that this phenomenological approach, which has also been studied by Kroll,[19] yields a value for ℓ roughly twice as big as an ε-expansion based on D = 4. Nicoll[20] has found that the ε-expansion, with suitable refinements and evaluated for ε = 1 (i.e., D = 3) does not give an ℓ significantly larger than for D = 4. In the absence of a rigorous calculation for the D = 3 model it is not possible to decide on purely theoretical grounds which approach is more reliable. We therefore turn now to the experimental situation for an indication.

Unfortunately the total scattering intensity was not measured in the most recent experimental study of light scattering in liquid ^{4}He, that of Tarvin, Vidal, and Greytak.[21] Equivalent information is contained, however, in their measurements of $\gamma - 1$, the ratio of the strengths of the low frequency and high frequency (i.e., Brillouin) parts of the spectrum. $\gamma = c_P/c_V$ where c_V is the constant volume specific heat, related to c_P by

$$c_V = c_P - T \, \alpha_T^2/\beta_T, \qquad (2.15)$$

where α_T is the thermal expansion coefficient. Equation (2.15) enables us to eliminate c_V from γ and reexpress it in the form

$$\gamma - 1 = \frac{\alpha_T^2}{c_V \, \beta_T/T} = \frac{\alpha_T^2}{c_P \, \beta_T/T - \alpha_T^2} . \qquad (2.16)$$

It is convenient to rewrite Eq. (2.1) in the abbreviated form

$$c_P = A_o L \tag{2.17}$$

where

$$L = \nu \ln(t_o/t) \tag{2.18}$$

is equal to $C_{TH}(y)$, up to a constant. The other thermodynamic functions appearing in Eq. (2.16) have similar logarithmic critical behavior. Allowing for differences in the additive constants, we have

$$\alpha_T = -A(L + B) \tag{2.19}$$

and

$$\beta_T/T = A'(L + B'), \tag{2.20}$$

with the constants B and B' to be determined empirically. The multiplicative coefficients are, however, constrained by

$$A_o A' = A^2 \tag{2.21}$$

by virtue of Eq. (2.15). Because of Eq. (2.21) the L^2 terms cancel from the denominator of Eq. (2.16). A small amount of algebra yields

$$\gamma - 1 = \frac{B''}{B^2} \left[L + \frac{(B + B'')^2}{L - B''} + 2B + B'' \right], \tag{2.22}$$

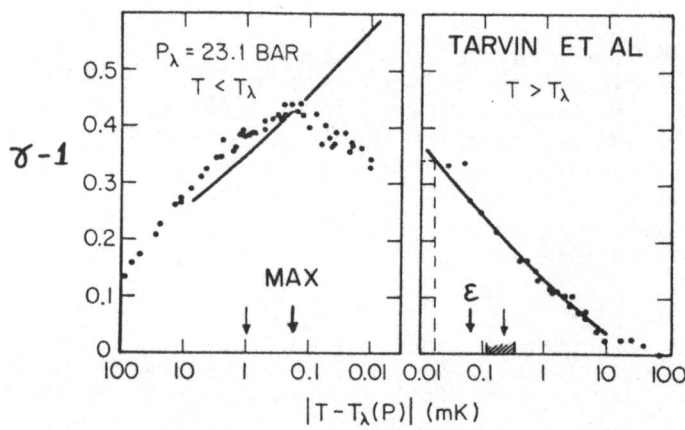

Fig. 3.   Temperature dependence of the nonlocal specific heat ratio, $\gamma - 1$. Data of Tarvin, Vidal, and Greytak at 23.1 atmospheres and $k = 1.79 \times 10^5$ cm^{-1}. The local thermodynamic limit is shown by the solid curves.

where

$$B" = B^2/(B' - 2B) \qquad (2.23)$$

The thermodynamic limit of $\gamma - 1$ has been plotted by Tarvin et al.[21] and is shown as the solid curves in Fig. 3. The second term in Eq. (2.22), proportional to $(L - B")^{-1}$, contributes the curvature to the semi-log plot. As $T \to T_\lambda$, nonlocality enters and $\gamma - 1$ approaches the $y = \kappa/k = 0$ limiting value of 0.34. The construction of horizontal and vertical dashed lines, as in Fig. 1, indicates that at the temperature $T = T_\lambda + \Delta T$, where $\Delta T = 0.0172$ mK, $\kappa = \kappa_{eff} = k/\ell$. The value of ΔT at which κ is equal to k is therefore larger by the factor $\ell^{1/\mathcal{N}} = \ell^{3/2}$. With ℓ in the phenomenological range $3.9 \leqslant \ell \leqslant 7.6$, this yields the temperature range shown by the shading along the horizontal axis. The criterion of Hohenberg et al.[22] for this temperature is indicated by the downwards pointing arrow and falls within the phenomenological range found here. On the other hand, the ε-expansion results of Nicoll give a small value of ΔT, as shown by the downwards pointing arrow with the label "ε." It will clearly be of great value to have detailed measurements of the total intensity itself, so as to be able to avoid the indirect route via the specific heat ratio.

We close this section by alluding briefly to the left-hand portion of Fig. 3, where the theory becomes more complicated than above the λ-point. Stephen[23] has employed an Ornstein-Zernike approximation for discussing the maximum in $\gamma - 1$. Bray[8] has argued on general grounds that a maximum must occur for $T < T_\lambda$. A full theory of this effect re-remains to be worked out. We note, however, purely empirically, that the maximum gives a natural indication of the temperature at which the inverse correlation length matches the scattering wave number. This is illustrated by the downwards pointing arrow labeled "MAX." It gains additional support from the temperature dependence of the dynamics, as discussed in the next section. The Hohenberg et al.[22] criterion, shown by the leftmost downwards pointing arrow, seems to be much too far below the λ-point. The experimental data indicate a much narrower critical region.

III. DYNAMICS

In a recent review of critical dynamics[4] it was stated, "The presently available experimental data...do not show any temperature dependence... , in striking contrast to the dynamic scaling predictions... Thus...the overall situation remains quite unclear." In this section we wish to review recent developments[6,7,25] which we believe have clarified the picture referred to in this quotation. We attempt this with a minimum of theoretical equipment, while emphasizing the basic physical ideas. In contrast to the statics, where there is no obvious trouble with scaling, the situation in regard to dynamics is indeed, at first sight, quite perplexing. The clue to understanding the situation is the presence of strong background contributions to the dynamics. These are quite important in the experimental range, and tend to mask the true critical behavior.

The critical portion of the thermal conductivity above the λ-point according to simple kinetic theory, is composed of a sum of contributions from particles of various momenta p according to

$$\lambda_c \propto \int dp \; \tau_p \;\; = \;\; \int \frac{dp}{\gamma_p} \propto \int \frac{dp}{p^2 D_\psi} \tag{3.1}$$

τ_p is the lifetime of a particle of momentum p. The decay rate is $\tau_p^{-1} = \gamma_p = 2p^2 D_\psi$, which we have factored into the thermodynamic force (proportional to p , the reciprocal of the order parameter correlation function, or susceptibility) times the kinetic, or Onsager coefficient D_ψ. Far from the λ-point the Onsager coefficient loses its critical variation and assumes the background value B_ψ. In this "van Hove," or precritical region, the thermal conductivity acquires the critical temperature dependence

$$\lambda_c \propto \frac{1}{B_\psi} \int \frac{dp}{p^2} \propto B_\psi^{-1} \; \kappa^{-1}, \tag{3.2}$$

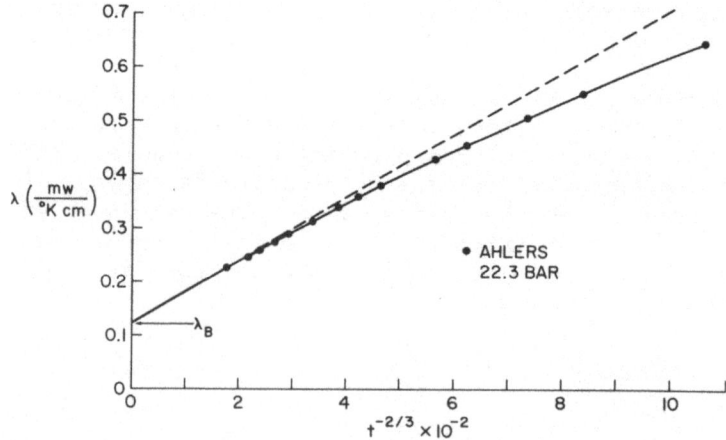

Fig. 4. Thermal conductivity vs. correlation length. The intercept and slope of the dashed line determine the background values of the thermal and particle Onsager coefficients, respectively.

with the critical exponent ν. (The integral has a lower cutoff proportional to κ.) The data of Ahlers[25] reveals this behavior, as
shown in Fig. 4 of λ vs. $t^{-2/3}$. The intercept is the background
thermal conductivity (originating from noncritical microscopic processes) of $\lambda_B = 0.13$ mw/$^\circ$K cm. The slope of the dashed line determines $B_\psi = 1.3 \times 10^{-4}$ cm^2/sec, close to the characteristic diffusion
constant for the problem $\hbar/m = 1.6 \times 10^{-4}$ cm^2/sec, where m is the
Helium atom mass and $2\pi\hbar$ is Planck's constant. This value results
from taking the particle mean free path equal to the De Broglie wavelength.

As D_ψ begins to rise, λ_c drops below Eq. (3.2), as shown by the
curvature of the plot of the Ahlers data in Fig. 4. Finally, close to
the λ-point, when the rise is enough bigger than the background that
the latter can be neglected, we enter the scaling region where both
Onsager coefficients scale with the same critical exponent, according
to $D_\psi \propto p^{-\frac{1}{2}}$ and $\lambda \propto \kappa^{-\frac{1}{2}}$. Dividing the latter by c_p and going to the
extreme nonlocal limit (i.e., $\kappa \to 0$) gives the wave number dependent
thermal diffusion coefficient plotted in Fig. 5 as the upper solid
curve. The lower solid curve shows D_ψ as a function of k. The dashed
curves show the scaling solution, with D_ψ one order of magnitude
smaller than the thermal diffusion coefficient. This small ratio,
found first by De Dominicis and Peliti[5] from an ε-expansion to two-

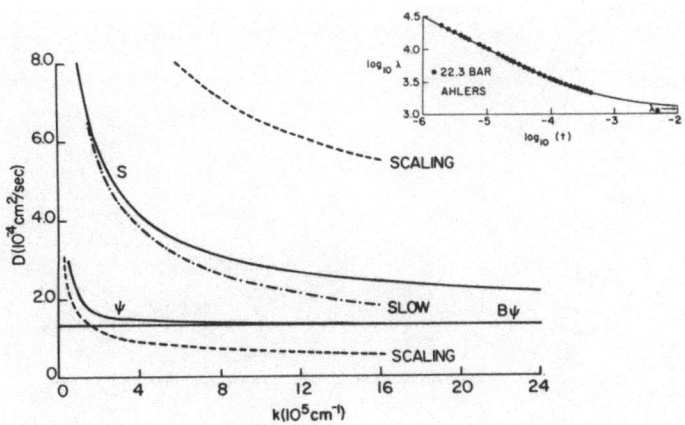

Fig. 5. Nonlocal thermal diffusion (S) and particle relaxation (ψ)
vs. wave number. The dashed curves show the scaling solution while the solid curves include background and transient
effects. The inset shows the good fit to Ahlers' data.

loop order, and subsequently confirmed in various ways by Dohm and Ferrell[26] and by the present authors,[27] corresponds in the conceptual framework of this paper to the well-known kinetic effect of velocity persistence.

The intermediate region between the precritical and the scaling regions is described by the transient solutions to the coupled integral equations. The slow transient enters the thermal diffusion with negative amplitude and causes the effective critical exponent to exceed the dynamic scaling value. The effect of the slow transient alone is illustrated by the middle (dot-dash) curve in Fig. 5 labeled "SLOW." The inset in the upper right-hand corner of Fig. 5 shows the excellent fit to Ahlers' data attained by adding the transient solutions to the scaling solutions.

Now, to discuss the light scattering data it is necessary to have a frequency-dependent thermal conductivity. Fortunately, fairly general theoretical ideas suffice to make possible a clear prediction of the frequency dependence. It is necessary simply to add a frequency term to γ_p in the denominator of Eq. (3.1), with the consequence that a lower momentum cutoff to the integral occurs either because of κ or because of the frequency, depending upon which is relatively larger. λ_c is therefore reduced in either of two ways:

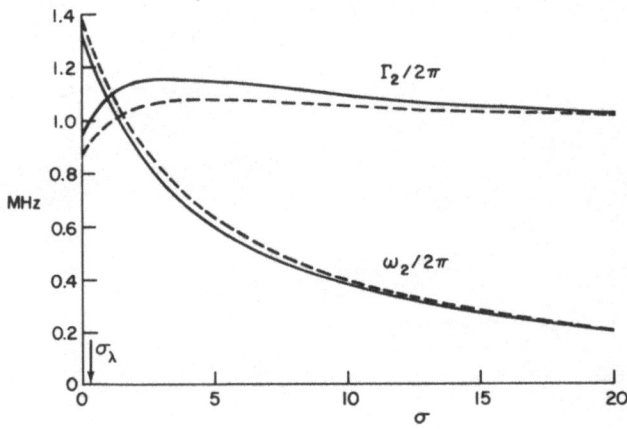

Fig. 6.   Predicted temperature dependence of the two-Lorentzian parameters. σ is proportional to $T - T_\lambda$. The calculation is carried out using two different approximations (solid and dashed curves).

1) by a temperature increase, or 2) by a finite frequency. In either
case we can describe the behavior of the thermal conductivity function by the Ahlers curve, with the variable for the horizontal axis
defined in such a way to include the effects of both temperature and
frequency. With this simplification of the theory, it is easy to
compute the fluctuation spectrum as a function of frequency for any
$T > T_\lambda$. Furthermore the theory can be so manipulated[25] as to provide
the two-Lorentzian form that was used for fitting the experimental
data.[21] The result of the calculation is that the spectra are
Lorentzian far from the λ-point, where the background is constant in
temperature and independent of frequency. As the λ-point is approached, frequency dependence sets in. The distorted spectrum is no
longer pure Lorentzian. It develops a deficiency at low frequencies
which can be fit by a two-Lorentzian expression for the spectrum.
The computed width and splitting parameters, Γ_2 and ω_2, respectively,
are shown in Fig. 6. It will be noted that, as the temperature is
lowered, Γ_2 at first increases, but then ω_2 also starts to increase.
The resulting splitting takes over and broadens the spectrum by separating the two Lorentzians, with the limiting value at the λ-point of
$\omega_2/2\pi = 1.3$ MHz. The width of the separated Lorentzians stays roughly temperature independent, at $\Gamma_2/2\pi \approx 1$ MHz. The experimental measurements of Γ_2 are shown by the data points in the right-hand portion of Fig. 7. The observed absence of temperature dependence is

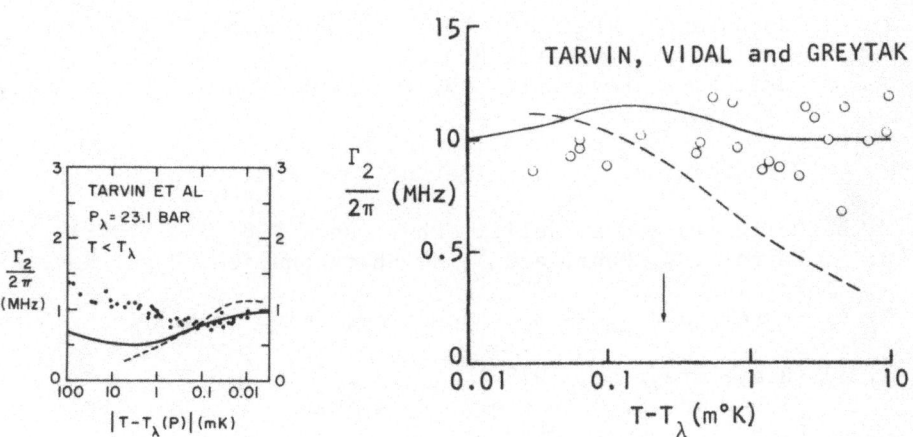

Fig. 7. Comparison of theory (solid line) and experiment (data
         points) for the width parameter, above and below the λ-
         point. The dashed curves show the earlier theoretical pre-
         diction neglecting background.

reasonably well accounted for by the rather flat behavior of the theoretical curve. The dashed line shows the previous calculation of Hohenberg, Siggia, and Halperin,[28] which did not include the background effects, and which consequently drops off away from the λ-point. The fall-off of Γ_2 for $T < T_\lambda$ is also well accounted for by the theory, as shown in the left-hand portion of Fig. 7, except far from the λ-point, where the predicted background strength seems too small by a factor of two. Again, the dashed curve shows the previous calculation[28] of Γ_2, which falls off too rapidly.

SUMMARY

The total intensity of scattered light and its frequency spectrum near the λ-point of liquid ^{4}He have been discussed. It has been described how the observed temperature dependence of the spectral parameters can be reconciled with the dynamic scaling predictions by including the effects of the noncritical background. The corrections to scaling caused by these noncritical terms also give a good account of the variation of the thermal conductivity with temperature. In discussing the statics, we have seen how further measurements of the total intensity can be expected to provide needed information concerning the scale of the correlation length.

REFERENCES

1. L. D. Landau and G. Placzek, Phys. Z. Sowjun. 5 172 (1934).
2. R. A. Ferrell, N. Ményhard, H. Schmidt, F. Schwabl, and P. Szépfalusy, Phys. Rev. Lett. 18, 891 (1967); Ann. Phys. (N.Y.) 47, 565 (1968).
3. P. C. Hohenberg and B. I. Halperin, Phys. Rev. 177, 952 (1969).
4. P. C. Hohenberg and B. I. Halperin, Rev. Mod. Phys. 49, 435 (1977).
5. C. De Dominicis and L. Peliti, Phys. Rev. B18, 353 (1978).
6. R. A. Ferrell, V. Dohm, and J. K. Bhattacharjee, Phys. Rev. Lett. 41, 1818 (1978).
7. R. A. Ferrell and J. K. Bhattacharjee, Univ. of Md. Tech. Report #79-075.
8. A. J. Bray, preprint, 1978.
9. M. E. Fisher and R. J. Burford, Phys. Rev. 156, 583 (1967).
10. R. A. Ferrell, Journal de Physique 32, 85 (1971).
11. R. A. Ferrell and D. J. Scalapino, Phys. Rev. Lett. 29, 413 (1972).
12. R. A. Ferrell and D. J. Scalapino, Phys. Lett. 41A, 371 (1972).
13. S. K. Ma, Phys. Rev. Lett. 29, 1311 (1972) and Phys. Rev. A7, 2172 (1973).
14. R. Abe, Prog. Theor. Phys. 48, 1414 (1972).

15. R. A. Ferrell and D. J. Scalapino, Phys. Rev. Lett. $\underline{34}$, 200 (1975).

16. R. A. Ferrell, "Theory of Light Scattering in Condensed Matter," ed., B. Bendow, J. L. Berman, and V. M. Agranovich, pp. 509-516 Plenum Press, New York (1976).

17. R. A. Ferrell and J. K. Bhattacharjee, Univ. of Md. Tech. Report #79-083.

18. A. J. Bray, Phys. Rev. Lett. $\underline{36}$, 285 (1976) and Phys. Rev. $\underline{B14}$, 1248 (1976).

19. D. M. Kroll, Zeit. Phys. $\underline{B31}$, 309 (1978).

20. J. F. Nicoll, Univ. of Md. IPST Tech. Report #BN-903.

21. J. A. Tarvin, F. Vidal, and T. J. Greytak, Phys. Rev. $\underline{B15}$, 4193 (1977).

22. P. C. Hohenberg, A. Aharony, B. I. Halperin, and E. D. Siggia, Phys. Rev. $\underline{B13}$, 2986 (1976).

23. M. J. Stephen in "The Physics of Liquid and Solid Helium," ed. by K. H. Benneman and J. B. Ketterson, Wiley, New York (1976), Vol. I., Chap. IV.

24. G. Ahlers in "The Physics of Liquid and Solid Helium," ed. by K. H. Benneman and J. B. Ketterson, Wiley, New York (1976) Vol. 1, Chap. II.

25. R. A. Ferrell and J. K. Bhattacharjee, Univ. of Md. Tech. Report #79-099.

26. V. Dohm and R. A. Ferrell, Phys. Lett. $\underline{A67}$, 387 (1978).

27. R. A. Ferrell and J. K. Bhattacharjee, J. of Low Temp. Phys. $\underline{36}$ (1979).

28. P. C. Hohenberg, E. D. Siggia, and B. I. Halperin, Phys. Rev. $\underline{B14}$, 2865 (1976).

TRANSITION TO TURBULENCE

IN COUETTE-TAYLOR FLOW*

Harry L. Swinney

Department of Physics
The University of Texas
Austin, Texas   78712

INTRODUCTION

The Couette-Taylor problem is described in this section and
our laser Doppler velocimetry studies of transitions in Couette-
Taylor flow are presented in the next section, which also includes
a summary of three other recent studies of flow spectra in the
transition region. In the final section the experimental results
will be compared with the behavior found in a numerical studies of
a finite-dimensional mathematical model.

In the Couette-Taylor system a fluid is contained between
concentric cylinders with one or both cylinders rotating; we
will be concerned with the case with the outer cylinder at rest.
For this problem it is convenient to define the Reynolds number as
$R=\Omega(b-a)/\nu$, where Ω is the rotation rate of the inner cylinder,
ν is the kinematic viscosity, and a and b are the radii of the
inner and outer cylinders respectively.

In 1923 the instability of the basic flow, which is purely
azimuthal, was observed and then calculated by G. I. Taylor[1] in
a classic study in hydrodynamics. He investigated the stability
of the basic azimuthal flow against infinitesimal perturbations.
Solving the resultant linearized equation for the perturbation,
he found that the basic flow becomes unstable at a critical Reynolds
number R_c, which depends on the radius ratio. For $R>R_c$ there
evolves a horizontal (time-independent) torodial vortex pattern,

---
*This research is supported by the National Science Foundation.

now known as Taylor vortex flow. Taylor used dye to visualize the
flow, and the measured values of R_c agreed well with those determined from his linear stability analysis.

As R is increased above R_c a second well-defined critical
Reynolds number R_c' is reached. At this instability traveling
azimuthal waves become superimposed on the horizontal vortices.
The wavy vortex flow was studied photographically in 1965 by
Coles,[2] who visualized the flow by suspending small flat flakes
(aluminum paint pigment) in the fluid.

A linear stability analysis of Taylor vortex flow analagous
to that done by Taylor for the basic flow is not possible since
the solution of the Navier-Stokes equation for the Taylor vortex
flow (about which a perturbation would be considered) is unknown
except as an infinite series of interacting modes. A half century
passed between Taylor's work and the monumental <u>nonlinear</u> analysis
by Davey, DiPrima, and Stuart[3] of the stability of Taylor vortex
flow. They found that R_c' for radius ratios near unity is only
5% greater than R_c.

How does the fluid ultimately become turbulent as R is increased beyond R_c'? The stability analyses have proved to be too
difficult in practice to extend to $R > R_c'$. The only detailed
experimental investigation of the flow at $R > R_c'$ prior to those
described in the next section was Coles' photographic study.[2] He
found that ω_1/Ω (where ω_1 is the wave frequency) decreased from
0.5 at R_c' to 0.34 at $R/R_c' = 23$, where the waves became lost in the
noise. As R was increased beyond R_c' the flow gradually began to
show small scale irregularities, and finally "the flow can only
be described as fully turbulent";[2] however, the time-dependence of
the flow was not measured so the transition process could not be
described quantitatively.

SPECTROSCOPIC STUDIES OF THE TRANSITION TO TURBULENCE

At City College we used the laser Doppler velocimetry technique to measure the radial component of the velocity in a
Couette-Taylor system. Velocity values measured in successive
time intervals were recorded in a computer and then Fourier-trans-
formed to obtain velocity power spectra. The high resolution
velocity spectra obtained from long data records were supple-
mented by flow photographs of the type obtained by Coles. The
experiments we will describe were performed by P. R. Fenstermacher
and the author; the initial phase of this research was done in
collaboration with J. P. Gollub.[4] Our experimental results will
be briefly summarized here; the details have been published
elsewhere,[4] but the comparison of our results with other recent
experiments (see Table 1) has not been previously published.

The velocity power spectrum in the range $R_c' < R < (10.1)R_c$ contains only a single fundamental frequency component (the wave frequency ω_1) and its harmonics, as shown in Fig. 1(a); the flow is strictly periodic. The amplitude of the fundamental is more than 5 orders of magnitude above the instrumental noise level.

As the Reynolds number is increased above $R/R_c = 10.1$ a second fundamental frequency appears in the spectrum. The amplitude of this component increases continuously from zero for $R/R_c > 10.1$; hence this is a continuous transition, analagous to a second order phase transition. This second fundamental frequency of the steady state flow has been designated ω_3 since a transient component ω_2 is observed at lower Reynolds number. The frequency ratio ω_3/ω_1 increases continuously (within the experimental resolution) with increasing R. Therefore, the frequencies ω_1 and ω_3 are incommensurate and the flow is quasiperiodic.

A broad weak component appears in the spectrum at $R/R_c \approx 12$; this component is labeled B in Fig. 1(b). The essential change in the qualitative character of the flow marked by the appearance of B should be emphasized. Unlike a flow characterized by two or any number of discrete frequencies, the behavior of the flow described by a broad spectral component is no longer predictable at distant future times. The flow must now be described as chaotic (or "turbulent") even though more than 99% of the spectral energy remains in the discrete spectral lines.

With further increase in Reynolds number the components ω_1 and ω_3 disappear at $R/R_c = 19.3$ and 21.9, respectively, leaving a spectrum with only the component B and a continuous background. No further transitions are observed.

Walden and Donnelly[5] have studied the transitions in Couette-Taylor flow by measuring the time dependence of the ion current between the inner cylinder and an electrode embedded in the outer cylinder. Their ion current spectra obtained for a system with a height to gap ratio of 20 contain the same frequency components, ω_1, ω_3, and B, that were observed in our velocity power spectra, and the Reynolds numbers obtained in the two experiments agree within the experimental uncertainty. Walden and Donnelly have also obtained ion current spectra for cylinder height to gap ratios ranging from 18 to 80. They find that for height to gap ratios greater than 25 another spectral component, called ω_r, appears in the spectrum in the range $28 \lesssim R/R_c \lesssim 36$; one of their spectra is shown in Fig. 1(c). The flow is certainly turbulent, as is clear from the presence of the component B and from the noisy appearance of the photographs of the flow; ω_r must arise from some large scale periodic structure. In this connection it should be noted that the axial periodicity corresponding to the Taylor vortex structure persists even at the largest Reynolds number

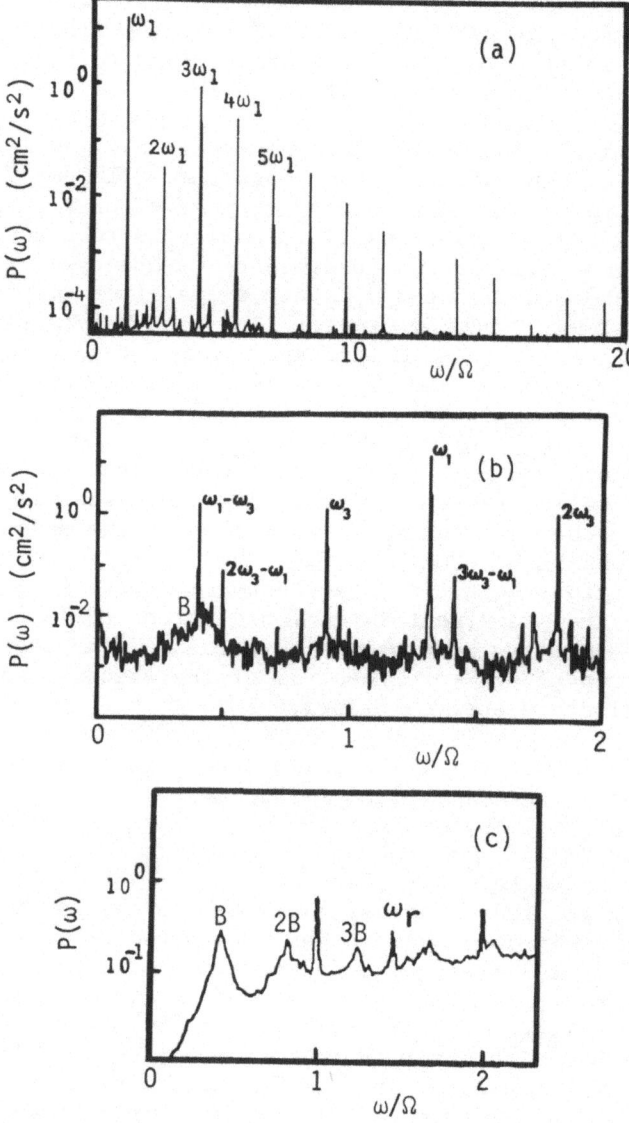

Fig. 1.   (a) A velocity power spectrum at R/R_c=5.6.[4] All spectral
          lines are harmonics of the azimuthal wave frequency ω_1.
          (b) A velocity power spectrum at R/R_c=15.1.[4] All discrete
          components are harmonics and linear combinations of ω_1 and
          ω_3.   (c) An ion current power spectrum at R/R_c=28.[5] The
          peaks at ω/Ω = 1 and 2 are instrumental artifacts.

studied, $R/R_c > 1,000$, where the flow is highly turbulent.[6]

Cognet and Bouabdallah have measured the time dependence of the velocity gradient at a point on the inner cylinder wall by using an electrochemical technique.[7] Their power spectra appear to be consistent with those obtained in the experiments described above, although their spectra have much lower resolution. They have explored in particular the variation of the spectra with axial position, and they find that turbulence appears to originate at the vortex boundaries corresponding to fluid outflow.

Mobbs, Preston, and Ozogan have obtained power spectra of the time-dependent flow from measurements by a fourth technique, hot film anemometry.[8] In the experiments we have described above only the transitions marked by changes in the character of the power spectra have been discussed. In addition to those transitions there are transitions between flows with different spatial states (i.e., different numbers of axial vortices and azimuthal waves); these transitions generally occur without a change in the character of the spectrum, although the frequencies of the spectral components of course change. Mobbs et al.[8] find that in the vicinity of a change in azimuthal wavenumber from m to m±1 the spectrum is always dominated by a component at a frequency corresponding to an azimuthal wave with m=1.

DISCUSSION

A remarkably consistent and simple picture of the transition to turbulence in Couette-Taylor flow emerges from the four independent studies that were described in the previous section and are summarized by Fig. 2 and Table 1.[9] As R is increased beyond the onset of wavy vortex flow, the system is initially periodic, then quasiperiodic with two frequencies, and finally at $R/R_c \approx 12$ a chaotic element appears in the flow (that is, the spectra contain a broad component). We have examined photographs of the flow and found that $R/R_c \approx 12$ is also the Reynolds number at which the photographs begin to show irregular small scale structure. Thus the experiments consistently indicate that the flow begins to become chaotic at $R/R_c \approx 12$. However, the flow at this point is still largely ordered; only at much larger Reynolds number does the flow become strongly chaotic in the sense that all of the energy is in broadband spectral components.

The observation that Couette-Taylor flow is characterized by only two discrete frequencies before the system becomes chaotic suggests that it may be possible to construct a realistic model of Couette-Taylor flow with only a small number of interacting modes. This approach has been followed by Yahata,[10] who Fourier-analyzed the velocity field in the axial and azimuthal directions, and then each Fourier coefficient in this double Fourier series was

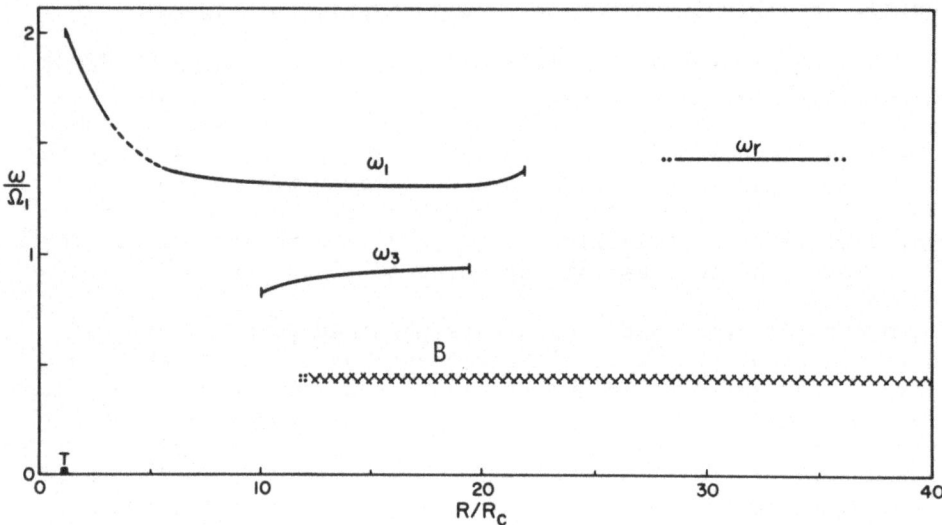

Fig. 2.   Fundamental frequencies observed in Couette-Taylor Flow.
          "T" indicates the region where there are stable time-
          independent Taylor vortices, $R_c < R < R_c'$.

Table 1.   Summary of the Spectral Studies of the Transition to
           Turbulence in Couette-Taylor Flow (radius ratio $\simeq 7/8$)[a]

Investigators	Cylinder height to gap ratio	Maximum R/R_c studied	R/R_c for noise appearance	R/R_c for wave disappearance
Fenstermacher et al.[4]	20	45	~12	21.9
Walden & Donnelly[5]	18–80	67	~11	21–25[b]
Cognet & Bouabdallah[7]	40	70	~12	19.5
Mobbs et al.[8]	65	320	~5–8[c]	21

[a]The transition marking the appearance and disappearance of the
   second fundamental frequency (ω_3 in Fig. 2) were studied in detail
   only in the experiment by Fenstermacher et al.
[b]The Reynolds number at which the wave disappeared increased
   monotonically with the cylinder height to gap ratio.
[c]In this range "the flow appears visually to be chaotic," but above
   this range "there is a return to a more orderly flow."

expanded in a complete set of functions in the radial variable. The series were then severely truncated; only the fundamental axial mode, the $m=0$ and $m=4$ azimuthal modes, and several radial modes were retained. The resultant model consisted of 32 ordinary coupled differential equations for the mode amplitudes. Velocity power spectra calculated for this model at four Reynolds numbers showed: (1) At $R/R_c=7.97$ the spectrum contained a single component at $\omega/\Omega=1.5$; presumably this corresponds to ω_1. (2) At $R/R_c=15.94$ the spectrum contained a second sharp frequency component at $\omega/\Omega=1.1$; presumably this corresponds to ω_3. (3) At $R/R_c=22.31$ the component at $\omega/\Omega=1.1$ had disappeared but the component at $\omega/\Omega=1.5$ remained. (4) At $R/R_c=23.91$ the component at $\omega/\Omega=1.5$ had become broad. Thus the behavior of this simple finite-dimensional model system is in accord with most of the experimental observations, except for the appearance of noise at $R/R_c \simeq 12$.

A more general approach to the study of the dynamics of nonlinear systems is provided by dynamical systems theory, which is concerned with the phase space topology of solutions to the equations of motion. Although this approach cannot give specific Reynolds numbers for transitions or values of characteristic frequencies, dynamical systems theorems may provide powerful insights into the qualitative behavior of nonlinear systems. A theorem of Newhouse, Ruelle, and Takens[11] is interesting in connection with the experiments on Couette-Taylor flow. The theorem implies that infinitesimal perturbations of a system with three (or more) characteristic frequencies may lead to chaotic behavior. If the theorem is applicable to fluid mechanics, it would mean that a quasiperiodic flow with three or more frequencies may be unobservable. It is interesting to note that Couette-Taylor flow and other flows that have been investigated so far are characterized by not more than two temporal frequencies.

The transition to turbulence has been studied in detail for very few systems. The transitions observed in Rayleigh-Bénard convection in boxes with lateral dimensions not more than a few times the height are described by the same dynamical regimes that are observed in Couette-Taylor flow: periodic, quasiperiodic with two frequencies, and chaotic.[12,13] However, for boxes with large lateral dimensions the system becomes chaotic at the first instability.[12] This strong dependence on the shape of the box is not understood.

Recently Michael Gorman in our laboratory at The University of Texas has found that in Couette-Taylor flow with the two cylinders counter-rotating there is a rich variety of different types of transitional behavior; the particular transitional sequence observed depends on the ratio of the inner to outer cylinder rotation rates and on which cylinder rotation rate is being increased while the other is held fixed.

In conclusion, although significant progress towards the understanding of the transition to turbulence has been achieved through the use of new mathematical approaches and modern experimental techniques, it is clear that much more experimental and theoretical work is needed before an understanding of the transitional problem in general is realized.

REFERENCES
1. G. I. Taylor, Phil. Trans. Roy. Soc. (London) A 223, 289 (1923).
2. D. Coles, J. Fluid Mech. 21, 385 (1965).
3. A. Davey, R. C. Di Prima, J. T. Stuart, J. Fluid Mech. 31, 17 (1968).
4. J. P. Gollub and H. L. Swinney, Phys. Rev. Lett. 35, 927 (1975); H. L. Swinney, P. R. Fenstermacher, and J. P. Gollub, in "Synergetics, a Workshop," H. Haken, ed., Springer, New York, 1977, p. 60; H. L. Swinney and J. P. Gollub, Physics Today 31, No. 8, 41 (August 1978); P. R. Fenstermacher, H. L. Swinney, and J. P. Gollub, J. Fluid Mech., to be published (1979).
5. R. W. Walden and R. J. Donnelly, Phys. Rev. Lett. 42, 301 (1979).
6. A Townsend, private communication.
7. G. Cognet and A. Bouabdallah, Taylor Vortex Flow Working Party, Leeds, 1979; see also G. Cognet, J. Mecanique 10, 65 (1971).
8. F. R. Mobbs, S. Preston, and M. S. Ozogan, Taylor Vortex Flow Working Party, Leeds, 1979; see also A. Barcilon, J. Brindley, M. Lessen, and F. R. Mobbs, J. Fluid Mech., to be published (1979).
9. It should be noted, however, that all these experiments were done for a radius ratio a/b=7/8. The transition sequence can be quite different for significantly smaller radius ratios.
10. H. Yahata, to be published in Prog. Theor. Phys. (1979).
11. S. Newhouse, D. Ruelle, and F. Takens, Commun. Math. Phys. 64, 35 (1978).
12. G. Ahlers and R. P. Behringer, Phys. Rev. Lett. 40, 712 (1978).
13. J. P. Gollub and S. V. Benson, Phys. Rev. Lett. 41, 948 (1978).

HYDRODYNAMIC INSTABILITIES AND TURBULENCE

P. C. Hohenberg

Bell Laboratories

Murray Hill, New Jersey   07974

Abstract

When a classical fluid is subjected to external stress it
undergoes transitions to flow states which become more and more
disordered, or turbulent, as the stress is increased. Two limiting cases are of particular interest, the region of weak stress
or onset of chaotic motion, and the region of large stress or
fully-developed turbulence. In the first case a classical theory
due to Landau and Hopf describes the onset of disorder as the
pile-up of a large number of instabilities, with modes of motion
at mutually incommensurate frequencies. More recently it has been
realized both experimentally and theoretically, that chaotic motion
can also result from non-linear interactions among a small number
of modes, after the appearance of only two or three instabilities.
The most striking experimental demonstration of these effects
occurs in the study of Rayleigh-Bénard convection and Couette-
Taylor flow. In the region of large external stress (fully developed turbulence), a statistical description of short-scale velocity correlations is sought, with the hope of finding certain
universal features. Various phenomenological theories have been
proposed, beginning with the famous 1941 prediction of Kolmogorov,
that the energy spectrum as a function of wavenumber will vary as
$k^{-5/3}$ for large k. Experimental techniques for studying fluid
turbulence will be briefly surveyed.

Outline

I. INTRODUCTION: WEAK AND STRONG TURBULENCE

A. Statement of the Problem

Consider a classical fluid, described by deterministic hydrodynamic equations. Apply a constant external stress, characterized
by a dimensionless strength R [in the case of a stirred fluid for
example, $R = UL/\nu$, where U, L, and ν are typical values of the
velocity, the length scale, and the viscosity, respectively].
What flow states are obtained as the stress is increased?

B. Classification of Flow States

Typically, one observes some or all of the following types
of behavior with increasing R:
$\sim$ laminar (no flow, steady flow, or periodic flow)
$\sim$ quasi-periodic (flow with various incommensurate frequencies)
$\sim$ chaotic (no well-defined periodicities)
$\vdots$
$\sim$ fully developed turbulence (isotropic and homogeneous in small
   scales).

References: Martin (1975); Swinney and Gollub (1978,1979); Riste
(1975); Nelkin (1978); Joseph (1976).

C. Experimental Examples

$\sim$ Rayleigh-Bénard convection: buoyancy driven flow of a
fluid confined between parallel plates and heated from below.
$\sim$ Taylor-Couette flow: fluid between concentric rotating
cylinders.
$\sim$ Pipe flow and flow past a solid body.

References: Chandrasekhar (1961); Normand et al. (1977); Swinney
and Gollub (1978,1979).

D. Experimental Methods

$\sim$ bulk measurements: total heat flux or torque.
$\sim$ local measurements: velocity v and its derivatives
$\partial v_i/\partial x_j$. Use of hot-wire anemometery or laser-Doppler velocimetry.

References: Comte-Bellot (1976 a,b); Durrani and Greated (1977);
Buchhave et al. (1979); AGARD (1976).

II. ONSET OF TURBULENCE: CHAOTIC MOTION IN TIME

A. Experimental Example: Rayleigh-Bénard Convection

Measurement of Nusselt number (N = total heat flux/conductive
heat flux) or local velocity, as a function of time. Periodic,
quasi-periodic, and chaotic flows are observed, as R increases.
Chaotic motion is signaled by the appearance of broad band contributions to f(ω), the power spectrum of N(t) or v(t). The succession of instabilities is observed to depend on aspect ratio
(height/lateral dimension). Spatial order seems to persist to
rather high R. More work is needed to understand spatial correlations.

References: Ahlers and Behringer (1978); Gollub and Benson (1978);
Bergé and Dubois (1976).

B. Theory

The Landau-Hopf picture involves an infinite sequence of instabilities to modes of motion with incommensurate frequencies.
E. N. Lorenz (1963) has shown numerically that chaotic motion can
result from nonlinear interactions between a small number of modes.
Later mathematical and numerical work has reproduced the succession of flow states discussed in I-B above, and has yielded qualitative agreement with many of the experiments. A number of features still remain unexplained, however, even qualitatively.

References: Landau and Lifshitz (1959); Lorenz (1963); Ruelle
and Takens (1971); Martin (1975); Curry (1978); Rabinovich (1978).

III. FULLY DEVELOPED TURBULENCE: THE SEARCH FOR UNIVERSALITY

A. Theory

The "mean-field" theory of Kolmogorov (1941) assumes a local
cascade of energy, with rate ε, from large scales L, down to a
dissipation scale η. Universal velocity correlations may then be
found by dimensional analysis. The energy spectrum is given by
$E(k) \sim \varepsilon^{2/3} k^{-5/3}$, for wavenumbers in the inertial range
$L^{-1} < k < \eta^{-1}$. This range is only sizeable for R >> 1, since
$L/\eta \sim R^{4/3}$. Deviations from Kolmogorov (1941), due to fluctuations,
lead to intermittency, and change the exponents of velocity correlations in the inertial range.

References: Landau and Lifshitz (1959); deGennes, in Riste (1975);
Rose and Sulem (1978); Nelkin (1978);Monin(1978);Frish et al(1978).

B. Experiment

The Kolmogorov prediction for E(k) has been verified in a number of fluids. To see deviations one must look at high-order velocity correlations. These are difficult to measure accurately, since large values of R and good spatial resolution are needed. The experimental situation is unclear at present.

References: Van Atta and Park (1972); Champagne (1978).

References

The references are intended primarily to provide an entry into the vast literature on hydrodynamic instabilities and turbulence, to those unfamiliar with the field.

AGARD, 1976, Applications of non-intrusive instrumentation in fluid flow research, NATO Advisory Group for Aerospace Research and Development, Conference Proceedings CP193.

Ahlers, G. and R. P. Behringer, 1978, Evolution of turbulence from the Rayleigh-Bénard instability, Phys. Rev. Lett. 40, 712.

Bergé, P. and M. Dubois, 1976, Time-dependent velocity in Rayleigh-Bénard convection: a transition to turbulence, Optics Commun. 19, 129.

Buchhave, P., W. K. George, Jr., and J. L. Lumley, 1979, The measurement of turbulence with the Laser-Doppler anemometer, Ann. Rev. Fluid Mech. 11, 443.

Champagne, F. H., 1978, The fine-scale structure of the turbulent velocity field, J. Fluid Mech. 86, 67.

Chandrasekhar, S., 1961, Hydrodynamic and hydromagnetic stability, Oxford Univ. Press, Oxford.

Comte-Bellot, G., 1976a, Les méthodes de mesure physique de la turbulence, J. Physique (Paris) 37 C1-67.

Comte-Bellot, G., 1976b, Hot-wire anemometry, Ann. Rev. Fluid Mech. 8, 209.

Curry, J. H., 1978, A generalized Lorenz system, Comm. Math. Phys. 60, 193.

Durrani, T. S., and C. A. Greated, 1977, Laser systems in flow measurements, Plenum, New York.

Frisch, U., P. L. Sulem, and M. Nelkin, 1978, A simple dynamical model of intermittent fully developed turbulence, J. Fluid Mech. 87, 719.

Gollub, J. P., and S. V. Benson, 1978, Chaotic response to periodic perturbation of a convecting fluid, Phys. Rev. Lett. 41, 948.

Joseph, D. D., 1976, Stability of fluid motions, Vols. I and II, Springer, New York.

Landau, L. D., and E. M. Lifshitz, 1959, Fluid mechanics, Chap. 3:
        Turbulence, Pergamon, New York.
Lorenz, F. N., 1963, Deterministic nonperiodic flow, J. Atoms. Sci.,
        20, 130.
Martin, P. C., 1975, The onset of turbulence: a review of recent
        developments in theory and experiment, in Proc. Internat.
        Conf. on Stat. Phys., Budapest, 1975, ed. by L. Pal and
        P. Szépfalusy, North Holland, Amsterdam.
Monin, A. S., 1978, On the nature of turbulence, Usp. Fiz. Nauk
        125, 7 [Soviet Phys. Usp. (to be published)].
Nelkin, M., 1978, Universality and scaling in fully-developed tur-
        bulence, in Proc. Intern. Conf. on Stat. Phys., Haifa,
        1978, p. 236.
Normand, C., Y. Pomeau, and M. G. Velarde, 1977, Convective insta-
        bility: a physicist's approach, Rev. Mod. Phys. 49, 581.
Rabinovich, M. I., 1978, Stochastic oscillations and turbulence,
        Usp. Fiz. Nauk 125, 123 [Soviet Phys. Usp. (to be publi-
        shed)].
Riste, T., 1975, editor, Fluctuations, instabilities, and phase
        transitions, Plenum, New York.
Ruelle, D., and F. Takens, 1971, On the nature of turbulence,
        Commun. Math. Phys. 20, 167.
Rose, H. A., and P. L. Sulem, 1978, Fully developed turbulence and
        statistical mechanics, J. Phys. (Paris), 39, 441.
Swinney, H. L., and J. P. Gollub, 1978, The transition to turbu-
        lence, Physics Today, August, p. 41.
Swinney, H. L., and J. P. Gollub, 1979, editors, Hydrodynamic
        instabilities and the transition to turbulence, Springer-
        Verlag Topics in Current Physics (to be published).
Van Atta, C. W., and J. Park, 1972, Statistical self-similarity
        and inertial subrange turbulence, in Statistical Models
        and Turbulence, edited by M. Rosenblatt and C. Van Atta,
        Springer, New York, p. 402.

LIGHT SCATTERING FROM GELS AND A SINGLE POLYMER CHAIN

NEAR PHASE TRANSITIONS

Toyoichi Tanaka, Amiram Hochberg, Izumi Nishio,
   Shao-Tang Sun, & Gerald Swislow

Department of Physics, Center for Materials Science
   & Engineering, Massachusetts Institute of Technology
Cambridge, MA 02139

INTRODUCTION

In this report, we describe light scattering studies of phase transitions in polyacrylamide gels and in single polyacrylamide chains in solution. A gel is a cross-linked polymer network immersed in a gluid medium. In polyacrylamide gels, the crosslinks are permanent covalent bonds. The polymer chains constituting the network are quie flexible and consequently constantly undergo random thermal motions. Thus, the network concentration fluctuates in space and time, creating inhomogeneities in the refractive index of the gel which scatter light. In 1973, Tanaka, Hocker, and Benedek were able to observe such concentration fluctuations using the technique of laser light scattering spectroscopy.[1] In this report, we shall show how the technique can provide information fundamentally important to the understanding of the physics of gels, especially near the gel phase transition, wherein a gel separates into two gel phases having higher concentration and lower concentration than the original gel. Using the same technique, we shall also demonstrate, for the first time, the existence of the coil-globule transition in a single polyacrylamide chain.

EQUATION OF MOTION

Dynamic properties of a gel such as swelling, shrinking and concentration fluctuations in the network are described by the equation of motion for the displacement vector, $\vec{u}(\vec{r},t)$, which represents the displacement of a point in the network from its average position. The equation of motion is given by[1]

$$\frac{\partial \vec{u}}{\partial t} = \frac{K}{f} \Delta \vec{u} \tag{1}$$

where K is the bulk modulus of the network* and f is the friction force per unit volume of network as it passes through the fluid with unit velocity. Eq. (1) indicates that the network 'diffuses' with a diffusion coefficient $D \equiv K/f$. It has been shown that Eq. (1) successfully describes both microscopic concentration fluctuations in the gel and the macroscopic swelling of the gel.[1,2]

LIGHT SCATTERING FROM CONCENTRATION FLUCTUATIONS IN GELS[1]

The correlation function of the electric field E_s scattered from concentration fluctuations can be calculated using Eq. (1) and the equipartition principle. The result is

$$\langle E_s(\vec{q},t)\, E_s(\vec{q},0)\rangle \propto (kT/K)\, \exp\, (-Kq^2 t/f) \tag{2}$$

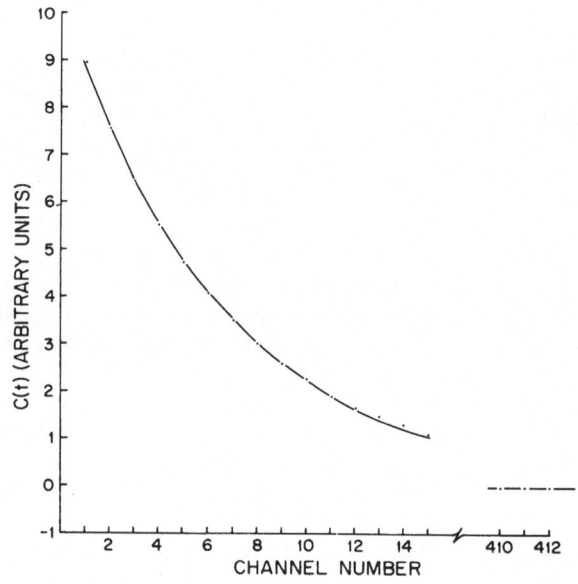

Fig. 1. The measured correlation function of a 5% polyacrylamide gel at 25°C at a 90° scattering angle. The solid line is the best single expotential fit to the data. One channel corresponds to 10μsec.

---

*More precisely, K is replaced in Eq. (1) and (2) with $K + (4/3)\mu$, the longitudinal modulus of the network, where μ is the shear-modulus. However, $\mu \ll K$, except near the spinodal line.

where $\vec{q}$ is the scattering wave vector. Thus the intensity measurements of the scattered light can be used to obtain the bulk modulus
K. In addition, the correlation time of the scattered light determines the ratio K/f. In Fig. 1, a typical correlation function is
shown; it is indeed an exponential as Eq. (2) requires. Fig. 2
shows the linear dependence of the decay rate on the square of q.
The diffusion coefficient K/f thus obtained is 2.5 x 10^{-7}cm^2/sec.

KINETICS OF SWELLING OF GELS[2]

Eq. (1) also governs the kinetics of swelling or shrinking of
gels. Fig. 3 shows the change in the radius of spherical acrylamide gel when immersed in water. The characteristic time of the
swelling should be proportional to the radius of the gel. This is
clearly demonstrated in Fig. 4. From the slope, it is possible to
obtain the diffusion coefficient K/f = 3 x 10^{-7}cm^2/sec. It is remarkable that the values obtained by microscopic light scattering
measurements and by macroscopic swelling experiments give quantitative agreement.

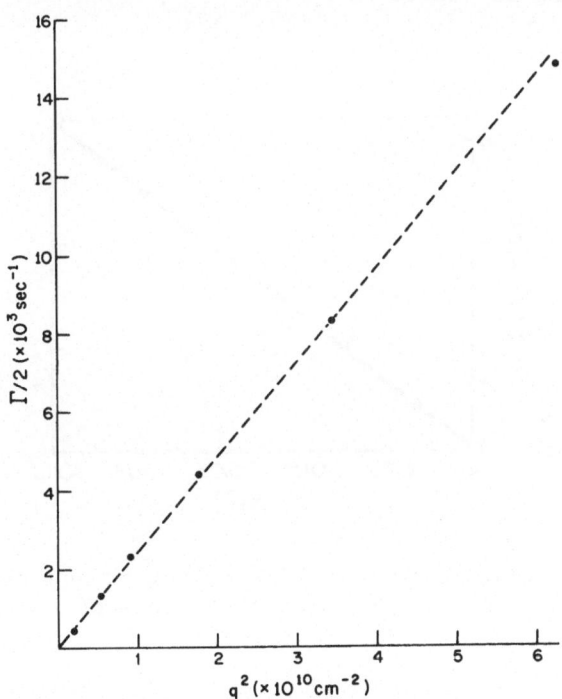

Fig. 2. The q^2 dependence of the relaxation rate $\Gamma = (K/f)q^2$ of
the time correlation of scattered light in a 5% polyacrylamide gel at 25°C.

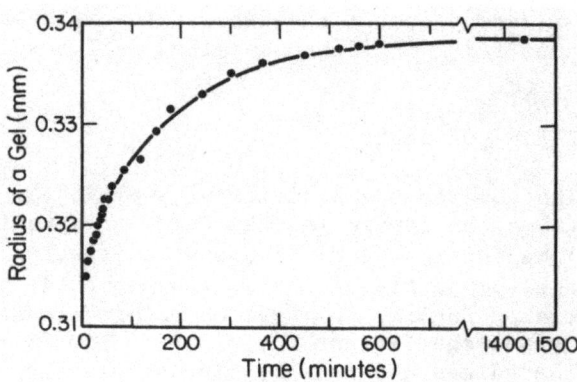

Fig. 3. Time dependence of the radius of a 5% polyacrylamide gel
         swelling in water.

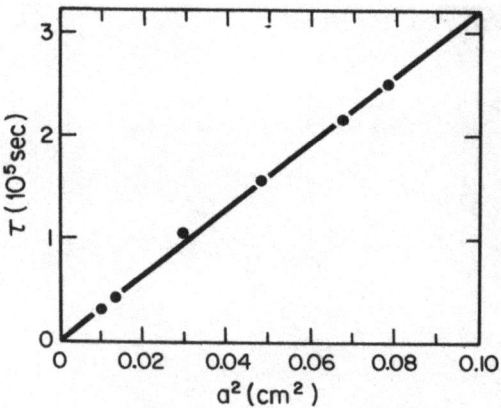

Fig. 4. Characteristic time, τ, of swelling of spherical polyacryla-
         mide gels as a function of the square of the final radius a.

PHASE DIAGRAM OF A GEL

Recently, it was discovered experimentally that a gel undergoes a phase separation upon either lowering the temperature or decresing the solubility of the gel network by varying the composition of the gel fluid.[3,4] The phase diagram was calculated using a mean field theory and is shown in Fig. 5. The vertical axis represents either temperature or solvent composition. The horizontal axis represents the network concentration. The phase boundary consists of two lines: one is the coexistence curve and the other is the volume curve at which the osmotic pressure of the gel is zero. Above these, the gel is stable. In between these two curves, the gel is unstable and separates into domains of two different gels having volume fractions determined by the intercepts of the T = constant line (or solvent composition = constant) with the coexistence curve. The two coexisting gel phases merge at the maximum of the coexistence curve, which corresponds to the critical point. Below the volume curve, the osmotic pressure of the gel is negative, and the network shrinks, increasing the network concentration until the gel reaches the state of zero osmotic pressure on the volume curve.

Also plotted in Fig. 5 is the calculated spinodal line on which the bulk modulus K of the network vanishes. On this line, the concentration fluctuations in the network diverge and become infinately slow. In the regions between the coexistence curve, the volume curve, and the spinodal curve, the gel is metastable. It is possible to enter the metastable region experimentally by approaching it while

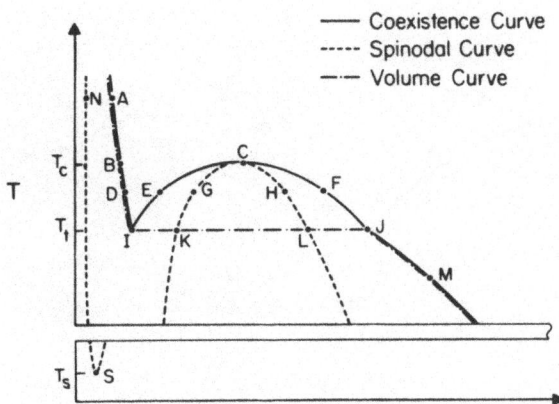

Fig. 5.   The phase diagram of polyacrylamide gel. The vertical axis represents either increasing temperature or decreasing poorness of the solvent (gel fluid). The horizontal axis represents increasing network concentration.

avoiding mechanical disturbances to the system.

The volume curve, characterized by zero osmotic pressure, can be determined by measuring the equilibrium concentration of the gel when it is immersed in a large volume of fluid.[3] Equilibrium between the gel and fluid is reached when the osmotic pressure of the network becomes equal to that of the fluid, which is, of course, zero. The volume curve in Fig. 5 has a discrete change, which indicates that by changing temperature or solvent composition infinitesimally, we can cause a finite change in the gel volume. In Fig. 6 are the volume curves of gels as a function of temperature and solvent composition. We see a discrete change in the gel concentration. The volume curve sometimes shows hysterisis. When the temperature is carefully lowered or increased the gel can go into the metastable state.

CRITICAL BEHAVIOR

The spinodal curve was determined by light scattering measurements. Fig. 7 shows the intensity and relaxation rate of light

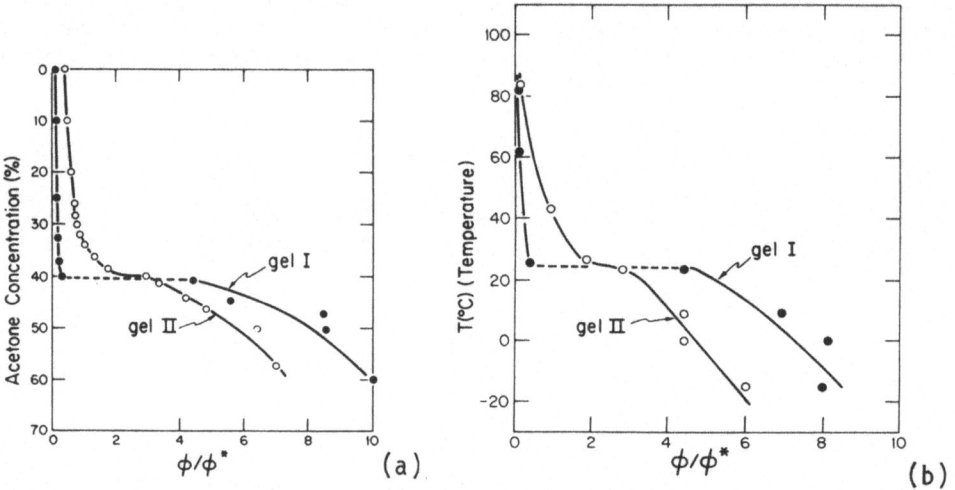

Fig. 6. Volume curves of polyacrylamide gels as function of
         (a) acetone concentration of the gel fluid, and (b) temperature. Here ϕ^* and ϕ are the volume fractions of the network before and after swelling. The number of free branches in the gel network was reduced in Gel I by curing it in the presence of the polymerization initiators for 30 days. Gel II was cured for 3 days only.

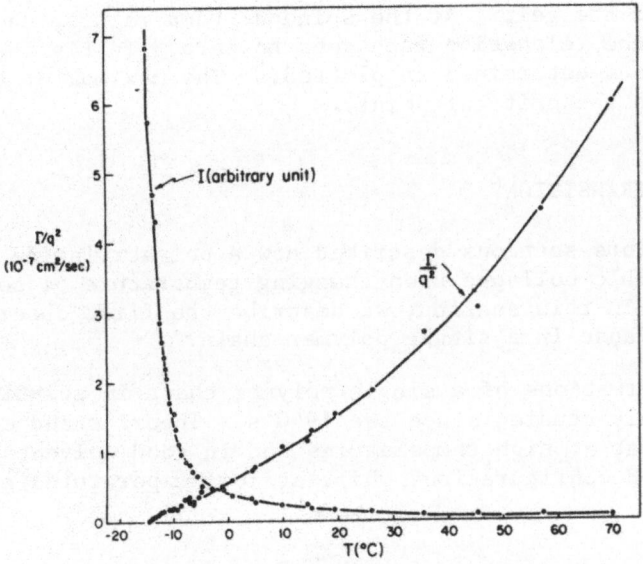

Fig. 7.   The intensity, I, and the reciprocal relaxation rate
$\Gamma = (K/f)q^2$ of laser light scattered by a 2.5% polyacrylamide gel. Here Γ is divided by the square of the scatter-
ing vector $|\vec{q}|$.

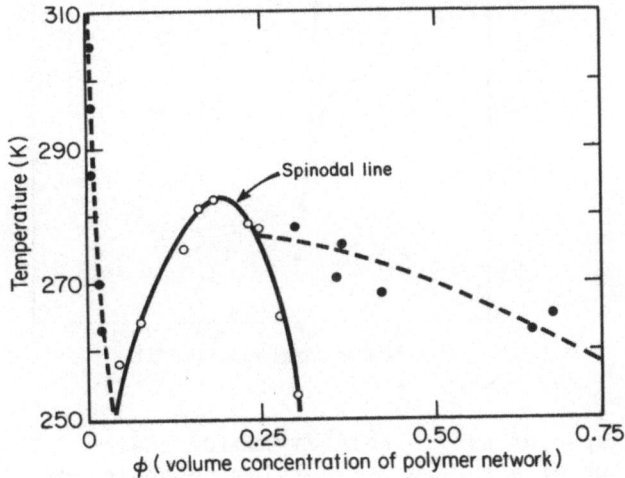

Fig. 8.   The spinodal line of a polyacrylamide gel determined by
light scattering measurements. The solid circles donote
the equilibrium concentrations of the gel when immersed in
a 44% acetone-water mixture.

scattered from the gel.[3] At the spinodal temperature, the intensity diverges and the relaxation rate goes to zero. In Fig. 8, the spinodal curve thus determined is plotted.[5] The maximum of the curve corresponds to the critical point.

COIL-GLOBULE TRANSITION[6]

The previous sections described how a polyacrylamide gel undergoes a reversible collapse upon changing temperature or solvent composition. In this section, we describe the first observations of such a collapse in a single polymer chain.

The conformations of a single polymer chain in solution have been extensively studied since the 1940's. Theories and experiments established that at high temperatures and in good solvents a polymer has an extended configuration, while at low temperatures and in poor

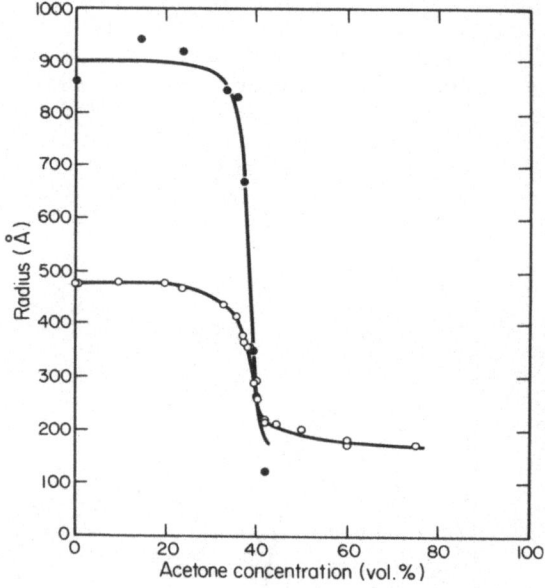

Fig. 9. Collapse of single polyacrylamide chains. The hydrodynamic radius of a single polyacrylamide chain, determined by laser light scattering spectroscopy is plotted (open circles) as a function of the acetone concentration of the acetone-water mixtures used as solvent. The radius of gyration of the polyacrylamide chain determined from the angular dependence of the scattered light intensity is also plotted (solid circles).

solvents, a polymer is in a collapsed state. The transition between the extended and collapsed state as the temperature or solvent composition is varied was thought to be smooth and continuous, and experiments supported this idea. However, in 1968, Lifshitz suggested the transition between the two configurations was discrete, calling it the coil-globule transition.

We have made the first observation of the coil-globule transition in a single polyacrylamide chain dissolved in acetone-water mixtures. Acetone is a poor solvent for polyacrylamide, whereas waster is a good solvent for the molecule. By gradually changing the composition of the mixture, the polyacrylamide molecules can be brought through the transition between the regions of good solvent and poor solvent.

In the experiment, monodisperse polyacrylamide polymers of molecular weight $5 - 6 \times 10^6$ Daltons were used. A single chain of the polymer consists of approximately 8×10^4 acrylamide monomers and has a length of about 24µm. The concentration of the polymer in solution was less than 10µg/ml. At this concentration, the mean distance between adjacent polymers, on the order of 1µm was much larger than the average polymer size, thus interpolymer entanglements were avoided. The decay rate of the correlation function of the scattered light intensity determines the diffusion coefficient, D, of the polymer, which is related to the hydrodynamic radius, ξ, of the polymer through the Stokes-Einstein relation, $D = kT/(6\pi\eta\xi)$, where k is the Boltzman constant, T is the absolute temperature, and η is the viscosity of the solvent.

The results of the measurements of the hydrodynamic radius of the polymer as a function of acetone concentration are plotted is Fig. 9 (open circles). At low acetone concentrations, the radius is large, approximately 500Å. Near an acetone concentration of 39%, the polymer shows a drastic decrease in hydrodynamic radius to about 200Å. With further increases in acetone concentration, the polymer radius remains constant. The transition was found to be reversible and reproducible.

The radius of gyration of the chain was also determined using measurements of the angular dissymmetry of the scattered light intensity. The results are also shown in Fig. 9 (solid circles). The coil-globule transition is seen to occur at exactly the same acetone concentration as in the curve of the hydrodynamic radius.

CONCLUSION

Using laser light scattering techniques, we have been able to observe the study the phase transition in gels and in a single polymer chain. The sharp change in volume observed in gels and in single polymer chains upon changing temperature or solvent composition may eventually be utilized to perform functions of amplification, switching and memory. This would have tremendous applications in biology, medicine, and chemical engineering. We also hope that these studies can provide an improved perspective on the interdisciplinary field where polymer science and the physics of phase transitions merge.

ACKNOWLEDGMENT

This work has been supported by NSF, CHE77-26924, NIH, EY02433 and the Whitaker Health Sciences Fund, MIT.

REFERENCES

1. T. Tanaka, L.O. Hocker, & G.B. Benedek, J. Chem. Phys. 59, 5151 (1973).
2. T. Tanaka, & D.J. Fillmore, J. Chem. Phys. 70, 1214 (1979)
3. T. Tanaka, Phys. Rev. Lett., 40 820 (1978)
4. T. Tanaka, S. Ishiwata & C. Ishimoto, Phys. Rev. Lett. 38, 771, (1977).
5. A. Hochberg, T. Tanaka & D. Nicoli, submitted for publication.
6. I. Nishio, S-T. Sun, G. Swislow & T. Tanaka, submitted for publication.

INTERFEROMETRIC STUDIES OF THICK FILM CRITICAL BEHAVIOR

W. J. O'Sullivan, B. A. Scheibner, M. R. Meadows
and R. C. Mockler
Department of Physics and Astrophysics
University of Colorado
Boulder, Colorado 80309

INTRODUCTION:

Consider a film of thickness $L \simeq 1$ μm, consisting of a critical binary fluid mixture trapped between parallel flat surfaces. Scaling theory supplies a number of predictions regarding the effects of restricted geometry on the critical properties of such a film.[1] These are based on the ansatz that the relevant scale of thickness for the film is determined by the bulk correlation length $\xi(T)$ which scales with the 3d critical exponent ν. ξ diverges, and becomes comparable to L as the film temperature approaches the bulk critical temperature, $T_{C\infty}$; at which point the presence of the walls introduces a constraint which lowers the symmetry of the system.

Then, a transition from 3d to 2d Ising scaling should occur at a crossover temperature $T_X(L)$ such that $\xi(T) \simeq L$. A bulk property $Y_\infty(T)$, which varies as $A_3 \epsilon^{\psi_3}$ as $\epsilon \to 0$ ($\epsilon = (T-T_{C\infty})/T_{C\infty}$), should have a corresponding film property $Y(T,L)$ behaving as $A_2(L) \epsilon_2^{\psi_2}$ near the film critical temperature $T_C(L)$. (ψ_3 and ψ_2 are the relevant 3d and 2d exponents and $\epsilon_2 = (T - T_C(L))/T_{C\infty}$.) There should be a generalized law of corresponding states for films of the form $Y(T,L) \sim \epsilon_2^{\psi_3} X(L\epsilon_2^\nu)$, where $X(L\epsilon_2^\nu)$ is a scaling function. Thus, if $\epsilon_2^{-\psi_3} Y(T,L)$ is plotted against $L\epsilon_2^\nu$ for different values of L, a universal curve results. The 2d amplitude should vary as $A_2(L) \sim L^{(\psi_3 - \psi_2)/\nu}$. The crossover temperature should have a thickness dependence given by $T_X(L) - T_{C\infty} \sim L^{-1/\nu}$, and in the absence of an overall constraint, the film critical temperature shift should also vary as $T - T_C(L) \sim L^{-1/\nu}$.

None of these predicted finite size modifications of film critical behavior have been verified experimentally for Ising class systems.

In this paper we discuss the results of index of refraction-coexistence curve measurements on trapped films of 2,6-lutidine + water near that system's <u>lower</u> critical point. These measurements confirm the predicted effects of finite size on the critical properties of Ising class fluid films.

EXPERIMENT:

Figure 1 is a schematic representation of the experiment. The films are formed between a pair of coated optical flats which constitutes a vertical wedge plate interferometer. The fluid filled interferometer is mounted in a sample cell whose temperature is controlled to $\pm$ 0.1mK with an absolute calibration of $\pm$ 10mK. Both the mirror separation and wedge angle are adjustable with external controls. The films are prepared from fluid in a reservoir surrounding the interferometer, with the temperature in the <u>one phase region</u> $\sim$12C below $T_{C\infty}$ = 33.98C. The film preparation must be carried out with great care to insure that each film has the critical composition. The mirror separation is then reduced to the target L value, and the wedge angle adjusted to give 3 to 5 vertical multi beam Fizeau fringes when expanded and attenuated HeNe laser light illuminates the interferometer. The positions of these fringes are measured with a telescope riding on a micrometer stage.

The temperature is then raised through $T_C(L)$ to induce phase separation. For L > 15μm, entrance into the two phase region produces separated upper and lower phases. For L < 15μm, the two phases remain interdispersed in small drops. In either case, phase separation results in a splitting of each interference fringe into two parts, and measurements of the splitting lead to values for $\Delta n(L,T)$, the index of refraction difference between coexisting phases.

In previous work[2] we showed how measurements of Δn can be related to the volume fraction difference, $\Delta\phi$, which seems to be the preferred order parameter for the 2,6-lutidine + water mixture.[3] As a result, measurements of $\Delta n(L,T)$ taken at various film thicknesses as a function of temperature, enable us to study the structure of fluid film coexistence curves from L in the submicron range to L $\simeq$ 300μm, where bulk critical properties should be manifested.

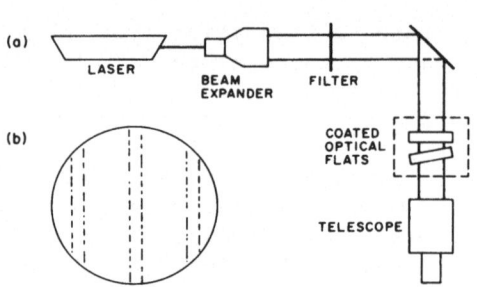

Fig. 1 a) Experimental set up (schematic); b) view through the telescope showing three pairs of split fringes.

Films trapped between Ag
coated flats and SiO$_2$ overcoated
dielectric mirrors were studied.
The data taken with the Ag
mirrors consist of twenty nine
coexistence curves for 0.46μm
≤L≤5.9μm and three at L≈14μm.
Each curve consists of about
forty Δn(T) measurements. Phase
separation temperatures (PST)
were determined for each spacing
and eleven additional PST were
measured for films of thickness
6μm≤L≤182μm. The data set with
the dielectric coatings consists
of thirty two PST for L from
8μm to 272μm. PST were
determined to ±1mK.

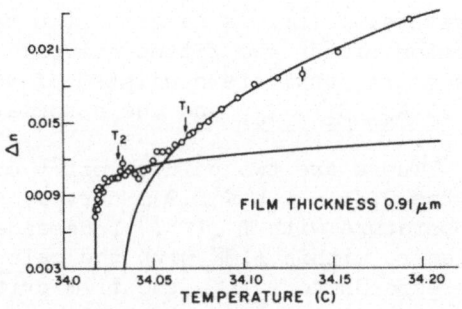

Fig. 2 Plot of Δn values for
L=0.91μm, as a function of temperature. Three points at T>34.2C
are not included. The solid lines
are theoretical fits to the data as
discussed in the text.

ANALYSIS:

The three coexistence curves for L≈14μm show 3d scaling behavior
to within ±0.1mK of the film critical temperature, the limit set by
our temperature control. The twenty nine curves for L≤5.9μm each
have a definite two part structure as seen in the L=0.91μm results
shown in Fig. 2.

The coexistence curves were analyzed by fitting the outlying
data to the form

$$\Delta n = A_3 (T-T_{c3}(L))^{\beta_3} \tag{1}$$

and the data close to the PST by

$$\Delta n = A_2 (T-T_{c2}(L))^{\beta_2} \tag{2}$$

The analysis began by estimating $T_{c3}(L)$ from an extrapolation
of the high temperature data to Δn = 0, and by choosing a conserva-
tively high value for T_1, the expected <u>lower</u> limit for 3d behavior.
A least squares analysis of the T>T_1 data was performed, using the
$T_{c3}(L)$ estimate. The resulting A_3, β_3 pair with $T_{c3}(L)$ were used
as starting values in a weighted non linear grid search, in which
A_3, β_3 and T_{c3} were allowed to vary while the routine sought a
minimum χ^2. Then the data set was expanded by adding the next Δn
value nearer the critical point. New parameter values were returned
and this process repeated until all data were used or χ^2 diverged,
signifying entrance into the crossover region. A similar process
was followed, using Eq.2, for the data between $T_{c2}(L)$ and T_2, the
expected <u>upper</u> limit of 2d behavior. Finally, the crossover tem-

perature, $T_x(L)$, is taken as the temperature defined by the intersection of the two fitted curves. Thus, the output from the
preceding analysis consisted of values for (T_{c2}, A_2, β_2),
(T_{c3}, A_3, β_3), T_x, and the associated χ^2 for each L.

There are two points worthy of stress at this juncture. First,
in the $0.46\mu m \leq L \leq 5.9\mu m$ curves, we identify the phase separation
temperature with $T_{c2}(L)$. Independent measured values for the PST
agree to within $\pm 1mK$ with the calculated $T_{c2}(L)$. Thus $T_{c2}(L)$
corresponds to $T_c(L)$, the film critical temperature as it appears
in the scaling predictions discussed earlier. For $L \geq 6\mu m$ there is
no resolvable 2d regime, and T_{c2} is indistinguishable from T_{c3}.

The second point concerns the interpretation of $T_{c3}(L)$.
Naively, T_{c3}, the extrapolated 3d region critical temperature for all
the curves, should correspond to the bulk critical temperature $T_{c\infty}$.
In the case of an idealized film, the film surface is represented
by a boundary at which the fluid is truncated without otherwise
changing physically. Here, as long as L is much greater than ξ_0,
there is a unique $T_{c\infty}$ toward which the 3d length ξ approaches
divergence. Scaling theory for such ideal films does not bear on
the possibility that the virtual divergence temperature for ξ can be
a function of L. However, our data clearly reveal an effect due to
the presence of the mirrors, which is manifested in a simple translation of each film's coexistence curves in temperature by an amount
dependent upon L. Anticipating results to be discussed later, we
find that T_{c3} varies with L, and that the $T_{c3}(L)$ extracted from the
coexistence curves join with the eleven Ag mirror PST values to
define a single power law in L, spanning a range from $0.46\mu m$ to
$182\mu m$. We interpret this translation as due to an as yet unspecified
interaction between the film and mirrors, and, as unrelated to the
particular finite size effect considered by scaling theory for free
(non interacting) films. The necessary modification required to
bring our data into correspondence with scaling theory for free films
is the replacement of $T_{c\infty}$ by $T_{c3}(L)$. That is, the temperature shifts
associated with the sought for finite size effects are measured with
respect to the temperature at which ξ tends toward divergence, irre-
spective of the shift of this point as a function of L due to physical boundary effects. This position can be supported by arguments
based on the generalized homogeneity approach to scaling. However,
invoking such arguments introduces no new physics nor added rigor.

RESULTS:

We find the weighted means, $\beta_3 = 0.332 \pm 0.003$ and $\beta_2 = 0.126 \pm 0.005$.
The β_3 result represents an average over thirty two coexistence
curves and β_2 an average over twenty eight such curves. The quoted
uncertainties are one standard deviation of the mean $(\bar{\sigma} = \sigma/\sqrt{N})$.

The β_3 result agrees with values found in bulk binary fluids, and β_2 agrees with the 2d Ising value of 0.125.

In Fig. 3 we plot $\epsilon_2^{-\beta_3} \Delta n(T,L)$ versus $L\epsilon_2^{\nu}$. Since about 1300 points are involved, we have reduced the point density by a factor of ten. These results, which include a sample of over twenty different L values, are persuasive evidence for the existence of a law of corresponding states for critical films.

The 3d amplitude A_3 is found to be independent of L. Upon conversion of A_3 using the Lorentz-Lorenz relation, we find $A = 1.91\pm0.003$ where $\Delta\phi = A\epsilon^{\beta_3}$. This can be compared to the result[3] $A=1.95\pm0.70$ determined from the coexistence curve data of Loven and Rice.[4]

The critical temperature shift in our "interacting films" should scale as

$$T_{c2}(L) - T_{c3}(L) = KL^{-1/\nu} \qquad , K \text{ const.} \qquad (3)$$

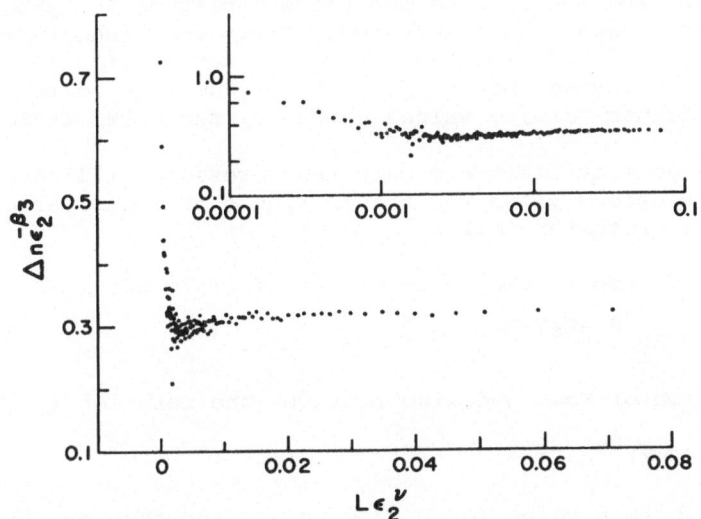

Fig. 3 Smoothed plot of $\epsilon_2^{-\beta_3} \Delta n(T,L)$ vs. $L\epsilon_2^{\nu}$, and of $\ln\left[\epsilon_2^{-\beta_3} \Delta n(T,L)\right]$ vs. $\ln(L\epsilon_2^{\nu})$ (inset). Each plotted point represents 10 data points.

We find $\nu=0.64\pm0.09$ and $\kappa=-0.0129\pm0.0012$.

The width of the 2d region should behave as,

$$T_x(L)-T_{c2}(L) = JL^{-1/\nu} \qquad , \text{ J const.} \qquad (4)$$

we find $\nu=0.61\pm0.06$ and $J=0.0264\pm0.0016$.

The equation

$$T_x(L)-T_{c3}(L) = ML^{-1/\nu} \qquad , \text{ M const.} \qquad (5)$$

embodies the expectation that crossover should occur when ξ becomes comparable to L. Our results are $\nu=0.66\pm0.06$ and $M=0.0127\pm0.0010$.

These results support scaling theory predictions of finite size critical temperature shifts in films. Our values for ν agree with the RG value[5] of 0.630 ± 0.008. Within precision limits, $|\kappa|=M$ and $J/M=2$, indicating that $T_{c3}(L)$ is near the middle of the 2d domain.

The crossover is expected to occur when $L \simeq \xi$. If we set $L=m\xi$ and use Eq. 5 to eliminate L, we find $m=M^{\nu}/\xi_0 T_{c\infty}^{\nu}$, where ξ_0 is the correlation length amplitude. Using $\xi_0=2.9\text{Å}$ for 2,6-lutidine + water[6], we find $m=4^{+3}_{-2}$. The two phase amplitude ξ_0' should be used, but this hasn't been measured. Tarko and Fisher[7] have calculated $\xi_0/\xi_0'\simeq1.96$ for the 2nd moment correlation length, while Fisher[8] has suggested that $\xi_0/\xi_0'\simeq1.5$ for the true correlation range. In either case, m values larger by the cited factors result.

The large scatter in our data near crossover militates against drawing conclusions vis a vis the sharpness of crossover and the width of the crossover region.

The predicted L dependence of the 2d amplitude $A_2(L)$ is

$$A_2(L) = EL^{(\beta_2-\beta_3)/\nu} \qquad , \text{ E const.} \qquad (6)$$

The derivation of Eqs. 3-6 also provides the relation

$$E = A_3(J+\kappa)^{\beta_3}/J^{\beta_2} \qquad (7)$$

We can determine a value for E from Eq. 7, and thus can "predict" $A_2(L)$. All ten films for $L\leq2.2\mu m$ have A_2 values randomly scattered about the corresponding predictions. Results from seven films of greater L also agree in the above sense, but eleven films for $L>2.2\mu m$ show A_2 values larger than predicted by 25% to 60%. While the majority of our data are consistent with the scaling prediction, we cannot account for the large values of the remainder. Thus, we are unable to claim confirmation of Eq. 6.

In the case of the Ag mirrors, the data set from which the L
dependence of the virtual 3d region critical temperature was
extracted consists of thirty two $T_{c3}(L)$ values from coexistence
curve fits, plus eleven PST values for $T_{c3}(L)$ for $6\mu m < L \leq 182\mu m$.
We find,

$$T_{c3}(L) - T_{c\infty} = (0.047 \pm 0.002) L^{-(0.80 \pm 0.08)}$$

In addition, we measured thirty two PST for films bounded by the
dielectric material. Thicknesses ranged from $8\mu m$ to $272\mu m$. We
find, for the dielectric mirrors,

$$T_{c3}(L) - T_{c\infty} = -(0.24 \pm 0.08) L^{-(0.81 \pm 0.09)}$$

The power law exponents are equal for the two kinds of mirrors, but
a change in the mirror material altered both the direction and
amplitude of the $T_{c3}(L)$ shift.

The foregoing is strong evidence for the role of a film-mirror
interaction in activating the coexistence curve displacements.
However, the restricted geometry effects which scaling theory has
addressed, namely those which arise from the constraint on the
evolution of ξ in one dimension, are independent of this interaction.
We base this conclusion on our results: β_2, β_3 and A_3 are
independent of L and have the expected values, and the thickness
dependences of $T_{c2}(L)$ and $T_x(L)$ conform to free film scaling
predictions when they are referenced to the apparent divergence
temperature for ξ, $T_{c3}(L)$.

SUMMARY:

Except for our inability to verify the thickness dependence of
$A_2(L)$, our results on the 2,6-lutidine + water system confirm scaling
theory predictions of finite size effects in Ising class critical
fluid films. In addition, we have found evidence of an interaction
between the fluid films and the bounding surfaces, which is revealed
as a thickness dependent displacement of the coexistence curves in
temperature.

A major concern of ours, that critically wetted surface films[9]
might influence our results, did not materialize. The lack of any
systematic dependence of A_3 upon either L or temperature is evidence
that, if such layers form in the 2,6-lutidine + water films, their
influence is submerged in the scatter of our data.

In earlier work[10] our group reported the observation of mean
field behavior in coexistence curves for films of a critical mixture
of methanol + cyclohexane. We are repeating measurements on films

of this system and preliminary evidence indicates that the formation of wetted surface layers may play an important role.

ACKNOWLEDGEMENT:

    This work was supported in part by the U. S. Department of Energy, Contract DE-A502-76ER02203.

REFERENCES:

1. M. E. Fisher, J. Vac. Sci. Technol 10, 665 (1973); Critical Phenomena, ed. M.S. Green (Academic, New York, 1971).
2. D. T. Jacobs, D. J. Anthony, R. C. Mockler and W. J. O'Sullivan, Chem. Phys. 20, 219 (1977).
3. A. Stein and G. F. Allen, J. Phys. Chem. Ref. Data 2, 443 (1973).
4. A. W. Loven and O. K. Rice, Trans. Faraday Soc. 59, 2723 (1963).
5. S. C. Greer, Accts. Chem. Res. 11, 427 (1978).
6. E. Gülari, A. F. Collings, R. L. Schmidt and C. J. Pings, J. Chem. Phys. 56, 6169 (1972).
7. H. B. Tarko and M. E. Fisher, Phys. Rev. Lett. 31, 926 (1973).
8. M. E. Fisher, Fluctuations in Superconductors, W. S. Goree and F. Chitton, eds. (Stanford Research Institute, Stanford, Calif., 1968).
9. J. W. Cahn, J. Chem. Phys. 66, 3667 (1977).
10. D. T. Jacobs, R. C. Mockler and W. J. O'Sullivan, Phys. Rev. Lett. 37, 1471 (1976).

MELTING AND LIQUID CRYSTALS IN TWO DIMENSIONS

B. I. Halperin and D. R. Nelson

Lyman Laboratory of Physics
Harvard University
Cambridge, Massachusetts 02138

Abstract

The consequences of a dislocation-mediated theory of two-dimensional melting have been worked out for triangular lattices. Dissociation of dislocation pairs first drives a transition into a "hexatic" liquid crystal phase with exponential decay of translational order, but power law decay of six-fold orientational order. A subsequent dissociation of <u>disclination</u> pairs at a higher temperature then produces an isotropic fluid. Physical systems where the theory may be applicable include phase transitions in suspended smectic films and melting of the recently observed two-dimensional solid formed by electrons on a liquid helium surface. Light-scattering is potentially a very useful probe in the latter case, as the electron lattice spacing is comparable to the wavelength of light. Light scattering is also useful in liquid crystal-films if tilted molecules lead to a local birefringence of the film. We discuss the possible phases when both solidification and tilt transitions occur.

I. INTRODUCTION

The theory of melting in two-dimensions, and related problems of liquid-crystal phases in very thin films have received much attention recently. In this present talk, I will review some of the predictions of the theory of dislocation-mediated melting, whose details have been recently worked out by David Nelson and myself,[1-3] and by A. P. Young,[4] based on the proposals of Kosterlitz and Thouless.[5,6]

The theory of two-dimensional melting has a number of possible applications to experimental system. One system where light-scattering may be an important tool is the two-dimensional solid formed by electrons on the surface of liquid helium, at suffic-iently low temperatures and sufficiently high electron density.[7] This solid has been recently observed for the first time by Grimes and Adams, using a radio-frequency resonance technique.[8,9] Appli-cation of the dislocation theory of melting to the two-dimensional electron solid has been discussed by Thouless[10] and Morf.[11] Be-cause the electron densities employed are of the order of $4 \times 10^8/cm^2$, the lattice constant of the solid is comparable to the wavelength of visible light. Furthermore, in the solid phase, at least, each electron should be accompanied by a shallow depression or "dimple" in the surface of the helium.[9,12] The very weak scattering of light by these dimples gives one a possibility, at least in prin-ciple, of studying correlations near the melting transition.

Other systems of great experimental interest include various smectic liquid crystals, which can be prepared as free suspended films, with two or more molecular layers in the sample. A number of these films undergo transitions which may be described as the formation of a two-dimensional solid.[13] The degree to which light-scattering will be useful in the study of these transitions is not yet clear. In the simplest cases, the solid phase has a hexagonal symmetry (triangular lattice) giving an isotropic dielectric tensor in the plane of the sample. In such cases, there is no dramatic change in light scattering behavior as one passes through the melting temperature. In more complicated cases, however, the molecules of the substance are tilted away from the normal to the film. Such a film will be locally biaxial, and light scattering effects will be associated with fluctuations in the orientation of the tilted molecules. Indeed, light scattering has been observed in films as thin as two smectic layers, and has been used to study the transition between tilted (Smectic C) and non-tilted (Smectic A) phases [14,15]. The orientation of the molecules will couple to the local orientation of the "bonds" between molecules, and, under suitable conditions, one may be able to derive important informa-tion from the changes in light scattering in the neighborhood of the melting transition.

Phase transitions of various kinds have also been observed in lipid layers, floating on a water surface. Some of these transi-tions may be manifestations of two-dimensional melting.

The theory of two-dimensional melting is also applicable, with some modifications, to melting of a layer adsorbed on a crys-talline substrate, provided that the lattice constant of the adsor-bate is not locked to that of the substrate. Indeed, the calcula-tion of Reference 2 suggests that some aspects of the dislocation unbinding transition may be preserved even in the limit of infi-nitely strong substrate potentials, which is a crude way of model-ing chemisorption. Certain aspects of the commensurate-in-commensurate transition for an adsorbed solid layer can also be

related mathematically to the melting transition. Transitions
similar to melting may also occur in the reconstruction or distortion of a clean solid surface.

II. MELTING OF A TRIANGULAR LATTICE

We shall first review the theory of two-dimensional melting,
developed in Refs. (1)-(3). Consider the properties of a two-
dimensional triangular solid on a smooth substrate. By definition,
the solid has non-zero long wavelength elastic constants. The
structure factor exhibits[16] power law singularities,

$$S(\vec{q}) \sim |\vec{q}-\vec{G}|^{-2+\eta_{\vec{G}}} \quad , \tag{1}$$

near a set of reciprocal lattice vectors $\{\vec{G}\}$, with exponents $\eta_{\vec{G}}$
related to the Lamé elastic constants $\mu_R(T)$ and $\lambda_R(T)$ by

$$\eta_{\vec{G}} = k_B T |\vec{G}|^2 (3\mu_R + \lambda_R)/4\pi\mu_R(2\mu_R + \lambda_R) \quad . \tag{2}$$

These singularities, which replace the δ-function Bragg peaks found
in three-dimensional solids, reflect power law decay at large dis-
tances of the correlation function $\langle e^{i\vec{G}\cdot[\vec{u}(\vec{r}) - \vec{u}(\vec{0})]}\rangle$, where $\vec{u}(\vec{r})$
is the lattice displacement at point $\vec{r}$. One can also define an
order parameter (analogous to $e^{i\vec{G}\cdot\vec{u}}$) for bond orientations, namely

$$\psi \equiv e^{i6\theta} \quad , \tag{3}$$

where $\theta(\vec{r})$ is the orientation relative to the x-axis of a bond between two nearest neighbor atoms at $\vec{r}$. In a solid, θ is given in
terms of the displacement field,

$$\theta = \tfrac{1}{2}(\partial_y u_x - \partial_x u_y) \quad . \tag{4}$$

The solid phase exhibits long range orientational order, since
$\langle\psi*(\vec{r})\ \psi(\vec{0})\rangle$ approaches a nonzero constant at large $\vec{r}$.[17]
If melting is indeed characterized by an unbinding of dis-
location pairs at a temperature T_m, one expects that a density
$n_f(T)$ of free dislocations above T_m will lead to exponential decay
of the translational order parameter $e^{i\vec{G}\cdot\vec{u}}$, with a correlation
length

$$\xi_+(T) \approx n_f^{-\frac{1}{2}} \quad . \tag{5}$$

This length diverges as $T \to T_m^+$ [see (13) below]. The structure factor $S(\vec{q})$ is now finite at all Bragg points, and the Lamé coefficients vanish at long wavelengths. We shall see, however, that orientational order persists, in the sense that bond-angle correlations now decay algebraically,

$$<\psi*(\vec{r}) \ \psi(\vec{0})> \ \sim 1/r^{\eta_6 (T)} \quad . \tag{6}$$

This phase can be described as a liquid crystal, similar to a two dimensional nematic, but with a six-fold rather than two-fold anisotropy. The exponent $\eta_6(T)$ is related to the Frank constant $K_A(T)$, which is the coefficient of $\frac{1}{2}|\vec{\nabla}\theta|^2$ in the free energy density, by

$$\eta_6(T) = 18k_B T/\pi K_A(T) \quad . \tag{7}$$

We find that K_A is infinite just above T_m, but decreases with increasing temperatures, until a temperature T_i, where dissociation of _disclination_ pairs drives a transiiton into an isotropic phase in which both the translational and orientational order decays exponentially.

   The liquid-crystal phase is isomorphic to a d = 2 superfluid, except that ±60° disclinations play the role of vortices. The transition at T_i should belong to the same universality class as the superfluid transition, and we except, in particular, that $\eta_6(T_i) = \frac{1}{4}$. Although disclination pairs are very tightly bound in the solid phase, screening by a gas of free dislocations produces a weaker logarithmic binding for $T_m < T < T_i$. It is interesting to note that an isolated dislocation can itself be regarded as a tightly bound disclination pair,[18] separated by one lattice constant. Evidence for the existence of the new liquid crystal phase, which we denote the "hexatic phase," has been found recently in numerical simulations of a two-dimensional gas with 6-12 potentials, by Frenkel and McTague.[19]

   To see the origins of these results, let us decompose the displacement field of a solid into a smoothly varying phonon field $\vec{\phi}(\vec{r})$, and a part due to dislocations.[5] The Hamiltonain $\mathcal{H}_E$ for the solid, within continuum elasticity theory,[18] then breaks into two parts, $\mathcal{H}_E = \mathcal{H}_0 + \mathcal{H}_D$, with

$$\mathcal{H}_0/k_B T = \frac{1}{2} \int \frac{d^2 r}{a_0^2} [2 \ \bar{\mu}\phi_{ij}^2 + \bar{\lambda}\phi_{ii}^2] \quad , \tag{8}$$

$$\mathcal{H}_D/k_B T = - \frac{K}{8\pi} \sum_{R \neq R'} \left[\vec{b}(\vec{R}) \cdot \vec{b}(\vec{R}') \ell n(|\vec{R}-\vec{R}'|/a) \right.$$

$$\left. - \frac{\vec{b}(\vec{R}) \cdot (\vec{R}-\vec{R}')\vec{b}(\vec{R}') \cdot (\vec{R}-\vec{R}')}{|\vec{R}-\vec{R}'|^2} \right]$$

$$+ (E_c/k_B T) \sum_{\vec{R}} |\vec{b}(\vec{R})|^2 \quad . \tag{9}$$

In (8), ϕ_{ij} is related to the smooth part of the displacement
field $\phi_{ij} = \frac{1}{2}(\partial_i \phi_j + \partial_j \phi_i)$, and $\bar{\mu}$ and $\bar{\lambda}$ are "reduced" elastic constants, given by the usual Lamé coefficients μ and λ multiplied
by the squared lattice spacing a_0^2 and divided by $k_B T$. In (9),
$\vec{b}(\vec{R})$ is a dimensionless dislocation Burger's vector of the form
$\vec{b}(\vec{R}) = m(\vec{R})\vec{e}_1 + n(\vec{R})\vec{e}_2$, where $m(\vec{R})$ and $n(\vec{R})$ are integers, and $\vec{e}_1$
and $\vec{e}_2$ are unit vectors spanning the underlying Bravais lattice.
Here, we restrict ourselves to the triangular lattice $(\vec{e}_1 \cdot \vec{e}_2 = \frac{1}{2})$.
The sums in (9) are over, say, a square mesh with spacing $\underline{a}$ of
sites in physical space, and the $\vec{b}(\vec{R})$ must satisfy a vector charge
neutrality condition, $\sum_{\vec{R}} \vec{b}(\vec{R}) = 0$. The quantity K is given by
$K = 4\mu(\bar{\mu}+\bar{\lambda})/(2\bar{\mu}+\bar{\lambda})$, and E_c is the core energy of a dislocation.

If dislocations only exist in bound pairs at low temperatures,
one expects that they can be ignored, and that the long wavelength
properties of the solid will simply be given by (8), with suitably
renormalized elastic constants. The properties of the solid phase
quoted above follow directly from this observation.

The properties of H_D are studied using the renormalization
group approach developed by Kosterlitz[6] for the two-dimensional
superfluid transition. Recursion relations for K and for
$y \equiv e^{-E_c/k_B T}$ can in fact be obtained rather straightforwardly, by
considering the renormalization of elastic constants due to dislocation pairs, in analogy to calculations of the effect of vortices on the superfluid density in a ^{4}He film.[20] Integrating
over mesh sizes between $\underline{a}$ and $\underline{a}\, e^{\ell}$, we obtain partially dressed
parameters $\bar{\mu}(\ell)$, $\bar{\lambda}(\ell)$, $y(\ell)$, and $K(\ell)$, which satisfy, to $O[y^2(\ell)]$,

$$\frac{d\mu^{-1}}{d\ell} = 3\pi y^2 e^{K/8\pi} I_0 \left[\frac{K}{8\pi} \right] \quad . \tag{10}$$

$$\frac{d[\overline{\mu} + \overline{\lambda}]^{-1}}{d\ell} = 3\pi y^2 e^{K/8\pi}\left[I_0\left(\frac{K}{8\pi}\right) - I_1\left(\frac{K}{8\pi}\right)\right] \quad , \tag{11}$$

$$\frac{dy}{d\ell} = \left(2 - \frac{K}{8\pi}\right) y + 2\pi y^2 e^{K/16\pi}I_0\left(\frac{K}{8\pi}\right) \quad , \tag{12}$$

where $I_0(x)$ and $I_1(x)$ are modified Bessel functions of the first and second kind. We find that $K^{-1}(\ell) = \frac{1}{4}\{\overline{\mu}^{-1}(\ell) + (\overline{\mu}(\ell) + \overline{\lambda}(\ell))^{-1}\}$ for all ℓ, so that its recursion relation can be obtained trivially from (10) and (11).

The parameter $y(\ell)$ is driven to zero at large ℓ, for all temperatures below a critical value T_m. Above T_m, $y(\ell)$ is ultimately driven toward large values and $K(\ell)$ is driven towards zero, an instability we associate with dislocation pair unbinding.

We determine the behavior near T_m by studying (3-5) near the critical value $K_c = 16\pi$. We identify the correlation length $\xi_+(T)$ with $a\,e^{\ell*}$, with $\ell*$ chosen such that $K(\ell*) \approx \frac{1}{2}K_c$. In this way, we find that

$$\xi_+(T) \approx a \exp[b/(T/T_m - 1)^{.36963\cdots}] \tag{13}$$

as $T \to T_m^+$, where b is a constant, and $.36963\ldots$ can be expressed in terms of the roots of a quadratic equation with Bessel function coefficients. The specific heat exhibits only an essential singularity, $C_p \approx \xi_+^{-2}$, while the structure factor at the Bragg points is given by $S(\vec{G}) \approx \xi_+^{2-\eta_{\vec{G}}}$. Taking over the discussion for the superfluid density in Ref. 20, we find that the reduced shear modulus in the solid phase is $\overline{\mu}_R(T) = \lim_{\ell\to\infty}\overline{\mu}(\ell)$. It follows from Eqs. (10)-(12) that $\mu_R(T)$ approaches a finite limiting value as $T \to T_m^-$. Just below T_m we find

$\mu_R(T) = \mu_R(T_m)[1 + \text{const.}(T_m - T)^{.36963\cdots}]$ with a similar result for $\lambda_R(T)$. There is a universal relationship involving the elastic constants at T_m,

$$\lim_{T\to T_m^-}\left\{\frac{1}{\mu_R(T)} + \frac{1}{\mu_R(T) + \lambda_R(T)}\right\} = \frac{a_0^2}{4\pi k_B T_m} \tag{14}$$

This corresponds to the critical value $K_c = 16\pi$, and is also suggested by the "entropy argument" of Kosterlitz and Thouless.[5]

The results for orientational correlations above T_m follow from a calculation of the Frank constant K_A,

$$k_B T/K_A = \lim_{q \to 0} q^2 \langle \hat{\theta}(\vec{q}) \, \hat{\theta}(-\vec{q}) \rangle$$

$$= \lim_{q \to 0} \frac{q_i q_j}{q^2} \langle \hat{b}_i(\vec{q}) \, \hat{b}_j(-\vec{q}) \rangle \, a_0^2 \qquad (15)$$

where $\hat{\theta}(\vec{q})$ and $\hat{b}_i(\vec{q})$ are the Fourier transformed orientational and Burger's-vector fields, respectively. The second line of (15) follows because the contribution of $\vec{\phi}(\vec{r})$ to K_A^{-1} is zero and because the dislocation part of $\hat{\theta}(\vec{q})$ is just [2,18] $\hat{\theta}(q) = ia_0 q_j \hat{b}_j(\vec{q})/q^2$. To estimate K_A just above T_m, we use its transformation properties under the renormalization group, $K_A[K(0), y(0)] = e^{2\ell} K_A[K(\ell), y(\ell)]$. Choosing $\ell = \ell* = \ln(\xi_+/a)$, we can evaluate K_A using Debye-Hückel theory, which treats $\vec{b}(\vec{R})$ as a continuous vector field, rather than restricting it to discrete points on a Bravais lattice. Upon Fourier transformation, $\mathcal{H}_D$ becomes

$$\mathcal{H}_D/k_B T = \frac{1}{2} \sum_{\vec{q}} \frac{1}{2} \left[\frac{K}{q^2} \left(\delta_{ij} - \frac{q_i q_j}{q^2} \right) + \frac{2E_c a^2}{k_B T} \delta_{ij} \right] \hat{b}_i(\vec{q}) \, \hat{b}_j(-\vec{q}) \, . \qquad (16)$$

Since the term proportional to the transverse projection operator in (16) does not contribute to (15), one obtains $K_A[K(\ell*), y(\ell*)] \approx 2E_c(\ell*) = O(k_B T_m)$. It follows that the physical Frank constant is $K_A \approx \xi_+^2(T)$. The algebraic decay of orientational order above T_m, and the relationship between $\eta_6(T)$ and $K_A(T)$, are straightforward consequences of this result.

It should be emphasized that we have only explored consequences of the dislocation model of melting perturbatively in $y = e^{-E_c/k_B T}$. Although the theory is stable and self-consistent, we cannot rule out other mechanisms for melting, perhaps leading to a first order transition. A "premature" unbinding of disclinations (before dislocations dissociate) might constitute such a mechanism.

III. TILTED MOLECULES

In order to describe a phase in which there are tilted molecules, we introduce an order parameter $\Phi(r) = e^{i\phi(r)} \sin \gamma$ where γ is the magnitude of the local tilt angle, and $\phi(r)$ describes the orientation of the projection of the molecular axis onto the x-y plane. As usual we may neglect fluctuations in the magnitude of the order parameter (sin γ) except that we must keep track of disclinations, where $\Phi(r)$ vanishes. Coupling between the tilt axis and bond orientations is expressed by a term in the free energy of form

$$-\lambda \, \mathrm{Re}[\psi^*\phi^6] \, \propto \, -\lambda \cos(6\phi-6\theta) \qquad (17)$$

A positive value of λ means that a molecular tilt axis tends to line up along one of the six nearest neighbor bond directions, while a negative value of λ favors a tilt towards the midpoint between two nearest neighbors.

We now discuss some of the phases which can occur in a thin smectic film, when the local tilt angle γ is non-zero. A fuller discussion and derivation will be given elsewhere.[21]

(1). The two correlation functions, $\langle\psi^*(\vec{r}) \, \psi(0)\rangle$ and $\langle\phi^*(\vec{r}) \, \phi(0)\rangle$ may both fall off exponentially with r, (short-range order). This we identify with the smectic A phase, or an isotropic two-dimensional fluid.

(2). The order parameter ψ may show quasi-longrange order, as in Eq. (6) above, while ϕ has only short-range order. This is identical to the hexatic phase, described above.

(3). The molecules may form a triangular solid while ϕ has only short-range order. This is the solid phase described above, with finite shear modulus, power-law Bragg peaks and $\langle\psi\rangle \neq 0$ (true longrange order in the bond-angle field). Recent x-ray measurements on thin films[13] and bulk samples[22] of one material indicate that this solid phase should be identical with the smectic B phase.

(4). There can be a solid phase, in which both ψ and ϕ show true longrange order. The relative orientation between the order parameters is such that $\langle\psi^*\rangle\langle\phi\rangle^6$ is positive or negative, depending on the sign of λ in Eq. (17). Coupling between the tilt and bond orientations will inevitably produce some shear in the crystal, so that the lattice will no longer possess perfect hexagonal symmetry. We identify the present state with the smectic H phase.

(5). There can be a liquid crystal phase which we identify with the smectic C phase, in which both ψ and ϕ show quasi-longrange order and the relative orientations of ψ and ϕ are "locked." Let us define exponents η_1 and η_x by

$$\langle\phi^*(\vec{r}) \, \phi(0)\rangle \sim 1/r^{\eta_1} \qquad (18)$$

$$\langle\psi^*(\vec{r}) \, \phi^6(0)\rangle \sim \pm \, 1/r^{\eta_x} \qquad (19)$$

[The sign in Eq. (19) is determined by the sign of λ in Eq. (17).]

The exponents in this locked phase are related by

$$\eta_6 = \eta_\times = 36\eta_1 \tag{20}$$

where η_6 is defined by Eq. (6). There is a single Frank constant
which gives the free-energy cost of spatial variations in orientation, where ψ and Φ vary simultaneously. The exponents η_6, etc.,
are related to the Frank constant by expressions analogous to Eq.
(7). The exponent η_1 must be less than ¼, for the phase to be
stable against the formation of 360° disclinations, in the combined
ψ and Φ fields.[23,20]

(6). There can exist a phase in which the ψ field and Φ show
"independent" quasi-longrange order. Now there are no simple rela-
tions among the exponents η_1, η_6 and $\eta_\times$, although various inequal-
ities must be satisfied for stability of the phase. There are
three independent Frank constants which describe the energy cost
of independent gradients in the orientations of ψ and Φ.

(7). In a simplified model, there exists a solid phase (long-
range order in ψ) in which the Φ field shows quasi-longrange order,
as in Eq. (18). This is the X-Y like phase discussed by José et
al.,[24] for a two-dimensional X-Y magnet in the presence of six-fold
crystalline anisotropy. This phase is indeed a possibility for
orientable molecules in a rigid lattice, or for molecules adsorbed
on a three-dimensional crystalline substrate. For a free standing
film however, the XY-like solid phase appears to be unstable with
respect to shear of the lattice. The probable result is a uniaxial crystal, with Φ locked to ψ, and true longrange order in both
fields -- i.e. the smectic H phase described in (4) above.

It is worth emphasizing that there cannot exist a phase with
longrange order or quasi-longrange order in the Φ field but only
short-range order in the ψ-field. Roughly speaking, we may say
that a non-zero value of Φ in some region will generate via Eq.
(17) an effective field which couples linearly to ψ. This will
tend to align ψ and induce an order in the ψ field, unless the ψ
field has already acquired a quasi-longrange order of its own as
in phase (6) above.

We may remark that phases (1)-(5) should have counterparts
in bulk smectic phases,[25] although the hexatic phase has not been
established.[13,22] [In bulk systems, of course, quasi-longrange
order is replaced by true longrange order.] Phase (6),
with independent tilt and hexatic bond orientations, cannot exist
in bulk.

The nature of the transitions between the various two-
dimensional phases described above, and the possible phase diagrams
that can result, will be explored in a future publication.[21]

Acknowledgments

The authors are grateful for helpful discussions with R. J. Birgeneau, D. S. Fisher, C. C. Grimes, J. D. Litster, D. E. Moncton, R. Morf, P. Pershan, R. Pindak and A. P. Young. Research has been supported in part by the National Science Foundation through the Harvard Material Research Laboratory and through Grant No. DMR77-10210.

References

1. B. I. Halperin and D. R. Nelson, Phys. Rev. Lett. $\underline{41}$, 121 (1978); E $\underline{41}$, 519 (1978).
2. D. R. Nelson and B. I. Halperin, Phys. Rev. B$\underline{19}$, 2457 (1979).
3. D. R. Nelson, Phys. Rev. B$\underline{18}$, 2318 (1978).
4. A. P. Young, Phys. Rev. B$\underline{19}$, 1855 (1979).
5. J. M. Kosterlitz and D. J. Thouless, J. Phys. C$\underline{6}$, 1181 (1973).
6. J. M. Kosterlitz and D. J. Thouless, Prog. in Low Temp. Phys. (to be published); J. M. Kosterlitz, J. Phys. C$\underline{7}$, 1046 (1974).
7. For a recent review of electrons on helium, see C. C. Grimes, Surf. Sci. $\underline{73}$, 379 (1978).
8. C. C. Grimes and G. Adams, Phys. Rev. Lett. $\underline{42}$, 795 (1979).
9. D. S. Fisher, B. I. Halperin and P. M. Platzman, Phys. Rev. Lett. $\underline{42}$, 798 (1979).
10. D. J. Thouless, J. Phys. C$\underline{11}$, L189 (1978).
11. R. Morf (Harvard preprint).
12. Yu. P. Monarkha and V. B. Shikin, Zh. Eksp. Teor. Fiz. $\underline{68}$, 1423 (1975) [Sov. Phys. JETP $\underline{41}$, 710 (1976)].
13. D. E. Moncton and R. Pindak, Bull. Am Phys. Soc. $\underline{24}$, 251 (1979); preprint; and private communication.
14. C. Y. Young, R. Pindak, N. A. Clark, and R. B. Meyer, Phys. Rev. Lett. $\underline{40}$, 773 (1978).
15. C. Rosenblatt, R. Pindak, N. A. Clark, and R. B. Meyer, Phys. Rev. Lett. $\underline{42}$, 1220 (1979).
16. See, e.g., Y. Imry and L. Gunther, Phys. Rev. B$\underline{3}$, 3939 (1971); B. Jancovici, Phys. Rev. Lett. $\underline{19}$, 20 (1967).
17. N. D. Mermin, Phys. Rev. $\underline{176}$, 250 (1968).
18. F. R. N. Nabarro, Theory of Dislocations, (Clarendon, New York 1967).
19. D. Frenkel and J. P. McTague (UCLA preprint).
20. D. R. Nelson and J. M. Kosterlitz, Phys. Rev. Lett. $\underline{39}$, 1201 (1977).
21. D. R. Nelson and B. I. Halperin, (manuscript in preparation).
22. P. Pershan, G. Aeppli, R. Birgeneau, and D. Litster (private communication). A review of recent experiments on smectic liquid crystals, including this work, has been given by J. D. Litster, R. J. Birgeneau, M. Kaplan, C. R. Safinya, and

J. Als-Nielsen, Lecture notes from NATO Advanced Study Institute: "Ordering in Strongly-Fluctuating Condensed Matter Systems," Geilo, Norway, April 17-28, 1979 (Plenum Press, to be published).

23. Cf. R. A. Pelcovits and B. I. Halperin, Phys. Rev. B19, (1979) (in press).

24. J. José, L. P. Kadanoff, S. Kirkpatrick, and D. R. Nelson, Phys. Rev. B16, 1217 (1977).

25. R. J. Birgeneau and J. D. Litster, J. Phys. Lett. (Paris) 39, L 399 (1978).

LIGHT SCATTERING STUDIES OF MOLECULAR ORIENTATION FLUCTUATIONS IN TWO DIMENSIONS

Noel A. Clark

Department of Physics and Astrophysics
University of Colorado
Boulder, Colorado 80309

The static and dynamic behavior of the two dimensional orientation field of molecules in the plane of thin, freely-suspended smectic C liquid crystal films is effectively studied by light scattering. Films of 1 to 10 layers in thickness (30Å per layer) have been studied. Thin smectic C (SC) films are a nearly ideal physical representation of the Classical Continuous XY Model. Light scattering with well-defined wave-vector has been used to study the dispersion relations for the two allowed orientation fluctuation modes in two dimensions (bend and splay). These exhibit Frank-like elastic behavior with, in ferroelectric SC films, an additional term in the bend mode dispersion relation due to polarization space charge interaction effects. Exploitation of this latter term allows the absolute measurement of the polarization, elastic constants, and viscosities. The Frank elastic constants decrease with increasing temperature to values comparable to those predicted for the Kosterlitz-Thouless transition at which point a phase transition to an isotropic (smectic A) film is found to occur. Measurements of the 2-d elastic constants vs. film thickness shows that, 3-d Frank elasticity is dispersive (nonlocal) at wave vectors $q \sim 10^6 cm^{-1}$. Light scattering studies which employ the collection of a broad distribution of wave vectors have also been carried out. These allow the direct visualization of the 2-d molecular orientational field, a feature unique to these systems. The static and dynamic behavior of spontaneous and field induced point and line defects has been investigated using this method. It has also been used to study directly the fluctuations in the average local orientation which show contributions of modes having inverse wavevectors ranging from the size of the averaging area to the size of the sample.

MULTIPHONON BOUNDARY OF THE EXCITATION SPECTRUM

L. P. Pitaevsky

Institute of Physical Problems
Academy of Sciences
Moscow, USSR

1.   In this paper we investigate the creation of elementary excitations by neutrons or light inelastic scattering when the conservation laws only permit the creation of a large number of excitations.

It is now well known (see, for example, Ref.1) that the excitation spectrum in helium has the form shown in Fig.1: At normal pressure the spectrum near the origin goes above the sonic line

$$\varepsilon \;=\; up \tag{1.1}$$

(u is the sound velocity), and is described by the expression

$$\varepsilon \;=\; up + \gamma p^3 \;. \tag{1.2}$$

However, at some distance from the origin the excitation velocity starts to decline and at $p = p^*$ the spectrum intersects the sonic line and then goes below it.

Suppose now, that in inelastic scattering the liquid acquires the energy ε and momentum p. This process cannot occur by the creation of phonons if the point ε, p is below the sonic line. For points above this line, the necessary number of phonons increases as the sonic line is approached and the difference

$$\delta\varepsilon \;=\; \varepsilon - up \tag{1.3}$$

is decreased. The probability of scattering for the given ε and p is described by the dynamic form-factor of the liquid. For $\delta\varepsilon \to 0$ this form-factor must vanish. We wish to find the way in which it vanishes along the whole sonic line (1·1).

Let us find the smallest number of phonons among which the energy ε and momentum p can be distributed. Clearly, the momenta of the phonons should be equal, since the total momentum for the given energy will then be largest. If n phonons are produced with energy, ω, and momentum, k, so that $\omega = \varepsilon/n$, $k = p/n$, we obtain from Eq.(1·2)

$$\frac{\varepsilon}{n} = u\,\frac{p}{n} + \gamma\,\left(\frac{p}{n}\right)^3$$

or

$$n = \frac{p^{3/2}\,\gamma^{1/2}}{(\delta\varepsilon)^{1/2}}\,.$$ (1·4)

Note that $n \to \infty$, if $\delta\varepsilon \to 0$. All of our calculations will be based on the assumption that $n \gg 1$.

Up to now we have only discussed liquid helium and we shall only consider this case in our paper. However, it is clear that the same problem arises in a solid, if its phonon spectrum has the form shown in Fig.1. Our method of solution can be used for a solid also, but our expression for the phonon emission probability may change when anisotropy is taken into account.

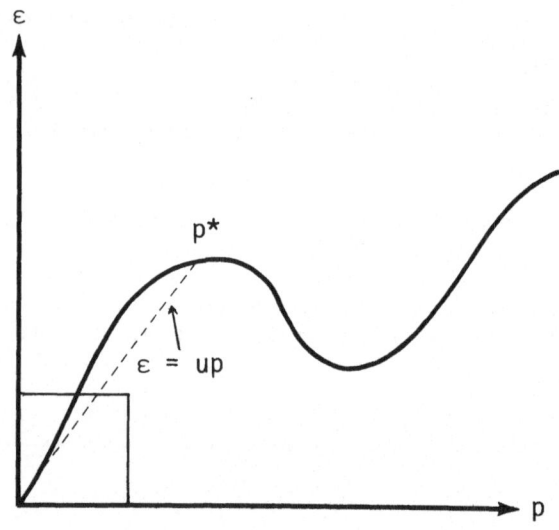

Fig. 1.

2. It is easy to obtain the probability of a two-phonon scattering in first order perturbation theory when the anharmonicity is cubic. However, to obtain the probability for an n-phonon process it is necessary to use n-th order perturbation theory in the cubic anharmonicity or to take into account higher order anharmonicities. For large n it is practically impossible to solve the problem in this way. The number of various Feynman graphs becomes very large and it is very difficult to evaluate the whole sum. It may not converge and the signs of the different terms vary.

Clearly, for large n, we should be able to use a semiclassical approach. We shall see that in this limit the calculations are rather simple. It is very instructive to consider first a simpler case: a system with one degree of freedom. Let us calculate the probability of exciting a slightly anharmonic oscillator with a high-frequency external field. The potential energy of the oscillator is

$$U(x) = \frac{m \omega_0^2}{2} x^2 + gx^3 \quad . \tag{2·1}$$

In the presence of a weak external field, described by the Hamiltonian

$$V = -Fx(e^{-i\varepsilon t} + e^{i\varepsilon t}) \ ,$$

the probability for exciting the oscillator is given, in the quadratic approximation, by

$$w = 2\pi\delta(\varepsilon - \varepsilon_n + \varepsilon_0) \ |M_{n0}|^2 F^2 \ ,$$

$$M_{n0} = - \int \psi_n^* x \psi_0 dx \quad . \tag{2·2}$$

Let us assume that the field frequency ε greatly exceeds the eigen frequency of the oscillator ω_0:

$$\varepsilon \approx (n + \frac{1}{2})\omega_0, \qquad n \gg 1. \tag{2·3}$$

In this limit, the first non-vanishing contribution to the matrix element M_{n0} is obtained in n-th order perturbation theory in the interaction constant g. When the condition (2·3) is satisfied, matrix element (2·2) is exponentially small. It can be evaluated

by Landau's method (see Ref.3, §52). According to Landau the value of the matrix element is defined essentially by the analytical continuation of the semiclassical wave function to the singular point of the potential energy in the complex plane(x_0). If there are several singular points we must take the one which gives the largest value for the matrix element.

When the potential energy has the form (2·1) and no finite singular points, we must let $x_0 \to \infty$. So with the exponential accuracy (omitting the phase factors) we obtain

$$M_{n0} \sim \exp \left\{ \int^{\infty} (\sqrt{2m(U(x)-\varepsilon)} - \sqrt{2mU(x)})dx \right\} . \qquad (2·4)$$

We shall assume that the anharmonicity constant g is small in the sense that $\varepsilon \ll \varepsilon_g$, where $\varepsilon_g \sim m^3 \omega_0^6 g^{-2}$ is the value of energy, for which the anharmonic effects are of the order of unity (see Fig.2). The main contribution to the integral (2·4) then comes from distances x for which

$$\varepsilon \ll U(x) \ll \varepsilon_g . \qquad (2·5)$$

Expanding in ε (which is small according to (2·5)) we have from (2·4)

$$M \sim \exp(-\varepsilon\tau) , \qquad \tau = \int^{\infty} \sqrt{\frac{m}{2U(x)}} \, dx . \qquad (2·6)$$

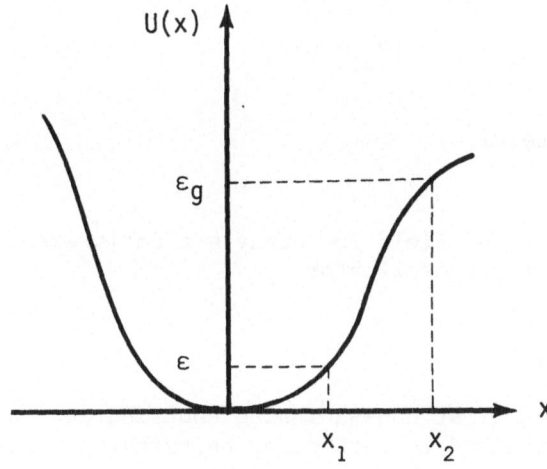

Fig. 2.

This formula may be interpreted as follows: In the classically forbidden region (2·5) the momentum of the oscillator is purely imaginary $p(x) = i\sqrt{2m(U-\varepsilon)}$. We may say that the particle is moving with an imaginary velocity, and interpret τ as the imaginary time needed for the particle to reach infinity. In the domain producing the essential contribution we can neglect anharmonicity and put $U = (m\omega_0^2/2)x^2$ cutting off the logarithmically divergent integral at the value $x_1 \sim (\varepsilon/m\omega_0^2)^{1/2}$ below, and at the value $x_2 \sim (m\omega_0^2/g)$ above (where the anharmonicity is of the order of unity). To logarithmic accuracy we then have

$$\tau \approx \frac{1}{\omega_0} \ln \frac{x_2}{x_1} = \frac{1}{\omega_0} \ln \{\frac{(m\omega_0^2)^{3/2}}{\varepsilon^{1/2} g}\}$$

whence the probability of excitation is

$$w \sim |M_{n0}|^2 \sim \exp \{- \frac{2\varepsilon}{\omega_0} \ln(\frac{\varepsilon g}{\varepsilon})^{1/2}\} = \frac{\varepsilon^n g^{2n}}{(m\omega_0^2)^{3n}} . \tag{2·7}$$

This very simple calculation depends essentially on the large value of the logarithm in τ. It is clear from (2·7) that this logarithm gives a dependence on the interaction constant g, consistent with n-th order perturbation theory.

Let us return to our main problem —— the instantaneous emission of n phonons. We shall show that the equivalent of the formula (2·6) for the matrix element is valid for a many-body system.

If a system has generalized coordinates x_i, its semiclassical wave function with the energy ε (essentially the part of this function of interest) is:

$$\psi \sim \exp[iS_0(x_i,\varepsilon)] ,$$

where $S_0(x_i,\varepsilon)$ is the "abbreviated action", that is, the quantity related to the total action $S(x_i,t)$ of the system by

$$S(x_i,t) = S_0(x_i,\varepsilon) - \varepsilon t . \tag{2·8}$$

Landau's arguments show that the matrix element for the transition from the ground state is

$$M \sim \exp\{i[S_0(x_i^0,\varepsilon) - S_0(x_i^0,0)]\} \approx \exp[i(\frac{\partial S_0}{\partial \varepsilon})_{\varepsilon=0}\varepsilon] \; .$$

It is easy to show that the derivative $\frac{\partial S_0}{\partial \varepsilon} = t(x_i^0)$, where $t(x_i^0)$ is the time needed for the system to reach the point x_i^0 from some point in a classically accessible region. Indeed, from (2·8) and the identity for $S(x_i,t)$

$$dS = p_i dx_i - \varepsilon dt$$

it follows that

$$dS_0 = p_i dx_i + td\varepsilon \; .$$

When the interactions are weak, we must take values of x_i^0 for which the anharmonicity is of order unity. These x_i^0 will be situated in the classically forbidden region. The equivalent time $t = i\tau(x_i^0)$ is then imaginary, and the matrix element is given by

$$M \sim \exp(-\varepsilon\tau) \tag{2·9}$$

which coincides with (2·6).

From Eq.(2·9) we may calculate the probability of a neutron creating phonons whose total energy and momentum is near the sonic line.

We take for generalized coordinates of our system the density variation at each point of the fluid:

$$\rho'(x) = \rho(x) - \bar{\rho}$$

or the Fourier components of this variation. The conjugate momenta will be the Fourier components of the fluid velocity potential (e.g., Ref.3, §24). The fluid-neutron interaction can be described by an interaction potential of the form

$$V = \frac{2\pi a}{m\mu} \rho'(x) \tag{2·10}$$

(a is the scattering length, m is the mass of the atom, and μ is the reduced mass of atom and neutron). So the problem is reduced to calculating the matrix element of the Fourier component of the fluid density with wave number p between the ground state and the state with the energy ε and the momentum p:

$$M \sim < \varepsilon \mid \rho_p \mid 0 > .$$

The character of the final state is clear from physical considerations. Since phonons are created with almost equal momenta
(see Ref.1), this state must be a wave packet sharply peaked in
momentum space about

$$k_0 = \frac{p}{n} , \qquad\qquad n = \frac{p^{3/2} \gamma^{1/2}}{(\delta\varepsilon)^{1/2}} \qquad\qquad (2\cdot11)$$

with total energy ε and total momentum p. (This normalization
corresponds to the energy-dependent cutoff in the integral for the
oscillator problem. As we shall see below, the energy dependence
of τ is logarithmic).

Now the solution of the problem is given directly by the
expression (2·9), where τ represents the imaginary time required
for a density perturbation in the wave packet to reach a value of
the order of the undisturbed density $\bar\rho$. It is obvious that, if
$\rho' \sim \bar\rho$, the anharmonicity effects will be of the order of unity.
The shape of the wave packet is chosen to minimize the time.

Let us perform the necessary calculation. The Fourier components of the fluid density can be written in the form:

$$\rho_n = Af(\frac{k_{||} - k_0}{\Delta_{||}} , \frac{k_\perp}{\Delta_\perp}) ,$$

where the function $f(\alpha,\beta)$ is normalized by the condition $f(0,0)=1$.
A is the normalization constant, $\Delta_{||}$ and $\Delta_\perp$ are the widths of the
packet in the directions along and perpendicular to $\vec{p}$, and
$k_{||}$ and $k_\perp$ are the appropriate projections of the wave vector. With
the accuracy assumed in our method we only need to know to an order
of magnitude the dependence of A, $\Delta_{||}$ and $\Delta_\perp$ on the parameters of
the problem. We shall therefore omit numerical constants in the
calculations that follow. The connection between A and the
quantities $\Delta_{||}$ and $\Delta_\perp$ is determined from the expression for the
energy in the harmonic approximation:

$$\varepsilon = \frac{u^2}{\bar\rho^2} \int |\rho_k|^2 \frac{d^3k}{(2\pi)^3} ; \text{ or approximating} \qquad\qquad (2\cdot12)$$

$$\varepsilon \simeq \frac{u^2}{\bar\rho^2} A^2 \Delta_{||} \Delta_\perp^2 .$$

Thus

$$A \sim (\bar{\rho}\varepsilon/u^2 \Delta_{||} \Delta_{\perp}^2)^{1/2}$$

and

$$\rho_k \sim \frac{\bar{\rho}}{u}^{-1/4} \varepsilon^{1/2} \Delta_{||}^{-1/2} \Delta_{\perp}^{-1} f(\frac{k_{||}-k_0}{\Delta_{||}}, \frac{k_{\perp}}{\Delta_{\perp}}).$$

The time dependence of the Fourier component is given by the factor $\rho_k \sim \exp(i\omega_k t) \sim \exp(u k \tau)$. Consequently, the τ-dependence of the fluid density in the center of the wave packet is

$$\rho(r=0) \sim \frac{\bar{\rho}}{u}^{-1/2} \varepsilon^{1/2} \Delta_{||}^{1/2} \Delta_{\perp} \exp(\omega_{k_0}\tau), \quad \omega_{k_0} = \frac{\varepsilon}{n} \qquad (2\cdot13)$$

and the condition that, $\rho'(r=0) \sim \bar{\rho}$ implies that

$$\tau \approx \frac{\varepsilon}{n} \ell n \frac{u\bar{\rho}^{-1/2}}{\varepsilon^{1/2}\Delta_{||}^{1/2}\Delta_{\perp}} . \qquad (2\cdot14)$$

To make τ small, the wave packet should not be very narrow, i.e., the quantities $\Delta_{||}$ and $\Delta_{\perp}$ must not be very small. On the other hand, a large spread in the widths would be inconsistent with the normalization condition for the momentum analogous to $(2\cdot12)$,

$$p = \frac{u^2}{\bar{\rho}} \int \frac{k}{\omega(k)} |\rho_k|^2 \frac{d^3k}{(2\pi)^3} .$$

For small $k_{||} - k_0$ and $k_{\perp}$ we obtain

$$\frac{k_{||}}{\omega(k)} \approx \frac{1}{u}(1 - \frac{k_{\perp}^2}{k_0^2} - \frac{\gamma}{u} k_{||}^2$$

and, correspondingly

$$\frac{\varepsilon-pu}{\varepsilon} = \frac{\delta\varepsilon}{\varepsilon} \sim \frac{\gamma k_0^2}{u} + \frac{\gamma}{u} \Delta_{||}^2 \frac{\Delta_{\perp}^2}{k_0^2} . \qquad (2\cdot15)$$

From this relation we see that in order to make the smearing of $\delta\varepsilon/\varepsilon$ less than its mean value, it is necessary to have $\Delta_{||}^2 \ll u\delta\varepsilon/\gamma\varepsilon$ and $\Delta_{\perp}^2 \ll u\delta\varepsilon^2/\gamma\varepsilon^2$. Since the dependence of τ on $\Delta_{\perp}$ and $\Delta_{||}$ is logarithmic, we can, with logarithmic accuracy, use these bounds as estimates:

$$\Delta_{||} \sim (\frac{u}{\gamma})^{1/2}(\frac{\delta\epsilon}{\epsilon})^{1/2} \quad , \qquad \Delta_{\perp} \sim (\frac{u}{\gamma})^{1/2}\frac{\delta\epsilon}{\epsilon} \quad , \tag{2.16}$$

Inserting (2·16) into (2·14), we obtain in conventional units:

$$\tau \approx \frac{n}{\epsilon\hbar} \ell n \, (\frac{\hbar^3\bar{\rho}^{1/2}u^4}{\epsilon^3\gamma^{1/2}} \, n^{5/2}) \quad . \tag{2.17}$$

Finally, using (2·9), we obtain asymptotic expression for the multiple phonon emission probability (i.e., for the tail of the dynamic form-factor):

$$w \sim \exp(-2\epsilon\tau) \sim (\frac{\epsilon^6\gamma}{\hbar^3\bar{\rho}u^8})^n \, \exp(-5n \, \ell n \, n) \quad . \tag{2.18}$$

The relation of n to $\delta\dot{\epsilon}$ is given by (1·5). As we have explained above the procedure is only justified when the logarithm in (2·17) is large. Moreover, we wish to stress that since the formulas (2·17) and (2·18) have only been obtained to logarithmic accuracy, the numerical factor under the logarithm is undetermined.

One might ask if the expression (2·18) should contain some statistical weight factor, to account for different distributions of phonons with the given energy and momentum. There can be no such factor since it would alter the energy and momentum of the wave packet (2·11) and they have been appropriately normalized. Taking into account variations in the shape of the packet for a given width and length would affect the result only beyond the stated accuracy.

It should be noted that according to (2·16) the width of the packet in coordinate space is much smaller than its length since, in the absence of dispersion, a spread in phonon momenta along $\vec{p}$ (in contrast with a transverse spread) does not modify the total energy and displace the point in the ϵ, p plane from the sonic line. Correspondingly, in coordinate space the wave packet is a "pancake" perpendicular to $\vec{p}$, with a thickness along $\vec{p}$ that tends to zero as $\delta\epsilon \to 0$.

A full version of this paper is given in S. V. Iordansky and L. P. Pitaevsky, ZhETF 76, 769 (1979).

REFERENCES

1. H. J. Maris and W. E. Massey, Phys. Rev. Lett. <u>25</u>, 220 (1970).
2. S. V. Iordansky and L. P. Pitaevsky, JETP Letters <u>27</u>, 621 (1978).
3. L. D. Landau and E. M. Lifshitz, "Quantum Mechanics" (Nauka, Moscow, 1974).

RAMAN SCATTERING SPECTRA OF PROUSTITE AND PYRARGYRITE

CRYSTALS IN LOW-TEMPERATURE PHASES

K.E. Haller and L.A. Rebane
Institute of Physics of the Estonian SSR
Academy of Sciences,
Tartu 202400, USSR

Crystals of proustite (Ag_3AsS_3) and pyrargyrite (Ag_3SbS_3) are of considerable interest as new nonlinear materials. The structure of both crystals at room temperature belongs to the trigonal space group $C_{3v}^6(3\,m)$ with two molecules per unit cell and differences in lattice parameters within a few per cent (1). Low-temperature phase transitions were found in proustite at 56 K (second order) and at 24 - 28°K (first order) on the basis of NQR spectra (2), electrical constants (3) and optical absorption (4) behaviour. For pyrargyrite a second-order phase transition was found at 9.7°K (5). The Raman and infrared absorption spectra of the crystals were first measured by Byer et al. (6) at 65°K and found to correspond well to the trigonal structure. New data on the Raman spectra of proustite above 60°K displaying some more lines were interpreted in (7) on the basis of monoclinic $C_2(m)$ structure and an additional phase transition C_{3v}-C_S at 210°K was assumed. We present the Raman scattering spectra of prousitite and pyrargyrite crystals at low temperatures and discuss their structures and phase transitions taking into account ordering in the cation sublattice.

In the case of trigonal structure we have to deal with Raman-active phonons, which are all polar: $6A_1$ normal vibrations which induce polarization along $C_3(z)$ and 13 pairs of E-type vibrations which create a polarization E(-x) parallel and E(y) perpendicular to hexagonal axes. To interpret the Raman spectra it is necessary to use the generalized expression for the scattered light intensity (8):

$$I = \{ \sum_{\substack{\rho,\sigma,\tau= \\ =x,y,z}} e_i^{\sigma} R_{\sigma\rho} \, (\alpha\xi^{\tau} + \beta q^{\tau}) \, e_s^{\rho} \}^2 \, , \qquad (1)$$

where ξ^{τ} and q^{τ} are the components of the vectors of lattice displacement and phonon polarization, respectively, and β is proportional to certain electro-optical coefficients.

The measurements were carried out on synthetic crystals oriented along the C_3 and hexagonal axes of the trigonal phase. The exciting lines from CW He-Ne and krypton lasers were 6328, 6471 and 6764 Å which are in the region of in band absorption (4). To clarify the situation all possible 90° geometries were used.

Figure 1 shows the Raman spectra of proustite at 65°K in three orientations that allow us to compare the scattering from E(-x) and E(y) lattice displacements and that from "c" (xy) and "d" (yz) components of the scattering tensors. Table 1 contains the frequencies and relative intensities of E_c and E_d lines in the spectra of x(zx)y and x(yx)z orientations.

ν, cm^{-1}	$\dfrac{E_d(-x)}{E_c(-x)}$	ν, cm^{-1}	$\dfrac{E_d(-x)}{E_c(-x)}$	
19	0.094	122	0.16	Table 1. The fre-
23	> 50	142	4.7	quencies and related
27	10.8	192	0.16	intensities from E_c
33	< 0.02	228	0.26	and E_d normal vib-
37	> 50	278	0.53	rations in the
42	< 0.02	337	0.77	Raman-scattering of
				proustite at
				65°K.
50	0.45	344	0.15	
68	> 50	348	0.02	
105	< 0.05	360	> 50	

The spectra of E modes contain 18 lines, from which 5 appear from c- and 5 from the d-components of scattering tensors. The single lines at 122, 142, 192 and 228 cm^{-1} display no shifts or splittings (within 2 cm^{-1}) that one could expect resulting from anisotropy and the influence of long-range electric forces (8). The spectra of E(-x) and E(y) vibrations are much alike, which demonstrates the degeneracy of these vibrations with respect to the C_3 axes. The extra number of E-modes may also be the result of near-resonant conditions. The spectrum of A_1-modes is shown in Fig. 2a, curve 3. It contains 6 lines in accordance with the trigonal structure (5). The frequencies of these modes are very

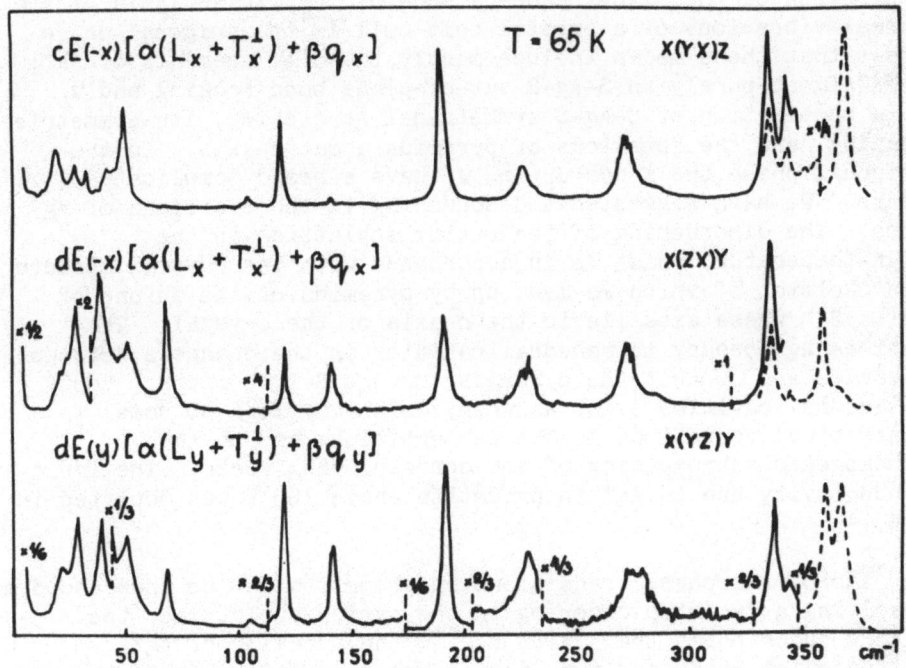

Fig. 1. The Raman spectra of proustite at 65 K measured in the
scattering configurations x(yx)z, x(zx)y and x(yz)y with the
resolution 1 cm⁻¹. The excitation 6471 Å. Dashed lines
obtained with 6328 Å excitation.

near to those in the spectrum of E-modes. The small difference in
the frequencies of the modes with displacements along and
perpendicular to the optical axis indicate the low anisotropy of
the local field. These vibrations should be considered as having
a mixed A + E symmetry.

The spectra of proustite in three different phases at 100, 36
and 5°K are shown in Fig. 2. The most essential modification of
the spectra by the phase transition occurs in two regions: in the
region of low-frequency external modes ν_1, $\nu_7 - \nu_{10}$, and in
the region of the high-frequency mode of ν_4. An analysis of the
normal vibrations of a crystal unit cell in the trigonal phase
shows that these modes include mostly the displacements of Ag^+:
ν_1 is almost purely an S-Ag-S out-of-plane bond rocking and ν_4
is a combination of S-Ag-S symmetrical stretching, its symmetrical
bending, and the rotations of pyramids about c-axis. In the
trigonal phase the modes ν_1 and ν_4 have a broad complicated
shape. We have suggested a disordering in the positions of Ag^+
ions. The disordering of the cation sublattice in the
high-temperature phase is in accordance with the crystal structure
the skeleton of which is made up by pyramids of the anions of
$As(Sb)S_3^{3-}$ whose axes lie in the c-axis of the crystal. The
cations Ag^+ occupy tetrahedral cavities in the channels between
pyramids and connect the pyramids through S-Ag-S bonds. Empty
octahedral cavities exist as well, which may lead to some
statistical deviations in the Ag^+ positions and to the
inhomogeneous broadening of the corresponding modes. The ionic
conductivity due to Ag^+ in proustite above 100 K was reported in
(10).

Two of the phase transitions in proustite may be understood as
involving a two-step ordering in the cation sublattice: the
second-order phase transition due to the freezing of Ag^+
diffusion at 56 K, and the first-order transition due to the
freezing of Ag^+ temperature-activated tunnelling at 28 K. These
processes do not need any breaking of molecular bonds, which is in
agreement with the rather low temperatures of the transitions.
The idea of insignificant distortions of the anion skeleton in the
phase transition at 56°K is supported by the behaviour of the
Raman spectra in the region of internal modes. The modes ν_2
and ν_6, which are attributed to the symmetric bending and
stretching vibrations of the pyramids (5), remain unchanged in
their number and positions.

If in the intermediate phase the Ag^+ ion is involved in
tunnelling in a two-well potential, between S-Ag...S, it can bring
about some distortions in symmetry without destroying the crystal
structure. From Fig. 2a, curve 2 one can see that the only
changes in A_1-spectra are the passing of some E modes (ν_{10},
ν_{11} and ν_{12}) into parallel polarization and the appearance of

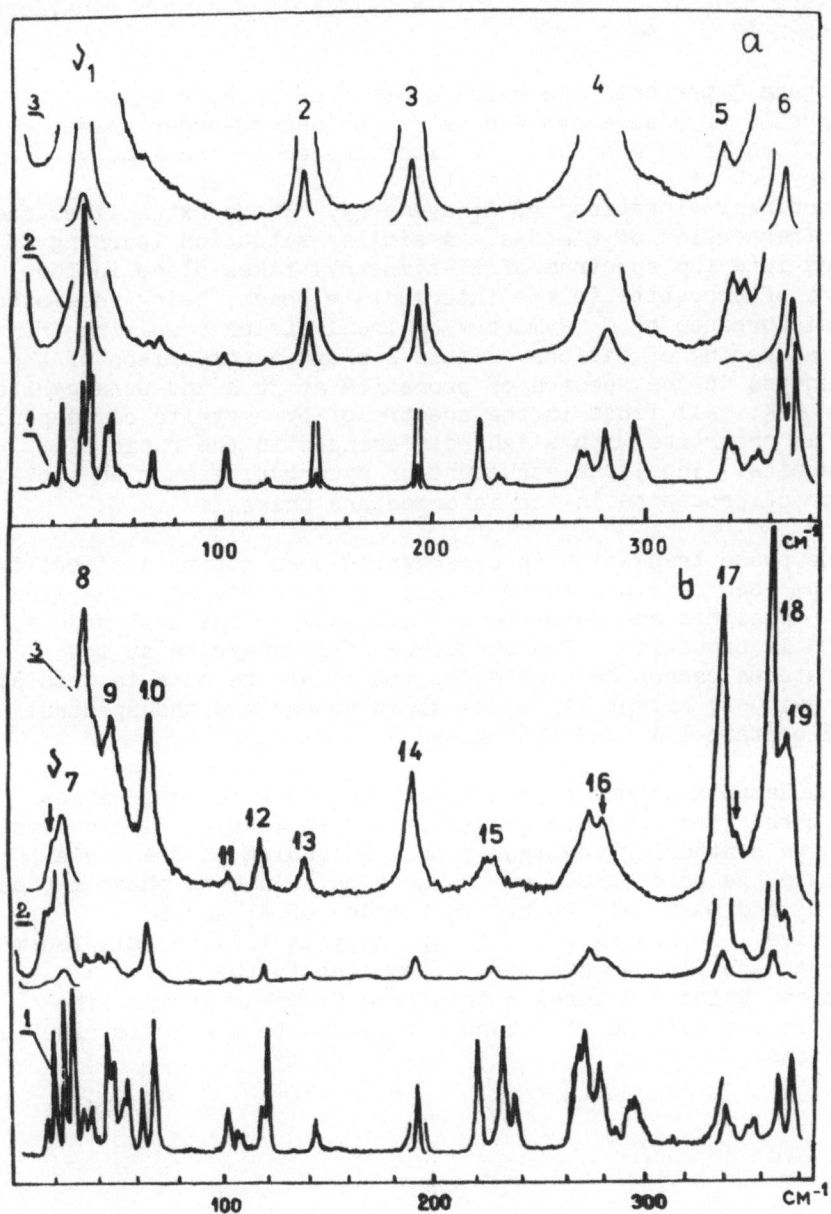

Fig. 2. The Raman spectra of proustite in the scattering
configurations x(zz)y (a) and x(zx)y (b) in three phases at
temperatures 5 K (1), 36 K (2) and 100 K (3). The resolution is 1
cm^{-1}. The intensities are comparable. The breaks indicate a
scale change of 1/10. The notation is that given in (5).

one extra mode at 10 cm^{-1}. This mode displays a soft behaviour near 56°K (7).

Figure 3 presents the Raman scattering spectra of a pyrargyrite crystal above and below the second-order phase transitions at 12 and 5°K. 16 lines appear in the x(zz)y geometry (curve 1a) instead of the allowed 6 transverse extraordinary vibrations of A_1 symmetry. These extra lines seem to be frequencies of E modes. A similar situation (passing of E-modes into the spectrum of A_1-symmetry) takes place in the spectra of proustite in the intermediate phase, being understood as a disturbance of C_3 symmetry by local fields resulting from the disordering of cations. Table 2 shows a comparison of the frequencies in the spectra of proustite at 36 K and pyrargyrite at 12 and 5 K. All lines in the spectra of pyrargyrite correspond to lines in proustite with slight differences in the ratio of frequencies. Thus, the structure of pyrargyrite must be similar to that of proustite in the intermediate phase.

The phase transition in pyrargyrite does not much affect the spectrum (see curves a and b in Fig. 3 and Table 2). The line at 16 cm^{-1} vanishes and seems to be analogous to the soft mode at 10 cm^{-1} in proustite. The structure of pyrargyrite at low temperatures cannot be considered monoclinic because in this case the previously silent $7A_2$ modes would appear and the spectrum should contain $19A'$ and $20A''$ modes.

The Raman spectra of proustite below 28 K becomes rather complicated (see (10) and curves 1a and 1b in Fig. 2) and does not depend on scattering configuration. According to the Curie principle the point group of the unit cell in this phase may be C_s or C_1 and must lead to two soft modes of A_1 and E symmetries. According to an X-ray analysis (11) the displacement vector at the first-order transition lies in the < 11.0 > direction, which indicates a triclinic C_1 space group. The change in the anionic structure may be due to the redistributions of cations.

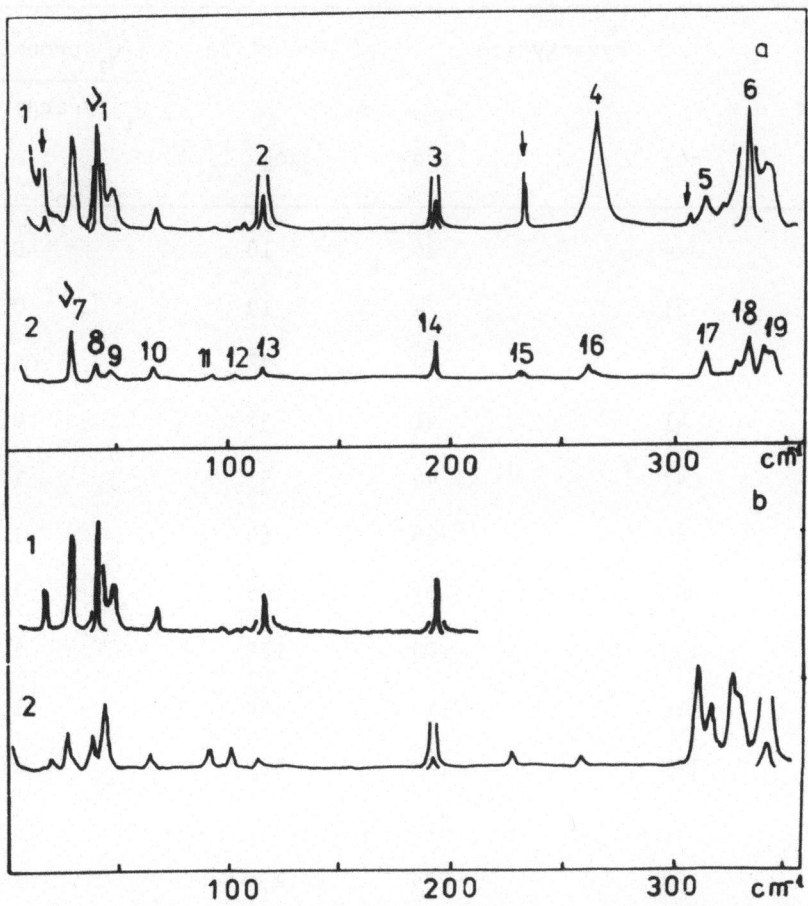

Fig. 3. The Raman spectra of pyrargyrite in the scattering configurations x(zz)y (a) and x(zx)y (b) in two phases at temperatures 12 K (1) and 5 K (2). The resolution is 1 cm⁻¹. The intensities are comparable. The breaks indicate a scale change of 1/10.

Table 2

Phonon frequencies in the Raman spectra of pyrargyrite at 12 and 5 K. Comparison with the phonon frequencies of proustite in the intermediate phase at 36 K is given.

Mode index	Pyrargyrite		Proustite	$\dfrac{\nu_i \text{ proustite}}{\nu_i \text{ pyrargyrite}}$
	$5^\circ K$	$12^\circ K$	$36^\circ K$	
-	- -	16	10	0.6
-	21	-	19	0.9
ν_7	29	29	28	0.97
ν_1, ν_8	41	41	38	0.94
ν_9	47	46	51	1.10
ν_{10}	66	66	69	1.05
ν_{11}	93	93	106	1.15
ν_{12}	103	102	122	1.2
ν_2, ν_{13}	116	115	144	1.26
ν_3, ν_{14}	193	194	193	1.00
ν_{15}	-	230		-
	233	234	230	1.00
ν_4, ν_{16}	262	261	276	1.05
ν_5, ν_{17}	313	313	338	1.08
ν_{18}	318	321	346	1.08
	329	332	360	1.08
ν_6	338	338	369	1.09
ν_{19}	341	342	372	1.09

REFERENCES

1. D. Harker, J. Chem. Phys., $\underline{9}$, 381, 1936.
2. D.F. Baisa, N.E. Eibinder, A.V. Bondar, Ukr. Fiz. Zh., $\underline{20}$, 154 (1975) (in Russian).
3. N.D. Gavrilova, V.A. Koptsik, V.K. Novik, T.V. Popova, Kristallografia, $\underline{23}$, 1067, (1978).
4. Ja. O. Dovgi, V.N. Korolishin, E.G. Moros, V.V. Turkewitz, Fiz. tverdogo tela, $\underline{13}$, 202, (1971) (in Russian).
5. H.H. Byer, L.C. Bobb, I. Lefkowitz, B.S. Deaver, Ferroelectrics, $\underline{5}$, 207, (1973).
6. D.F. Baisa, A.V. Bondar, A. Ja. Gordon, Fiz. tverdogo tela, $\underline{19}$, 1273, (1977).
7. G.A. Smolenskij, I.G. Sinij, E.G. Kuzminov, A.A. Godovikov, Proc. of Raman Scattering Conference, p. 253, Moscow, (1978).
8. R. Loudon, Adv. Phys., $\underline{13}$, 423, (1964).
9. P.H. Davies, C.T. Elliott, K.F. Hulme, J. Appl. Phys., $\underline{2}$, 165, (1969).
10. K. Haller, L. Rebane, Yu. Vysochanskij, B. Slivka, ENSV Tead. Akad. Toimetised, Fuusika * Matem., $\underline{27}$, 112, (1978).
11. B.A. Abdikamalov, V.I. Ivanov, V. Sh. Shechtman, and I.M. Shmitko, Fiz. tverdogo tela, $\underline{20}$, 2963, (1978).

FINITE FIELD LOCAL FIELD CATASTROPHE--

APPLICATION TO THE SPECTRA OF KCN_xCl_{1-x}

C. M. Varma

Bell Laboratories
Murray Hill, NJ 07974

The oscillator strength of the CN molecular exciton in KCN_xCl_{1-x} varies by about two orders of magnitude for a nominal variation in x or temperature. The enhancement in pure KCN as a function of temperature appears below the orientational phase transition of the CN molecules. The enhancement is explained by a finite frequency version of the Lorentz local field catastrophe. The conditions for the occurrence of such phenomena are explored. The crucial role is played by the background susceptibility at the exciton frequency, due to the interband polarizability. As this background susceptibility rises above a certain value, rapid transfer of the oscillator strength from the interband to the exciton peak occurs. It is suggested that the rise of the background susceptibility is connected with the ordering of the CN^- dipoles as the temperature is decreased. A mean field theory of the phase transition in pure KCN is also presented. For details of the theory, please see: A. J. Holden, V. Heine, J. C. Inkson, C. M. Varma and M. A. Bosch, J. Phys C, 12 (1035), 1979. The reflectivity spectra in KCN is presented by M. A. Bosch and G. Zumofen, Phys. Rev. Letters 41, 590 (1978).

THE EFFECT OF LONG RANGE FLUCTUATIONS IN IMPURITY POTENTIAL

ON THE ELECTRON LIGHT SCATTERING IN HEAVILY DOPED SEMICONDUCTORS

V. A. Voitenko, I. P. Ipatova and A. V. Subashiev

A. F. Ioffe Physical-Technical Institute
Academy of Sciences of the USSR
Leningrad, USSR

The light scattering from one particle electronic excitations in solids is studied mainly in heavily doped semiconductors. The heavy doping condition

$$a_B \gg \bar{R} \qquad \text{or} \qquad N a_B^3 \gg 1 \qquad (1)$$

means that the impurities are ionized and the degenerated electron gas is a nearly ideal one with respect to electron-electron interaction. Here a_B is the electron Bohr radius, N is the concentration of impurities, $\bar{R}$ is the average separation of impurities. The effect of impurities themselves is usually taken into account through the momentum relaxation time resulting from the elastic scattering of electrons by impurities.

But there is another considerable effect of impurities caused by the long-range fluctuations of impurity potential. This impurity potential effects the frequency dependence of the light scattering cross section.

When there are no impurity potential fluctuations, the light cross section for free carriers is constrained by conservation laws of the energy and the momentum

$$\omega_I - \omega_S = \omega, \qquad \vec{K}_I - \vec{K}_S = \vec{q}, \qquad (2)$$

where ω_I, ω_S and $\vec{K}_I$, $\vec{K}_S$ are frequencies and wave vectors of incident and scattered photons. ω and $\vec{q}$ are the frequency and the wave vector of electronic excitation. The single particle light scattering spectrum of interest occurs when $K_I \cdot r \ll 1$, where r is the electron screening radius.[1] It follows from (2) that in the

degenerate electron gas at zero temperature only the electrons in
a spherical layer with depth $\hbar\omega/v_F$ near the Fermi surface
contribute to the differential cross section of the light, v_F
being the electron Fermi velocity. As the number of electrons
increases linearly with ω, the light cross section also increases
linearly with ω, and then vanishes at

$$\omega = q \cdot v_F .$$

(3)

When $T \neq 0$ this edge becomes broadened over a range determined by
the temperature. In heavily doped semiconductors there is another
reason for broadening — the interaction of electrons with the
meansquare fluctuation impurity potential. The effect resembles
the edge broadening of Burstein-Moss effect studied by Dyakonov
et al.[2]

§1. THE LIGHT SCATTERING CROSS SECTION

We have studied the effect of the impurity potential on two
main nonscreened processes of light scattering from single
particle electronic excitation. The first is the light scattering
from intervalley fluctuations of the electron density. Since a
number of experimental data concern n-Si[3-5] we study the n-Si type
semiconductor with 6 valleys along (100) directions. The cross
section in this case was first obtained by Platzman.[1,6] When the
wave vector $\vec{q}$ is directed along the (111) axis of the crystal,
the light cross section has the form

$$\frac{d^2\Sigma}{d\omega d\Omega} = (-) \frac{e^4}{\pi c^4} \left[1 - \exp\left(-\frac{\hbar\omega}{T} \right) \right]^{-1} \times$$

$$\times \frac{1}{6} \sum_{\ell < \ell'} \left[(m_\ell^{-1})_{ik} e_i^I e_k^S - (m_{\ell'}^{-1})_{ik} e_i^I e_k^S \right]^2 \mathrm{Im}\, F^1(\vec{q},\omega)$$

(4)

Here $\vec{e}^I$, $\vec{e}^S$ are polarization vectors of incident and scattered
light, m_ℓ^{-1} is the electron inverse mass tensor in ℓth valley,
$F^1(\vec{q},\omega)$ is the electron polarization operator. $F^1(\vec{q},\omega)$ does not
depend on ℓ for our geometry and is equal to

$$F^1(\vec{q},\omega) = (-i) \int_0^\infty dt\, e^{i\omega t} < [\, \hat{\rho}_{\vec{q}}(t),\, \hat{\rho}_{-\vec{q}}(0)\,] >,$$

(5)

where $\hat{\rho}_{\vec{q}}$ is the electron density operator for all the valleys.

The second nonscreened process of light scattering is the scattering from electron spin density fluctuations. We consider the direct gap semiconductor when $\hbar\omega_I \sim E_g$. Hamilton and McWhorter,[9] Blum and Davies[10,11] have shown that the cross section has the form

$$\frac{d^2\Sigma}{d\omega d\Omega} = -\frac{e^4}{m^2 c^4} \frac{1}{\pi} \left[1 - \exp\left(-\frac{\hbar\omega}{T}\right) \right]^{-1} \times$$

$$\times \ B^2 (\vec{e}^I \times \vec{e}^S)^2 \ \mathrm{Im} \ F(\vec{q},\omega), \tag{6}$$

where

$$B = \frac{2P^2}{3m} \hbar\omega_I \left[\frac{1}{E_g^2 - (\hbar\omega_I)^2} - \frac{1}{(E_g + \Delta)^2 - (\hbar\omega_I)^2} \right]. \tag{7}$$

Here $P \equiv i < x|\hat{P}_x|s>$ is the Kane model parameter, and m is the free electron mass. Platzman has shown[6] that the polarization operator $F^1(\vec{q},\omega)$ for the multivalley semiconductor is related to the isotropic operator, $F(\vec{q},\omega)$, as given in (5) by the following substitutions[1]

$$m^* \rightarrow m_{/\!/}^{1/3} m_\perp^{2/3} \ ; \qquad n \rightarrow \frac{n}{6} \ ;$$

$$q \rightarrow m_{/\!/}^{1/6} m_\perp^{1/3} \sqrt{q_i (m^{-1})_{ik} q_k} \ , \tag{8}$$

where $m_{/\!/}$, $m_\perp$ are eigenvalues of the m^{-1} matrix, m^* is the electron isotropic effective mass. Therefore the problem of the light scattering by one-particle excitation is reduced to the calculation of $F(\vec{q},\omega)$.

When calculating $F(\vec{q},\omega)$ for heavily doped semiconductor one should take into account the electron scattering by the mean square impurity fluctuation potential.

§2. ELECTRONIC STATES IN HEAVILY DOPED SEMICONDUCTOR

We consider a semiconductor in which the condition of heavy doping (3) is realised. The distribution of donors is assumed random. The mean-square fluctuation of impurity concentration, in a sphere of radius R is equal to

$$\sqrt{<\Delta N^2>} \ \cong \ (NR^3)^{\frac{1}{2}} \ . \tag{9}$$

Fluctuations of ionized impurity concentration create potential fluctuations. Thus, there appears a mean-square impurity fluctuation potential

$$\gamma(R) \cong \frac{e^2}{\varepsilon_0 R} (NR^3)^{\frac{1}{2}},\tag{10}$$

where ε_0 is the dielectric susceptibility. Due to screening effects, the radius R is restricted by the screening radius, r:

$$R < r \cong (\frac{E_F \varepsilon_0}{ne^2})^{\frac{1}{2}},\tag{11}$$

where n is the electron concentration, E_F is the Fermi Energy. Substitution of Eq.(11) into Eq.(10) gives the following average depth of impurity fluctuation potential wells:

$$\gamma(r) = E_F(\frac{N}{n})^{3/4}(Na_B{}^3)^{-1/4} \ll E_F.\tag{12}$$

The linear superposition of these potential wells creates, according to Shklovski and Efros,[13] a long-range impurity fluctuation potential. Due to the condition of linear screening $r \gg N^{-1/3}$ these potential wells contain a large number of impurity atoms.

It follows from inequality (12) that the electrons near the Fermi surface have wavelength $\lambda \ll r$. Therefore $\gamma(\lambda) \ll \gamma(r)$ and the short-range fluctuations, of range λ, are of no importance. Hence the quasi-classical approach can be used for description of the electrons near the Fermi surface.

Since
$$\frac{a_B}{r} = (\frac{n}{N})^{1/6}(Na_B)^{1/6} \gg 1,\tag{13}$$

there are no bound state of the electron in the impurity atom. Thus the scattering of electrons by the single impurity potential well should be considered in the Born approximation.

On the other hand, the large value of N makes all the processes of simultaneous electron scattering by groups of impurities ——— consisting of two, three and more donors ——— equally important. All these interfering processes should be taken into account. Their contribution is proportional to some power of the parameter $N\sigma\lambda \gg 1$, where σ is the electron cross section of the single impurity.

§3. LIGHT CROSS SECTION WITH ACCOUNT OF THE LONG-RANGE
FLUCTUATION POTENTIAL

The diagram technique for calculation of the one-particle electronic Green function has been developed by Efros.[14] We applied this technique to calculate the polarization operator $F(\vec{q},\omega)$. To take into account the finite temperatures we used temperature Green functions (see, e.g., Ref.15, Chap.III).

The temperature polarization operator, $\mathcal{F}(\vec{q},\omega_m)$, corresponding to diagram in Fig.2 has the form

$$\mathcal{F}(\vec{q},\omega_m) = 2T\hbar \sum_n \frac{V}{(2\pi\hbar)^3} \int d^3p\; G_{\vec{p}}(\varepsilon_n) G_{\vec{p}+\hbar\vec{q}}(\varepsilon_n + \hbar\omega_n)$$

$$\times\; \mathcal{T}(\vec{p},\varepsilon_n;\; \vec{p} + \hbar\vec{q};\; \varepsilon_n + \hbar\omega_m;\; \vec{q};\; \omega) \quad . \qquad (14)$$

Here $\varepsilon_n = (2n + 1)\pi T$, $\hbar\omega_m = 2m\pi T$, n, m are integral numbers, $\mathcal{T}$ is the temperature vertex, and $G_{\vec{p}}(\varepsilon_n)$ is the exact temperature one-particle Green function. The simplest diagrams for $\mathcal{T}$ with respect to impurity potential are shown in Fig.3.

The inequality Eq.(13) enables us to neglect all the diagrams for $\mathcal{T}$ of the type Fig.3b. They correspond to the multiple scattering of electrons by the single impurity atom. All the interference diagrams of the type Figs.3a,c corresponding to the simultaneous scattering by one, two, three and more donors should be summed. Due to the smooth, quasi-classical behaviour of the long-range impurity fluctuation potential only the zero harmonic in Fourier expansion of $\gamma(R)$ is of importance. To take it into account explicitly one may take the Green function $D^0(\vec{k})$ corresponding to broken line in Fig.3 in the following form:

$$D^0(\vec{k}) = \frac{1}{2}\gamma^2(2\pi)^3\, \delta(\vec{k}) \quad . \qquad (15)$$

Equation (15) holds within the accuracy $\hbar/(r\rho_{\overline{F}})=(Na_B{}^3)^{-1/6}(N/n)^{1/6} < 1$.

Using Eq.(15) one can sum up the perturbation expansion for $\mathcal{T}$. Making an analytical continuation[15] one can get

$$F(\vec{q},\omega) = \frac{2V}{\gamma}\frac{\hbar}{\sqrt{\pi}} \int_{-\infty}^{+\infty} du\; e^{-\frac{u^2}{\gamma^2}} \int d^3p\; \frac{1}{(2\pi\hbar)^3}$$

$$\times\; \frac{n(\varepsilon_{\vec{p}} - u) - n(\varepsilon_{\vec{p}+\hbar q} - u)}{\hbar\omega - \varepsilon_{\vec{p}+\hbar\vec{q}} + \varepsilon_{\vec{p}} + i\hbar\delta} \quad , \qquad (16)$$

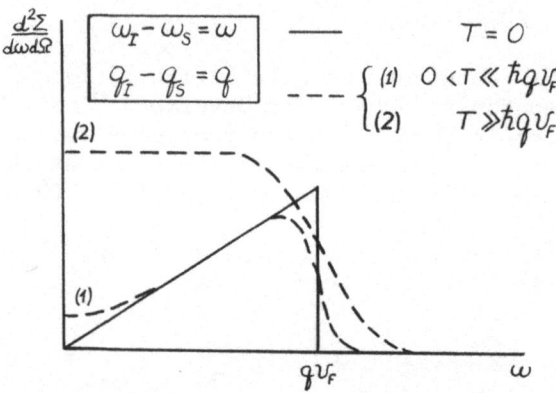

Fig. 1.   The qualitative frequency dependence of the light cross
          section from degenerate free electron gas.

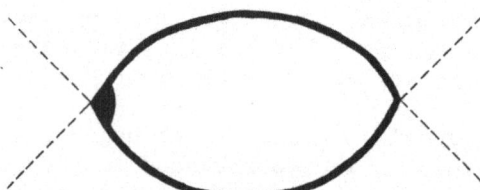

Fig. 2.   The diagram for polarization operator $\mathcal{F}(\vec{q}, \omega_m)$.   The solid
          line represents the electron Green function.

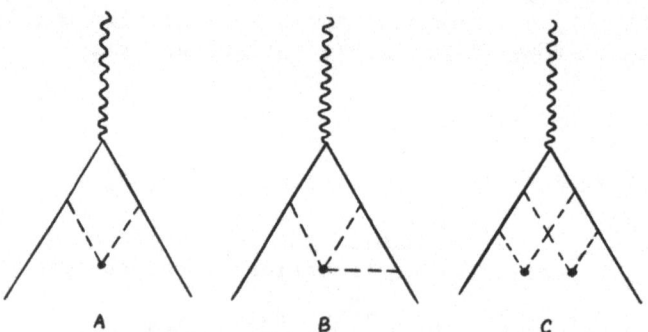

Fig. 3.   The perturbation expansion for the vertex $\mathcal{T}$.   The solid
          line represents the electron Green function, the broken
          line represents the electron scattering by impurities.

where

$$n(\varepsilon_p - u) = \frac{1}{e^{\frac{\varepsilon_{\vec{p}} - u - \mu}{T}} + 1} \quad , \tag{17}$$

μ is the electron chemical potential, and ε_p is the electron energy.

Since $\mu \gg T$ at any temperature of interest, the integral in Eq.(16) should be taken within the accuracy $\exp(-\mu/T) \ll 1$ (see, e.g., Ref.12, p.190). Then we get

$$F(\vec{q},\omega) = \frac{2V\hbar}{\gamma\sqrt{\pi}} \int \frac{d^3p}{(2\pi\hbar)^3} \frac{(\vec{v}_p \cdot \vec{q})}{\omega - (\vec{v}_p \cdot \vec{q}) + i\delta} \; I(\varepsilon_{\vec{p}} - \mu), \tag{18}$$

where

$$I(\varepsilon_{\vec{p}} - \mu) = \exp\left[- \frac{(\varepsilon_{\vec{p}} - \mu)^2}{\gamma^2} \right]$$

$$- T \frac{\partial}{\partial\mu} \int dz \; \frac{\exp\left[-\frac{(\varepsilon_{\vec{p}}-\mu+T_z)^2}{\gamma^2}\right] - \exp\left[-\frac{(\varepsilon_p - \mu - T_z)^2}{\gamma^2}\right]}{(e^z + 1)}$$

$$\tag{19}$$

Equation (19) applies to an isotropic electron energy spectrum. The anisotropic case follows from the substitution of Eqs.(6). Taking the integral in (18), one gets

$$\text{Im } F(\vec{q},\omega) = (-)\frac{Vm^*}{\pi^2\hbar} (3n\pi^2)^{1/3} \frac{\omega}{qv_F}$$

$$\times \left\{ \frac{1}{2}\left[1 + \Phi\left[1 - (\frac{\omega}{qv_F})^2 \right)\frac{\mu}{\gamma} \right] \right] + g\left[\frac{T}{\gamma}; \frac{\mu}{T} \left(1 - (\frac{\omega}{qv_F})^2\right) \right] \right\} , \tag{20}$$

where $\Phi(x) = \frac{2}{\sqrt{\pi}} \int_0^x e^{-y^2} dy$ is the probability integral, and

$$g[\alpha,x] = \frac{\alpha}{\sqrt{\pi}} \int_0^\infty dz \frac{e^{-\alpha^2(x+z)^2} - e^{-\alpha^2(x-z)^2}}{(e^z + 1)} \quad . \tag{21}$$

We introduce for convenience the dimensionless polarization operator:

$$f[(\frac{\omega}{qv_F})^2] = - \frac{\pi^2 \hbar}{Vm^*} (3n\pi^2)^{-1/3} \frac{qv_F}{\omega} \text{ Im } F(\vec{q},\omega) \quad . \tag{22}$$

At low temperatures one can expand the integrand of Eq.(21) in powers of the small parameter $\alpha = T/\gamma \ll 1$. Termwise integration gives

$$f[(\frac{\omega}{qv_F})^2] = \frac{1}{2} \{1 + \Phi[\frac{\mu}{\gamma}(1 - (\frac{\omega}{qv_F})^2)]\} - \frac{2}{\sqrt{\pi}} e^{-(\frac{\mu}{\gamma})^2(1-(\frac{\omega}{qv_F})^2)^2}$$

$$\times \sum_{K=1}^\infty \frac{2^{2K-1}-1}{(2K)!} (\pi\frac{T}{\gamma})^{2K} B_K H_{2K-1}[\frac{\mu}{\gamma}(1-(\frac{\omega}{qv_F})^2)] \tag{23}$$

Here B_K is the Bernoulli numbers, H_{2K-1} is the Hermite polynomial. The light cross section dependence on the frequency is defined by f through the factor $\frac{\mu}{\gamma}(1-(\frac{\omega}{qv_F})^2)$. Thus the spectrum edge is broadened over a range of the order

$$\Delta\omega = qv_F \frac{\gamma}{\mu} \quad . \tag{24}$$

When $T \ll \gamma \sqrt{\gamma/\mu} \ll \gamma$ the expansion Eq.(23) converges rapidly and it is sufficient to keep its first term. Then we have

$$f[(\frac{\omega}{qv_F})^2] = \frac{1}{2} \{ 1 + \Phi[\frac{\mu}{\gamma}(1 - (\frac{\omega}{qv_F})^2)]\}$$

$$- \frac{1}{\sqrt{\pi}} \frac{\pi^2}{3} e^{-(\frac{\mu}{\gamma})^2(1-(\frac{\omega}{qv_F})^2)^2} [1 - (\frac{\omega}{qv_F})^2] \frac{\mu}{\gamma}(\frac{T}{\gamma})^2 \quad . \tag{25}$$

The arguments of the Hermite polynomials are usually large
($\mu/\gamma \gg 1$). When $T/\gamma \approx \sqrt{\gamma/\mu} < 1$ one should keep several terms
of expansion. Near the spectrum edge where $|1-(\frac{\omega}{qv_F})^2|\frac{\mu}{\gamma} \ll 1$
there is a linear dependence of f on this parameter

$$f\left[(\frac{\omega}{qv_F})^2\right] = \frac{1}{2} + \frac{1}{\sqrt{\pi}}\frac{\mu}{\gamma}\left[1 - \frac{\pi^2}{3}(\frac{T}{\gamma})^2\right]\left[1 - (\frac{\omega}{qv_F})^2\right]. \qquad (26)$$

The results of numerical calculations of $f\left[(\frac{\omega}{qv_F})^2\right]$ for two sets
of parameters ($n \approx 0.6 \times 10^{19} cm^{-3}$, T = 40K and $n \approx 10^{20} cm^{-3}$,
T = 65K) are presented in Figs. 4 and 5.

In the opposite case of high temperatures, the light cross
section is weakly changed by the long-range impurity fluctuation
potential. Near the edge, when

$$|\Delta\omega| \ll qv_F \frac{\gamma}{\mu} \quad , \qquad (27)$$

f is equal to

$$f\left[(\frac{\omega}{qv_F})^2\right] \simeq \frac{1}{2} + \frac{\mu}{4T}\left(1 - \frac{3}{4}(\frac{\gamma}{T})^2\right)\left(1 - (\frac{\omega}{qv_F})^2\right). \qquad (28)$$

Far from the edge, where

$$\left|1 - (\frac{\omega}{qv_F})^2\right|\frac{\mu}{T} \gg 1 , \qquad (29)$$

f equals

$$f\left[(\frac{\omega}{qv_F})^2\right] \simeq \left\{e^{-\frac{\mu}{T}\left[1 - (\frac{\omega}{qv_F})^2\right]} + 1\right\}^{-1}$$

$$+ sgn(\frac{\omega}{qv_F} - 1)\frac{\gamma^2}{4T^2}e^{-\frac{\mu}{T}\left|1 - (\frac{\omega}{qv_F})^2\right|}. \qquad (30)$$

The first terms in Eqs.(28)-(30) incorporate the effects of
temperature only. The second terms represent the small correction
from the long-range potential which enhances the edge broadening.

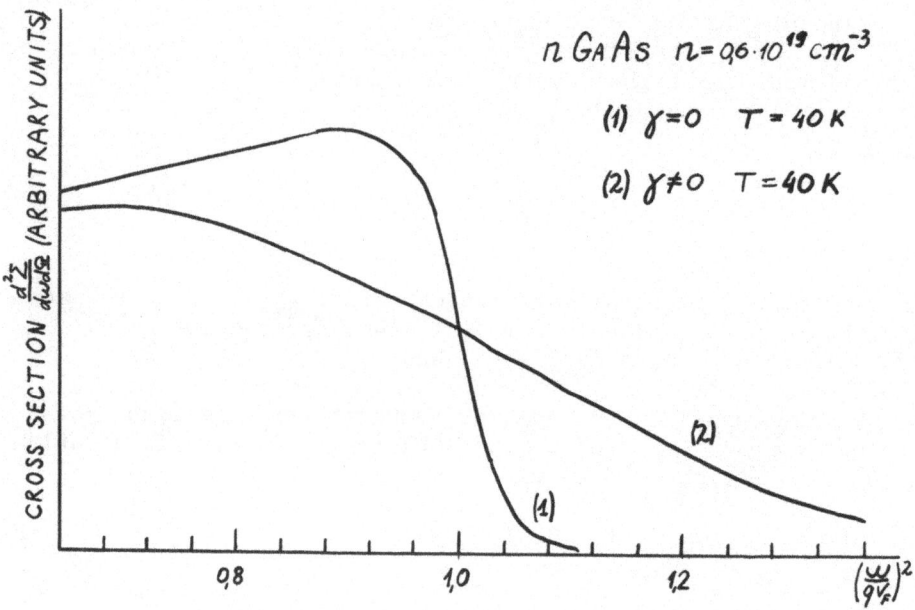

Fig. 4.   The frequency dependence of the electron cross section
near the edge $\omega = qv_F$ for $n \approx 0.6 \; 10^{19} \mathrm{cm}^{-3}$, T = 40K.

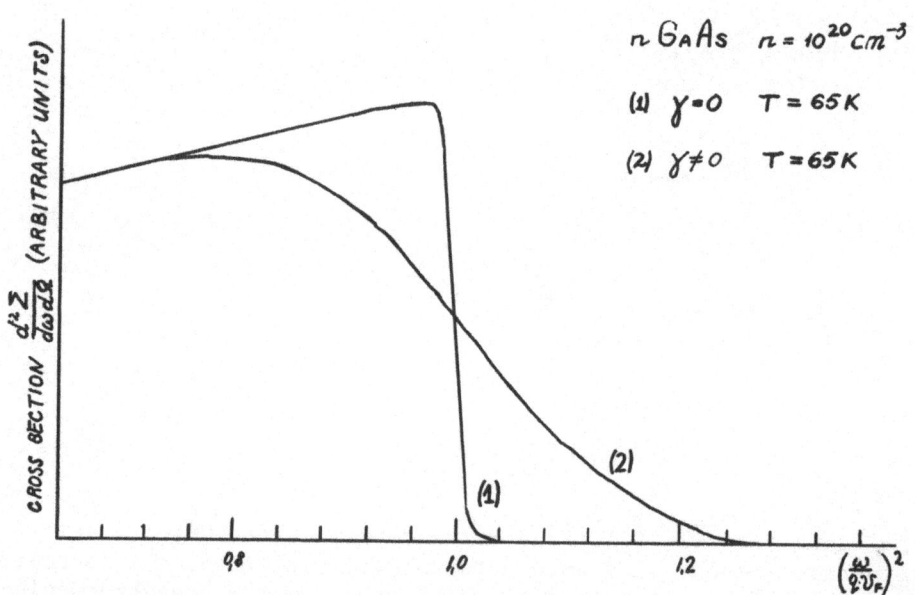

Fig. 5.   The frequency dependence of the electron cross section
near the edge $\omega = qv_F$ for $n \approx 10^{20} \mathrm{cm}^{-3}$, T = 65K.

It is seen that the long-range potential broadening should be taken into account when discussing the low temperature experimental data. It should be noted that the theory above does not contain quantum corrections which could be important at low frequencies.

REFERENCES

1. M. V. Klein, in "Light Scattering in Solids" edited by M. Cardona (Springer-Verlag, 1975), p.148.

2. M. I. Dyakonov, A. L. Efros and D. L. Mitchell, Phys. Rev. 180, 819 (1969).

3. M. Jouanne, R. Beserman, I. Ipatova and A. Subashiev, Solid State Commun. 16, 1947 (1975).

4. K. Jain, S. Lai and M. V. Klein, Phys. Rev. B13, 5448 (1976).

5. M. Chandrasekhar, M. Cardona and E. O. Klein, Phys. Rev. B16, 3579 (1977).

6. P. M. Platzman, Phys. Rev. 193A, 379 (1965).

7. A. Mooradian, in "Light Scattering Spectra of Solids" edited by G. B. McWhorter (Springer-Verlag, 1969), p.285.

8. A. Pinczuk, L. Brillson, E. Burstein and E. Anastassakis, Phys. Rev. Lett. 27, 317 (1971).

9. D. C. Hamilton and A. L. McWhorter, in "Light Scattering Spectra of Solids" edited by G. B. Wright (Springer-Verlag, 1969), p.309.

10. F. A. Blum, Phys. Rev. B1, 1125 (1970).

11. F. A. Blum and R. W. Dawies, Phys. Rev. B3, 3270 (1977).

12. L. D. Landau and E. M. Lifshitz, "Statistical Physics" (Moscow, Nauka, 1978).

13. B. I. Shklovski and A. L. Efros, "Electronic Properties of Heavily Doped Semiconductors" (Moscos, Nauka, 1979).

14. A. L. Efros, JETP 59, 860 (1970).

15. A. A. Abrikosov, L. P. Gor'kov and I. E. Dzialoshinski, "Quantum Field Theory Methods in Statistical Physics" (Moscow, Fizmatgis, 1962).

RESONANT SCATTERING AND TRAPPING OF 29 cm^{-1} ACOUSTIC PHONONS IN RUBY CRYSTALS

A. A. Kaplyanskii, S. A. Basoon and V. L. Shekhtman

Ioffe Physico-Technical Institute
Academy of Sciences of the USSR
Leningrad, USSR

I. INTRODUCTION

The phenomenon of resonant radiation trapping caused by
multiple scattering is of general physical interest. In an
optically dense resonant medium with the photon mean free path
small compared to the size of the system, photons escape from the
medium after several reabsorption events resulting in an increase of
the photon residence time in the medium. This phenomenon has been
well known to exist in gases, the mechanism of trapping having
been considered by Holstein [1] and Biberman [2]. A distinctive
feature of radiation trapping in gases consists in the existence
of both Doppler and collisional broadening of spectral lines. As
a result, photon frequency may vary over the line profile in each
absorption and reemission event. Photons reemitted at line wings
are absorbed only weakly in the medium and escape freely from the
bulk, while those at line center undergo repeated absorption.
Because of such a frequency transformation, radiation emerges from
the bulk primarily at the wings of the line, the emission line
revealing a dip at the center (self-absorption). This illustration demonstrates the importance of the nature of the scattering
event and of the character of line broadening for the trapping
process. In gases, because of the interaction of atoms with the
bath the secondary radiation spectrum does not depend on the
excitation event and corresponds to thermalized luminescence.

A different situation may exist in a solid in the case where
a two-level system interacts _only_ with radiation and the homogeneous line width is strictly "radiative". In this case the photon
frequency does not change in a scattering event, i.e. a purely

elastic scattering takes place ("resonant fluorescence"[3]).
The mechanism of trapping will differ from that of Holstein-
Biberman (HB). This alternative mechanism of trapping connected
with spatial diffusion of quanta has been recently considered
theoretically by Levinson [4] and Malyshev and Shekhtman [5] (LMS).

The present report deals with an experimental study of radiation
trapping described by the LMS theory. We have studied the trapping
in crystals not of optical radiation (photons) but rather of
"mechanical" radiation (phonons) more accurately, of acoustic
phonons, which is caused by multiple scattering of "phonon"
resonant fluorescence.

The ruby crystals used in the work had the composition
$Al_2O_3:0.05\%$ Cr^{3+}. We studied the resonant interaction of phonons
with the Cr^{3+} ions in the excited metastable ^2E state. The energy
gap between the sublevels of this state, $\bar{E} - 2\bar{A}$, is $\Delta = 29$ cm^{-1}
(Fig.1). The $\bar{E} \rightarrow 2\bar{A}$ transitions involve the absorption of 29 cm^{-1}
phonons belonging to the acoustic branches of the lattice. In the
reverse $2\bar{A} \rightarrow \bar{E}$ transition, the 29 cm^{-1} acoustic phonons are
generated in the lattice. The probability of this spontaneous one-
phonon transition being $T_1^{-1} \approx 10^9 s^{-1}$ [6,8]. The phonon interaction
with the $\bar{E}$, $2\bar{A}$ levels was studied optically by the fluorescence
lines R_1 and R_2 corresponding to transitions from $\bar{E}$, $2\bar{A}$ to the
ground state $^4A_2$ (with the times $\tau_R \approx 10^{-3} s$). Study was made of
non-equilibrium 29 cm^{-1} phonons generated in different ways in a
crystal maintained at low temperature.

II. HEAT PULSE EXPERIMENTS

The first experiments employed the technique of optical
detection of nonequilibrium phonons proposed in 1971 by Renk [7]
(see Fig.1). A ruby crystal is maintained at 1.8 K. On its
surface is deposited a thin metal film "h" heated by short current
pulses ($\Delta t_0 = 100$ ns). As a result, heat phonon pulses with a
continuous quasi-Planckian frequency distribution are injected into
the crystal. At low temperature, phonons propagate ballistically
in the crystal with the group velocity of sound. Inside the crystal,
an excited volume "d" of cylindrical shape is produced by steady-
state laser pumping (Ar, $\lambda = 5145$ A) via upper broad bands of ruby.
In this volume, part of the Cr^{3+} ions reside on the lower sublevel $\bar{E}$ of the excited ^2E state ($\Delta << kT$). This volume emits a strong
luminescence line $R_1(\bar{E} \rightarrow {}^4A_2)$. When a ballistic pulse reaches the
volume "d", the 29 cm^{-1} phonons induce the resonant transition
$\bar{E} \rightarrow 2\bar{A}$, so that an R_2 luminescence pulse from the upper $2\bar{A}$ sublevel
appears.

Figure 2a shows R_2 luminescence pulses produced by first longitudinal (sound velocity $v_L = 11.4 \times 10^3$ m/s), and then transverse
($v_T = 6.6 \times 10^3$ m/s) phonons entering the volume "d". The pulses

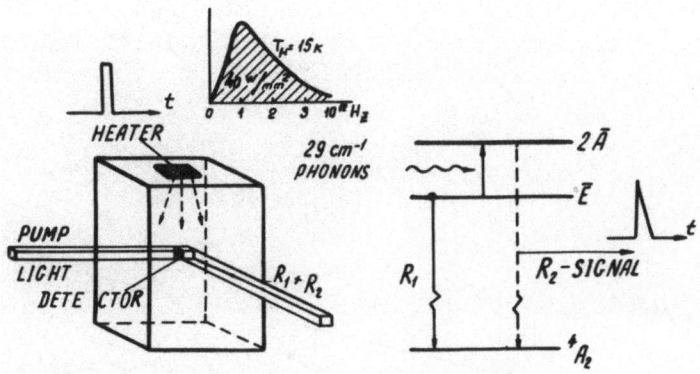

Fig. 1. The scheme of heat pulse experiments.

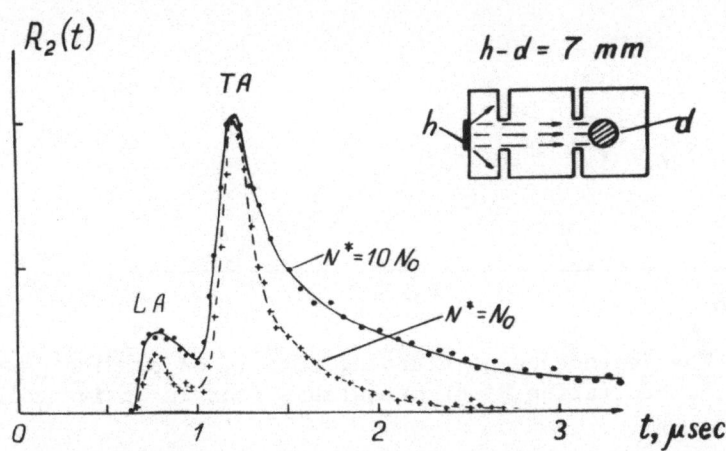

Fig. 2. The R_2 fluorescence pulses for different concentration N^* of metastable Cr ions ($N_0 \approx 10^{16} cm^{-3}$). The distance between heater "h" and detector "d" is 7 mm. The side cuts around the sample walls collimate the phonon beam.

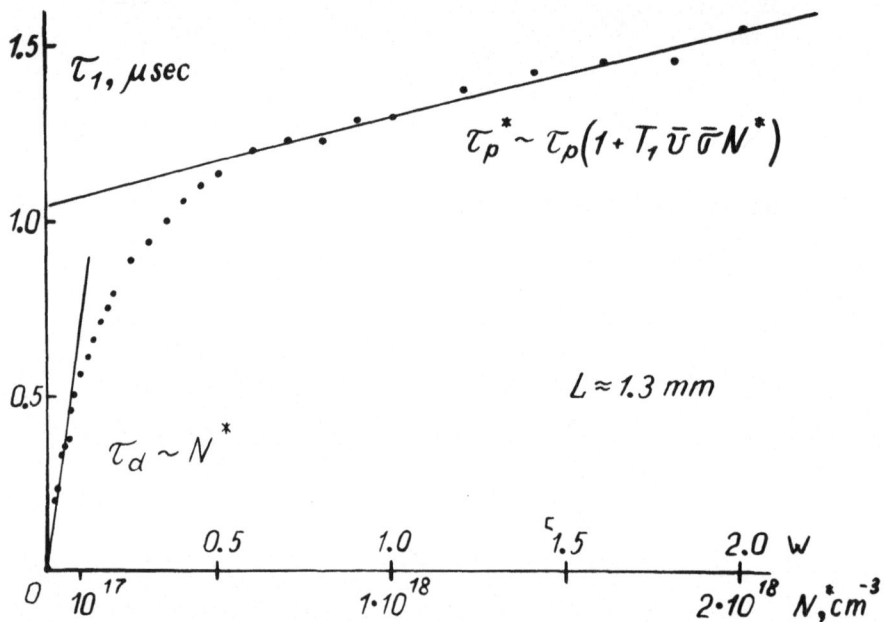

Fig. 3. The dependence of trapping time τ_1 on pumping power W and
concentration N* of metastable ions in active volume.

were obtained at moderate pumping of the excited volume with the excited Cr^{3+} ion concentration, $N* \approx 10^{16}$ cm^{-3}. The pulses are narrow (Δt=200 ns), their width being determined by the current pulse duration and by the spread of the phonon ballistic flight time which depends on geometric factors (the size of the active volume and heater etc.). Now when the pump power is increased, i.e. when the $N*$ concentration is increased, the R_2 luminescence pulses become progressively longer (Fig.2b).

Figure 3 shows typical experimental dependence of the R_2-pulse time delay τ_1 on the concentration $N*$ in the active volume.*) The tailing of the R_2 pulse has been observed for the first time [7] for an active volume in the immediate vicinity of the heater and attributed to phonon delay in the active volume because of trapping.

III. PHONON TRAPPING MECHANISM

The 29 cm^{-1} phonon trapping in an excited ruby was studied optically [7, 9-20]. Recently, we have carried out a number of new experiments. The results obtained permit a reliable identification of the phonon trapping micromechanism.

Note that an excited Cr^{3+} ion represents a practically ideal scatterer for the 29 cm^{-1} phonons. Indeed, since $T_1^{-1} \approx 10^9$ s^{-1} and $\tau_R \approx 3 \times 10^{-3}$s, after the $\bar{E} \to 2\bar{A}$ transition involving phonon absorption the ion will predominantly return to the $\bar{E}$ state with the emission of a resonant phonon.**) At a low concentration $N*$, the reemitted phonon will escape from the volume. At high $N*$, multiple scattering of phonons results in their being trapped in the volume thus bringing about a tailing of the R_2 pulse. Thus, out of the continuous spectrum of the phonon heat pulse passing through the excited volume the latter segregates the 29 cm^{-1} phonons which become imprisoned in it.

The mechanism of trapping is connected ultimately with the nature of the broadening of the 29 cm^{-1} $\bar{E} \to 2\bar{A}$ phonon line. The predominant contribution to the homogeneous line width at 1.8 K comes from the "radiative" 2A level broadening caused by the finite life-time with respect to spontaneous emission of the $2\bar{A} \to \bar{E}$ phonon: $\Gamma = T_1^{-1} \approx 10^9s^{-1} \approx 0.01$ cm^{-1}. The cross relaxation broadening of

---

*)The time τ_1 is determined as the difference between the first moments $< t >$ of the experimental pulse $R_2(t)$ (Fig.2b) and the "instrumental" pulse $R_2(t)$ at $N* \to 0$ (Fig.2a)

**)The quantum efficiency of conversion of an absorbed 29 cm^{-1}phonon into an R photon is small: $\eta \equiv T_1/\tau_R = 10^{-7}$. Despite this, the process is observed reliably by the photon counting technique.

the E, 2A levels at 1.8K caused by the Raman two-phonon processes
responsible for the broadening of the optical R_2 and R_1 lines at
higher temperatures is negligible [21]. Inhomogeneous broadening
of the phonon line due to random strains is of the same order of
magnitude as the homogeneous one ($\approx 10^{-2}$ cm^{-1}) [13, 19]. Thus,
an elastic scattering of 29 cm^{-1} phonons takes place and the LMS
model of trapping is valid.

The phonon mean free path depends on frequency within the
phonon line: $\ell^{-1}(\omega) = K(\omega) = N^*\sigma(\omega)$, where $\sigma(\omega)$ and $K(\omega)$ are the
cross section and coefficient of absorption for a phonon of
frequency ω respectively. The mean free path time
$\tau_{res}(\omega) = \ell^{-1}(\omega)\bar{v}$, where $\bar{v}$ is the averaged phonon velocity.

There are two ways for the trapped 29 cm^{-1} phonons to escape
from the excited volume [9, 10]:
(1) Spatial diffusion. At $\bar{\ell} \ll L$, where L is the smallest linear
size of the active volume, phonon propagation is diffusive in its
nature. The time for the diffusive phonon escape from the volume

$$\tau_d = (\tau_{res} + T_1) \times \frac{\alpha L^2}{\bar{\ell}^2} \qquad (1)$$

(duration of one scattering event multiplied by their number,
$\alpha L^2/\ell^2$), $\alpha \approx 1$ is a numerical coefficient depending on the shape of
the scattering volume. At small N*, $\tau_{res} \ll T_1$ and $\tau_d \sim \bar{\ell}^{-1} \sim N^*$.
(2) Anharmonic decay. Let τ_p be the anharmonic lifetime of a free
29 cm^{-1} phonon. When trapped, the phonon lifetime will be longer
because of the phonons residing as electronic excitations of the
Cr^{3+} ions when their decay is impossible:

$$\tau_p^* = \tau_p \left(1 + \frac{T_1}{\tau_{res}} \right) . \qquad (2)$$

The expression for τ_p^* contains a term linear in N* ($\tau_{res}^{-1} \sim N^*$).

The total inverse delay time of phonons of frequency ω in the
volume is approximately

$$\tau^{-1}(\omega) = \tau_d^{-1}(\omega) + \tau_p^{*-1}(\omega) . \qquad (3)$$

When calculating the observed R_2 pulse delay time associated with
phonon trapping, one has to carry out averaging over all frequencies
within the phonon line taking into account the conditions of phonon
injection.

The above considerations and formulas (1) − (3) explain
qualitatively the peculiar experimental dependence of the delay
time τ_1 on excited Cr^{3+} ion concentration N* (Fig.3). In the
region $N^* \leqslant 10^{17}$ cm^{-3} the phonons diffuse from the volume with the

time (1) $\tau_d \sim N*$. At $N* > 10^{17} cm^{-3}$ anharmonic decay of the phonons becomes noticeable. At $N* > 5.10^{17} cm^{-3}$, when the phonons are practically trapped in the volume, the trapping time is determined by the decay time $\tau*(2)$. The corresponding linear behaviour of $\tau_1(N*)$ up to $N* \approx 10^{19} cm^{-3}$ is in agreement with the trapping mechanism considered.

IV. THE EFFECT OF MAGNETIC FIELD ON 29 cm⁻¹ PHONON TRAPPING [14]

The excited volume bonbarded by heat pulses was placed in· a superconducting magnet with H // C_3, where C_3 is the trigonal crystal axis (Fig.4). Figure 4 shows R_2 luminescence pulses with and without a field applied. A narrow pulse is observed in zero field H = 0 at low pumping of the volume ($P_0 \approx 1$ mW) when there is practically no trapping. Increasing the pumping (P = 500 P_0) produces a tailing in the R_2 pulse resulting from the resonant phonon trapping. Applying a field H = 3 kGs cuts down the duration of the R_2 pulse considerably, i.e. reduces the trapping time τ_1.

Figure 5 displays an experimental magnetic field dependence of the trapping time ratio with and without the field applied, $\tau_1(H)/\tau_1(0)$. It has a resonant contour shape with a halfwidth $\Delta H \approx 400$ Gs revealing saturation at high field at a level $\sim 1/2$. Thus in the high field limit, the trapping time decreases by about a factor of 2. The $\tau_1(H)/\tau_1(0)$ curve does not depend markedly on the pumping power P which affects strongly $\tau_1(0)$ (Fig.3).

The effect of magnetic field is evidently accounted for by the Zeeman splitting of the $\bar{E}$, $2\bar{A}$ Kramers levels resulting in the splitting of the $\bar{E} \rightarrow 2\bar{A}$ phonon line (Fig.5). Neglecting weak side spin-flip transitions [6] one may consider the phonon line contour $K(\omega)$ to split in a field into a doublet, $K(\omega, H) = \frac{1}{2}[K(\omega - \varepsilon) + K(\omega + \varepsilon)]$ where $\varepsilon = \frac{1}{2}(g_1 - g_2)(\mu H/h)$. Thus, in the case of splitting the spectral absorption coefficient decreases, the transparency of the volume increases so that the trapping time τ_1 in a field decreases.

The experimentally observed phonon trapping time τ_1 is obtained by averaging the time $\tau_1(\omega)$ for a fixed frequency over the phonon line contour, $\tau_1 = \int I(\omega)\tau_1(\omega)d\omega$, where $I(\omega) \sim K(\omega)$ is the spectral form factor of the $\bar{E} \rightarrow 2\bar{A}$ transition. In the diffusion region $\tau_d \sim K(\omega)$, and in the region of anharmonic decay τ_p^* includes also a term $\sim K(\omega)$, so that τ_p is approximately proportional to the convolution of the line contour, $\tau_1 \sim \int |K(\omega)|^2 d\omega$. For a Lorentzian contour, $\frac{\tau(H)}{\tau(0)} = \frac{1}{2}[1 + (\Gamma^2/4)/(\varepsilon^2 + \Gamma^2/4)]$, yielding (in agreement with experiment) a reduction of τ_1 by one half at $H \rightarrow \infty$. From the experimental halfwidth $\tau_1(H)$, $\Delta H = 400$ Gs (Fig.5), one can evaluate the halfwidth of the phonon line $\Gamma \approx 0.01$ cm⁻¹ which agrees with the data of ref. [22].

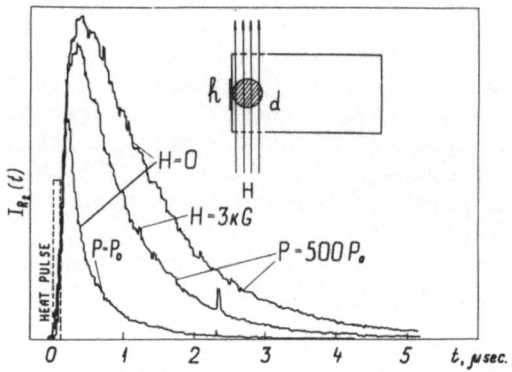

Fig. 4.   R₂-fluorescence pulses in magnetic field.

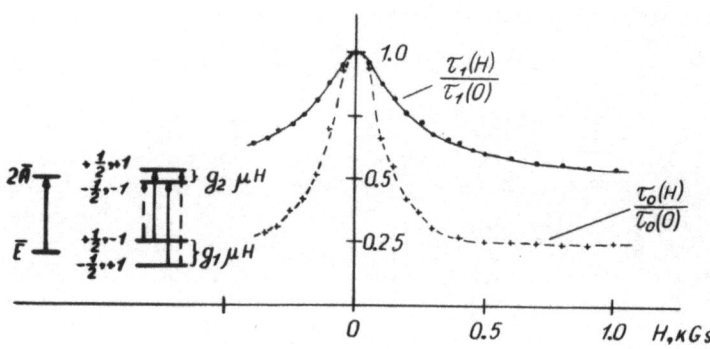

Fig. 5.   The magnetic field dependences of τ_1 and τ_0.

Experiments in a magnetic field indicate also a negligible role played in trapping at moderate N* by processes involving phonon frequency change in scattering events at H = 0. Indeed, a change of frequency in a scattering event can take place because of splitting of the $\bar{E}$, $2\bar{A}$ levels in random magnetic fields (Fig.5). A phonon absorbed in a $< + \frac{1}{2}, -1| \rightarrow | + \frac{1}{2}, +1 >$ transition can be emitted in another $< + \frac{1}{2}, +1| \rightarrow | - \frac{1}{2}, +1>$ transition with the frequency shifted by δ which is the splitting of the lower $\bar{E}$ level. In such a resonant Raman scattering of a phonon, a spin-flip and a non-spin flip transition pair is involved. The magnitude of the splitting of $\bar{E}$ level in the random fields is $\delta \approx 10^{-3}$cm^{-1} [23]. Since $\delta << \Gamma \approx 0.01$ cm^{-1}, the frequency transfer over the contour is of spectral diffusion type, a large number of scattering events $\sim(\Gamma/\delta)^2$ being required for transfer to the line wings [4, 13]. As seen from Fig.5, application of external field $H \approx 10$ Gs (comparable with H_{int}) practically does not affect the trapping time. This indicates spectral diffusion at H = 0 to be inefficient. At the same time, at fields H > 1 kGs when the splitting δ is comparable with linewidth, spectral transfer may turn out to be substantial [13].

Quantitative interpretation of heat pulse experiments meets with difficulties because nonequilibrium phonons are injected into the volume from outside. A homogeneous distribution of phonons occurs in the volume as a result of their extinction. Phonons with frequencies near the line center practically do not penetrate into the volume becoming concentrated in the surface layer. Another problem arising here is that of taking into account the inhomogeneity of concentration N* over the laser beam cross section, and the existence of optical excitation beyond the laser beam ("halo").

V. EXPERIMENTS WITH STEADY-STATE OPTICAL PHONON GENERATION

Trapping was also studied in experiments with nonequilibrium phonons generated optically in the active volume [13, 15] rather than entering it through injection of heat pulses. In ref. [15], resonant 29 cm^{-1} phonons were produced by pulsed optical excitation into the higher states of Cr^{3+} accompanied by nonradiative relaxation to $\bar{E}$ via the $2\bar{A}$ levels. Another simple version of the phonon trapping experiment involves their steady-state generation in the $2\bar{A} \rightarrow \bar{E}$ transitions [13]. This version was employed to study luminescence in ruby under steady-state optical excitation via upper broad bands and by measuring the relative line intensity R_2/R_1.

As follows from the balance equations,

$$\frac{R_2}{R_1} = \eta \frac{\tau_o}{\tau_R} , \qquad (4)$$

where $\eta = I_2/(I_1 + I_2) = 0.28$ [8] is the pumping factor of the upper level (I_1 and I_2 correspond to the pumping from above of the $\bar{E}$ and $2\bar{A}$ levels), τ_0 is the "effective" lifetime of the $2\bar{A}$ level. This time can be much longer than T_1 due to reabsorption $\bar{E} \rightarrow 2\bar{A}$ of the phonons generated in the lattice in $2\bar{A} \rightarrow \bar{E}$ transitions, which results in repeated population of the $2\bar{A}$ level. Obviously, $\tau_0 = T_1 (1 + M)$, M is the number of reabsorption events of the originally generated phonon during its lifetime (trapping time) in the active volume. In the region of low concentrations N* where phonons escape by diffusion, $M = \alpha L^2/\bar{\ell}^2$, and $\tau_0 \sim L^2 N^{*2}$. At high N* where phonons escape by anharmonic decay, $M = \tau_p/\tau_{res}$, and $\tau_0 \sim \tau_{res}^{-1} \sim N^*$.

Experiment confirms these considerations. Shown in Fig. 6 is the dependence of the steady-state intensity ratio R_2/R_1 on concentration N*. It was obtained from steady-state luminescence of a cylindrical volume excited by an Ar laser ($\lambda = 5145$ Å) at different laser powers. At low pumping levels R_2/R_1 is seen to vary by a quadratic law ($\sim N^{*2}$) while at high N* it reaches a linear region extending to the highest values of N* (a similar linear region was observed earlier [13]). The region of N* where $R_2/R_1 \sim N^{*2}$ corresponds approximately to the concentration region where spatial diffusion of phonons from the volume is essential in heat pulse experiments(the R_2 luminescence delay time is shown in Fig.3). The $\tau_0 \sim N^{*2}$ dependence in the diffusive trapping region was confirmed by study of R_2/R_1 ratio in magnetic fields. It is seen from Fig.5 that $\tau_0(H)$ dependence is different from that for $\tau_1(H)$. The resonant shape of $\tau_0(H)/\tau_0(0)$ is narrower than of $\tau_1(H)/\tau_1(0)$ and the saturation level at $H \rightarrow \infty$ is not $\sim 1/2$ but $\sim 1/4$ (see also [13]). Since in trapping diffusive region $\tau_0(\omega) \sim K^2(\omega)$ the measured mean time $\tau_0 = \int I(\omega)\tau_0(\omega)d\omega$ is proportional to $\int |K(\omega)|^3 d\omega$ which gives indeed the value 1/4 for $\tau_0(\infty)/\tau_0(0)$.

In the high concentration region the ratio $(R_2/R_1)(N^*)$ was studied also by varying the diameter L of the cylindrical volume of the laser beam of fixed power W = 2W (by properly focusing the beam). In this case N* varies with pump density as $P = W/L^2$; one should take into account the nonlinear saturation effect caused by depletion of the ground state at strong pumping. One obtained the dependence of R_2/R_1 on L characteristic for the total blocking of phonon diffusion and their escape by anharmonic decay $(R_2/R_1 \sim N^*)$.*)

---

*)In ref. [17] similar measurements yielded $R_2/R_1 \sim 1/L$ which is probably a result of not having taken saturation into account. The dependence of R_2/R_1 on the volume diameter obtained in ref. [17] at fixed P (Fig.1 [17]) corresponds probably to the region where spatial trapping is still substantial.

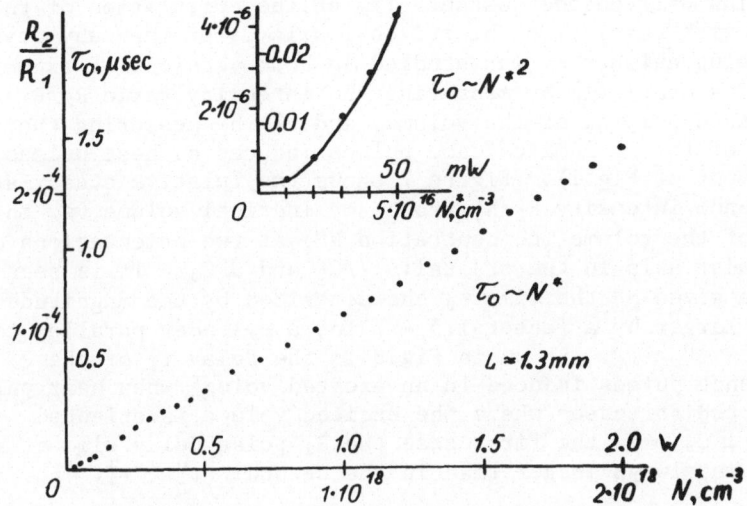

Fig. 6. The steady-state intensity ratio R_2/R_1 as a function of
concentration $N*$ of metastable Cr ions.

Measurements of $(R_2/R_1)(N*)$, Fig.6, yield a direct dependence
on $N*$ of the effective lifetime τ_0 of the $2\bar{A}$ level under conditions
of phonon trapping. The time τ_0 is the lifetime of a 29 cm⁻¹
excitation quantum as an electronic $2\bar{A}$ excitation. It depends
only on the number of reabsorption events M (and not on the time
τ_{res}) and is connected with the delay time $\tau_1 = (\tau_{res} + T_1)\cdot M$ of
phonons in the active volume through the formula

$$\tau_0 = \frac{\tau_1}{1 + \dfrac{\tau_{res}}{T_1}} \qquad (5)$$

It is essential that in the diffusion trapping region τ_0 and τ_1
not only differ in magnitude but also have a qualitatively
different concentration dependence, $\tau_0 \sim N*^2$, $\tau_1 \sim N*$. An
experimental observation of this fact supports in a convincing way
the validity of the trapping mechanism considered.

VI. TRAPPING ANISOTROPY

Recent experiments have culminated in the discovery of a
new peculiar phonon trapping mechanism caused by anisotropy of
the system and predicted in ref. [24]. It was found that if an
optically excited volume in a ruby crystal has a strongly
anisotropic shape (e.g. a cylindrical laser beam), then the degree

of trapping will depend sustantially on the orientation of this cylinder with respect to the trigonal axis C_3 of the ruby crystal. The trapping anisotropy was studied in both versions of experiment described above: (1) by measuring the intensity ratio R_2/R_1 at a steady-state pumping of the volume, and (2) by measuring the duration of the R_2 luminescence pulses induced by heat pulses (arrangement of Fig.1). Figure 7 shows the relative steady-state luminescence intensity R_2/R_1 from a cylindrical volume vs. the pumping of the volume (concentration N*) at two orientations of the cylinder axis in the crystal: $//C_3$ and $\perp C_3$. It is seen that at a given N* the time τ_0 characterized by the magnitude of R_2/R_1 is larger by a factor 1.5 - 2 for a cylinder parallel to the axis: $\tau_0^{\parallel} > \tau_0^{\perp}$. Shown in Fig.8 is the delay τ_1 of the R_2 luminescence pulses induced in an excited volume when heat pulses are injected for cases where the excited volume is oriented $//C_3$ and $\perp C_3$. In the first case the R_2 pulse delay time at a given N* is always longer than in the second: $\tau_1^{\parallel} > \tau_1^{\perp}$.

These results indicate that the degree of trapping in a cylindrical volume $//C_3$ at a given N* is higher than in a cylinder $\perp C_3$. Hence there is an "anisotropic" channel of spatial phonon escape which is effective in experimental arrangements where the small size of the volume (cylinder radius) is oriented along the trigonal axis. We believe this escape channel to be associated with the anisotropy in the interaction of phonons with chromium ions (phonon emitter anisotropy) and with the mode conversion of scattered phonons. Indeed, while frequency does not change in a scattering event, the mode i.e. the wave vector $\vec{q}$ and the polarization of a phonon TA, LA undergoes conversion. TA and LA phonons have different mean free paths $\bar{\ell}$ and anharmonic decay times τ_p^*.*)

"Axial" LA phonons with vectors $\vec{q}$ oriented close to the crystal axis C_3 deserve special attention. As shown experimentally [25], LA phonons with $\vec{q}$ $//C_3$ cannot interact with the $\bar{E} \rightarrow 2\bar{A}$ transition in Cr^{3+} at all (forbidden by selection rules) and are absorbed (or emitted) by ions only when q deviates from C_3, this being the stronger, the larger is the angle $\theta = \angle qC_3$. Hence, "axial" LA phonons have an anomalously large mean free path ℓ and, when generated in scattering, escape from deep in the excited volume without absorption. They escape effectively inside the cone with the axis C_3 and opening angle θ_{max} such that $\bar{\ell}(\theta_{max}) \approx L_{\parallel}$, where $L_{\parallel}$ is the size of the volume along C_3. This

---

*)The trapping parameters considered above ($\bar{\ell}$, τ_p and others) represent effective (mean) quantities depending on the corresponding parameters of individual modes and the intermode conversion factor for a scattering event.

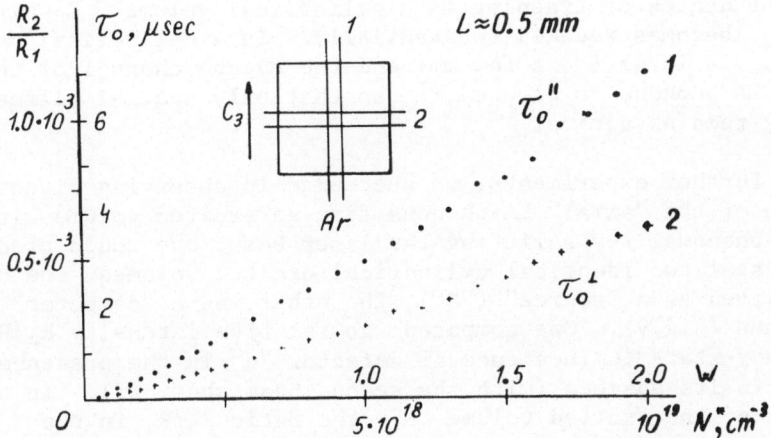

Fig. 7. The concentration dependence of R_2/R_1 ratio (time τ_0) for two orientations of laser beam in crystal (1 - $//C_3$, 2 - $\perp C_3$).

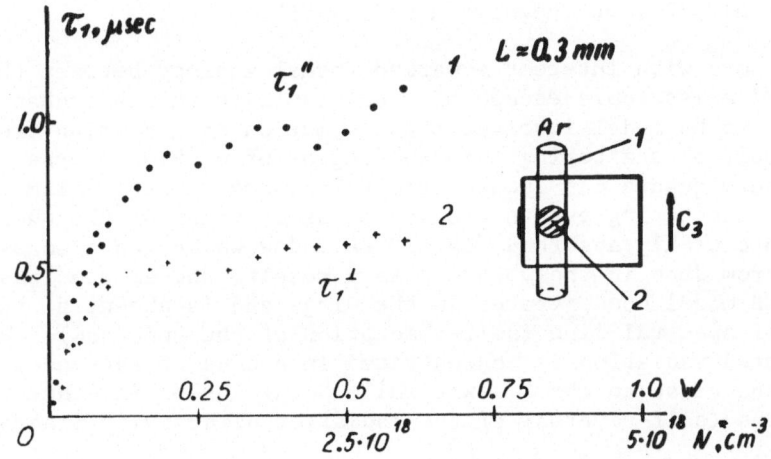

Fig. 8. The concentration dependence of trapping time τ_1 for two orientations of laser beam in crystal (1 - $//C_3$, 2 - $\perp C_3$).

escape channel is the more effective, the smaller is the ratio $L_{//}/L_\perp$. This is exactly what is observed experimentally (Figs.7,8) where the degree of trapping in a cylindrical volume $\perp C_3$ with small $L_{//}$ becomes reduced substantially. In a cylindrical volume $//C_3$, $L_{//}$ is large ($\approx$ a few mm) and the escape channel of the "axial" LA phonons is blocked,*) so that only spatial diffusion trapping remains active.

In further experiments, we succeeded in observing directional emission of the "axial" LA phonons from an excited volume with trapped phonons. By splitting the laser beam, one could produce in a crystal two identical cylindrical excited volumes, one of which served as a "source" ("S"), the other, as a "detector" ("d") of phonons (Fig.9). One compared the relative intensity R_2/R_1 of the steady-state luminescence of detector "d" in the presence of "S" and in its absence (with the second beam shut off). In the presence of the excited volume "S", the ratio R_2/R_1 in the luminescence of "d" was found to increase by a few percent. This indicates an increase in the number of phonons in volume "d" because of injection into it of phonons emitted by volume "S". This increase $\Delta(R_2/R_1)$ is the largest when the line connecting "S" with "d" coincides with the C_3 axis and decreases when the line "S" -"d" deviates from C_3, the angular halfwidth of the corresponding dependence $\Delta(R_2/R_1)$ (see Fig.9) making up $\Delta\theta \leqslant 20°$. Hence phonon emission from the "S" volume does indeed form a sufficiently narrow cone along C_3. Such a focussed nature of the phonon emission is confirmed also by a relatively weak dependence of the increase $\Delta(R_2/R_1)$ on the distance from "S" to "d".

We note with interest a marked formal analogy between the above effect of anisotropic escape of radiation with the HB mechanism. Indeed, the HB model involves transformation of radiation frequency as a result of scattering into the region of weakly absorbed line wings where quanta can escape freely from deep in the volume. Here multiple scattering events produce transformation of the phonon modes into weakly absorbed "axial" LA modes which can also escape freely from deep in the volume. As a result, the emitted radiation in the HB model concentrates in the wings and is absent at the center of spectral line (self-absorption of the spectrum). Here the emitted radiation is concentrated in a cone of easy escape directions close to the C_3 axis while being absent in other directions (self-reversal of the radiation directivity diagram).

---

*) Therefore a cylindrical volume $//C_3$ was used in studies of steady-state trapping (R_2/R_1) and heat pulse trapping, $R_2(t)$ (Figs.3, 6).

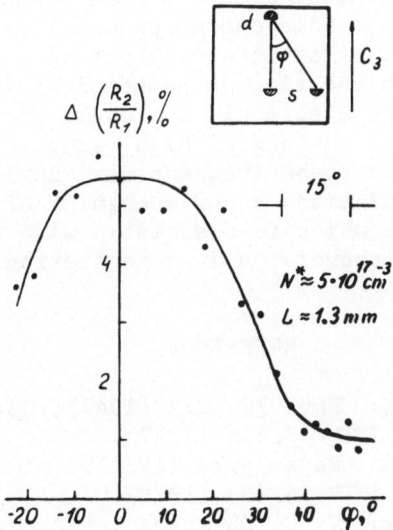

Fig. 9.   The relative increase of detector "d" steady state
          luminescence ratio R_2/R_1, induced by phonons emitted
          from volume "s" at different angles between C_3 and s-d
          direction.

VII. CONCLUSION

     The heat pulse experiments (the dependence of trapping time
τ_1 on N*, the effect of magnetic field on τ_1) and steady-state R_2
luminescence measurements (dependence of τ_0 on N*) are described
well within the framework of the LMS theory of spatial diffusion
trapping under conditions of finite anharmonic phonon lifetime.
These experiments provide a possibility to evaluate a number of
parameters affecting the trapping of 29 cm⁻¹ phonons, the mean
cross section for resonant scattering $\sigma = 2 \times 10^{-15}$cm², the line
width $\Gamma = 0.01$ cm⁻¹, the anharmonic phonon decay time $\tau_p \approx 2\mu$s.
These parameters ensure an extremely high degree of trapping with
the number of phonon reabsorption events from M = 10² at
N* = 10¹⁷cm⁻³ in the diffusion region (L ≈ 0.5 mm) to M = 10⁴ at
N* ≈ 10¹⁹cm⁻³ in the region of anharmonic decay. Due to trapping,
the ratio R_2/R_1 reaches in the steady-state at T = 1.8 K a value
of 10⁻³ or 10⁷ times that of the thermal equilibrium value, which
corresponds to an equivalent "temperature" of trapped 29 cm⁻¹
phonons of T ≈ 6 K (the bottlenecking effect).

     The above results show that trapping of the 29 cm⁻¹ phonons
between the $\bar{E} \to 2\bar{A}$ levels of excited ruby turns out to be a good
experimental model for the LMS theoretical mechanism of resonance
fluorescence trapping. This trapping mechanism representing an

alternative to the HB model occurs in the absence of cross relaxation processes when the homogeneous phonon line broadening has a radiative nature. It is of interest that the time delay in trapping is connected with the time of phonon free flight τ_{res} and finite lifetime of the electronic 2A level. Spectral diffusion within the line in the absence of high magnetic fields and at not too large N* turns out to be inessential. Besides the diffusive escape of trapped radiation, a new mechanism of "anisotropic" escape was discovered which is associated with the anisotropy of scattering, and mode conversion in a scattering event.

REFERENCES

1. T. Holstein, Phys. Rev. 72,1212 (1947); 83,1159 (1951).
2. L. N. Biberman, JETP 17,416 (1947).
3. V. Weisskopf, Ann. Phys. 9,23 (1931).
4. I. B. Levinson, JETP 75,234 (1978).
5. V. A. Malyshev and V. L. Shekhtman, Fiz. Tverd. Tela 20,2915 (1978).
6. M. Blume, R. Orbach, A. Kiel, and S. Geschwind, Phys. Rev. 139,A314 (1965).
7. K. F. Renk and J. Deisenhofer, Phys. Rev. Lett. 26,764 (1971).
8. J. E. Rives and R. S. Meltzer, Phys. Rev. B16,1808 (1977).
9. K. F. Renk, "Light Scattering in Solids" Flammarion, Paris (1971), p.12.
10. K. F. Renk and J. Peckenzell, Jour. Phys. No.10, Suppl. C-4, 103 (1972).
11. A. A. Kaplyanskii, S. A. Basoon, V. A. Rachin, and R. A. Titov, Pisma JTF 1,628 (1975).
12. A. A. Kaplyanskii, S. A. Basoon, V. A. Rachin, and R. A. Titov, Fiz. Tverd. Tela 17,3661 (1975).
13. J. T. Dijkhuis, A. van der Pol, and H. W. de Wijn, Phys. Rev. Lett. 37,1554 (1976).
14. A. V. Akimov, S. A. Basoon, A. A. Kaplyanskii, R. A. Titov, V. L. Shekhtman, Fiz. Tverd. Tela 19,3704 (1977).
15. R. S. Meltzer and J. E. Rives, Phys. Rev. Lett. 38,421 (1977).
16. G. Pauli and K. F. Renk, Proc. Intern. Conf. Lattice Dynamics M. Balkanskii, ed., Flammarion, Paris (1978), p.232.
17. G. Pauli and K. F. Renk, Phys. Lett. 67A,410 (1978).
18. A. P. Abramov, I. N. Abramova, I. Ja. Gerlovin, and I. K. Rasumova, Pisma JETP 27,3 (1978).
19. A. A. Kaplyanskii, Colloq. Intern. CNRS N.255, Lyon (1976), p.137.
20. G. Pauli, G. Klimke, H. J. Krenzer, and K. F. Renk, to be published.
21. D. E. McCumber and M. Sturge, J. Appl. Phys. 34,1682 (1963).
22. H. Lengfellner, G. Pauli, W. Heisel, and K. F. Renk, Appl. Phys. Lett. 29,566 (1976).

23.   S. Geshwind, G. E. Delvin, R. L. Cohen, and S. R. Chinn,
      Phys. Rev. <u>137</u>,A1087 (1965).
24.   V. A. Malyshev and V. L. Shekhtman, Opt. i spekt. <u>46</u>,800 (1979).
25.   A. A. Kaplyanskii, S. A. Basoon, V. A. Rachin, and R. A. Titov,
      Pisma JETP <u>21</u>,438 (1975).

SOME ASPECTS OF THE THEORY OF SURFACE POLARITONS

V. M. Agranovich

Institute of Spectroscopy
Academy of Sciences of the USSR
Troitsk, Moscow obl., USSR

I. INTRODUCTION

Studies of the physics of surfaces, thin films, and two-dimensional systems, and the great interest in these studies, have called for further development of various optical methods for the investigation of surface properties. The use of the technique of surface polariton spectroscopy opens up some new possibilities.

Surface polaritons commonly explored by optical methods (e.g. by ATR method) are macroscopic waves (their penetration depth considerably exceeds the lattice constant). Therefore, the surface polariton (SP) dispersion law is primarily determined by the dielectric permeability of the media which are in contact. Primarily, but not only! The SP dispersion law depends also on the so-called transition layer which is always present at the media interface. The properties of the transition layer, which are determined by the peculiarities of the spectrum of surface excitations (surface phonons, excitons, electrons, magnons) may differ appreciably from bulk properties of the media which are in contact. Since the SP dispersion law can be explored at present by various methods, theoretical analysis of the transition layer effect on SP dispersion is timely (see also[1]).

In view of the above I will try in the present paper to touch upon two aspects of the theory of surface polaritons, which at present seem to me the most worthy of consideration.

The first refers to the influence of the transition layer on the surface polariton dispersion. This problem is treated in the second section of the present paper. Since the first experiments

have already been made along this direction, I will also speak
about some results obtained.

In the third section of the report the results of the
theoretical study of surface polariton scattering by the
fluctuations of the order parameter near the points of bulk and
surface phase transitions are discussed. Experimental observations
of such scattering are difficult and will evidently require great
efforts. Nevertheless, I feel that "the game is worth the candle"
because such experiments could offer new possibilities for the
study of phase transitions in both transparent and opaque media
(e.g. in metals; in this connection it is also of interest to note
the possibility of analyzing boundary conditions for the order
parameter).   Also for studies of phase transition in the transition layer, and in particular, for observations of surface (i.e.
two-dimensional and quasi-two-dimensional) analogs of ferromagnetism,
ferroelectricity, piezoelectricity, superconductivity, superfluidity,
etc.).

II. DISPERSION OF SURFACE POLARITONS IN THE REGION OF RESONANCE
WITH VIBRATIONS IN THE TRANSITION LAYER (ADDITIONAL WAVES AND
ABC)

1. The effect of the transition layer is especially pronounced
in the case when the frequency ω_0 of dipole oscillations in the
transition layer*) falls within the "rearrangement region" of SP
frequencies. In this case, as shown in[2](see also[1]) a gap Δ is
formed in the spectrum of SP frequencies with a width of the order
of $(d/\lambda_0)^{\frac{1}{2}}$, $\lambda_0 = 2\pi c/\omega_0$, d is the thickness of the transition layer
This splitting of the SP dispersion curve as well as the square
root dependence of Δ on d were first observed in [3] for the IR
spectra region. SP propagation along the surface of a sapphire
substrate with a LiF film (at d = 100 Å, the value of $\Delta \approx 20$ cm^{-1})
was studied there. The gap width increases considerably for the
visible spectra region.[4] In that work the splitting effect was
observed for SP propagating along the surface of aluminium with
silver films (d = 20–60 Å). In this case the splitting value Δ
at d = 26 Å turns out to be approximately 0.5 ev according to the
theory.

Resonance of oscillations in the transition layer with SP is
apt to be a rather common phenomenon. In particular, the
possibility of it has to be taken into account when analyzing
light reflection spectra of molecular crystal surfaces (e.g.
anthracene[5]) and also while studying (see [6]) Fermi resonance with
SP.

---

*)Layers of such kind can also be obtained artificially, e.g.
  by coating various substrates with very thin films.

In this connection further analysis of SP dispersion in the presence of resonance with oscillations in the transition layer is called for and, in particular, the analysis of possible effects conditioned by the inclusion of spatial dispersion. Such analysis was carried out in [1] for the nonresonance case (see also [7] where energy dissipation in the transition layer was taken into account within the framework of a particular model).

In [1], in particular, it was shown that in the region of the Coulomb frequency ω_s of a surface polariton at the interface with vacuum (the frequency ω_s satisfies the condition $\varepsilon(\omega_s) = -1$, $\varepsilon(\omega)$ is the substrate permeability) a dependence of $\omega_s(k)$ linear in k appears under the influence of the transition layer there (k is the wave vector of the surface polariton). With due regard to retardation this gives rise to an additional surface electromagnetic wave.

However, in the region of frequencies $\omega \approx \omega_s$ considerable damping occurs which should prevent appreciable propagation of the additional surface wave. It should be noted in this connection that for SP propagation along dielectric surfaces considerable damping occurs not only at $\omega \simeq \omega_s$, but at $\omega < \omega_s$ as well, i.e. for the whole SP spectral region. The situation is, generally speaking, different for SP propagation along metal surfaces. Since the surface polariton field penetrates considerably into metal for waves with the frequency $\omega \approx \omega_s = \omega_p/\sqrt{2}$ (ω_p is the frequency of the bulk plasmon), the SP in this spectral region is appreciably damped. In the frequency region $\omega \ll \omega_p/\sqrt{2}$ however, the penetration of the surface wave field into the metal is slight, so SP damping is weak and its propagation length is macroscopically large, i.e. of the order of several cm, see [8]. A review of recent experiments can be found in [9]. As will be shown below damping will be relatively small in many cases in the frequency region of resonance with SP oscillations in the transition layer, if: the frequency of these oscillations $\omega_0 \ll \omega_p/\sqrt{2}$ and the transition layer is sufficiently thin. Therefore, the detection and investigation of additional surface waves will evidently be most possible just under the condition when such waves are propagating along metal surfaces.

2. We assume that an isotropic medium (II) with dielectric permeability occupies a space region z<0 and has a boundary with vacuum (I) along the plane z=0. If the transition layer of thickness $d \ll \lambda$ (λ is the wavelength) is taken into account, instead of the usual boundary conditions for the fields following from the Maxwell equations at the sharp boundary, the following boundary conditions at the surface, which are correct to linear terms in d/λ are to be used.

$$D_3(\text{II}) - D_3(\text{I}) = i\gamma \vec{k}_t \cdot \vec{E}_t(\text{I}) \ ,$$

$$\vec{E}_t(II) - \vec{E}_t(I) = -i\mu\vec{k}_t E_n(I) + ik_0 d[\vec{n}\vec{H}_t(I)],$$

$$\vec{H}_t(II) - \vec{H}_t(I) = -id\vec{k}_t H_n(I) - ik_0\gamma[\vec{n}\vec{E}_t(I)],$$

$$H_n(II) - H_n(I) = id\vec{k}_t \cdot \vec{H}_t(I), \tag{1}$$

where n and t denote vector components normal and tangential with respect to the plane z=0, $k_0 \equiv \omega/c$, ω is the field frequency.

The phenomenological values γ and μ, present in (1), are determined by the properties of the transition layer. If this layer can be considered as a macroscopic one, then $\gamma = d\tilde{\epsilon}$, $\mu = d/\tilde{\epsilon}$, where $\tilde{\epsilon}$ is the layer permeability. But if the thickness d is of the order of the lattice parameter then the determination of the values γ and μ requires a microscopic theory. The important thing here is that the resonances of γ and μ correspond in general to different values of   frequency. In their vicinity, for sufficiently weak damping, it is sufficient to retain only resonance terms in (1). In particular, for the frequency region $\omega \approx \omega_0$, $\mu(\omega_0) = \infty$ (experiments made in [3,4] correspond to this very case) it can be assumed that only the value $\vec{E}_t$ is discontinuous, and

$$\vec{E}_t(II) - \vec{E}_t(I) = -iuE_n(I)\vec{k}_t. \tag{2}$$

Assuming that in the frequency region considered $\omega \approx \omega_0$ the permeability $\epsilon(\omega) < 0$, we obtain the following dispersion law for surface waves

$$F(\omega, k) \equiv \frac{\kappa}{\epsilon} + \kappa_1 + \mu k^2 = 0, \tag{3}$$

where $\vec{k}$ is the two-dimensional wave vector of the surface wave, $\kappa = \sqrt{k^2 - \omega^2\epsilon(\omega)/c^2}$, $\kappa_1 = \sqrt{k^2 - \omega^2/c^2}$. If we set $\mu=0$ in Eq.(3) we obtain the well-known relationship:

$$k^2 = \frac{\omega^2}{c^2} \frac{\epsilon}{\epsilon+1}. \tag{4}$$

But if $\mu \neq 0$, $\mu = -Ad/(\omega^2-\omega_0^2)$, where A is a positive value weakly depending on ω in the resonance region, then at $\omega \approx \omega_0$, the dispersion law for surface waves changes appreciably. In what follows, we shall consider the dispersion of surface waves which correspond to the conditions in the experiments.[4] Namely, we assume that medium II corresponds to a metal with the plasma frequency $\omega_p \gg \omega_0$ and that the transition layer is macroscopic and is obtained by coating the surface z=0 with a thin film of another metal with the plasma frequency ω_0.

Since for electrons on the Fermi Surface $k_F \approx 10^8$ cm^{-1}, the above film can be considered as a macroscopic one on condition that the inequality $dk_F \gg 1$ is satisfied which we will assume to be valid. Assuming moreover the case of normal skin-effect we take

$$\varepsilon(\omega) = 1 - \omega_p^2/\omega(\omega + i\Gamma),$$

$$\tilde{\varepsilon}(\omega) = 1 - \omega_0^2/\omega(\omega + i\tilde{\Gamma}),$$ (5)

where Γ and $\tilde{\Gamma}$ are collision frequencies of electrons in the metal (II) and in the transition layer respectively. Substitution of (5) into (3) taking into account that $\mu = d/\tilde{\varepsilon}$ gives a relationship which permits us to determine the dispersion of surface waves in the presence of damping. We are interested here only in the situation when the surface wave frequency is real, corresponding to that of the pumping source. In this condition the wave vector k becomes complex, $k = k'+ik''$, and tne mean free path of the surface wave is $L = (2k'')^{-1}$. If the surface wave damping is sufficiently weak (i.e. if $k' \gg k''$) then as a first approximation damping can be totally omitted when determining the dispersion law. In this case the dispersion law for the surface wave, i.e. the dependence $\omega_s(k)$ is determined from the equation:

$$F(k, \omega) = -cdk^2\omega^2/(\omega^2 - \omega_0^2),$$ (6)

where

$$F(k, \omega) = \frac{\omega^2\sqrt{k^2c^2 + \omega_p^2 - \omega^2}}{\omega^2 - \omega_p^2} + \sqrt{k^2c^2 - \omega^2}.$$ (6a)

If $\omega_{0s}(k)$ is the surface polariton frequency at the sharp boundary (i.e. when the presence of the transition layer is neglected), then $F(k, \omega_{0s})=0$ and in the frequency region $\omega \approx \omega_{0s}(k)$

$$F(k, \omega) \approx \left(\frac{\partial F}{\partial \omega^2}\right)_0 \left[\omega^2 - \omega_{0s}^2(k)\right].$$

From (6a) it follows that at $\omega^2 \ll \omega_p^2$ and $k^2c^2 \ll \omega_p^2$

$$F(k, \omega) \approx -\frac{\omega^2}{\omega_p^2} + \sqrt{k^2c^2 - \omega^2}$$

so that

$$\omega_{0s}^2 (k) = k^2c^2 - k^4c^4/\omega_p^2 + \dots ,$$ (7a)

$$\left(\frac{\partial F}{\partial \omega^2} \right)_0 \approx - \frac{\omega_p}{2\omega^2} \tag{7b}$$

and, therefore,

$$F(k, \omega) \approx - \frac{\omega_p}{2\omega_{0s}^2(k)} \left[\omega^2 - \omega_{0s}^2(k) \right] . \tag{8}$$

Thus, for the frequency region $\omega \approx \omega_0 = \omega_{s0}(q_0)$ equation (6) can be written as

$$(\omega^2 - \omega_0^2) \left[\omega^2 - \omega_{0s}^2(k) \right] = 2cdk^2\omega^2\omega_{0s}^2(k)/\omega_p.$$

Solving this equation two solutions for $\omega^2 = \omega_{1,2}^2(k)$ are obtained:

$$\omega_{1,2}^2(k) = \frac{1}{2} \left[\omega_0^2 + \omega_{0s}^2(k) + 2cdk^2 \frac{\omega_{0s}^2(k)}{\omega} \right] \pm$$

$$\pm \frac{1}{2} \sqrt{ \left[\omega_0^2 - \omega_{0s}^2(k) \right]^2 + 4cdk^2\omega_{0s}^2(k) \left[\omega_0^2 + \omega_{0s}^2(k) \right] \omega_p^{-1} } \tag{9}$$

In (9) relatively small terms proportional to d^2 are omitted under the radical.

At $k = q_0$ where the frequency $\omega_{0s}(k) = \omega_0$ splitting of the branches arises. In fact, at $k = q_0$ for frequencies $\omega_{1,2}(q_0)$ we obtain

$$\omega_{1,2}^2(q_0) \approx \omega_0^2 (1 \pm \sqrt{2cdq_0^2\omega_p^{-1}})$$

so that the splitting $\Delta \equiv \omega_1(q_0) - \omega_2(q_0)$ is determined by the relation

$$\Delta \approx \omega_0 (\frac{2d\omega_0^2}{\omega_p c})^{\frac{1}{2}} \tag{10}$$

or in wavelengths*)

---

*) In the determination of Δ damping was neglected. Therefore, relation (10) is applicable, if the value Δ is large compared to the spectral width of the polariton line.

$$\frac{d\lambda}{\lambda_0} = 2 \left(\frac{\pi d}{\lambda_p} \right)^{\frac{1}{2}} \frac{\lambda_p}{\lambda_0} , \tag{10a}$$

where $\lambda_p = 2\pi c/\omega_p$, $\lambda_0 = 2\pi c/\omega_0$. Along with the splitting $\Delta = \omega_1(q_0) - \omega_2(q_0)$ the value $\Delta_1 = \omega_1$ (min) $- \omega_2$ (max) may also be introduced. In this relationship ω_1 (min) is the minimum value of the frequency at the lower branch, corresponding to that value $k_{min} = \omega_{min}/c$ where the upper branch comes to the asymptotic straight line $\omega = ck$ (see (7a)). From (6) it follows that at $\omega_0 \ll \omega_p$

$$\omega_1 \text{ (min)} - \omega_0 \approx \frac{1}{2} d\omega_0\omega_p/c$$

Note also that $\Delta_1 < \Delta$ (see Fig.1). The splitting value decreases with increase of ω_p, $\Delta \sim \omega_p^{-\frac{1}{2}}$. This is due to the fact that with increase of ω_p the value of the electric field strength at $z \simeq 0$ in the surface wave decreases and, accordingly, the wave interaction with the transition layer also decreases. If on the contrary, the value $\omega_p \to 0$ (this case obviously corresponds to a metal film in vacuum), eq.(6) becomes:

$$2\sqrt{k^2c^2 - \omega^2} = \frac{cdk^2\omega^2}{\omega_0^2 - \omega^2} . \tag{11}$$

From this equation it follows that nonradiative waves discussed here appear only at such values of ω and k, for which $kc > \omega$, $\omega < \omega_0$. No splitting of the surface wave spectrum occurs. Regarding the dependence $\omega(k)$, it coincides in this case with that for two-dimensional systems (see [10,11]).

However, let us go back to consider the dispersion of surface waves in case $\omega_p \gg \omega_0$. Note first of all that at $k \gg q_0$ when $kc \gg \omega_p$, i.e. in the nonrelativistic limit, eq.(6) can be simplified and becomes:

$$\frac{\omega_p^2 - 2\omega^2}{\omega^2 - \omega_p^2} = \frac{kd\omega^2}{\omega^2 - \omega_0^2} \tag{12}$$

It should be borne in mind that this nonrelativistic equation is valid only when the inequality $\omega_p d/c \ll 1$ is fulfilled. Only in this case will the transition to the nonrelativistic limit not contradict the inequality $kd \ll 1$ used in the boundary conditions (1).

It follows from (12) that for the upper branch of frequencies at large k (see also [1,7])

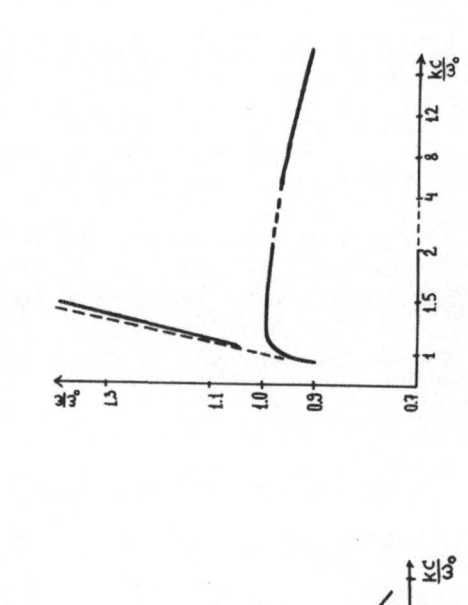

Fig. 2. Dispersion law for
the surface polariton in case
of y_p =45; α=0.01.

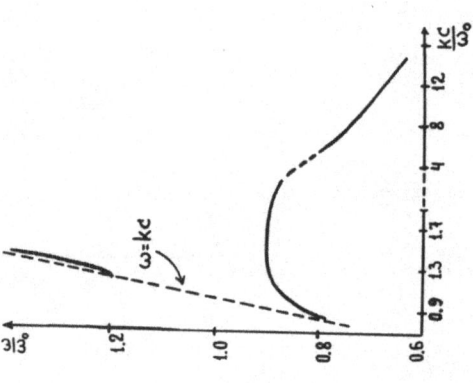

Fig. 1b. Dispersion law for
the surface polariton in case
of y_p=15; α=0.1.

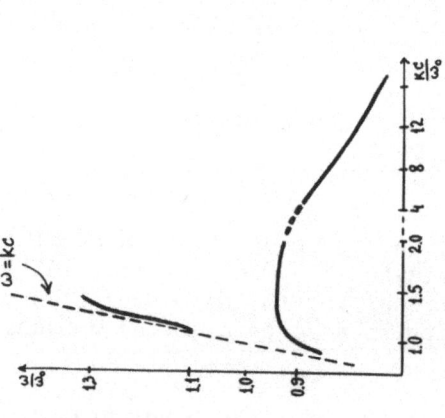

Fig. 1a. Dispersion law for
the surface polariton in case
of y_p =15; α=0.05.

$$\omega_1(k) \approx \frac{\omega_p}{\sqrt{2}} \; (\; 1 + \frac{kd}{4} \;) \tag{13a}$$

while for the lower one

$$\omega_2(k) \approx \omega_0 (1 - \frac{kd}{2} \;) \; . \tag{13b}$$

The fact that for the lower branch of SP frequencies a linear dispersion law with a negative slope is valid leads to the appearance of an additional (see below) surface wave in the frequency region $\omega \approx \omega_0$. In this frequency region not one but two surface polaritons may exist with one and the same frequency but with different values of the wave vector.

The dispersion law of polaritons in the case considered (metal film on a second metal) and in which damping is neglected is represented in Fig.1 for various values of the parameter $\alpha \equiv \omega_0 d/c$ and $y_p \equiv \omega_p^2/\omega_0^2$. Note that for aluminium coated with a silver film we have $y_p = 15.2$ and $\omega_0/c = 2.10^5$ cm^{-1}. For LiF film on a silver substrate $y_p = 45$ and $\omega_0/c = 4.10^3$ cm^{-1}. In the cited work[4] the splitting in the polariton spectra for Ag/Al pair was studied for film corresponding to the values of the parameter $\alpha = 5.10^{-2}$, 8.10^{-2} and 12.10^{-2}.

3. With regard to energy dissipation in the film and substrate the values ε and μ in (3) are complex (see (5)) even for real $\omega : \varepsilon = \varepsilon' + i\varepsilon''$, $\mu = \mu' + i\mu''$, $\varepsilon'' > 0$, $\mu'' < 0$. In this case the relationship (3) allows determination of the real and imaginary parts of k, $k = k' + ik''$ as functions of ω. Since the frequency of electron collisions in metal Γ, $\tilde{\Gamma} \sim 10^{14}$ sec^{-1} (see [12]), for the frequency region considered $\omega \sim \omega_0 \sim 10^{15}$sec^{-1}, $\Gamma/\omega_p << 1$, $\tilde{\Gamma}/\omega << 1$ so that $|\varepsilon'(\omega)| >> \varepsilon''(\omega)$ and $|\mu'(\omega)| >> |\mu''(\omega)|$. This means that for the determination of k'' in the case discussed the first approximation for ε'' and μ'' can be used.*) In accordance with the above we obtain from (3):

$$k''(\omega) = - \frac{k^2\mu'' - (\kappa/\varepsilon^2 + \omega^2/2\varepsilon\kappa c^2)\varepsilon''}{k(1/\varepsilon\kappa + 1/\kappa_1 + 2\mu)} \tag{14}$$

where the value $k = k(\omega)$ determined by the dispersion law (3), (6) as well as the values ε and μ should now be considered real (with Γ, $\tilde{\Gamma} = 0$).

At $\mu = 0$, i.e. when the transition layer is neglected

---

*) As the evaluation shows a similar situation also occurs in most cases of dielectric films and dielectric substrates e.g. for LiF films on sapphire).

$$k'' = \frac{\varepsilon''(\kappa/\varepsilon^2 + \omega^2/2\varepsilon\kappa c^2)}{k(1/\varepsilon\kappa + 1/\kappa_1)} \qquad (15)$$

and, taking into account (5) for frequencies where the value
$k'' = \frac{1}{2}\frac{\Gamma}{c}(\frac{\omega}{\omega_p})^2$.

With due regard for the transition layer in the frequency region $\omega \lesssim \omega_0$ each value of ω corresponds, as has already been stated above, to two surface polaritons (the usual and the additional one) (see Fig.1) and the relationship (14) allows us to compare their corresponding damping lengths L. Let us evaluate first of all the value $k_2''(\omega)$ which corresponds to the additional solution, which begins with the frequency region $\omega < \omega_0$ where the nonrelativistic approximation (13b) is already applicable. Since in this spectral region $\kappa \approx \kappa_1 \approx k$, we find from (14), taking account of inequality $|\varepsilon| \gg 1$

$$k_2''(\omega) = -\mu''(\omega)k^2/3 + \varepsilon''k/3\varepsilon^2$$

and, using (5) and (13b) we obtain

$$k_2''(\omega) \approx \tilde{\Gamma}/4\omega_0 d \qquad (16)$$

In this case of large d (but still k<1/d) the surface polariton field practically does not penetrate into the metal. Therefore, it is not surprising that (15) does not include substrate characteristics, and that this relationship can also be obtained (when damping is included) for a metal sheet in vacuum using the dispersion relationship (11) taken at $k \gg \omega/c$. At $\tilde{\Gamma}/\omega_0 = 3.10^{-2}$ and d=30A, we have $k_2'' = 2.10^4 \text{cm}^{-1}$, corresponding to propagation lengths L of the order of tenths of a micron.

For the evaluation of propagation lengths of both the usual and the additional polaritons, which correspond to small k, the general relationship (14) should be used. This relationship can be rewritten in terms of dimensionless quantities as

$$\frac{ck''}{\omega_0} = A(x,y)/B(x,y),$$

$$A(x,y) = \alpha\tilde{\nu}xy^{3/2}(y-1)^{-2} + \nu y_p y^{-3/2}\varepsilon^{-1}[(x-y\varepsilon)^{1/2}\varepsilon^{-1}$$
$$+ y/2(x-y\varepsilon)^{1/2}],$$

$$B(x,y) = \sqrt{x}\,[\frac{1}{\varepsilon}(x-y\varepsilon)^{-1/2} + (x-y)^{-1/2} + 2\alpha y(y-1)^{-1}],$$

where

$$x = c^2 k^2 / \omega_0^2,$$

$$y = \omega^2 / \omega_0^2,$$

$$y_p = \omega_p^2 / \omega_0^2,$$

$$\nu = \Gamma / \omega_0,$$

$$\tilde{\nu} = \tilde{\Gamma} / \omega_0,$$

$$\alpha = \omega_0 d / c,$$

$$\varepsilon = 1 - y_p / y.$$

Since according to assumption, $y_p \gg 1$, in the region $x>1$ and $y \simeq 1$, where $|\varepsilon| \approx y_p/y \gg 1$, the expressions for $A(x,y)$ and $B(x,y)$ simplify:

$$A(x,y) \simeq \alpha \tilde{\nu} x y^{3/2} (y-1)^{-2} + \nu \left(\frac{y}{y_p}\right)^{1/2},$$

$$B(x,y) \approx \sqrt{x} \left[(x-y)^{-1/2} + 2\alpha y (y-1)^{-1} \right].$$

For example, at $\alpha = 10^{-2}$ and $y_p = 45$ (see Fig.2) the value $y = 0.98$ ($\omega = 0.98\omega_0$) corresponds to $x=1.1$ (the usual wave) and $x=1.9$ (additional wave). In this case for the usual wave $k_1'' = \frac{1}{16} \frac{\omega_0}{c}$, for the additional one $-k_2'' = 0.1 \frac{\omega_0}{c}$ and $L_1 \approx L_2$. But if $y = 0.89$, then $x_1 = 0.9$ and $x_2 = 20$ so that $k_1'' = 10^{-3} \frac{\omega_0}{c}$, $k_2'' = 0.2 \frac{\omega_0}{c}$. Therefore, while moving away from resonance, the mean free path L_2 of the additional polariton decreases abruptly and in the given case (i.e. at $y = 0.89$) $L_1/L_2 \approx 200$.

The evaluations given here for the propagation lengths of surface polaritons indicate that as in the case of bulk polaritons the additional wave can only be observed for special choice of substrates and films.

For metal surfaces thin dielectric films make the mean free path L of the surface polariton only slightly shorter even in the resonance region in the film (see [13]). Additional surface waves are likely to be found just for these conditions.

Before proceeding to the discussion of ABC problems one remark should be made about the damping of surface waves in the case of metal transition layer (metal films). In the region of very small thickness metal films are usually not solid, but have a

granular structure. In this case along with damping of waves due
to non-Hermitian property of the permeability tensors, the Landau
damping mechanism may come into action.

As shown in [14], for metal drops of radius R Landau damping
leads to broadening

$$\Gamma \sim \frac{e^2}{R}$$

so that at $R \sim 10 \overset{\bullet}{A}$ the value Γ is of the order of 1 ev. This is
likely to mean that in experiments[4] done for very small thickness
of the silver coatings, no splitting in SP spectra was observed
because in these conditions the films were not solid, but consisted
of grains with characteristic dimensions of the order of the coating thickness. The contribution to damping from surface scattering of electrons must be taken into account in the same
situation.[15]

4. It was stressed in [1] that for the observation of optical
effects arising from the additional surface wave the excitation
of surface polariton by wedge light diffraction of laser light
for example can be used. See Fig.3. Recently (see [16]) such a

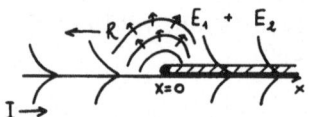

Fig. 3.   Incident (I), reflected (R), transmitted (E_1, E_2)
          surface waves and cylindrical wave.

method of excitation was realized for the IR range on a metal wedge. Perhaps, the further development of the experiments of the type described in [16] will permit us to study also the effects caused by the interference of additional and usual surface waves of the same frequency.

When the additional surface wave is taken into account, the usual boundary conditions are insufficient for the determination of field amplitudes. Similar to the case of bulk crystal optics, the problem of additional boundary conditions (ABC) arises.

The form of the ABC should, generally speaking, depend on the type of film and the character of those dipole oscillations in it which cause resonance with the surface polariton. In this connection we will confine our further attention to the discussion of the form of ABC for the case considered in §2 for obtaining the dispersion relationship (3).

Note first of all that the correction to the boundary condition (2) is not caused by the film polarization along the normal to it. Therefore, when determining ABC in the case of a dielectric film we may assume that the transition dipole moment is directed along the axis z. Since the field variation along the film width is neglected in the approximation linear in d/λ, the film can generally be considered as two-dimensional. If such a two-dimensional system (a two-dimensional crystal) is finite along the x axis and if deformation of the molecules on the very boundary of such crystal is neglected, then similar to the three-dimensional case (see [17]), the boundary condition for polarization at x=0 is

$$P_z(x=0) = 0, \tag{17}$$

where $P_z(x,y)$ is the film polarization per its unit area. If the field obeys condition (17), then at $x \approx 0$ the film cannot lead to the breakup of $\vec{E}_t$ of the form (2). This follows because at x=0

$$\vec{E}_t(II) - \vec{E}_t(I) = -i\vec{k}_t \int_0^d E_n(z, \ x=0)dz \tag{18}$$

where $E_n(z, \ x=0)$ is the normal component $\vec{E}$ in the film and, since in the film $P_n=0$ at x=0 (see (17)) we conclude that $E_n(z,x=0) \approx D_n(z,x=0)$. Accordingly:

$$\int_0^d E_n(z,x=0)dz = \int_0^d D_n(z,x=0)dz = dD_n(I,x=0) = dE_n(I,x=0).$$

Thus, at x=0 on the right-hand side of (18) there appears no term of the order of $\mu=d/\tilde{\varepsilon}$ with a resonance at $\tilde{\varepsilon}=0$. Recalling that when (2) was derived from (1) nonresonant terms were omitted, at x=0 we should consider that

$$\vec{E}_t(II) - \vec{E}_t(I) = 0 . \qquad (19)$$

Comparing now (19) and (2) we conclude that the condition (17) for polarization in the film gives the sought after ABC in the form

$$E_n(I,x=0) = 0 . \qquad (20)$$

Because $E_n(I) = D_n(II)$ (see (1) at $\gamma = d\tilde{\varepsilon} \approx 0$, $\tilde{\varepsilon} \approx 0$), the relationship (20) can be changed to an equivalent condition

$$D_n(II,x=0) = 0 . \qquad (20a)$$

The ABC for $E_t(I,x=0)$ for dielectric films in the vicinity of the resonance of $\tilde{\varepsilon}(\omega)$ can be found in a similar way. If, however, this resonance is due to a two- or three-dimensional Wannier-Mott exciton, then the introduction of a "dead layer" of thickness $\ell \approx 2\tau_b$ (τ_b is the Bohr radius of the exciton) may turn out to be important similarly to the ABC theory for bulk waves.[18]

The ABC problem for metal films requires additional special consideration.

III. SCATTERING OF SURFACE POLARITONS NEAR PHASE-TRANSITION POINTS

Because the dispersion of surface polaritons is determined by the dielectric properties of the media which are in contact, in the vicinity of a phase-transition point, owing to the fluctuations of the order parameter, the dielectric constant also fluctuates. This uncovers new possibilities of transforming a surface electromagnetic wave by emission of a surface polariton or by Raman scattering of a surface polariton.[19] Far from phase-transition points these processes also take place due to SP scattering, e.g. by static surface roughness. However, the intensity of SP scattering by the fluctuation of order parameter can have a characteristic temperature dependence. This very circumstance as well as the possibility of making sufficiently smooth surfaces may play a decisive role in the experimental investigation of phase transitions by the SP spectroscopy method.

In the development of the theory for SP scattering by the order parameter fluctuations in the transition layer the results of Mills[20] can be used. Mills considered the influence of static surface roughness on the polariton mean free path. He defined the deviation of the surface from a plane (z=0) by a random function $\xi(x,y)$ with $\langle\xi(x,y)\rangle > 0$. The presence of these deviations is formally equivalent to the presence at z=0 of a film whose polarizability per unit area

$$\alpha = \frac{\varepsilon - 1}{4\pi}\,\xi(x,y) \tag{21}$$

is a function of x and y (ε is the dielectric constant of the medium).

When we deal with SP scattering by order parameter fluctuations in the transition layer (e.g. in a thin dielectric film on a metal substrate), then its polarizability α (per 1 cm^2; the film thickness D is far less than the polariton wavelength) can also be considered as a random function of x and y. Its dependence on the order parameter is

$$\alpha = \alpha_0 + \frac{D}{4\pi}\left(\frac{d\tilde{\varepsilon}}{d\eta}\right)_0 \delta\eta(x,y) \;, \tag{22}$$

where α_0 is the dc component of the polarizability which causes no scattering, $\delta\eta = \eta - \eta_0$, $\langle\eta\rangle = \eta_0$, while $\tilde{\varepsilon}(\eta)$ is the film-material dielectric constant corresponding to the value of the order parameter η. Relation (22) means that we are dealing here with a phase transition in which the η-dependent part of $\tilde{\varepsilon}(\eta)$ is linear in η at small values of η.

In fact, the order parameter varies also in time, thus leading to SP nonelastic scattering processes. However, in the vicinity of the phase transition point, due to the soft mode, time fluctuations of η are slow and they are neglected here.

It is shown in [20] that for surface polaritons with frequencies ω much lower than the plasma frequency of the metal (for these frequencies we have $|\varepsilon| \gg 1$), the intensity of the scattering by the roughness and the corresponding (partial) mean free path L_1 are determined by the relation

$$I \sim \frac{1}{L_1} = \frac{4}{3\pi}\frac{\omega^5 F_\xi(0)}{c^5|\varepsilon|^{\frac{1}{2}}} \;, \tag{23}$$

where $F_\xi(\vec{Q})$ is the Fourier component of the correlation function

$$F_\xi(\vec{Q}) = \int d^2\vec{\tau}_{\parallel}\langle\xi(\vec{\tau}_{\parallel})\xi(0)\rangle e^{i\vec{Q}\cdot\vec{\tau}_{\parallel}} \;.$$

The principal process in this case is the scattering corresponding
to the termination of the surface waves and their conversion into
bulk waves. Comparing (21) with (22) and using (23), we find that
in our case

$$I \sim \frac{1}{L_1} = \frac{4\omega^5 D^2 (\frac{d\tilde{\epsilon}}{d\eta})^2 F_\eta(0)}{3\pi^2 c^5 |\epsilon|^{5/2}} \qquad (24)$$

so that at $T \simeq T_c$ the temperature dependence of the quantities I
and $1/L_1$ coincides with the temperature dependence of $F_\eta(0)$. On
the other hand, in crystals such as quartz, where $\alpha - \alpha_0 \sim \eta^2$ (at
$T < T_c$ we have $\eta^2 = \eta_0^2 + 2\eta_0 \delta\eta(x,y)$), the use of relation (23)
yields $I \sim 1/L_1 \sim \eta_0^2 F_\eta(0)$. Inasmuch as $\eta_0^2 \sim (T_c - T)$ in this
case, the role of the fluctuations becomes minor. In the Landau
theory of second-order phase transitions we have $F_\eta(0) \sim |T-T_c|^{-1}$
and this, in accordance with (24), corresponds to a sharp
decrease of the partial polariton mean free path L_1. The
temperature region in which the effect can be revealed by the
decrease of the mean free path $L(1/L = 1/L_0 + 1/L_1)$ is determined
by the inequality $L_1(T) < L_0 \simeq 1$ cm. If for some reason this
inequality is not satisfied anywhere (fluctuations of the order
parameters are suppressed, etc.), then observation of the effect
calls for direct measurements of the intensity of the light
produced by the surface wave in the contact region and scattering
by this region.

   If the SP frequency lies in the vicinity of oscillations in
the transition layer (i.e. at $\omega \simeq \omega_0$), the intensity of SP
scattering processes may increase considerably due to the increase
of the derivative $(d\epsilon_0/d\eta)$ here. In fact, in this frequency
region, for example, taking a linear dependence of $\tilde{\epsilon}_0$ on η

$$\left(\frac{d\tilde{\epsilon}_0}{d\eta}\right)_0 \approx A\tilde{\epsilon}(0)$$

where A is the frequency independent constant, we obtain for the
scattering intensity $I \sim I_0 \left(\frac{\Omega}{\omega-\omega_0}\right)^2$, where I_0 is the value
which equals approximately the scattering intensity far from
resonance, $\Omega^2 = (\tilde{\epsilon}_0 - \tilde{\epsilon}_\infty)\omega_0^2$. In using this relationship it
should naturally be borne in mind that SP frequency ω does not
lie in the gap region (see Figs.1,2). Therefore, at $\Delta > \delta$, where
δ is the polariton line width, the minimum value $|\omega - \omega_0|$ Δ.
But if $\delta \gg \Delta$, then, when evaluating the maximum increase of SP
scattering intensity, it should be considered that $I \sim I_0(\Omega/\delta)^2$.

Since in what has been said above, the results of (24) were used for obtaining the relationship (23), renormalization of the SP spectra due to the presence of a transition layer (a film) was not taken into account. The inclusion of this fact at $\omega \approx \omega_0$ leads to the additional increase of the scattering intensity. This increase is caused by the decrease of the SP group velocity at $\omega \rightarrow \omega_0$ and leads to a weaker effect than the one considered above.*)

The theory of SP scattering by fluctuations of the order parameter for the case when the phase transition occurs in the whole substrate volume and not with transition layer is considered in [21].

REFERENCES

1. V. M. Agranovich, Usp. Fiz. Nauk 115,199 (1975).
2. V. M. Agranovich and A. G. Mal'shukov, Opt. Commun. 11,169 (1974).
3. V. A. Yakovlev, V. G. Nazin, and G. N. Zhizhin, Opt. Commun. 15,293 (1975); Zh. Exsp. Theor. Fiz. 72,687 (1977).
4. T. Lopez-Rios, F. Abeles, and G. Vuye, Journ. de Phys. 39,645 (1978).
5. M. R. Philpott and J. M. Turlett, Journ. Chem. Phys. 64,3852 (1976).
6. V. M. Agranovich and I. I. Lalov, Opt. Commun. 16,239 (1976).
7. A. Ya. Blank and V. L. Beresinskii, ZhETF 75,2317 (1978).
8. J. Schoenwald, E. Burstein, and J. M. Elson, Sol. State. Commun. 12,185 (1973).
9. V. M. Agranovich and V. L. Ginzburg, "Spatial Dispersion in Crystal Optics and the Theory of Excitons" (II edition), Nauka, Moscow (1979).
10. V. M. Agranovich and O. A. Dubovskii, Pis'ma ZhETF 3,345 (1966).
11. V. M. Agranovich, "The Theory of Excitons," Nauka, M. (1968).
12. F. Abeles, "Optical Properties of Solids", F. Abeles, ed., NH Publ. Comp., Amsterdam, London (1972), p.93.
13a. G. N. Zhizhin, M. A. Moskaleva, E. V. Shomina, and V. A. Yakovlev, Pis'ma ZhETF 24,221 (1976) (JETP Lett. 24,196 (1976)).
13b. C. A. Ward, R. W. Alexander, and R. J. Bell, Phys.Rev. B14,856 (1976).
14. Yu. E. Lozovik and V. N. Nishanov, FTT 20,3654 (1978).
15. E. N. Economou, Phys. Rev. 182,539 (1969).
16. G. N. Zhinzhin, M. A. Moskaleva, E. V. Shomina, and V. A. Yakovlev, Pis'ma ZhETF 29,9 (1979).
17. S. I. Pekar, ZhETF 33,1022 (1957).
18. J. J. Hopfield and D. G. Thomas, Phys. Rev. 132,561 (1963).

*) If $\tilde{\varepsilon}_0(\omega,\eta) \simeq \tilde{\varepsilon}_0(\omega,0) + \alpha\eta^2$ at $\omega \approx \omega_0$ it is better to use $1/\tilde{\varepsilon} \simeq 1/\tilde{\varepsilon}(\omega,0) + \beta\eta^2$ and $(d\tilde{\varepsilon}_0/d\eta)_0 \approx A\tilde{\varepsilon}^2(\omega,0)$. In this case $I \sim \eta_0^2 F_\eta(0)[\tilde{\varepsilon}(\omega,0)]^4$ and at $T \simeq T_c$, $I \sim \eta_0^2|T-T_c|^{-1}(\frac{\Omega}{\omega-\omega_0})^4 \sim (\frac{\Omega}{\omega-\omega_0})^4$.

19. V. M. Agranovich, Pis'ma ZhETF 24,602 (1976). (JETP Lett.
     24,588 (1976)).
20. D. L. Mills, Phys. Rev. B12,4036 (1975).
21. V. M. Agranovich and T. A. Leskova, Solid State Commun.
     21,1065 (1977).

RECENT DEVELOPMENTS IN NON-LOCAL OPTICS

Joseph L. Birman, and
Deva N. Pattanayak
Department of Physics
City College of the City University of New York
New York, New York 10031

ABSTRACT

In this paper a brief account is given of results of some recent investigations of four classes of effects due to non-locality (or spatial dispersion) upon optical phenomena in bounded solids. These are: Resonant Inelastic (Brillouin or Raman) Scattering; Transient Reflectivity; Wave Propagation and Additional Boundary Conditions in Gyrotropic Media; Lateral Beam Shift - Goos Hänchen Effect.

I. INTRODUCTION

Since the pioneering work of Pekar and Ginzburg, spatial dispersion i.e. the wave vector dependence of the dielectric function has played a significant role in the field of crystal optics. Many aspects of non-local optics have recently been studied and are still being actively pursued both by experimentalists and theoreticians.[1] Novel phenomena in the Brillouin and Raman Scattering, reflection and refraction, Cerenkov and transition radiation and transient precursors have been investigated by various authors. In this paper we briefly report on the effects of spatial dispersion on:

transient optical reflectivity;
wave propagation in gyrotropic medium;
lateral beam shift near total internal reflection (Goos-Hänchen Effect);
resonant Raman and Brillouin Scattering.

These consequences of spatial dispersion all follow from the wave-vector and frequency dependence of the dielectric suscepti-bility. One very popular model for spatially dispersive effects,

the so-called "dielectric model" has the merit of exhibiting most
of the relevant physics in an explicit and even analytic form.
In this case in the medium with $b = \hbar\omega/M$, where m_e^* is the exciton mass:

$$\chi(k,\omega) = \chi_0 + F/(\omega_0^2 + bk^2 - \omega^2 - i\omega\Gamma)$$

while outside the medium $\chi = 0$. Owing in part to the inconsistency
of assuming a translationally invariant susceptibility $\chi(\vec{r}-\vec{r}')$ in a
system with broken translational symmetry this Ansatz violates exciton
energy conservation at the surface. Plane wave solutions of Maxwell's equations obey the dispersion $k^2 = (\omega/c)^2 \epsilon(k,\omega)$. The
features revealed by this model are: the multiwave (linear combination of plane waves) structure of the physical propagating polariton mode, with the correct linear combination determined by combination of both: usual Maxwell, plus Additional, Boundary Conditions.

In general of course the non-local dielectric coefficient will
not be of the form given above, and Ting, Frankel and Birman[2] have
given a form suitable for Wannier exciton-like media; Zeyher, Brenig
and Birman have given a form suitable for Frenkel excitons.[2]

II. RESONANT INELASTIC (BRILLOUIN, RAMAN) SCATTERING

Zeyher, Brenig and Birman[3] first gave a quantitative theory
of inelastic (Brillouin) scattering based on the non-local, polariton picture and the multi-wave description of exciton polaritons.
The kinematics of inelastic scattering is simple. Let the functions
$\omega_j(k)$ be the dispersion equations of the physical exciton-polaritons:
the index j accounts for multibranch effect for finite mass. An
inelastic scattering event can occur from $\omega_j(\vec{k}) \rightarrow \omega'_{j'}(\vec{k}')$ if the
frequency shift $\Delta\omega = \omega_j(\vec{k}) - \omega'_{j'}(\vec{k}')$ and pseudomomentum transfer
$\vec{q} = \vec{k}' - \vec{k}$ satisfy the dispersion equation of some other physical
crystal excitation $\Delta\omega(\vec{q})$. For example for Resonant Brillouin Scat-
tering (RBS) $\Delta\omega(\vec{q}) = c|\vec{q}|$, c is a sound velocity; for Resonant
Raman Scattering (RRS) $\Delta\omega(q) = \omega_R$: essentially independent of q.
The dispersion of the quasiparticle produced, "tunes" the scattering
process. Branch indices j,j' may or may not correspond. Most of
the practical cases require "backward" scattering so $|\vec{k}'| = |\vec{k}|$.
Certain predictions of this model can give dramatic evidence of the
extra propagating branches above ω_ℓ; and the onset of new channels
for inelastic scattering. The extra modes yield a Brillouin "octet"
instead of "doublet" which occurs for $\omega_L \ll \omega_\ell$. Since the initial
experimental verification of these predictions additional work has
revealed (for ω_L well above ω_ℓ) multiphonon, multistep Brillouin
and mixed Brillouin-Raman Scattering. Forbidden scattering has
been observed also.[4]

In addition to kinematic effects, the ZBB theory quantitatively
predicted the frequency dependence of the RBS cross section, and the
line shape of the Brillouin scattered radiation. The theory of
ZBB compared quantitatively the predicted results due to use of each
of three presently available abc - each one is specific to one form
of $\epsilon(r,r')$ - and demonstrated the possibility of distinguishing

the applicable abc, and thus determining the structure of the physical polariton mode.[3]

Since another paper in this Symposium[5] will discuss the current experimental developments in RBS, show how exciton mass and polariton branch dispersion above resonance can be determined from RBS as well as compare the theory with the present experiments, we defer further discussion to that paper.

III. TRANSIENT OPTICAL REFLECTIVITY

In the usual studies of transient pulse propagation, reflection of the incident pulse is neglected. However, for pulses which contain frequencies in the resonance region of the material, transient reflection will be appreciable, and may give rise to new effects. Transient reflectivity of a Heavyside sinusoidal pulse from a semiinfinite crystal taking into account only frequency dispersion has been studied by Elert[6]. In this paper we include spatial dispersion in addition to frequency dispersion in the analysis of reflectivity.

For simplicity we take the incident pulse to be a sinusoidal signal starting abruptly at time $t = 0$, $x = 0$. Thus:

$$\vec{E}^{In}(0,t) = \sin \omega_L t \; \theta(t) = - \, \text{Im} \, e^{-i\omega_L t} \, \theta(t) \tag{1}$$

We assume the pulse to be normally incident upon the semiinfinite medium occupying the space $L < x < + \infty$. The Fourier spectrum of the incident pulse then is

$$\vec{E}^{In}(o,\omega) = \lim_{\eta \to 0} \frac{1}{2\pi i} \left(\frac{1}{\omega - \omega_L + i\eta} \right) \tag{2}$$

The medium is characterized by the constitutive relation

$$\vec{D} = \vec{E} + 4\pi \vec{P}_{NL} \, , \tag{3}$$

where

$$\vec{P}_{NL}(z,\omega) = \frac{i\chi}{8\pi\mu} \int_0^\infty \left\{ e^{i\mu|z-z'|} - e^{i\mu|z+z'|} \right\} \vec{E}(z',\omega) dz' \tag{4}$$

is the non-local polarization, and

$$\chi = \frac{4\pi\alpha m_e^* \, \omega_e}{\hbar} \, , \quad \mu = \left(\frac{m_e^*}{\hbar\omega_e}(\omega^2 - \omega_e^2 + i\omega\Gamma) \right)^{\frac{1}{2}} \tag{5}$$

with Im $\mu > 0$. In Eq.(5) m_e^* is the total mass of the exciton, $\hbar\omega_e$ its energy, $4\pi\alpha$ is the oscillator strength associated with the exciton transition and Γ is a phenomenological damping constant. The normal incidence reflection coefficient $R(\omega)$ for such a non-local medium has been given[1]. An expression for $R(\omega)$ in the dielectric model was also given[7]. For the model considered here we have

$$R(\omega) = \frac{1- n^*(\omega)}{1+ n^*(\omega)} \quad , \tag{6}$$

where

$$n^*(\omega) = \frac{n_1 n_2 + 1}{n_1 + n_2} \quad ,$$

$$n_1(\omega) = + \left(a - (a^2 - b)^{\frac{1}{2}} \right)^{\frac{1}{2}} \quad , \tag{7}$$

$$n_2(\omega) = + \left(a + (a^2 - b)^{\frac{1}{2}} \right)^{\frac{1}{2}} \quad , \tag{8}$$

$$a = \frac{1}{2} \left(1 + \mu^2/k_0^2 \right); b = \frac{\mu^2}{k_0^2} - \frac{\chi}{k_0^2} \quad . \tag{9}$$

With this much information we are in a position to write down an integral expression for the transient reflectivity. The transient reflection is, due to linearity, a superposition of the reflected frequency components of the incident signal i.e.

$$\vec{E}^{(R)}(0,t) = \frac{Re}{2\pi} \lim_{\eta \to 0} \oint_C \frac{R(\omega)}{\omega - \omega_L + i\eta} e^{-i\omega(t - \frac{2L}{c})} d\omega \tag{10}$$

The contour C in Eq. (10) is shown in Fig. (1). In order to evaluate this integral we investigate the singularities of the integrand The singularities are shown in Fig. 1, and are:

(a) Simple pole at $\omega = \omega_L - i\eta$;
(b) Branch points at $\omega_1, \omega_2, \omega_3, \omega_4, \omega_5, \omega_6$,

where

$$\omega_{1,2} = - ib_1^+ \pm (b_1^{+2} + b_2 \omega_e^2)^{\frac{1}{2}} \quad ,$$

$$\omega_{3,4} = - ib_1^- \pm (b_1^{-2} + b_2 \omega_e^2)^{\frac{1}{2}} \quad ,$$

$$\omega_{5,6} = - i\Gamma/2 \pm (\omega_\ell^2 - \Gamma^2/4)^{\frac{1}{2}} \quad ,$$

$$b_1^{\pm} = b_2 \left(\frac{\Gamma}{2} \pm \left(\frac{4\pi\alpha\hbar\omega_e}{m_e^* c^2} \right)^{\frac{1}{2}} \omega_e \right) \ ; \ b_2 = \frac{1}{\left(1 - \frac{\hbar\omega_e}{m_e^* c^2} \right)}$$

$$\omega_\ell^2 = (1 + 4\pi\alpha) \omega_e^2 \tag{11}$$

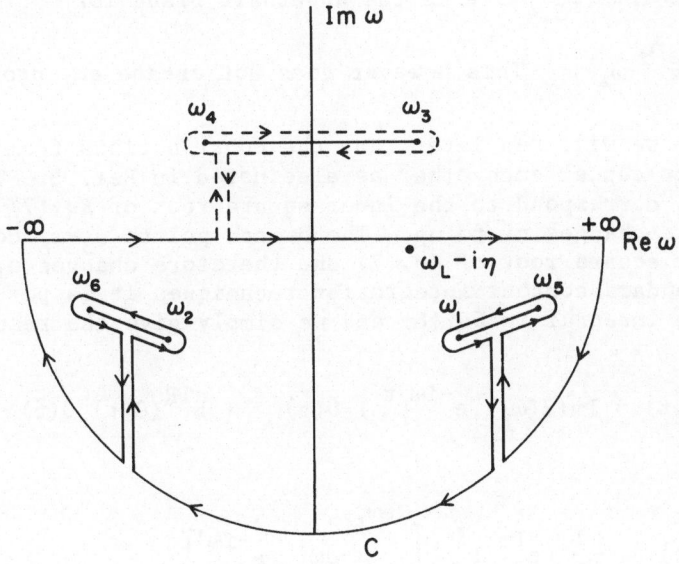

Fig. 1: The dotted portion of C does not contribute to transient
         reflection.

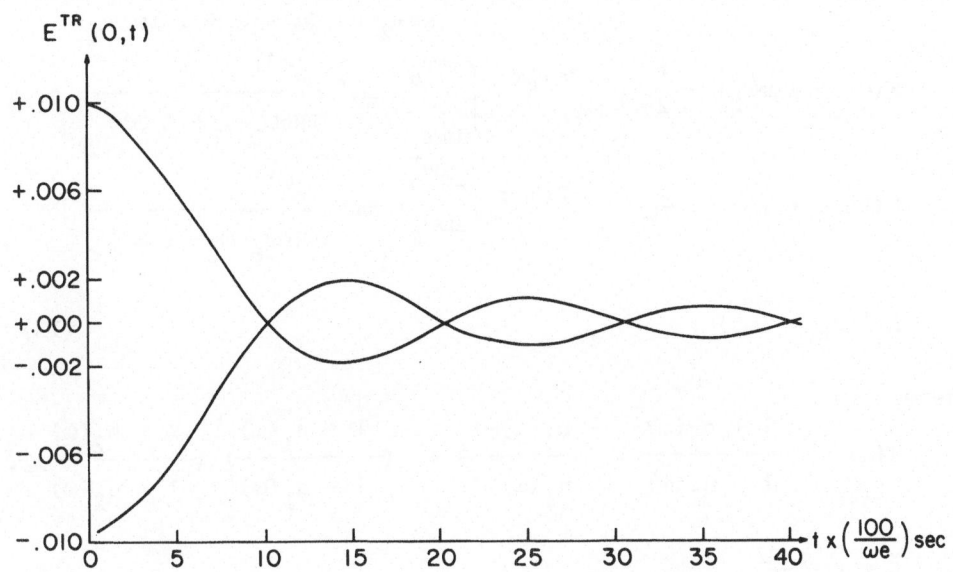

Fig. 2: Approximate numerical evaluation of $\vec{E}^{TR}(0,t)$, as in Eq.(13)
         for a model medium with the following parameters $\hbar\omega_e$ =
         2.53 eV; $4\pi\alpha$ = 0.125; $\hbar\Gamma$ = 5×10^{-5} eV; $\hbar\omega_L$ = 1.5 eV;
         m_e^* = 0.9 m, m being the mass of the electron.

The branch points $\omega_{3,4}$ lie in the upperhalf plane for

$$\frac{\Gamma}{2} < (\frac{4\pi\alpha\hbar\omega_e}{m^*_e c^2})^{\frac{1}{2}} \omega_e .$$ This however does not create any problem with

causality as we will see later that the contributions from these branch points cancel each other as also noted in Ref. 8. The branch points $\omega_1-\omega_4$ correspond to the inner square root of Eq.(7) and (8) and therefore changes n_1 to n_2. The branch points $\omega_5-\omega_6$ correspond to the outer square root of Eq.(7) and therefore changes n_1 to $-n_1$. By using standard contour integration techniques it is possible to evaluate the integral in Eq.(8) and we simply give the result here.

$$\vec{E}^{(R)}(0,t) = Im(R(\omega_L) e^{-i\omega_L \tilde{t}}) \theta(\tilde{t}) + \vec{E}^{TR}(0,t) \theta(\tilde{t}) \qquad (12)$$

where

$$\vec{E}^{TR}(0,t) \cong \frac{1}{2\pi} e^{Im\omega_1 \tilde{t}} \int_{Re\omega_1}^{Re\omega_5} d\omega' \frac{e^{-i\omega'\tilde{t}}}{(\omega'-\omega_L+ i\, Im\, \omega_1)} \times$$

$$K(\omega' + i\, Im\, \omega_1) + \frac{1}{2\pi} e^{Im\, \omega_2 \tilde{t}} \int_{Re\omega_2}^{Re\omega_6} d\omega' \frac{e^{-i\omega'\tilde{t}}}{(\omega'- \omega_L+ i\, Im\, \omega_2)} \times$$

$$K(\omega' + i\, Im\omega_2) + \frac{i}{2\pi} e^{-iRe\omega_5 \tilde{t}} \int_{Im\omega_1}^{Im\omega_5} d\omega' \frac{e^{\omega'\tilde{t}}}{(Re\omega_5-\omega_L+ i\, \omega')} \times$$

$$K(Re\omega_5 + i\omega') + \frac{i}{2\pi} e^{-iRe\omega_6 \tilde{t}} \int_{Im\omega_2}^{Im\omega_6} d\omega' \frac{e^{\omega'\tilde{t}}}{(Re\omega_6-\omega_L+ i\, \omega')} \times$$

$$K(Re\omega_6 + i\omega') ; \qquad\qquad\qquad\qquad\qquad\qquad\qquad (13)$$

where

$$K(\omega) = (\frac{1 + n_2(\omega)}{1 - n_2(\omega)}) (\frac{n_1(\omega)-1}{n_1(\omega)+1}) + (\frac{1 + n_1(\omega)}{1 - n_1(\omega)}) (\frac{1 - n_2(\omega)}{1 + n_2(\omega)}) ,$$

$$\tilde{t} = t - \frac{2L}{c} . \qquad\qquad\qquad\qquad\qquad\qquad\qquad\qquad (14)$$

The inequality sign in Eq.(13) implies that we have neglected terms which are of lower order. It can be shown that in the limit of $n_2 \to \infty$, our expression agrees with that due to Elert: in this case local optics is recovered. Analytical evaluation of (13) is difficult, and is not given here.

We made a computer calculation of these integrals and the result
is shown in Figure 2 for a typical set of parameters. We wish to
stress the qualitative aspect of the transient reflections. Soon
after the lapse of the "dead time" $\frac{2L}{c}$, the transient part has
its maximum amplitude and then it dies down rapidly. The transient
part arrives in pulses whose time period for the value of parameters
used is approximately 0.25 psec. These transients are superimposed
on the steady state reflection amplitude and may be detected experimentally perhaps by a heterodyne or beat technique.

The frequency content of this transient pulse lies in the
"stop gap" region of the crystal and therefore we conjecture that
in the case when surface polariton modes are excited by a transient
ATR configuration for example, there will be interference between
the surface polariton modes and this transient pulse.

IV. ADDITIONAL BOUNDARY CONDITIONS FOR GYROTROPIC MEDIUM NEAR EXCITON TRANSITION FREQUENCY

As has been pointed out by Ginzburg [9], near an exciton transition frequency in a gyrotropic crystal there will be three waves,
two of which are right polarized and the other left polarized or
vice versa. The dielectric function for a model gyrotropic medium
is assumed by Ginzburg to be

$$\varepsilon_{ij}^{-1} (\vec{k},\omega) = \varepsilon_0^{-1}(\omega) \, \delta_{ij} + i \, \varepsilon_1^{-1}(\omega) \, \varepsilon_{ij\ell} \, k_\ell, \tag{15}$$

The constitutive relation (15) is valid for an infinite medium.
The proper constitutive relation for a finite medium is not yet
known.

We propose a phenomenological constitutive relation for this
case. For this purpose we write Eq.(15) in coordinate space as

$$\vec{P} + \frac{4\pi\gamma(\omega)}{\varepsilon_b} \nabla \times \vec{P} = \alpha(\omega)\vec{E} - \gamma(\omega)\nabla \times \vec{E} , \tag{16}$$

where

$$\frac{4\pi\gamma(\omega)}{\varepsilon_b} = \varepsilon_0(\omega)\varepsilon_1^{-1}(\omega) \tag{17}$$

$$4\pi\alpha(\omega) = \varepsilon_0(\omega) - \varepsilon_b , \tag{18}$$

and ε_b is a background dielectric constant defined as

$$\vec{D} = \varepsilon_b \vec{E} + 4\pi\vec{P} \tag{19}$$

We neglect magnetization $\vec{M}$ i.e. $\vec{B} = \vec{H}$ for simplicity. The second

term on the right hand side of Eq.(16) arises due to spatial dispersion effects and is responsible for the existence of an additional wave. We further simplify Eq.(16) by assuming that we are
dealing with normal incidence, and we replace $\vec{\nabla}$ by $\hat{z}\frac{\partial}{\partial z}$. We then
get the following coupled set of equations

$$P_x (z,\omega) - \frac{4\pi\gamma}{\varepsilon_b} \frac{\partial P_y}{\partial z} = \alpha E_x - ik_0\gamma B_x \tag{20}$$

$$P_y (z,\omega) + \frac{4\pi\gamma}{\varepsilon_b} \frac{\partial P_x}{\partial z} = \alpha E_y - ik_0\gamma B_y \tag{21}$$

$$P_z = E_z = B_z = 0 . \tag{22}$$

In order to properly pose the boundary value problem for a bounded
system, we introduce a new function

$$\pi(z,\omega) = P_y(z,\omega) - iP_x(z,\omega) \tag{23}$$

It can then be shown that

$$\pi(z,\omega) + \frac{4\pi i\gamma(\omega)}{\varepsilon_b} \frac{\partial \pi(z,\omega)}{\partial z} = \alpha (E_y - iE_x)$$

$$- ik_0\gamma(B_y - iB_x) \tag{24}$$

Here, we follow [10]. Equation (24) however is not valid near the
surface. In the presence of the surface it is clear that γ and α
will depend upon distance from the surface. Then we must have on
the right hand side of Eq.(24) a term proportional to $\frac{\partial \gamma}{\partial z} \pi$. By
the addition of this term, and also on considering energy conser-
vation, it can be seen that for a bounded medium Eq.(24) may be
written as

$$\pi(z,\omega) + \frac{4\pi i}{\varepsilon_b} \frac{\partial}{\partial z} (\gamma(\omega,z)\pi) = \alpha (E_y - iE_x)$$

$$- ik_0\gamma(B_y - iB_x) \tag{25}$$

We can now obtain from a pillbox type of construction (as in the
case of obtaining Maxwell continuity conditions) that

$$\gamma(\omega,z = 0^+) \pi (z = 0^+,\omega) = 0 \tag{26}$$

From this we obtain the boundary condition

$$P_y(0) - iP_x(0) = 0 \tag{27}$$

In a similar fashion we also obtain from the conjugate of (25)

$$P_y(0) + iP_x(0) = 0. \tag{28}$$

One has to choose either the b.c. (27) or (28) depending upon
whether $\gamma > 0$ or $\gamma < 0$ respectively. We have solved the problem
of refraction and reflection based on the boundary conditions (27)
or (28) and will publish it elsewhere. Here we observe
that our way of solving the problem does not depend upon the model
situation   employed (for example) by Agranovich and Ginzburg
to solve the problem of refraction and reflection. The boundary
condition (27) and (28), however, is in qualitative agreement with
the proposed "effective" boundary condition of Agranovich and Ginzburg.
The possibility of a "helicity reversing" left→right circular
polarization Brillouin Scattering in this type of medium to detect
the third wave will be presented elsewhere.[11]

V. LATERAL BEAM SHIFT—GOOS-HÄNCHEN EFFECT

We have considered propagation of an optical beam in a bounded
spatially dispersive medium.

One of the interesting aspects of a beam propagating from a
local optically "dense" medium and impinging upon an interface
which separates it from an optically "rarer" medium is that near the
critical angle of incidence the beam suffers a lateral shift at
the interface. This effect is known as the Goos-Hänschen shift.

In case of a spatially dispersive medium, the beam may be
written as a superposition of angular spectrum modes[13], consisting
of a coupled set of waves. The critical angles for these consti-
tuent waves are different because each of these waves corresponds
essentially to a different refractive index associated with wave
propagation in the spatially dispersive medium. Our detailed
analysis of Goos-Hänchen shift at the interface of a spatially
dispersive dense medium and a local optically rarer medium is
more complicated. The analysis involves interface conditions
such as the "abc". The lateral shift will also be influenced by
surface roughness and the possible existence of an "exciton free"
dead layer. We present a very brief account of our presently
available results here.

Consider a special case of a Gaussian Beam and the special
geometry of Fig.(3). We write the incident Gaussian Beam at the
vacuum—medium interface as

$$\vec{E}^{(0)}(y_0 z_0) \cong \hat{x} \, e^{-ik_0 z_0} \, \exp\left(-\frac{y_0^2}{2\omega_0 z}\right) \tag{29}$$

We then write the fields inside the medium in terms of angular
spectrum representation appropriate to interfaces (1) and (2) of
Fig.(3). We then use the boundary conditions at interface (1) and
(2) and obtain general expressions for field amplitudes. We do

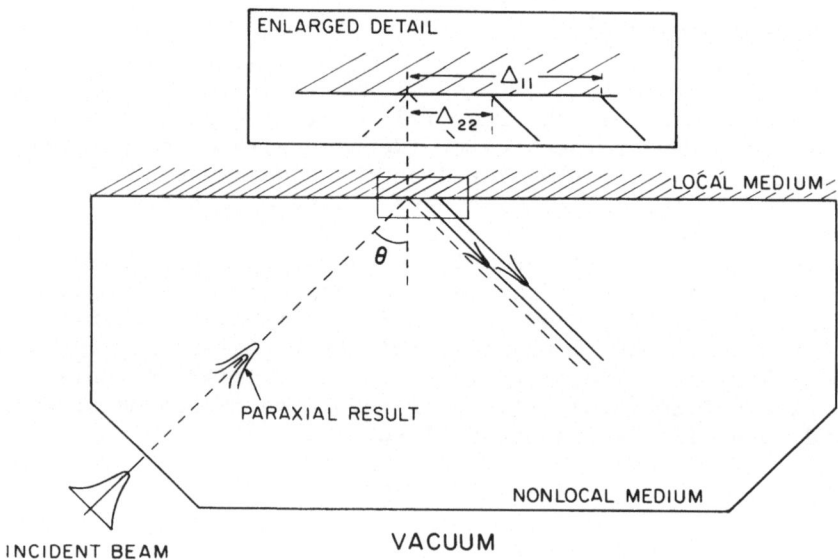

Fig. 3:   Schematic of the Goos-Hänchen shift of a beam at the interface of non-local and local media.

not give these expressions here   in detail. An approximate
expression for the internally reflected field at the interface (2),
which is valid for angles far   from the critical angle is

$$\vec{E}(y_r z_r) \cong \hat{x} \sum_{i=1}^{2} R_{ii}(0) \, e^{-ik_0 n_i z_r} \, e^{-(y_r - \Delta_{ii})^2} \tag{30}$$

where y_r and z_r refer to cartesian coordinates at the interface
(2); $R_{ii}(0)$ and Δ_{ii} are functions which depend upon refractive
indices, the angle θ and the form of abc. The approximate expression for Δ_{ii} is:

$$\Delta_{ii} = \frac{2 \sin\theta \, (1 + S_{ii}(0))}{k_0 (n_i^2 \sin^2\theta - n_T^2)^{\frac{1}{2}}} \tag{31}$$

The expression for the shift is similar in form to the local case,
if the term $S_{ii}(0)$ (which again we do not write here) is neglected.
Our expression (31) is not valid near the critical angles, where it
diverges.
     An expression for the shifts which does not diverge near
critical angles may be obtained following the method due to
Horowitz and Tamir[14] and will be given in detail elsewhere[15].

This work was supported in part by Army Research Office grant No.
DAAG29-79-G-0040, National Science Foundation grant No. DMR78-12399
and PSC:BHE award No. 13084.

REFERENCES

1. Pekar, S.I., JETP $\underline{33}$, 1022 (1957); Sov. Phys. JETP $\underline{6}$, 785 (1958);
    Ginzburg, V.L., JETP $\underline{34}$, 1593 (1958); Sov. Phys. JETP $\underline{7}$, 1096
    (1958). Earlier work is reviewed in: V.M. Agranovich and V.L.
    Ginzburg "Spatial Dispersion in Crystal Optics" (J. Wiley & Sons,
    New York, 1966), and "Polaritons" ed. E. Burstein, F. De Martini
    (Pergamon Press, Oxford 1974). A second edition of the first
    cited work has just appeared "Crystal Optics Taking into Account
    Spatial Dispersion and the Theory of Excitons", V.M. Agranovich
    and V.L. Ginzburg (Nauka 1979), in Russian.
2. Ting, C.S., Frankel, M.J., Birman, J.L., Sol. St. Comm. $\underline{17}$, 1285
    (1975); Brenig, W., Zeyher, R., and Birman, J.L., Phys. Rev. $\underline{B6}$,
    4613 (1972).
3. Zeyher, R., Brenig, W., and Birman, J.L., Phys. Rev. $\underline{B6}$, 4617
    (1972).
4. Weisbuch, C., and Ulbrich, R., Phys. Rev. Lett. $\underline{38}$, 865 (1977);
    Winterling, G., and Koteles, E., Sol. St. Comm. $\underline{23}$, 95 (1977);
    Yu, P.Y. and Evangelisti, F., Phys. Rev. Lett. $\underline{42}$, 1642 (1979).
5. Yu, P.Y.,"Light Scattering in Solids" (This volume) ed. J.L.
    Birman, H.Z. Cummins and K.K. Rebane Plenum Press, N.Y. (1979).
6. Elert, D., Ann. Phys. [5] $\underline{7}$, 65 (1932).
7. Birman, J.L. and Sein, J.J., Phys. Rev. $\underline{B6}$, 2482 (1972);
    Maradudin, A.A. and Mills, D.L., Phys. Rev. $\underline{B7}$, 2787 (1973);
    Agarwal, G.S., Pattanayak, D.N., and Wolf, E., Phys. Rev. $\underline{B10}$,
    1447 (1974).
8. Birman, J.L. and Frankel, M.J., Optics Comm. $\underline{13}$, 303 (1975);
    Phys. Rev. $\underline{A15}$, 2000 (1975).
9. Ginzburg, V.L., Sov. Phys. JETP $\underline{7}$, 1096 (1958).
10. Gakhov, F.D., "Boundary Value Problems" (Pergamon Press, Oxford
    1966) Section 39, p. 375.
11. Pattanayak, D.N. and Birman, J.L., to be published.
12. Goos, F. and Hänchen H., Ann. Phys. (6) $\underline{1}$, 333 (1947). For
    a review and more recent references on this subject, see H.K.V.
    Lotsch, Optik $\underline{32}$, (1970), p. 116–137, 189–204, 299–319, 553–
    569.
13. For a discussion of the angular spectrum representation, see,
    for example, P.C. Clemmow, "The Plane Wave Spectrum Represen-
    tations of Electromagnetic Fields", (Pergamon Press, Oxford,
    1966). In the context of spatial dispersion see G.S. Agarwal,
    D.N. Pattanayak and E. Wolf, Phys. Rev. $\underline{B10}$, 1447 (1974).
14. Horowitz, B.R. and Tamir, T., J.O.S.A. $\underline{61}$, 586 (1971).
15. Pattanayak, D.N. and Birman, J.L. (to be published).

RESONANT BRILLOUIN SCATTERING OF EXCITON POLARITONS

Peter Y. Yu

IBM Thomas J. Watson Research Center

Yorktown Heights, New York   10598

INTRODUCTION

Polaritons are coupled photon-polarization modes propagating in a dielectric medium. In this article the dielectric media of interest are semiconductors with zincblende and wurtzite crystal structures. Unless otherwise stated the polarization modes are excitons.[1]

Many techniques have been applied to study polaritons. These include reflectance[2,3], transmittance[4], photoluminescence[3] and non-linear techniques such as second harmonic generation[5], four-wave mixing[6] and hyper-Raman Scattering[7]. Several years ago Brenig, Zeyher and Birman[8] suggested a new technique, namely resonant Brillouin scattering (to be abbreviated as RBS in this article) for studying polaritons. It was pointed out by Brenig et al.[8] that RBS can measure not only the polariton dispersion and damping but also help to determine the additional boundary conditions necessary for describing the electrodynamics at the surface of non-local dielectric media.[9] In 1977, RBS of polaritons was observed by Ulbrich and Weisbuch[10] in GaAs. Since then this phenomenon has been reported in a number of other semiconductors such as CdS[11], CdTe[12], and CdSe[13]. Most of the experimental results have been explained by the original theory of Brenig et al.[8] Others are found to be explained by extensions of their theory to include effects such as wave vector dependent Brillouin scattering[14] and two-phonon processes[15]. The purpose of this article is to review these recent developments in the theoretical and experimental studies of RBS of polaritons. I shall concentrate mainly on GaAs and CdS since they have been studied in greater details and they are representative of semiconductors with similar crystal structures.

THEORY OF RESONANT BRILLOUIN SCATTERING OF POLARITONS

The kinetics of any scattering process is governed by two conservation laws, namely, those of energy and momentum:

$$\hbar\omega_i = \hbar\omega_s \pm \hbar\omega_p \qquad (1)$$

and

$$\hbar k_i = \hbar k_s \pm \hbar q \qquad (2)$$

$\hbar\omega_i$, $\hbar\omega_s$ and $\hbar\omega_p$ are respectively the energies of the incident photon, the scattered photon and the phonon; $\hbar k_i$, $\hbar k_s$ and $\hbar q$ are the corresponding momenta. The + (-) sign in Eqs. (1) and (2) corresponds to phonon emission (absorption). In Brillouin scattering the phonons of interest are the acoustic phonons. For q small compared to the reciprocal lattice vector the acoustic phonon dispersion is given by $\omega_p = V_s q$, where V_s is the appropriate sound velocity of the solid.

k_i and k_s are wave vectors of the photon inside the medium. If the dielectric function, $\varepsilon(\omega)$, of the medium is known, k_i and k_s can be calculated from the photon wave vectors in vacuo, k_{io} and k_{so}, using the relations $k_i = \sqrt{\varepsilon(\omega_i)}\, k_{io}$ and $k_s = \sqrt{\varepsilon(\omega_s)}\, k_{so}$. Using Eq. (2) and the experimentally measured values of ω_p one obtains the velocities of sound in the medium or conversely if the sound velocities are known, one can determine the dielectric constant of the medium. In RBS the incident photon energy is resonant with an excitonic transition in a semiconductor. For such photon energies the dielectric function of the medium is strongly dispersive. As a result the frequencies of the Brillouin peaks will also vary rapidly with the incident photon energies. Brenig et al. suggested that this dependence of the Brillouin frequencies on the photon energies can be used to deduce the polariton dispersion while the dispersion of the scattering efficiencies can be related to the additional boundary conditions.

It is well-known that the polariton dispersion corresponding to a dispersionless exciton level $\hbar\omega_e$ is given by :[16]

$$\varepsilon/\varepsilon_o = \frac{c^2 k^2}{\varepsilon_o \omega^2} = 1 + \frac{\omega_L^2 - \omega_e^2}{\omega_e^2 - \omega^2 - i\omega\Gamma} \qquad (3)$$

where $\hbar k$, $\hbar\omega_e$ and Γ are respectively the momentum, energy and damping constant of the polariton, ε_o is the dielectric constant of the medium without the exciton contribution and $\hbar\omega_L$ is the longitudinal exciton energy defined by $\varepsilon(\omega_L) = 0$ when $\Gamma = 0$. We are interested in dispersive excitons whose energy can be expanded as a function of k to second order as:

$$\omega_e(k) = \omega_T + \phi k + \frac{\hbar k^2}{2M}. \qquad (4)$$

$\hbar\omega_T$ is known as the transverse exciton energy. In most semiconductors ϕ is negligible and in crystals with inversion symmetry ϕ is identically zero. M is the effective mass of the exciton. The polariton dispersion is given by:[17]

$$\frac{\varepsilon(\omega, k)}{\varepsilon_o} \simeq 1 + \frac{\omega_L^2 - \omega_T^2}{\omega_T^2 + \frac{\hbar k^2}{M}\omega_T - \omega^2 - i\omega\Gamma} \qquad (5)$$

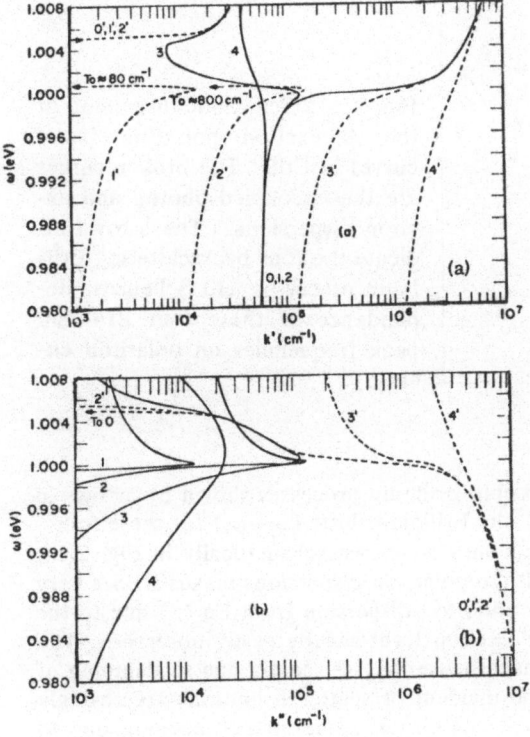

Fig. 1. Solutions of Eq. (5) with $\hbar\omega_T = 1.0\,\text{eV}$, $\hbar(\omega_L - \omega_T) = 5\,\text{meV}$, $M = 2 \times$ mass of free electron and various values of Γ : (0) and (0'), $\Gamma = 0\,\text{eV}$; (1) and (1'), $\Gamma = 10^{-5}\,\text{eV}$; (2) and (2'), $\Gamma = 10^{-4}\,\text{eV}$; (3) and (3'), $\Gamma = 10^{-3}\,\text{eV}$; (4) and (4)'), $\Gamma = 10^{-2}\,\text{eV}$ (from Ref. 18).

Tait[18] pointed out that there are in general two ways to display the solutions of Eq. (5). In one way the real and imaginary parts of k are plotted as a function of the real polariton frequency ω The dispersion curves obtained by Tait[18] are displayed as a function of Γ in Fig. 1. Note how the character of the solutions changes as Γ passes through a value Γ_c given by :[18]

$$\Gamma_c = 2[2\hbar\omega_T^2(\omega_L - \omega_T)/(Mc^2)]^{1/2}. \tag{6}$$

A solution is a true propagating mode inside the medium when the real part of k (to be denoted by k') is larger than its imaginary part (denoted by k''). Thus the term "polaritons" is meaningful only when Γ is smaller than Γ_c. When Γ exceeds Γ_c, the dispersion curves resemble more those of the uncoupled photon and exciton. For $\Gamma < \Gamma_c$ the polariton dispersion curves in Fig. 1(a) are the same as for $\Gamma = 0$. Thus, to first approximation the Brillouin peak frequencies can be assumed to be independent of Γ for $\Gamma < \Gamma_c$,[19] although the widths of the Brillouin peaks depend on Γ .

One-Phonon Scattering : Dispersion of Brillouin Peaks and Widths

Figure 2(a) shows schematically the polariton dispersion of the lowest exciton in CdS. An external photon with energy larger than $\hbar\omega_L$ can excite two polariton modes, one belonging to branch I and one to branch II. These two polariton states can be scattered by acoustic phonons into polariton states in either branch. Thus, for

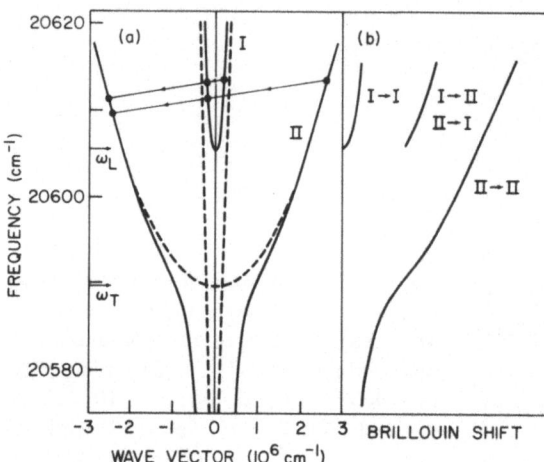

Fig. 2   (a)Schematic dispersion of the A exciton polariton (solid curve) in CdS. The broken curves are the uncoupled photon and exciton dispersions. The arrows indicate the four backscattering Brillouin processes. (b) Schematic dependence of these four Brillouin peak frequencies on polariton energy.

a given acoustic phonon there are four possible Brillouin processes shown by arrows in Fig. 2(a). As pointed out by Brenig et al.[8] the Brillouin shifts ($\omega_i - \omega_s$) for these four processes will depend on the polariton frequency as shown schematically in Fig. 2(b). The fact that these curves mirror quite well the polariton dispersions makes RBS a very useful technique because deviations in the polariton dispersion from Eq (5) due to the k-linear term or the presence of a second exciton level can be easily observed. The widths of the Brillouin peaks in non-resonant scattering are mostly due to damping of the acoustic phonons involved. In case the incident or scattered photons are strongly absorbed inside the medium, as in resonant scattering, additional broadening due to opacity results[20,21]. When the damping of the polaritons dominates over the damping of the phonons, the full width at half maximum (γ) of the Brillouin peaks is given by[8]

$$\gamma = 2V_s(k_i'' + k_s''), \qquad (7)$$

where k_i'' and k_s'' are respectively the imaginary part of the incident and scattered polariton wave vectors.

Brillouin Scattering Efficiencies

The calculation of the scattering efficiency, η, of photons by phonons (optical or acoustic) in semiconductors via polaritons is usually performed in two steps : (a) the transmission coefficients of the incident and scattered photons at the sample surface are evaluated and (b) the scattering efficiency of the polaritons by the phonon inside the sample is calculated using perturbation theory. The second step has been carried out by several authors and lead to the expression:[22,23]

$$\eta = \frac{1}{N_i} \frac{dN_s}{d\Omega} = \frac{(n_{ph}+1)}{8\pi^2\hbar^2} \left(\frac{\omega_s}{c}\right) \frac{|M_{is}|^2 |P_i|^2 |P_s|^2}{V_{gi} V_{gs}(k_i'' + k_s'')}. \qquad (8)$$

N_i is the number of incident photons per unit time, $dN_s/d\Omega$ is the number of scattered photons per unit time per unit solid angle Ω, V_s and P denote respectively the group velocity and the exciton polarization of the incident (subscript i) and scattered (subscript s) polaritons; $|M_{is}|^2$ denotes the exciton - phonon interaction matrix element; n_{ph} is the phonon occupation number and c is the speed of light. I shall first consider the exciton - phonon interaction and then the evaluation of the exciton polarizations.

Exciton-Acoustic Phonon Interactions

It is well-known that electrons in a crystal are perturbed by the strain field associated with acoustic phonons resulting in the deformation potential interaction (H_{dp}):[24]

$$H_{dp} = (D_e - D_h) \cdot \nabla U \tag{9}$$

where the second rank tensors D_e and D_h are respectively the deformation potentials for the electron and hole and U is the displacements of the atoms in the unit cell. In a piezoelectric crystal a strain field can also produce an electric field via the piezoelectric effect. The longitudinal component of this piezoelectrically induced electric field interacts strongly with excitons in a manner analogous to the longitudinal electric field of an optical phonon.[25] This piezoelectric exciton-phonon interaction H_{pe}, to the lowest order in q, is given by [26]

$$H_{pe} \propto q \cdot e \cdot (\nabla U), \tag{10}$$

where e is the third rank electromechanical tensor. The extra q factor in H_{pe} implies that wave vector dependent Brillouin scattering may be observed in piezoelectric crystals like CdS similar to wave vector dependent LO phonon scattering.[27]

Additional Boundary Conditions

To calculate the exciton polarizations P_i and P_s in Eq (8) requires a knowledge of the electrodynamics of the sample - vacuum interface. Due to the existence of two polariton branches in the sample there are three unknowns: E_r (the reflected wave electric field), E_1 and E_2 (electric field amplitudes of the polariton branches I and II respectively). Since only two boundary conditions are provided by Maxwell's Equations an additional boundary condition (ABC) is required.[9] A detailed discussion of the different ABC's is beyond the scope of this paper. It suffices to point out that there are two ABC's which have been discussed most extensively in the literature. The first ABC (to be denoted by ABC 1) was proposed by Pekar and has the form:[28]

$$P_1 + P_2 = 0 \tag{11}$$

at the sample surface. Another ABC (denoted by Zeyher et al.[9] as ABC 3) has the form:

$$\frac{E_1}{n_1 - n_e} + \frac{E_2}{n_2 - n_e} = 0 \tag{12}$$

at the surface. n_1, n_2 and n_e are defined respectively as $ck'_1/\omega, ck'_2/\omega$ and $(c/\omega)\{(\omega^2 - \omega_T^2 + i\omega\Gamma)/(\hbar\omega_T/M)\}^{1/2}$.

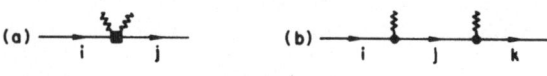

NOTATIONS

—•— POLARITON OF BRANCH i
 i   i, j, k = I OR II

ᴠᴠ ACOUSTIC PHONON

■ EXCITON-TWO PHONON INTERACTION

● EXCITON-ONE PHONON INTERACTION

Fig. 3.   Diagrammatic representations of the two-phonon Brillouin scattering processes.

The scattering efficiencies η calculated with these two ABC's have been shown to be qualitatively different.[29]

Two-Phonon Resonant Brillouin Scatterings

The one-phonon RBS theory can be extended easily to the case of two-honon scattering.[15]   As in two-phonon Raman scattering there are two-kinds of two-phonon Brillouin processes, shown diagrammatically in Fig. 3. In diagram (a) the two acoustic phonons are emitted simultaneously via the exciton-two phonon interactions.   In diagram (b) the two-phonons are emitted sequentially via the exciton-one phonon interaction. All the two-phonon processes of the second kind can be obtained by substituting i, j, and k in Fig. 3(b) with the possible permutations of polariton branches.   However, the relative intensities of the different processes depend strongly on the polariton frequency.   For example when ω_i and ω_s are well above ω_L the $I \rightarrow II \rightarrow I$ process is expected to be the strongest.   When ω_i and ω_s are in the vicinity of ω_L two other processes: $I \rightarrow II \rightarrow II$ and $II \rightarrow II \rightarrow II$ are also important.   The advantage of studying the two-phonon processes is that the phonon wave vectors are not restricted to zone center by momentum conservation so that one can study, for example, the wave vector dependence of the exciton-phonon interaction via the two-phonon scattering processes of Fig. 3(b).

RESULTS IN ZINCBLENDE SEMICONDUCTORS (GaAs)

Dramatic changes in the Brillouin spectra with excitation wavelength similar to those predicted by Brenig et al.[8] were first reported by Ulbrich and Weisbuch in GaAs.[10]

The experiment was performed in a backscattering geometry on high purity (impurity concentration $\lesssim 10^{15} cm^{-3}$) GaAs expitaxial layers. The sample was excited at low temperatures by a narrow bandwidth cw dye laser and the scattered radiation was analyzed by a double-grating spectrometer. The combined spectral width of this system is 0.02 meV.   The experimental results obtained from a GaAs [100] face are summarized in Figs 4(a), (b) and (c).   Qualitatively similar results have also been obtained from the [110] and [111] surfaces.

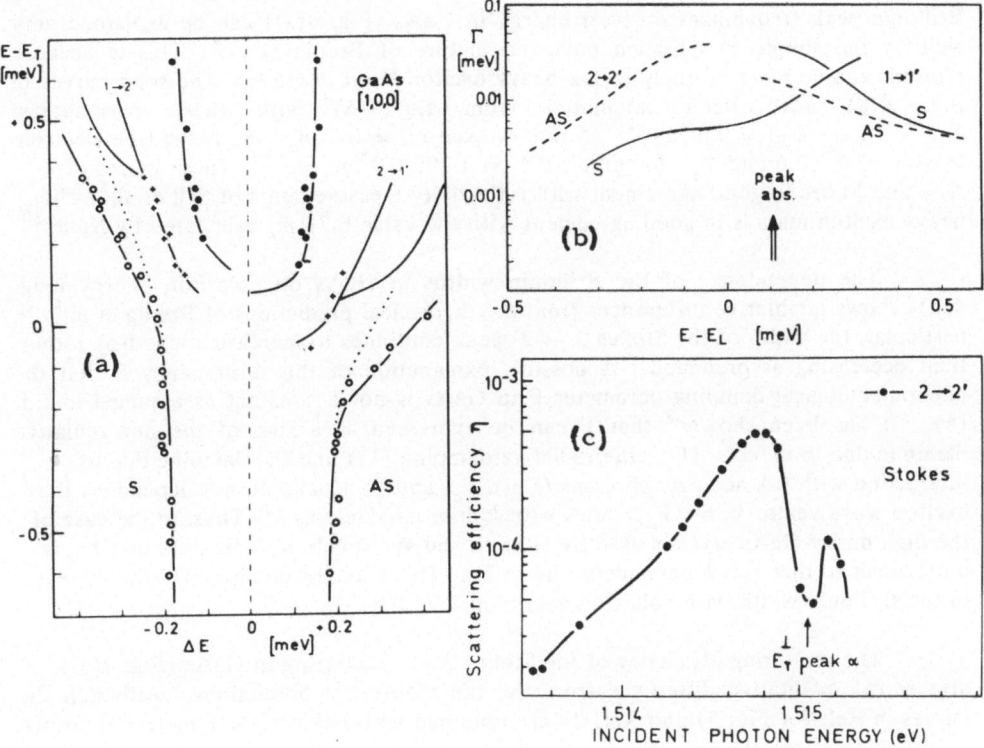

Fig. 4.   Dependence of the (a) peak frequencies, (b) full width at half maxima and (c) scattering efficiencies of the LA phonon in GaAs on polariton energy (from Ref. 12). The solid curves in (a) have been calculated using a two-branch polariton dispersion and the parameters given in the text.   S and AS stand respectively for Stokes and anti-Stokes scatterings.

To understand the results of Ulbrich and Weisbuch it is necessary to analyze the exciton dispersion in GaAs. The top valence bands of GaAs are well-known to be four-fold degenerate at the Brillouin zone center.[24] Away from the zone center these four levels are split into two doubly degenerate bands known as the heavy and light hole bands. These hole bands give rise to two heavy and light exciton bands whose dispersion (neglecting exchange interaction and k-linear terms) can be written as:[30]

$$E^{\pm}(k) = E_o + \frac{\hbar^2 k^2}{2M^*}(1 \pm \delta(k)), \qquad (13)$$

Where + and - correspond to the light and heavy exciton respectively. M* can be regarded as the average exciton mass of the two bands and δ as a measure of the difference between these masses.

Fishman[31] recently calculated the polariton dispersion curves for the heavy and light exciton bands in GaAs and showed there are three polariton branches instead of two. Thus, one may expect that the resonant Brillouin spectra in GaAs to be rather complicated. However, Ulbrich and Weisbuch found that the dependence of the

Brillouin peak frequencies on laser energy in GaAs (Fig. 4(a)) can be explained very well by the simple two branch polariton picture of Brenig et al.[8] This is because photons couple more strongly to the heavy exciton band in GaAs. The solid curves in Fig. 4(a) have been calculated from Eq. (5) with these parameters: $E_{LT} = \hbar(\omega_L - \omega_T) = 0.08$ meV   M (heavy exciton) $= 0.7\, m_o$ (m_o is the free electron mass), $\varepsilon_o = 12.6$ and V_s (longitudinal) $= 4.805 \times 10^5$ cm sec^{-1}. These values of E_{LT} and M are in good agreement with reflectivity measurements of Sell et al.[3] The heavy exciton mass is in good agreement with the value $0.76\, m_o$ calculated by Kane.[30]

The dependence of the Brillouin widths in GaAs on polariton energy (Fig. 4(b)) shows qualitative differences from the theoretical predictions of Brenig et al.[8] In particular, the width of the Stokes $2 \to 2'$ peak continues to increase above $\hbar\omega_L$ rather than decreasing as predicted. A possible explanation of this discrepancy is that the phenomenological damping parameter Γ in GaAs is not a constant as assumed in Eq. (5). It has been shown[32] that Γ can be expressed as a sum of the non-radiative damping due to defects (Γ_{nr}), the radiative damping (Γ_r) and the damping due to interaction with LA acoustic phonons (Γ_{ac}). Γ_{nr} and Γ_r usually do not depend on the exciton wave vector k, but Γ_{ac} varies with k approximately as k^2. Thus, in the case of the high purity GaAs crystals used by Ulbrich and Weisbuch, it is possible that Γ_{nr} is quite small so that Γ is k dependent due to Γ_{ac}. This explains qualitatively the increase in the Brillouin width with polariton energy in Fig. 4(b).

The scattering efficiency of the Stokes $2 \to 2'$ scattering in GaAs (Fig. 4(c)) also shows qualitative difference from the the theoretical predictions. Although the curves in Ref. 8 (Fig. 5) and Ref. 29 are obtained with polariton parameters appropriate for CdS rather than GaAs , the calculated curves for GaAs should be qualitatively similar. The only possible difference is that the two resonance peaks at ω_T and ω_L are not resolved in GaAs because of the smaller splitting between ω_T and ω_L. Even considering this difference, one finds that the experimental results disagree with the theory of Brenig et al. in two respects: the experimental enhancement in the $2 \to 2'$ scattering efficiency is weaker than the theoretical value by more than four orders of magnitude and secondly the theory does not predict a minimum above ω_L as in Fig. 4(c). Ulbrich and Weisbuch[33] pointed out that these discrepancies can be removed by using the experimentally measured absorption coefficient α of the polariton rather than the α calculated from Eq. (5) (α being equal to $2k''$). They suggested that the polaritons in their samples are attenuated predominantly by elastic scatterings with shallow impurities and this effect is not included in the theoretical α. This is supported by their observation that the dip in the scattering efficiency curve in Fig. 4(c) occurs at the same energy as the peak in experimental α and that thin GaAs samples have a strong fluorescence background. Thus, the failure of the theory of Brenig et al[8] in explaining the experimental Brillouin widths and efficiencies in GaAs can both be traced to the over-simplified assumption of a constant Γ in Eq. (5).

RESULTS IN WURTZITE SEMICONDUCTORS (CdS)

Exciton polaritons probably have been studied more extensively in CdS than any other material.[2,17,34] It is usually chosen as the test case for any theory on polaritons. Thus, it is not surprising that RBS was reported in CdS[11] soon after GaAs. In fact, dispersive Brillouin modes were reported by Bruce et al[35] in CdS even before

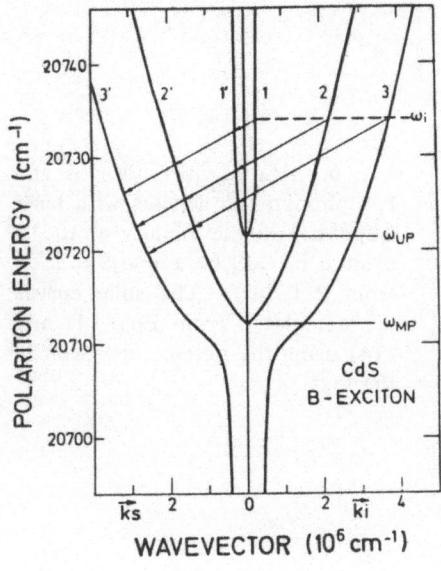

Fig. 5.   Dispersion of the B exciton polariton in CdS for $k \perp \hat{c}$ and $\hat{e} \perp \hat{c}$ (from Ref. 38). The arrows indicate the nine possible Brillouin processes.

GaAs. Instead of varying the laser frequencies, Bruce et al.[35] used the 488nm line of the Ar^+ laser and tuned the polariton energy by varying the sample temperature. Unfortunately, the limited tuning range obtainable did not allow a detailed comparison between theory and experiment.

The four-fold degenerate valence bands at. zone center in the zincblende materials are split in the wurtzite semiconductors into two doubly degenerate bands.[36] The resultant two exciton bands are known as the A (lower energy) and B (high energy) excitons. In CdS the A exciton is optically active only for polarization perpendicular to the c-axis ($\hat{e} \perp \hat{c}$) while the B exciton is optically active for both $\hat{e} \perp \hat{c}$ and $\hat{e} \parallel \hat{c}$. The dispersions of the A and B excitons in CdS have been calculated by Mahan and Hopfield.[37] The dispersion of the A exciton is parabolic, although the effective mass is not isotropic:

$$\omega_A(k) = \omega_A^o + \frac{\hbar}{2}[\frac{k_\perp^2}{M_{A\perp}} + \frac{k_\parallel^2}{M_{A\parallel}}] \tag{14}$$

where $\parallel$ and $\perp$ refer to the c-axis. Due to the k-linear term the B exciton dispersion depends on the polarization and direction of propagation. For $k \perp \hat{c}$ the dispersions are given by:[38]

$$(\hat{e} \parallel \hat{c}) \qquad \omega_B(k_\perp) = \omega_B^o + \frac{\hbar k_\perp^2}{2M_{B\perp}} \tag{15}$$

and

$$(\hat{e} \perp \hat{c}) \qquad \omega_B^\pm(k_\perp) = \omega_B^o \pm \phi k_\perp + \frac{\hbar k_\perp^2}{2M_{B\perp}^*} \quad . \tag{16}$$

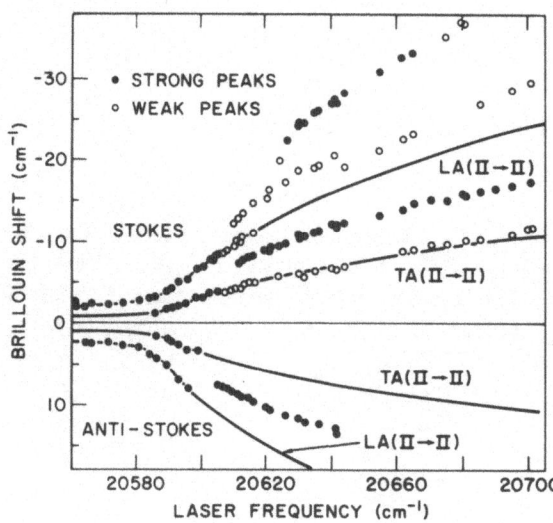

Fig. 6. Variation of the TA and LA phonon frequencies with laser frequency in the vicinity of the A exciton in CdS ($k \perp \hat{c}$ and $\hat{e} \perp \hat{c}$, from Ref. 40). The solid curves are calculated from Eqs. (4) and (14) using the parameters given in the text.

The two masses $M_{B\perp}$ and $M^*_{B\perp}$ are related by [38]

$$\left(M^*_{B\perp}\right)^{-1} = \left(M_{B\perp}\right)^{-1} + 2\phi^2/\Delta \qquad (17)$$

where Δ is the exchange energy. The polariton dispersion of the B exciton in CdS obtained from Eq. (16) for $k \perp \hat{c}$ is shown in Fig. 5. Note that the k-linear term in $\hat{e} \perp \hat{c}$ results in three polariton branches.

One-Phonon Resonant Brillouin Scattering in CdS: A Exciton

Resonant Brillouin scattering has been studied at both the A and B excitons in CdS by Winterling and Koteles,[11,14,38] and at the A exciton by Yu and Evangelisti.[15,23,39] At the A exciton, the observed dispersions of the Brillouin peaks for $k \perp \hat{c}$ and $\hat{e} \perp \hat{c}$ are shown in Fig. 6.[40] Unlike the case of GaAs, only two sets of Brillouin peaks are observed (in both Stokes and anti-Stokes scattering) when ω_i is in the vicinity of ω_T and ω_L. The higher frequency modes have been identified as two-phonon Brillouin modes and will be considered separately.[39] From the known phonon velocities in CdS, Winterling et al.[14] identified these two series of peaks as due to $\Pi \rightarrow \Pi$ scattering processes with emission of TA and LA phonons respectively. The solid curves in Fig. 6 have been calculated from Eq. (5) with these parameters for the A exciton $\omega_T = 20589.5$ cm^{-1}, $\omega_L - \omega_T = 15.4$ cm^{-1}, $M_{A\perp} = 0.89 m_0$ and $\varepsilon_0 = 9.3$. These values are in good agreement with those existing in the literature.[34]

In CdS the $I \rightarrow I$ scattering processes were not observed probably because of the poor experimental resolution (slit width ~ 0.6 cm^{-1}). Because of this limited resolution no dependence of the line width on laser frequency was reported, although broadening of the LA mode for $\omega_i > \omega_L$ was observed. The $\Pi \rightarrow I$ and $I \rightarrow \Pi$ processes for the LA phonon were unfortunately obscured by the TA phonon ($\Pi \rightarrow \Pi$) peak.

The TA mode is normally forbidden in the $\mathbf{k} \perp \hat{c}$ backscattering geometry. The presence of this mode in Fig. 6 can be explained by the wave vector dependent exciton-phonon interaction H_{pe}.[14] The explicit form of H_{pe} in CdS has been derived by Mahan and Hopfield.[41] For $\mathbf{k} \perp \hat{c}$ the LA mode is due entirely to H_{dp} while the TA mode is due entirely to H_{pe}. Because of the extra q^2 dependence in $|H_{pe}|^2$ as compared to $|H_{dp}|^2$ (Eqs. (9) and (10)), the ratio of the intensities of the TA peak and the LA peak is given approximately by:[14]

$$I_{TA}/I_{LA} \propto q_{TA}^2 \qquad (18)$$

where q_{TA} is the TA phonon wave vector. The experimental values of I_{TA}/I_{LA} obtained by Winterling et al.[14] are found to be in reasonable agreement with Eq. (18).

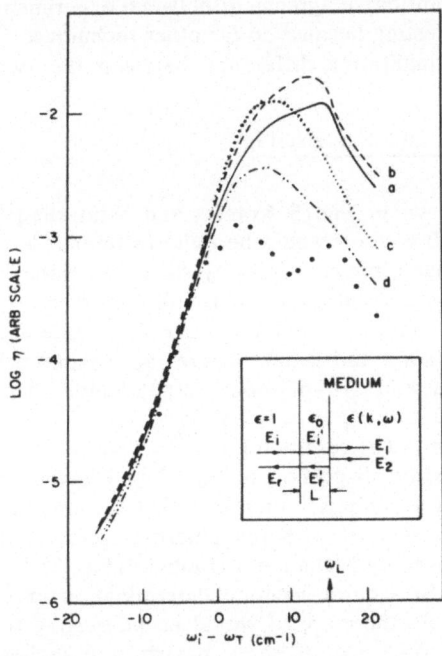
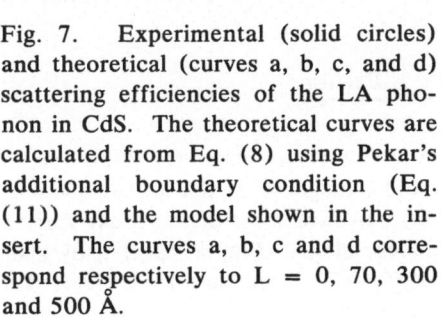
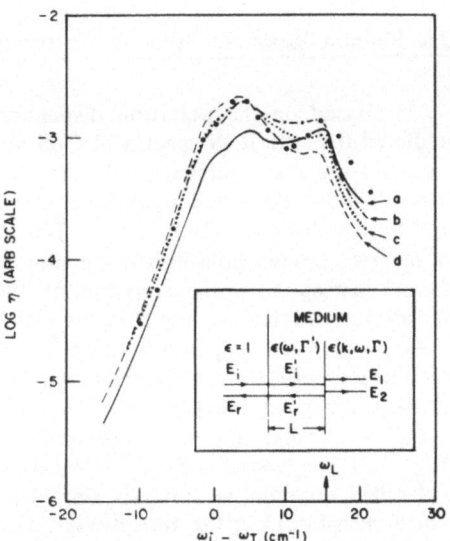

Fig. 7. Experimental (solid circles) and theoretical (curves a, b, c, and d) scattering efficiencies of the LA phonon in CdS. The theoretical curves are calculated from Eq. (8) using Pekar's additional boundary condition (Eq. (11)) and the model shown in the insert. The curves a, b, c and d correspond respectively to L = 0, 70, 300 and 500 Å.

Fig. 8. Same as in Fig. 7 except the theoretical curves are calculated with a different additional boundary condition (Eq. (12)) and the model shown in the insert. Curve a is obtained with L = 0 while curves b, c and d are calculated assuming L = 70 Å and Γ' = 20, 14 and 10 cm^{-1} respectively.

The dispersion of the LA phonon ($\Pi \rightarrow \Pi$) scattering efficiencies at the A exciton of CdS has been analyzed in detail by Yu and Evangelisti.[23] They compared the experimental scattering efficiency η with the theoretical values calculated from Eq. (8) using the ABC's represented by Eqs.(11) and (12). In addition their models for the semiconductor include a surface layer which is compatible with the particular ABC used.

For Pekar's ABC (Eq. (11)) an exciton-free layer as proposed by Hopfield and Thomas[17] was used. The curves calculated with this model (shown in insert of Fig. 7) for different thicknesses of the exciton-free layer are compared with the experimental points in Fig. 7. There is qualitative disagreement between theory and experiment even for unreasonably large values of L. For the second ABC (Eq. (11)) Yu and Evangelisti proposed a lossy layer on the sample surface due to surface enhanced recombination of excitons. This model is shown in the insert of Fig. 8. The exciton damping Γ' in this lossy layer was assumed to be larger than Γ_c (defined in Eq. (6)). The theoretical curves for $L = 0$ (curve a) and for $L = 70 \overset{\circ}{A}$ and three different values of Γ' are shown in Fig. 8. There is almost quantitative agreement between experiment and curve c ($\Gamma' = 14 cm^{-1}$). This result is interesting because so far other techniques such as reflectivity have failed to show a qualitative difference between the two ABC's.

One-Phonon Resonant Brillouin Scattering in CdS : B Exciton

Based on the polariton dispersion shown in Fig. 5 Koteles and Winterling[38] predicted that the RBS spectra in CdS at the B exciton would be quite different for $\hat{e} \parallel \hat{c}$ and for $\hat{e} \perp \hat{c}$. Similar predictions were made independently by Allen and Kane.[42] Figure 9 shows the Brillouin shifts as a function of ω_i at the B exciton of CdS measured by Koteles and Winterling[38]. The results for $k \perp \hat{c}$ and $\hat{e} \parallel \hat{c}$ are well explained by the two branch polariton dispersion. The solid and broken curves are theoretical curves corresponding to emission of the LA and TA phonons respectively. The polariton parameters for the B exciton obtained are: $\omega_B^0 = 20711.3 \ cm^{-1}$, $\omega_L - \omega_T = 7.7 \ cm^{-1}$, $\varepsilon_o = 8.9$ and $M_{B\perp}^* = 1.5 m_o$. The results for $\hat{e} \perp \hat{c}$ are more complicated reflecting the more complex polariton dispersion in Fig. 5 and have not yet been analyzed quantitatively. A preliminary analysis[38] indicated that the number of Brillouin peaks and their dispersion are consistent with the polariton dispersion ($M_{B\perp} = 1.2 m_o$ and $\phi = 5 \times 10^{-10}$ eV cm) proposed by Mahan and Hopfield (Fig. 5).[37] A detailed quantitative analysis should result in a more accurate determination of ϕ than is possible from the reflectivity spectra. Furthermore, it would be interesting to analyze the scattering efficiencies since the presence of three polariton branches necessitates two ABC's!

Two-Phonon Resonant Brillouin Scattering at the A Excitons

The higher frequency peaks which appear in CdS for ω_i above the A exciton frequency in Fig. 6 have been identified by Yu and Evangelisti as due to two-phonon Brillouin processes of the type shown in Fig. 3(b). In fact, most of the peaks can be quantitatively explained by the $I \rightarrow \Pi \rightarrow I$ process alone. Figure 10(a) shows an experimental spectrum obtained at ω_i equal to 20655 cm^{-1} (solid curve). The closed

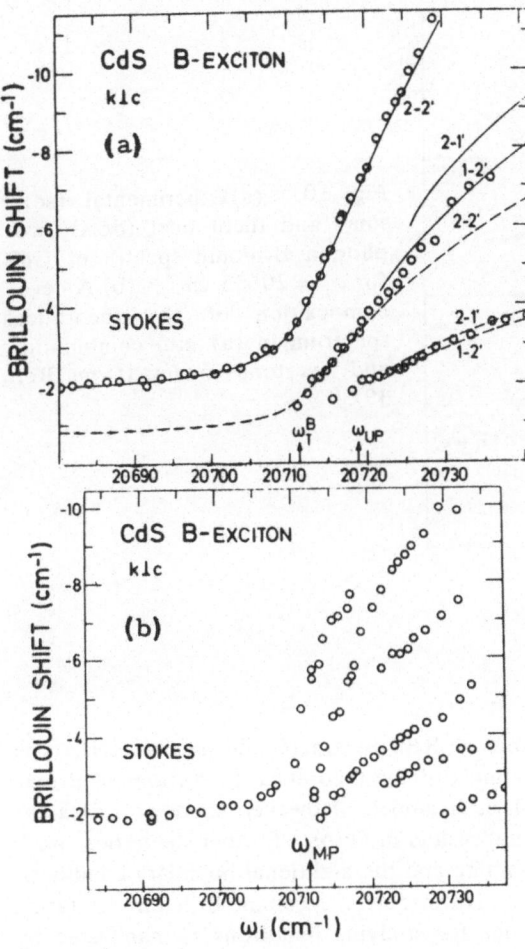

Fig. 9   Variation of the Brillouin shifts with laser frequency at
the B polariton of CdS for $k \perp \hat{c}$
and $\hat{e} \parallel \hat{c}$ (a) and $\hat{e} \perp \hat{c}$ (b). The
solid and broken curves are calcu-
lated from the two-branch polariton dispersion curve (Eqs. (4) and
(15)) assuming a LA and a TA
phonon are emitted respectively.
The B exciton parameters are giv-
en in the text (from Ref. 38).

circles are calculated by Yu and Evangelisti.[39] Other than an overall intensity normalization constant, there is only one unknown parameter $(M_{A\parallel})$ in this theory. $M_{A\parallel}$
was adjusted to be $(2.7 \pm 0.2)m_0$ to explain the positions of the two-phonon peaks.
This value of $M_{A\parallel}$ was confirmed by Winterling and Koteles by measuring the
one-phonon RBS for $k \parallel \hat{c}$.[43] The theoretical spectrum in Fig. 10(a) is decomposed
into three parts in Fig. 10(b) and the origin of the peaks is identified. It is interesting
to note that the 2 (quasi-TA) peak is rather sharp, in spite of the fact that it contains
contributions from the entire Brillouin zone. This results from the fact that the
anisotropy of the TA phonon velocity in CdS results in partial cancellation of the
broadening in the 2 (quasi - TA) peak due to the anisotropy in the exciton effective
mass. For $\omega_i \sim \omega_L$ Yu and Evangelisti[39] found that some of the observed peaks can be
explained by the $I \rightarrow \Pi \rightarrow \Pi$ two-phonon process. They were able to predict the
frequency and relative intensities of these peaks with no adjustable parameters.

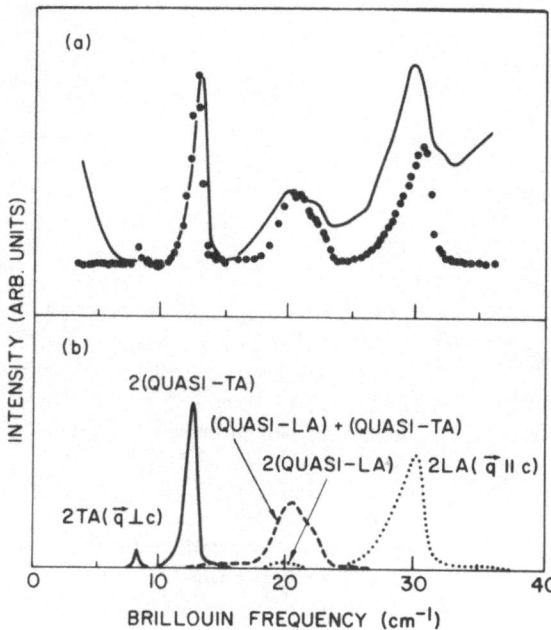

Fig. 10. (a) Experimental (solid line) and theoretical (dots) two-phonon Brillouin spectra of CdS for $\omega_i = 20655$ cm^{-1}. (b) A decomposition of the theoretical spectrum in (a) into combination and overtone modes (from Ref. 39).

CONCLUSION

Recent experimental investigations of RBS in zincblende and wurtzite-type semiconductors have verified qualitatively most of the theoretical predictions of Brenig et al. based on a simple two-branch polariton model. However, there are disagreements between theory and the experimental results in GaAs. Further theoretical work in the proper treatment of the exciton damping and the additional boundary conditions will probably remove these discrepancies. Nevertheless, resonant Brillouin scattering has proved to be a very powerful technique for studying polaritons as manifested by the recent work reviewed here. There are still many applications of this technique yet to be explored. An example is the study of surface exciton polaritons by RBS.

ACKNOWLEDGMENT

I wish to thank E. Koteles, G. Winterling, G. Weisbuch and R. Ulbrich for many discussions and for sending me preprints of their work. I am also grateful to D. R. Tilley for a preprint.

References

1.   A detailed treatment of exciton polaritons can be found in V. M. Agranovich and V. L. Ginzberg, "Spatial Dispersion in Crystal Optics and the Theory of Excitons", Interscience, New York (1966).

2.   E. F. Gross, S. Permogorov, V. Travnikov and A. Selkin, Solid State Commun. 10: 1071 (1972).

3.   D. D. Sell, S. E. Stokowski, R. Dingle, and J. V. DiLorenzo, Phys. Rev. B7:

4568 (1973).

4. V. A. Kiselev, B. S. Razbirin, and I. N. Ural'tser, Pis'ma Zh. Eksp. Teor. Fiz., 18 : 504 (1973)[JETP Lett. 18: 296 (1973)].

5. F. DeMartini, M. Colocci, S. E. Kohn, and Y. R. Shen, Phys. Rev. Lett. 38: 1223 (1977).

6. F. DeMartini, G. Giuliani, P. Mataloni, E. Palange, and Y. R. Shen, Phys. Rev. Lett 37: 440 (1976).

7. B. Hönerlage, A. Bivas, and Vu Duy Phach, Phys. Rev. Lett 41: 49 (1978).

8. W. Brenig, R. Zeyher and J. L. Birman, Phys. Rev. B6: 4617 (1972).

9. R. Zeyher, J. L. Birman, and W. Brenig, Phys. Rev. B 6: 4613 (1972).

10. R. G. Ulbrich and C. Weisbuch, Phys. Rev. Lett. 38: 865 (1977).

11. G. Winterling and E. Koteles, Solid State Commun. 23: 95 (1977).

12. R. G. Ulbrich and C. Weisbuch, in "Festkörperprobleme (Advances in Solid State Physics), Volume XVIII", page 217, J. Treusch, ed., Vieweg, Braunschweig (1978).

13. C. Hermann and P. Y. Yu, Solid State Commun. 28: 313 (1978).

14. G. Winterling, E. S. Koteles, and M. Cardona, Phys. Rev. Letters 39: 1286 (1977).

15. P. Y. Yu and F. Evangelisti, in "Physics of Semiconductors 1978", B. L. H. Wilson, ed. The Institute of Physics, Bristol and London (1979).

16. J. J. Hopfield, Phys. Rev. 112: 1555 (1958).

17. J. J. Hopfield and D. G. Thomas, Phys. Rev. 132: 563 (1963).

18. W. C. Tait, Phys. Rev. B5: 648 (1972).

19. Dependence of the Brillouin peak frequencies on Γ in RBS of polariton has recently been calculated. D. R. Tilley, (unpublished).

20. J. R. Sandercock, Phys. Rev. Lett 29: 1735 (1972).

21. A. Dervish and R. Loudon, J. Phys. C9 : L669 (1976).

22. E. Burstein, D. L. Mills, A. Pinczuk and S. Ushioda, Phys. Rev. Lett. 22: 348 (1969); E. M. Verlan and L. N. Ovander, Fiz. Tver. Tela 8 : 2435 (1966) [Sov. Phys. Solid State 8 : 1929 (1967)]; J. J. Hopfield, Phys. Rev. 182: 945 (1969) and B. Bendow and J. L. Birman, Phys. Rev. B1: 1678 (1970).

23. P. Y. Yu and F. Evangelisti, Phys. Rev. Lett. 42 : 1642 (1979).

24. See for example C. Kittel, "Quantum Theory of Solids", J. Wiley & Sons, Inc., New York (1966).

25. H. Fröhlich, Adv. in Phys. 3: 325 (1954).

26. A. R. Hutson, J. Appl. Phys. 32 (supplement): 2287 (1961).

27. R. M. Martin and T. C. Damen, Phys. Rev. Lett. 26: 86 (1971).

28. S. I. Perkar, Zh. Eksp. Teor. Fiz. 33 : 1022 (1957) [Sov. Phys. JETP 6: 785 (1958)]

29. P. Y. Yu, in "Proceedings of the Joint US - Japan Seminars on Inelastic Light Scattering in Condensed Matters", E. Burstein, ed., (in press).

30. E. O. Kane, Phys. Rev. B11: 3850 (1975); and M. Altarelli and N. O. Lipari, Phys. Rev. B15: 4898 (1977).

31. G. Fishman, J. Lum. 18/19 : 289 (1979).

32. P. Y. Yu and Y. R. Shen, Phys. Rev. B 12: 1277 (1975).

33. C. Weisbuch and R. Ulbrich, in "Lattice Dynamics", M. Balkanski, ed. Flamarion Sciences, Paris (1978) and private communications.

34. F. Evangelisti, A. Frova and F. Patella, Phys. Rev. B10: 4253 (1974).

35. R. H. Bruce, H. Z. Cummins, C. Frolivet, and F. H. Pollak, Bull. Am. Phys. Soc. 22: 315 (1977) and in Phys. Rev. (in press).

36.   J. J. Hopfield, J. Appl. Phys. 32: 2277 (1961).
37.   G. Mahan and J. J. Hopfield, Phys. Rev. 135: 428 (1964).
38.   E. Koteles and G. Winterling, J. Lum. 18/19 : 267 (1979).
39.   P. Y. Yu and F. Evangelisti, Solid State Commun. 27: 87 (1978).
40.   G. Winterling and E. Koteles, in "Lattice Dynamics", M. Balkanski, ed. Flamarion Sciences, Paris (1978).
41.   G. Mahan and J. J. Hopfield, Phys. Rev. Lett 12: 241 (1964)
42.   N. Allen and E. O. Kane, Solid State Commun. 29: 965 (1978).
43.   G. Winterling and E. Koteles, Bull. Am. Phys. Soc. 23: 247 (1978).

ULTRASLOW OPTICAL DEPHASING OF Pr^{3+}:LaF_3 *

R. G. DeVoe, A. Szabo,[†] S. C. Rand and R. G. Brewer

IBM Research Laboratory
5600 Cottle Road
San Jose, California 95193

ABSTRACT: Optical free induction dephasing times as long as 16 μsec, corresponding to an optical homogeneous linewidth of 10 kHz, have been observed for the $^3H_4 \leftrightarrow {}^1D_2$ transition of Pr^{3+} ions in LaF_3 at 2°K. Measurements are facilitated by a frequency-locked cw dye laser and a new form of laser frequency switching. Zeeman studies reveal a Pr-F dipole-dipole dephasing mechanism where the Pr nuclear moment is enhanced in both $^1D_2$ and $^3H_4$.

In this Letter, we report a new advance in the observation of extremely long optical dephasing times in a low temperature solid. Coherently prepared Pr^{3+} impurity ions in a LaF_3 host crystal exhibit optical free induction decay (FID) where the dephasing times correspond to an optical linewidth of only 10 kHz half-width half-maximum and a spectral resolution of 5×10^{10}. At this level of resolution, which represents a fifty fold increase over our previous measurements,[1] it is now possible to perform detailed optical studies of magnetic Pr-F dipole-dipole interactions in the ground and optically excited states. Heretofore, such weak relaxation effects could be detected only in the ground state by spin resonance techniques[2-4] or radio-frequency optical double resonance.[5,6]

---

*Work supported in part by the U.S. Office of Naval Research

To appear in Physical Review Letters June 4, 1979

[†]On leave from the National Research Council of Canada, Ottawa

159

The Pr^{3+} transition $^3H_4 \leftrightarrow ^1D_2$ monitored at 5925Å involves the lowest crystal field components of each state. These are singlet states where the 2J+1 degeneracy is lifted by the crystalline field due to the low Pr^{3+} site symmetry, perhaps C_2 or C_{2v}. The nuclear quadrupole interaction[7] of Pr^{3+} (I=5/2) splits each Stark level into three hyperfine components which are each doubly degenerate ($\pm I_z$), and to a first approximation, three equally probable optical transitions connecting these states occur, namely, $I_z'' \leftrightarrow I_z' = \pm 5/2 \leftrightarrow \pm 5/2$, $\pm 3/2 \leftrightarrow \pm 3/2$, and $\pm 1/2 \leftrightarrow \pm 1/2$. All three transitions overlap and can be excited simultaneously by a monochromatic laser field since the Pr^{3+} hyperfine splittings of order 10 MHz are considerably less than the inhomogeneous crystalline strain broadening of ~5 GHz. Weaker transitions of the type $5/2 \leftrightarrow 1/2$, $5/2 \leftrightarrow 3/2$, ... also occur among these hyperfine states because of a nonaxial field gradient at the Pr^{3+} nucleus which mixes the $|I_z\rangle$ wavefunctions slightly. As noted previously,[1,8] the weaker transitions redistribute the ground state hyperfine population distribution drastically in an optical pumping cycle, and play an important role in the optical dephasing measurements reported here.

Bleaney[9] has shown that when an electronic singlet of a rare earth ion admixes with close lying Stark split levels of a given J manifold, it produces in second order a pseudo-quadrupole moment and an enhanced nuclear magnetic moment

$$m_i = (g_N\beta_N - 2g_J\beta\Lambda_{ii})I_i \, , \tag{1}$$

where the notation is that of Teplov.[4] Here, the principal axes are labeled i=x,y,z, the nuclear and electronic g values are g_N and g_J, the electronic matrix element $\Lambda_{ii} = \sum\limits_{n \neq 0} A_J |\langle 0|J_i|n\rangle|^2 / (E_n - E_0)$ connects the lower state $|0\rangle$ with an excited state $|n\rangle$ removed in energy by $E_n - E_0$, and A_J is the Pr^{3+} hyperfine constant. Now imagine that a fluctuating local magnetic field $\tilde{H}_z$ exists at the Pr^{3+} site due to distant pairs of F nuclei participating in mutual spin flips, and ignore other dephasing mechanisms for the moment. This field modulates the optical transition frequency randomly through a Pr-F dipole-dipole interaction and produces a HWHM homogeneous optical linewidth

$$\Delta\nu = |\gamma_z''I_z'' - \gamma_z'I_z'|\tilde{H}_z/2\pi \tag{2}$$

where γ_z'' and γ_z' are the Pr^{3+} *enhanced* gyromagnetic ratios ($\gamma_z = m_z/\hbar I_z$) of $^3H_4$ and $^1D_2$. Because the Pr nuclear wavefunctions are mixed to some extent,[7] rigorously I_z is not a good quantum number. Nevertheless, to a good approximation[5,8] $I_z'' \rightarrow I_z'$ and as already mentioned, we expect three strong optical transitions $|\pm 5/2\rangle \rightarrow |\pm 5/2\rangle$, $|\pm 3/2\rangle \rightarrow |\pm 3/2\rangle$, and $|\pm 1/2\rangle \rightarrow |\pm 1/2\rangle$. Therefore, from (2) three different decay times should appear in an optical FID. We shall see that this idea is supported and that γ_z' for $^1D_2$ can be obtained since γ_z'' is known[5] and $\tilde{H}_z = 2\pi\Delta\nu_{rf}/\gamma_z''$ can be deduced from an rf-optical double resonance linewidth[6] of the $^3H_4$ state. Furthermore, these experiments offer a new way of testing ab initio calculations[10] of Λ_{ii} as well as the Pr^{3+} site symmetry, which remains controversial.[10,11]

The technique adopted for observing optical FID relies on laser frequency switching,[12] but in a new form. A cw dye laser radiates a beam at 5925Å which is linearly polarized at a power of ~4 mW. The beam passes through a lead molybdate acousto-optic modulator which is external to the laser cavity and oriented at the Bragg angle. The Bragg diffracted beam is focused to a 200 micron diameter in a $7\times7\times10$ mm^3 crystal of Pr^{3+}:LaF_3 (0.1 or 0.03 atomic % Pr^{3+}) which is immersed in liquid helium at 2°K, and the emerging laser and FID light, which propagates parallel to the crystal c axis, then strikes a PIN diode photodetector. The Pr^{3+} ions are coherently prepared while the modulator is driven continuously and efficiently at 110 MHz. FID follows when the rf frequency is suddenly shifted (100 nsec rise time) from 110 to 105 MHz, the duration of the switching pulse being 40 μsec. Note that the laser is switched 500 homogeneous linewidths. Figure 1 shows FID signals produced in this way where the dephasing time $T_2/(1+\sqrt{1+\chi^2 T_1 T_2}) \rightarrow T_2/2$ is independent of power broadening since $\chi^2 T_1 T_2 \ll 1$, χ being the Rabi frequency. The anticipated heterodyne beat of 5 MHz frequency is readily observed because the shifted laser and FID beams overlap due to the change in the Bragg angle (0.4 mrad) being less than the beam divergence (7 mrad). This type of extra-cavity laser frequency switching is compatible with laser frequency locking which we now consider.

To detect ultraslow dephasing times by FID, the laser frequency must remain fixed within the sample's narrow homogeneous linewidth $\Delta\nu = 1/(2\pi T_2)$ for an interval ~T_2 – a stability condition which is less stringent than in a linewidth measurement. In the present work, a frequency stability of ~10 kHz in a time of ~16 μsec is required. To this end, our laser is locked to an external reference cavity which provides an error signal in a servo loop of high gain for correcting slow frequency drift and high frequency jitter. The noise spectrum as seen from the error signal or a spectrum analyzer is not flat but is dominated by isolated jumps of 30 to 100 kHz in a 10 μsec period. At such times, the sample is prepared at two (or more) discrete frequencies which result in a deeply modulated FID pattern. This behavior agrees with a computer simulation of FID which assumes a bimodal spectrum. However, at other times frequency jumps do not occur, and the free induction decays monotonically as in Fig. 1. Under these conditions, a laser jitter of <10 kHz permits a reliable decay time measurement of these *single events* which are considerably longer-lived than the time-averaged value of

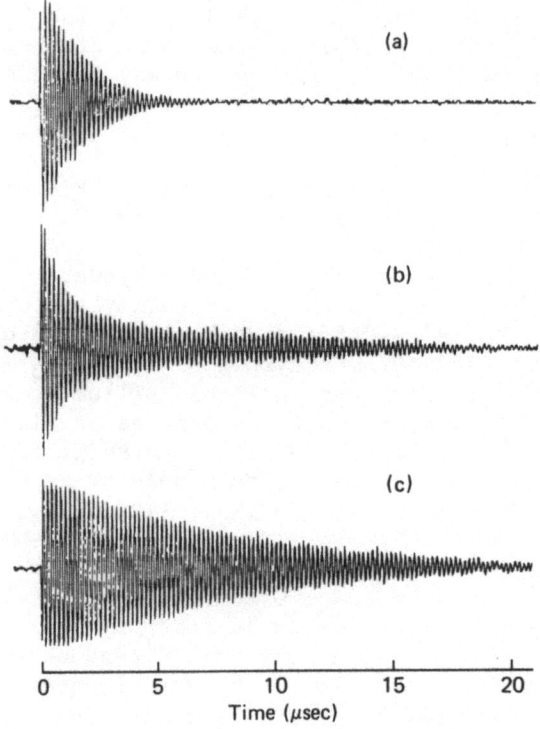

Fig. 1.   Free induction decay of
          0.1 atomic % Pr^{3+} in LaF_3
          at 2°K in the presence of
          an external magnetic field
          $H_0 \propto c$ axis. H_0 equals (a) 0.5G
          (earth's field), (b) 19% and
          (c) 76G. The optical heterodyne
          beat frequency is 5.005 MHz.
          Cases (a) and (c) are plotted
          in Fig. 2.

many decays. These signals are captured with a Biomation 8100
Transient Recorder and then reproduced on an X-Y chart recorder.

A key feature of the measurement is an optical pumping
absorption-emission cycle which transfers population from any
given hyperfine level of the $^3H_4$ ground state to its two
neighbors, for example from $|3/2\rangle$ to $|5/2\rangle$ and $|1/2\rangle$ within the
same inhomogeneous packet. As a result, each of the three $^3H_4$
hyperfine states excited (three packets) will be depleted and
FID cannot be detected. However, by sweeping the laser frequency
at a slow rate of $\leq$10 kHz/16 μsec so as not to influence the decay
rate, the pumping cycle can be reversed[1] and the hyperfine

population partially restored. The $^3H_4$ hyperfine population
distribution which results depends on the sweep rate and the
relative transition probability among the hyperfine states as
they decay from $^1D_2$ to $^3H_4$ via intermediate states. Therefore,
the pumping cycle dictates which of the three strong transitions
can be prepared to yield FID.

In Fig. 1, a dramatic variation in the FID occurs when a weak
external field H_0 is applied perpendicular to the crystal c axis.
The T_2 dephasing times for the three cases are (a) 3.6 μsec at
H_0=0.5G (earth's field), (b) 3.5 and 15.6 μsec at H_0=19G, and
(c) 15.8 μsec at H_0=76G. Note that case (c) corresponds to a
10 kHz HWHM linewidth which appears to be the *narrowest
homogeneously broadened optical transition detected in a solid.*
Its magnitude is comparable to NMR linewidths[2,4,6] which result
from a magnetic dipole-dipole dephasing process. Cases (a) and
(c) are single exponentials (Fig. 2), the ratio of the two decay
times being 4.6. The intermediate case (b) is dominantly a
biexponential and displays precisely the same two decay times
found in (a) and (c). It is significant that the decay time
ratio approximates 5 and that the magnitude of these decay times

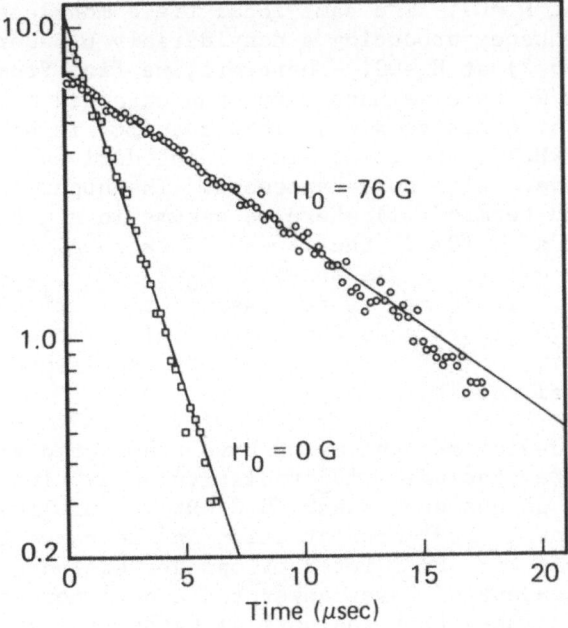

Fig. 2.   Semilog FID plots of the
          data of Figs. 1(a) and (c)
          showing a simple exponential
          decay.

is essentially independent of magnetic field. These results are consistent with Eq. (2) where we expect three decay times in the ratio 5:3:1, and we conclude that case (a) represents dephasing due to the $|5/2\rangle$ state, case (c) to the $|1/2\rangle$ state, and case (b) to both of these states with possibly a small contribution from $|3/2\rangle$ as well. We conclude that application of a weak magnetic field modifies the optical pumping cycle and the $^3H_4$ population distribution in a sensitive way by mixing the nuclear wavefunctions $|I_z\rangle$ further since the $^1D_2$ Zeeman and quadrupole energies[8] can be comparable. This model is also consistent with the zero field rf-optical double resonance observation[6,8] that the $^3H_4$ quadrupole transition $|5/2\rangle\leftrightarrow|3/2\rangle$ is more intense than the $|3/2\rangle\leftrightarrow|1/2\rangle$. More detailed calculations of the nuclear wavefunctions are needed to test these ideas further and will require determining the orientation of the principal axes x, y, z for both $^3H_4$ and $^1D_2$.

We now turn to Eq. (2) to determine the $^1D_2$ enhanced gyromagnetic ratio γ_z'. A fluctuating local dipolar field of $H_z=0.41$G at the Pr^{3+} site due to the fluorine nuclei can be deduced from the ground state value[5] $\gamma_z''/2\pi=23$ kHz/G and a ground state linewidth[6] of 9.5 kHz for the $^3H_4$ quadrupole transition $|5/2\rangle\leftrightarrow|3/2\rangle$ at $H_0=0$G. The same local field modulates the optical transition frequency producing a considerably broader linewidth of 44 kHz ($I_z=5/2$) at $H_0=0$G. Therefore, we find from (2) that $\gamma_z'/2\pi=20\pm4$ kHz/G where we have taken the enhanced moments of $^3H_4$ and $^1D_2$ to be of opposite sign. This quantity is bounded by $1.29<\gamma_z'/2\pi<19$ kHz/G, the lower limit being derived from the first term of (1), i.e., with no enhancement. The upper limit follows from the second term of (1) where we assume in Λ_{zz} the maximum matrix element $\langle1|J_z|0\rangle=2$, the lowest Stark level of $^1D_2$ mixes with the first excited state where $E_1-E_0=23$ cm^{-1}, $g_J=1$, and $A\sim1.093\times10^9$ Hz. If γ_z'' and γ_z' are assumed to be of the same sign, $\gamma_z'/2\pi=66$ kHz/G which exceeds the upper limit. In addition, ab initio calculations[10] of $\langle J_z\rangle$ are in serious disagreement with our experimental results.

Other broadening mechanisms we have considered appear to be negligible. They include a $^1D_2$ radiative decay time of 0.5 msec[13] (0.16 kHz) and phonon processes[8] (0.8 kHz). Our linewidths are also independent of Pr^{3+} concentration in the range 0.03 to 0.1 atomic % so that Pr^{3+}-Pr^{3+} interactions are excluded. Since the width is independent of laser power and a nutation signal is not detected, we estimate that the optical transition matrix element $\mu_{ij}\leq4.5\times10^{-5}$ Debye. This implies that only 10^{-5} of the $^1D_2$ ions return directly by radiative decay to the ground $^3H_4$ state; the remainder radiate to excited Stark split states of $^3H_4$ and other states[13] followed by rapid spontaneous phonon emission processes to the ground state. Clearly, the optical pumping cycle is not simple. The contribution of laser frequency jitter to the linewidth appears to be small since the decay time varies with external magnetic field in a predictable manner. We expect that a significantly higher spectral resolution can be achieved in the near future and will further improve precision measurements of this kind where ultraslow optical dephasing processes occur.

We are indebted to D. Horne for the design and construction of the laser frequency locking circuit and to K. L. Foster for technical assistance. We are pleased to acknowledge conversations with E. Wong, L. E. Erickson, C. S. Yannoni, I. D. Abella, E. L. Hahn, W. B. Mims, and A. Wokaun.

REFERENCES

1. A. Z. Genack, R. M. Macfarlane and R. G. Brewer, Phys. Rev. Lett. 37:1078 (1976); R. M. Macfarlane, A. Z. Genack, S. Kano and R. G. Brewer, Journal of Luminescence 18/19:933 (1979).

2. K. Lee and A. Shir, Phys. Rev. Lett. 14:1027 (1965).

3. W. B. Mims in: Electron Paramagnetic Resonance, ed. S. Geschwind (Plenum, NY, 1972), p. 263.

4. M. A. Teplov, Soviet Phys. JETP 26:872 (1968).

5. L. E. Erickson, Opt. Comm. 21:147 (1977).

6. R. M. Shelby, C. S. Yannoni and R. M. Macfarlane, Phys. Rev. Lett. 41:1739 (1978).

7. T. P. Das and E. L. Hahn, Nuclear Quadrupole Resonance Spectroscopy, (Academic, 1958).

8. L. E. Erickson, Phys. Rev. 16B:4731 (1977).

9. B. Bleaney, Physica 69:317 (1973).

10. S. Matthies and D. Welsch, Phys. Status Solidi B, 68:125 (1975).

11. E. Y. Wong, O. M. Stafsudd and D. R. Johnston, J. Chem. Phys. 39:786 (1963); V. K. Sharma, J. Chem. Phys. 54:496 (1971).

12. R. G. Brewer and A. Z. Genack, Phys. Rev. Lett. 36:959 (1976); A. Z. Genack and R. G. Brewer, Phys. Rev. 17A:1463 (1978).

13. M. J. Weber, J. Chem. Phys. 48:4774 (1978).

PICOSECOND RAMAN GAIN STUDIES OF MOLECULAR VIBRATIONS ON A SURFACE

J. P. Heritage and J. G. Bergman

Bell Telephone Laboratories
Holmdel, New Jersey   07733

We have developed a new picosecond Raman gain technique with
the capability of obtaining a vibrational spectrum of a molecular
monolayer. Very high sensitivity is required to detect the vibrations of so few molecules and the picosecond Raman gain technique
achieves the required sensitivity. In this paper we discuss the
new picosecond Raman gain technique, and we employ the technique
to obtain the first spectrum of a molecular monolayer using a
nonlinear optical technique. We present in this paper a vibrational spectrum of the surface enhanced Raman activity of a monolayer of cyanide on a silver surface. A surprising new result,
not obtainable by conventional Raman spectroscopy, emerges from
this work. The smooth continuum that accompanies the enhanced
Raman scattering is shown to be luminescence and not a Raman effect.

Gain is experienced by the stokes field when the difference
between the frequency of two optical fields (pump, stokes) is tuned
to a Raman active transition. It is instructive to clarify two
limiting cases of interest. The large gain limit (G>>1) occurs
when the intensity of a single applied field is sufficiently large
that spontaneous stokes scattering is amplified and efficient conversion of pump photons into Stokes-shifted photons occurs as a
result of strong nonlinear coupling. This result is well known as
stimulated Raman scattering and the nonlinear effects are very
important for understanding the evolution of the Stokes pulse shape.
The very small gain (G<<1) limit is called simply, Raman gain and
needs both the pump and Stokes fields to be applied externally in
order to be measurable. In this low gain regime, the effect of
the reshaping of the Stokes intensity profile by transient gain
and the effect of the Stokes pulse intensity growth with propagation may be neglected.

High sensitivity is obtained in Raman gain experiments by mod-
ulating a continuous pump laser and detecting the indiced change in
the continuous Stokes beam intensity with synchronous detection.
Since the Stokes (probe) laser is incident on the detecting photo
diode, laser power fluctuations in a band around the modulation
frequency present the principal source of noise. When the probe
laser power fluctuations are very small and with sufficiently high
probe laser power, amplifier noise may be overcome and shot noise
presents the ultimate limit. The important point is that at the
shot noise limit, extremely small changes in the power of the probe
beam may be detected.

Owyoung developed the technique of obtaining Raman gain spectra
with continuous single-mode lasers and has obtained high resolution
spectra in liquids[1] and gases[2]. Heritage[3] introduced time resolved
Raman gain spectroscopy using continuous modelocked lasers to
obtain vibrational dephasing dynamics in the liquid state.

A minimum detectable gain of $G=\ell=10^{-5}$ was reported in the
dephasing work and it was limited by laser fluctuations. In this
context we mean by gain the fractional change in probe laser power
measured at the photo diode that occurs as the result of the Raman
gain process where g is the gain factor and l the sample thickness.
We have improved the minimum detectable gain by a factor of 10^3 by
modulation and synchronous detection at 10 MHz. We detect a gain
as small as 10^{-8} in one second of integration with 10 mW of probe
power. This minimum detectable signal is comparable to that
achieved with single mode lasers[1], and approaches within a factor
of 4 of the shot noise limit. We achieve, however, an optimum
signal-to-noise ratio that, in the case of liquid benzene is
approximately 200 times larger than the single mode laser results.
This improvement comes about directly from the fact that the change
in the power of the steady state probe beam is proportional to the
product of the pump and Stokes intensity. This means that the
greatest gain is realized, for a given available energy flux, if
both the pump and probe energy is concentrated in two synchronized
pulses. The optimal gain is realized when ultrashort pulses are
used.

We illustrate the surface monolayer sensitivity of picosecond
Raman gain with a simple calculation. The steady state gain may be
estimated from the relation[4]

$$g = \frac{16\pi^2}{\omega^3} \frac{c^2}{n^2} \frac{N}{\Gamma \hbar} \frac{d\sigma}{d\Omega} \quad .$$

(5)

Using the differential cross-section for benzene $\frac{d\sigma}{d\Omega} = 7.85 \times 10^{-30}$
$\frac{cm^2}{sr}$ density $N = 8 \times 10^{21}$ cm^{-3}, $\omega = 3.14 \times 10^{15}$ sec^{-1}. We obtain
$g = 2.7 \times 10^{-3}$ cm/MW. We take 4 Å as a representative thickness
and obtain $G=g\ell=1.1 \times 10^{-10}$ cm^2/MW. The power of a 12 psec pump

pulse may exceed 100 W in a continuous modelocked laser and can easily be focused to a diameter of 10 μm, yielding an intensity of $= 1.3 \times 10^8$ W/cm^2. The expected gain for a single pass through a monolayer of benzene is then 1.4×10^{-8}. At the shot noise limit, the signal-to-noise ratio may be estimated from[1] $X/N = G\left(\dfrac{QP}{2\omega\Delta\nu}\right)^{1/2}$

where Q is the detector quantum efficiency, P the probe laser average power, ω is the stokes frequency, and $\Delta\nu$ the detection band width. With P = 40 mW, Q = 0.7 and $\Delta\nu = 0.01$ sec^{-1}, we find S/N = 28 for our idealized monolayer of benzene. We comment at this point that this calculation is only a crude estimate whose purpose is to illustrate the tremendous potential sensitivity of picosecond Raman gain applied to a monolayer. Local field corrections have not been taken into account in this estimate, but we have used the conservative vapor phase value of the differential scattering cross section. More detailed calculations have been performed by Levine et al.[6], and they arrive at surprisingly similar results.

A molecular monolayer with a Raman cross section much larger than benzene is useful to test the surface sensitivity of picosecond Raman gain. Recently, monolayers of molecules adsorbed on an anodized silver surface have been shown to exhibit a Raman cross section of $\cong 10^5$ larger than the isolated molecule[7]. Several molecular species have been shown to exhibit an enhanced Raman effect on silver[8]. Most experiments have been done with samples prepared and studied in an electrochemical cell under potentiostatic control. Two exceptions are CO and CN. CO has displayed an enhanced Raman cross section when adsorbed on silver at low temperatures in a vacuum[9]. CN, adsorbed on silver in solution, has been shown to give a strong Raman signals after the sample is rinsed and dried[10,11]. In general, one observes significant enhancement of Raman cross sections of well characterized molecular modes, along with certain low frequency vibrations associated with adsorbate-metal motion, and a broad, structureless continuum of significant intensity extending from 300 cm^{-1} to nearly 4000 cm^{-1}. The origin of the smooth continuum has not been determined. Given its structureless character, the techniques used in spontaneous Raman spectroscopy are unable to distinguish it from luminescence. A conventional Raman spectrum of cyanide on silver taken in air is presented in Figure 1. The CN stretch band at 2145 cm^{-1} is evident along with the continuum. The band at 1600 cm^{-1} has been assigned to a carbonate impurity that appears in the air spectrum[10].

We now turn to a discussion of the experimental arrangement. Two synchronously modelocked dye lasers provide continuous trains of picosecond pulses that are well synchronized and widely tunable[12]. In these experiments, the lasers are tuned near 5680 A (pump) and 6468 A (probe), corresponding to the Raman frequencies near 2145 cm^{-1}. The experimental arrangement is shown in Figure 2 and consists of a variable delay line and a dichroic mirror for

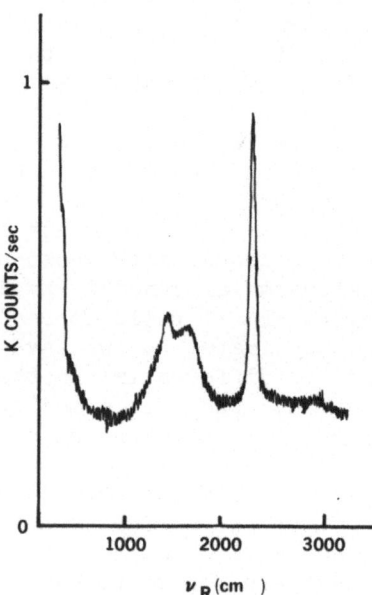

Fig. 1.   Conventional Raman spectrum of a monolayer of cyanide on
          silver. This spectrum, labeled in cm⁻¹, was obtained in
          air.

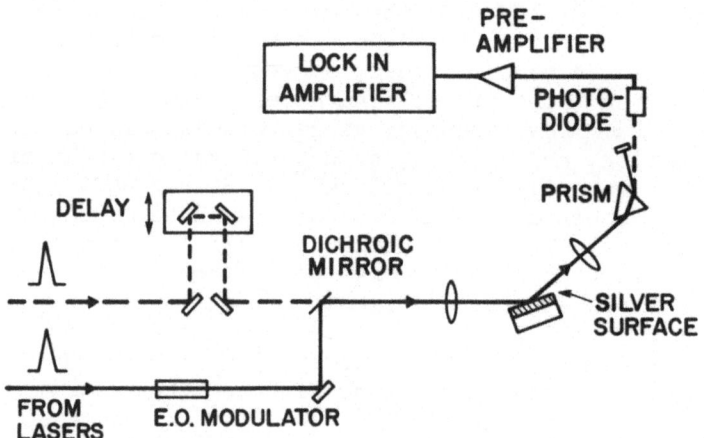

Fig. 2.   Experimental arrangement for picosecond Raman gain
          spectroscopy of silver surfaces.

combining the pump and probe pulse trains spatially and temporally. Raman gain spectra are obtained by tuning one laser at fixed temporal overlap of pump and probe pulse trains. We also obtain time resolved traces while exciting the CN vibration and while investigating the continuum.

Time resolved scans permit identification of a time-dependent background that turns out to be due to the influence of the pump pulse. This background is unique to the Raman gain approach and is definitely not related to the continuum. An unambiguous subtraction can be made to leave the pure Raman gain signal. The details of this technique are discussed elsewhere[13]. The Raman gain spectra and Raman gain time-resolved traces presented here have the background subtracted.

In Figure 3 we present a Raman gain spectrum of a monolayer of cyanide on silver. This sample was prepared by the methods described by Bergman et al.[11] who verified the monolayer coverage by radioactive Carbon 14 tracer measurements. This spectrum was obtained at a low enhancement. A spontaneous spectrum obtained from this sample, immediately after the Raman gain spectrum, revealed a peak at 2145 cm^{-1} only 20% higher than the continuum. This spectrum, which agrees with the conventional Raman spectrum, verifies the monolayer detection capability of the surface picosecond Raman gain technique.

We have investigated the continuum carefully by obtaining time resolved spectra in the continuum. In Figure 4 (solid line), we present a time-resolved trace obtained on the CN resonance ($\Delta \nu$ = 2145 cm^{-1}). This trace shows the expected Raman gain cross-correlation and establishes the detection of gain on that sample. Figure 4 (dotted line) is the time trace obtained in the continuum ($\Delta \nu$ = 2000 cm^{-1}). The gain disappears in the continuum. The fact that there is no gain at $\Delta \nu$ = 2000 cm^{-1} resonance proves, without need of further discussion, that the origin of this part of the continuum is luminescence.

Even though the absence of a Raman effect in the continuum has so far been verified only near 2000 cm^{-1}, we anticipate that this result will remain true throughout the featureless region of the continuum. There may, however, be weak Raman structure added to the continuum, especially when large multi-mode molecules are adsorbed on the surface. The question of the origin of the band of structure that lies near 1600 cm^{-1} remains open[10]. The sharp features in this region may be of Raman origin, but the adsorbate is not yet positively identified. The low frequency region below about 200 cm^{-1} rises steeply as $\Delta \nu$ approaches zero and we do not consider this region to be part of the continuum.

The excitation[14] and subsequent radiative decay of surface

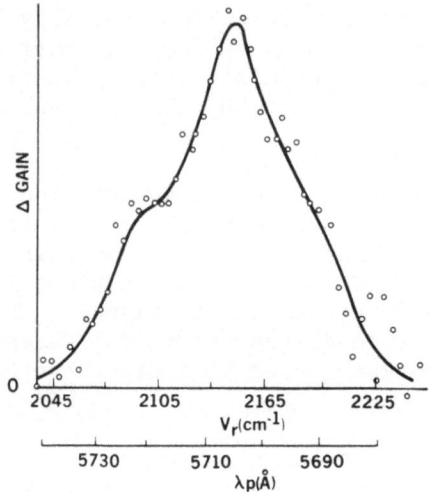

Fig. 3. Picosecond Raman gain spectrum of CN on silver.

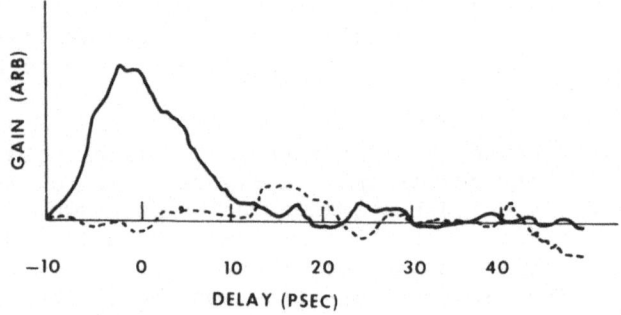

Fig. 4. Time resolved Raman gain trace. Solid line – $\Delta\nu = 2145$ cm^{-1} Dotted line – $\Delta\nu = 2000$ cm^{-1}.

enhanced Raman effect will probably include as well a detailed description of the origin of the luminescence continuum.

In conclusion, we have developed and demonstrated a sensitive picosecond Raman gain technique and have obtained the first Raman gain spectrum of a monolayer. Additionally, we have obtained the new result, not obtainable with conventional spectroscopic techniques, that the continuum associated with surface enhanced Raman scattering must be luminescence.

ACKNOWLEDGMENTS:

We gratefully acknowledge assistance and useful conversations with J. M. Worlock and A. Pinczuk. We thank C. V. Shank for stimulating discussion.

REFERENCES

1. A. Owyoung, Opt. Comm. 323 (1977); A. Owyoung and E. D. Jones, Opt. Lett. 1, 152 (1977); A. Owyoung, IEEE J. Quant. Elect. QE 14 (1978).

2. A. Owyoung, L. W. Patterson and R. S. McDowell, Chem. Phys. Lett. 59, 156 (1978).

3. J. P. Heritage, Appl. Phys. Lett. 34, 470 (1979).

4. M. Maier, Appl. Phys. 11, 209 (1976).

5. J. R. Nestor and E. R. Lippincott, J. of Raman Spect. 1, 305 (1973).

6. B. F. Levine, C. V. Shank and J. P. Heritage, to be published.

7. D. L. Jeanmarie and R. P. Van Duyne, J. Electroanal. Chem. 84, 1 (1977).

8. For a recent review, see R. P. Van Duyne, in "Chemical and Biochemical Applications of Lasers", ed. by C. B. Moore, Vol. 4, Chapt. 5 (1978).

9. T. R. Wood and M. V. Klein, to be published.

10. A. Otto, Surface Science, 75, 392 (1978).

11. J. G. Bergman, J. P. Heritage, A. Pinczuk and J. M. Worlock, to be published.

12. R. K. Jain and J. P. Heritage, Appl. Phys. Lett. 32, 41 (1978).

13. J. P. Heritage, J. G. Bergman, A. Pinczuk, J. M. Worlock, to be published.

14. E. Burstein, Y. J. Chen, C. Y. Chen, S. Lundgist and E. Tosatti, Sol. St. Comm. 29, 567 (1979).

BRILLOUIN-MANDELSTAM SCATTERING OF LIGHT IN ANTIFERROMAGNETIC CoCO$_3$

A.S. Borovik-Romanov, N.M. Kreines, V.G. Jotikov

Institute for Physical Problems, USSR Acad. of Sci., Moscow, USSR

1. INTRODUCTION

This paper reviews experimental investigations of spin-wave spectra and their relaxation rate in CoCO$_3$ by Brillouin-Mandelstam light scattering method (BMS). The first experiments on light scattering by spin waves was by Fleury and his colleagues (1,2) who observed one and two magnon scattering by magnons with a pronounced ($\sim$ 3-5 cm^{-1}) gap in the energy spectrum. Details of the investigation of such Raman scattering from magnons can be found in review articles (3-5). In the last few years, Sandercock and his colleagues have published a number of papers on BMS by the low frequency branch of the magnon spectrum (6,7). These experiments required an interferometer of high contrast, and Sandercock developed a multipass Fabry-Perot interferometer which was able to meet this requirement (8).

Easy plane, weak, ferromagnets are especially interesting objects for investigation using BMS method. Two of such magnets were previously investigated. Jantz, Sandercock and Wettling have studied FeBO$_3$ (9,10) and our group has investigated CoCO$_3$ (11,12). Both these substances are of rhombohedral structure with two molecules in the elementary cell (Fig. 1). At T$_N$=18.1K CoCO$_3$ becomes antiferromagnetic with spins lying in the base plane.

There are two branches of the spin wave spectrum, differing in the components of the sublattice magnetization vectors that

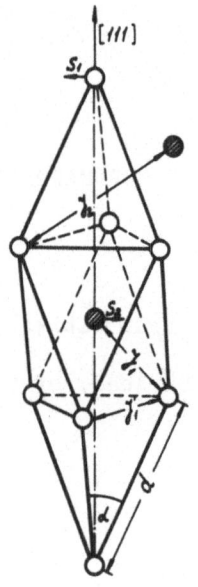

Figure 1. Magnetic structure of $CoCO_3$ and $FeBO_3$.

Figure 2. Experimental
apparatus: L_1, L_2 and
L_3, lenses; D_1, D_2 and
D_3, pinholes; II, polarizer;
A, analizer; $K_1 K_2$, collimation
systems; lF, interference filter;
PM, photomultiplier; PA, pre-
amplifier; PCS, photon counting
system.

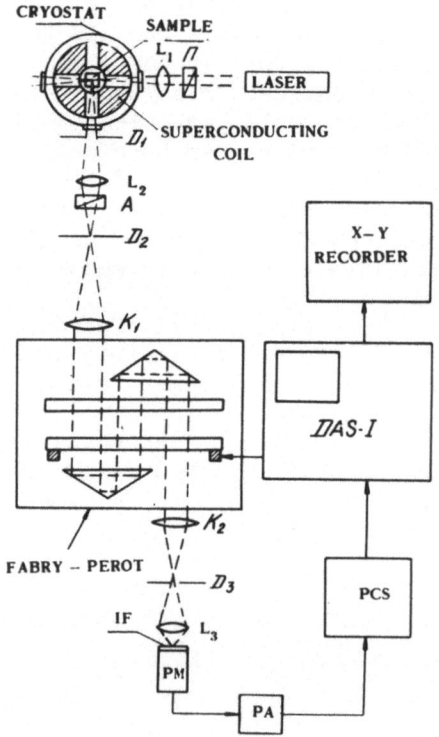

participate in the oscillations. It is expedient to replace the magnetization vectors M_1 and M_2 by the linear combinations:

$$1 = M_1-M_2; \quad m = M_1 + M_2 \tag{1.1}$$

If the coordinate system is chosen so that the z axis is directed along the trigonal axis and the x axis is in the direction of the magnetic field applied in the base plane of the crystal, then the components 1_x, m_y and m_z will oscillate in the first spin wave mode and the components 1_y, 1_z and m_x will oscillate in the second. These modes differ both in magnetic energy and anisotropy energy. Below, attention is restricted to the first mode. The dispersion law for the low frequency mode of spin waves propagating along the coordinate axis is:

$$(\nu/\gamma)^2 = H(H + H_D) + a_x^2 k_x^2 \tag{1.2}$$

$$(\nu/\gamma)^2 = H(H + H_D) + 4\pi\kappa(H + H_D)^2 + a_y^2 k_y^2 \tag{1.3}$$

$$(\nu/\gamma)^2 = H(H + H_D)(1 + 4\pi\kappa) + a_z^2 1_z^2 \tag{1.4}$$

where γ is the gyromagnetic ratio, ν is the frequency and k is the wave vector of the spin wave; H_D is the Dzyaloshinskii field, causing the canting of the spins; a, are the exchange constants; κ is the magnetic susceptibility in the base plane. This leads to additional terms associated with the dipole-dipole interaction.

In the next section we will describe experiments on BMS which proved the above relation. The third section is devoted to the investigation of spin wave relaxation by means of optical methods.

2. INVESTIGATION OF SPIN WAVE SPECTRUM BY MEANS OF BMS

2.1. The experimental set up and the sample. Experiments on $CoCO_3$ were carried out in 90° geometry. In this case the wave vector of the scattering quasiparticle k = 2.5 x 10^5 cm^{-1} at the incident light wavelength λ = 632.8 nm. The spectral composition of the scattered light was investigated using a high-contrast three-pass Fabry-Perot scanning interferometer, manufactured by Burleigh (USA), according to the design developed by Sandercock (8). The contrast was larger than 10^8. A DAS-1 system, also manufactured by Burleigh, was used for control of the interferometer and automatic adjustment. A diagram of the experimental apparatus is shown in Fig. 2.

The light scattering experiments were carried out both at room temperature (to investigate phonons) and at T = 1.5-2.0K in a bath of superfluid helium (to investigate magnons). The magnetic field was produced by superconducting coils.

In all experiments a single sample of $CoCO_3$ kindly provided
by Ikornikova and Egorov was used. This was a rectangular
parallelepiped (base 1x1.2 mm^2, height 1.8 mm). The axis (C_3)
was directed along the diagonal of the base. In the geometry of
Fig. 3, scattering by phonons and spin waves with $k = k_z$,
directed along the z-axis was observed. Rotation of the sample
through 90° allowed scattering by the quasiparticles propagating
in the base plane to be observed. For brevity, magnons travelling
along the x, y and z axis will be referred to henceforward, as x,
y and z magnons respectively.

2.2. Experimental results. Results obtained in the
experimental investigation of light scattering by magnons are now
considered (12).

An example of the light scattering spectrum for magnons
travelling along the C_3 axis is shown in Fig. 3. This is the
direction in which the scattering is the most intense. For all
spectra we have observed, the intensity of the Stokes and
anti-Stokes spectral components was the same.

In Fig. 4 the square of the magnon frequency is plotted
against the amplitude of the applied magnetic field for all three
directions of the wave vector, the absolute value of which was 2.5
10^5 cm^{-1}. The magnetic field values given in Fig. 4 take into
account the demagnetizing factor of the sample. The uncertainty
in the value of the demagnetizing factor is the main source of
errors in the spectrum of magnons. The continuous curves in Fig.
4 were plotted from Eqs (1.2) - (1.4) using the values of the
constants obtained by the least squares method. These values are
given in the Table. For comparison, the Table also gives the
parameters of the spin-wave spectrum for $FeBO_3$, determined by
Jantz, Sandercock and Wettling (19, 10), also from BMS.

The experiments on $CoCO_3$ were the first to allow observation
of scattering by magnons propagating in all three principal
directions. The agreement between the results of the present
experiments and Eqs (1.2) - (1.4) confirms that the dipole
component of the spin-wave energy of an easy-plane weak
ferromagnet has been correctly calculated (13,14). Fig. 5
presents spin-wave spectra constructed from Eqs (1.2) - (1.4)
using the values of the constants obtained in the present
experiments. It is evident from this Figure that, as a result of
the dipole-dipole interaction, the spectrum of magnons travelling
along the y axis has a gap of 24 GHz in zero magnetic field.
Recently Jantz and Wettling (10) have proved that in $FeBO_3$ the
dipole-dipole interaction cause a gap of 5 GHz.

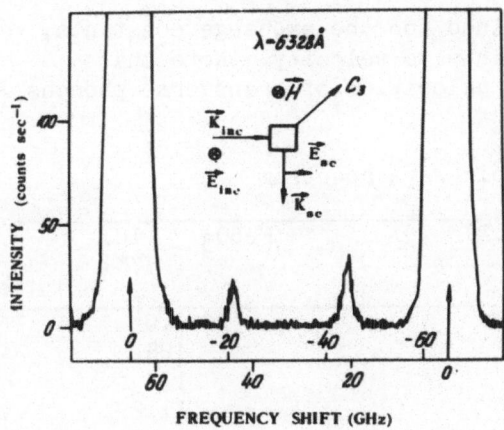

Figure 3. Spectrum of light scattered at 90° in $CoCO_3$ (T = 2K) by magnons travelling along the C_3 axis.

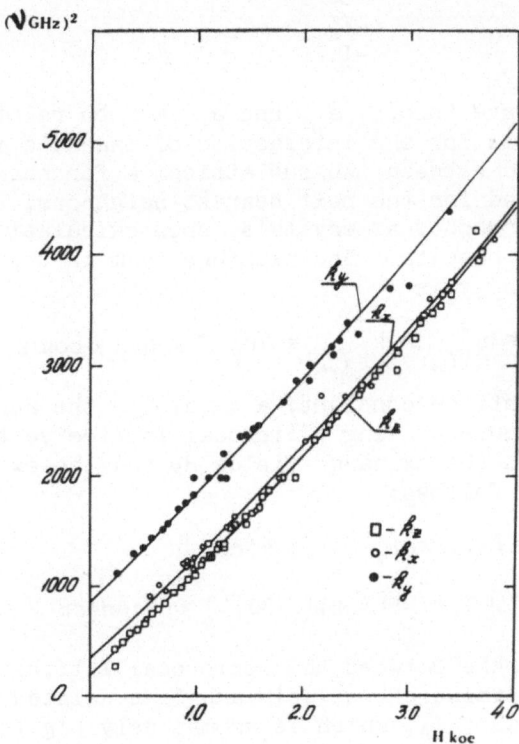

Figure 4. Dependence of square of magnon frequency on magnetic field for three directions of magnon propagation in $CoCO_3$: k_z along C_3 axis; k_x in the base plane along the magnetic field; k_y in the base plane perpendicular to the magnetic field; k = 2.5 x 10^5cm^{-1}. The continuous curves are plotted using Eqs (1.2) - (1.3).

Now consider the results obtained for the exchange constants, $a_{\parallel}$ and $a_{\perp}$, which determine the spin-wave velocity. Note that these velocities are close to the velocities of transverse phonons in the corresponding directions.

Parameters of spin-wave spectrum

Parameter	$CoCO_3$ (1.2) (T < 2K)	$FeBO_3$ (10) (T = 77K)
g-factor	4.1	2.0
H_D, kOe	27	105
$2\pi\gamma\, a_{\parallel}$, km/s	3.43	135
$2\pi\gamma\, a_{\perp}$, km/s	4.38	11.2
$H_E^{\parallel}$, kQe	240	1.990
$H_E^{\perp}$, kOe	310	1.650
J_1, GHz	-36	-268
J_2, GHz	-75	-29
$J_1^{'}$, GHz	-1.1	-17

Using spin-wave theory, $a_{\parallel}$ and $a_{\perp}$ may be related to the exchange integrals for the interaction of magnetic ions inside a sublattice $J_1^{'}$ and between two sublattices - for nearest neighbors, J_1, and for the next nearest neighbors, J_2 (see Fig. 1). For rhombohedral crystals, such calculations (9,15) lead to the following results. The exchange term in the expression for the spin wave energy is

$$v^{\parallel (\perp)} = a_{\parallel(\perp)}^{'} \times k_{\parallel (\perp)} = \gamma H_E^{\parallel(\perp)} \times d \times k \cos \varepsilon \qquad (2.1)$$

where d is the lattice constant; $a^{'} = \gamma a$; ε is the angle between the direct and the corresponding reciprocal lattice vectors. The relations between the exchange fields H_E and the exchange integrals are as follows:

$$H_E^{\parallel} = (1/\gamma) \times 2S(J_1 + J_2) \times Q \times \sec \beta \qquad (2.2)$$

$$H_E^{\perp} = (1/\gamma) \times 2S[J_1 + J_2)(J_1 + 4J_2 - 3J_1^{'})]^{\frac{1}{2}} Q \sqrt{8} \mathrm{cosec} \beta \qquad (2.3)$$

where β is the angle between the reciprocal lattice vector and the z axis; S is the value of the spin and Q is related to the anisotropy constant, K_A, which is anomalously big for $CoCO_3$:

$$Q = [1 - K_A/\{12(J_1 + J_2)\}]^{\frac{1}{2}} \qquad (2.4)$$

If the interaction with nearest neighbors is dominant, the spin wave velocities and the corresponding exchange fields must satisfy the relation:

$$a_{\parallel}/a_{\perp} = H_E^{\parallel}/H_E^{\perp} = \sqrt{8} \times \mathrm{tg}\, \beta \approx 1.3 \qquad (2.5)$$

For $FeBO_3$ the experimental results are close to this value, thus
confirming that the main cause of the anisotropy of the spin-wave
velocity in this material is the rhombohedral distortion of the
lattice. For $CoCO_3$ the anisotropy of the exchange fields is of
a different type. From Eqs (2.1) - (2.4) values may be obtained
for the following combinations of exchange integrals: $(J_1 + J_2)$ and $(J_2 - J_1^1)$. Using data (16) on two magnon scattering
in $CoCO_3$ we found the values of all the three integrals, which
are given in the Table.

3. INVESTIGATION OF SPIN-WAVE RELAXATION BY OPTICAL METHODS

3.1. Introduction. The aim of this section is to demonstrate
that optical methods offer important new opportunities for the
study of relaxation processes. First we describe experiments to
determine the number of magnons created in antiferromagnetic
resonance (AFMR) excitation. It is found possible to determine
the excess of magnons with various ν and k in comparison with the
equilibrium case. Finally the results of BMS on parametrically
excited quasiparticles will be given.

To perform the experiments described in this section we have
added to the optical part shown in Fig. 2 a simple microwave
spectrometer. The sample was placed in an 8-mm wave-guide with
appertures for the illumination of the sample and the observation
of transmitted or scattered light. A 5 mW klystron or continuous-
operation 1 W magnetron were used.

3.2. AFMR relaxation in $CoCO_3$. To elucidate the process of
AFMR relaxation three experiments were carried out. In the first
(17) we observed components shifted by ± 36 GHZ in the spectrum of
the light transmitted through a system consisting of a crossed
polarizer and analyzer on either side of the sample in which AFMR
is excited. The relative intensity of these components is
proportional to the square of the amplitude of the homogeneous
spin oscillations, i.e. to the number of the excited magnons with
$k = 0(N_0)$. Given the experimental results for the number of
spin waves $N_0 = 2 \times 10^{13}$ it is possible to determine the
relaxation rate of these spin waves from the relation

$$W/h\nu \ = \ N_o/\tau_o \qquad\qquad (3.1)$$

where W is the power absorbed by the sample. In the experiment
$W \sim 5$ mW, so that $\tau_o \sim 3 \times 10^{-11}$s.

The next step was to determine the total number of magnons
with $k \neq 0$ produced due to AFMR. This was done by measuring the
change in the magnetic birefringence at AFMR (18). Using a
modulation technique we have found that the total change of the

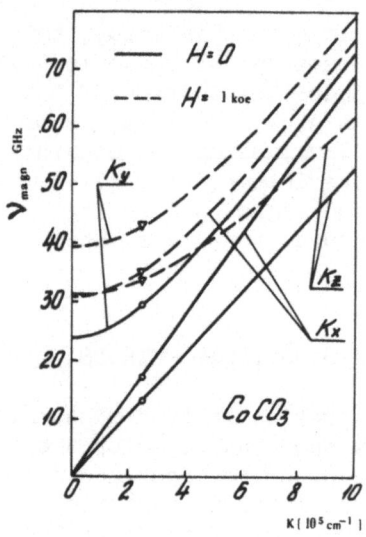

Figure 5. Low-frequency part of spin-wave spectrum for CoCO₃ plotted from Eqs (1.2) - (1.4).

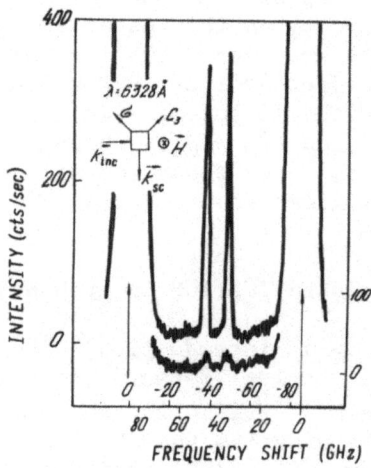

Figure 6. Spectra of light scattered at 90° in CoCO₃ by thermal magno nons (lower curve) and magnons excited at AFMR-amplified magnons (upper curve).

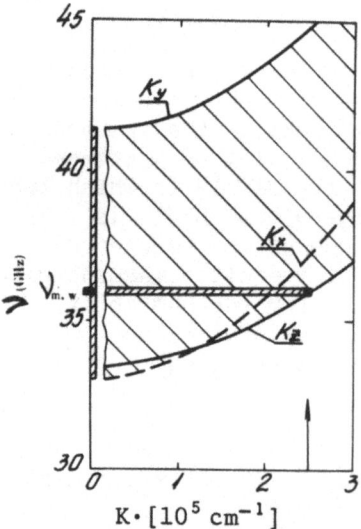

Figure 7. Magnons produced in AFMR relaxation by two-magnon process (horizontal shaded band).

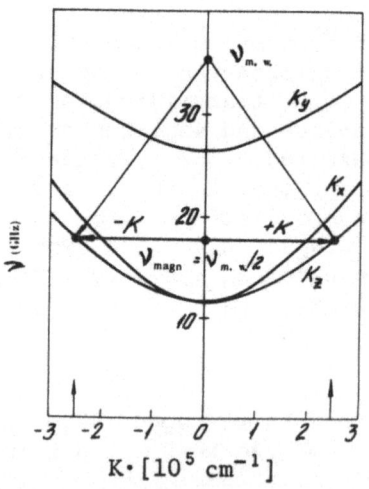

Figure 8. Parametric excitation of half-frequency magnons in CoCO₃ .

number of spin waves at AFMR was $N_k = 10^{17}$ (at the same
absorption rate ~ 5 mW. This means that the corresponding
relaxation time is at least three orders of magnitude larger
than τ_0. Thus in AFMR relaxation processes there may be a
phenomenon analogous to the so called "bottleneck" in EPR, but in
antiferromagnets there are no phonons that are overheated, but
rather spin waves.

To elucidate whether the spin-wave system is thermalized in
AFMR, or an isolated group of spin waves is overheated BMS was
used. It was found that if microwave power is fed to the sample,
considerable amplification of the magnon-peak intensity is
observed at the magnetic field corresponding to the existence of
magnons with frequency equal to the pumping frequency and k = 2.5x
10^5 cm^{-1} (see Fig. 6 and 7). This amplification was only
observed in a narrow range of magnetic fields $\pm$25 Oe, although
microwave absorption was observed in a field range of 600 Oe, due
to exciting of AFMR (either a homogeneous precession or one of the
magnetostatic modes). The amplification was observed both for z
and x magnons, but not for y magnons. These results can be
explained by the assumption that the two-magnon process is
dominant in relaxation, as illustrated in Fig. 7.

   3.3. Parametric Excitation of Magnons and Phonons in
Antiferromagnets. Under certain conditions one microwave photon
may create two magnons, each with opposite value of the wave
vector and frequency equal to half the photon frequency. This
process is called parametric excitation of spin waves. Like any
parametric process, it begins when the microwave power applied
exceeds some critical value corresponding to a certain value of
microwave magnetic field. This critical field bears a simple
relationship to the relaxation of the magnons. Until now
parametric excitation has only been observed by the appearance of
"additional" microwave absorption in magnetic fields less than
that corresponding to AFMR (19). Parametric excitation of
electronic spin waves has been observed only in two
antiferromagnets $MnCO_3$ and $CsMnF_3$. Comparison of the
relaxation rates obtained for these two substances with those
given in previous section for $CoCO_3$ indicated that parametric
excitation of magnons is also possible in $CoCO_3$.

The decay of a microwave photon to two magnons is shown
schematically in Fig. 8 from which it is evident that if the
frequency of microwave pumping is given, then for each value of
magnetic field in a given direction magnons with a fixed wave
vector will be parametrically excited. The appearance of an
excess number of magnons with a given wave vector and frequency
may be observed by BMS as an increase in the intensity of the
corresponding thermal magnon peak. An example of such an increase
is shown in Fig. 9, (20), in which the intense peaks correspond to

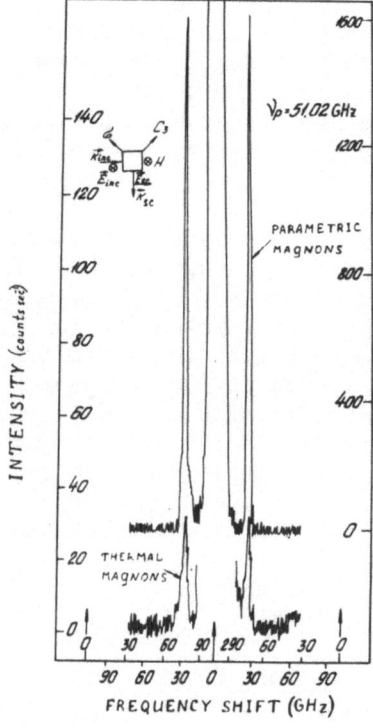

Figure 9. Spectrum of light scattered in CoCO$_3$ by magnons travelling along the z-axis with a 250 mW power supply at ν = 51 GHz.

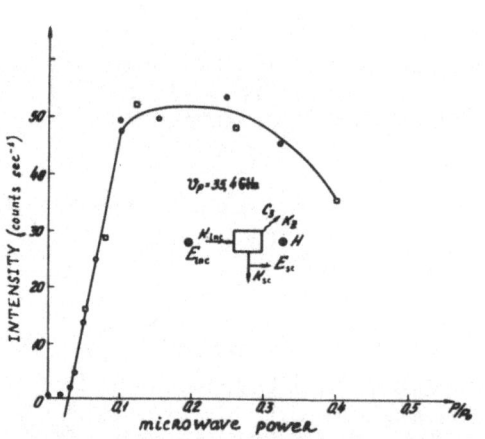

Figure 10. The dependence of the intensity of the parametric satellites on the pumping power.

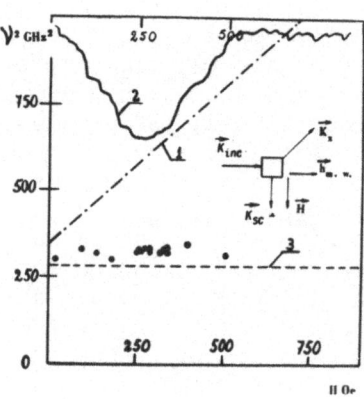

Figure 11. Dependence of square of frequency on magnetic field for thermal magnons (1) and pumped (points) quasiparticles travelling along the x-axis; curve 2 is the microwave absorption line, curve 3 is the frequency of transverse photons in the base plane at T = 100K.

parametrically excited z-magnons with frequency equal to half of
ν_p= 51.02 GHz. These intense peaks are only observed in a
narrow range of fields $\pm$10 Oe, in accordance with the diagram in
Fig. 8. In this case for a microwave power input of 500 mW the
intensity of the parametric peaks is 50 times the intensity of the
thermal peaks. If, as usual (19), one assumes that the excited
magnons are monochromatic to the order of 1 MHz, this experiment
implies that the number of excited parametric magnons exceeds the
number of thermal magnons by a factor 25,000. Similar results for
z-magnons were obtained at a pumping frequency ν_p= 35.4 GHz.
Intense parametric peaks corresponding to x magnons were observed
only for pumping frequency ν_p= 51.02 GHz. All these satellites
satisfied the polarization condition $E_i \perp E_s$, which showed that
the particles observed were magnons. Investigating the dependence
of the satellite intensity on microwave pumping power (see Fig.
10) we obtained the values of critical fields h_c and calculated
the relaxation rates of the magnons under consideration. It was
found that the relaxation time is of the order of 1 microsecond.
This coincides with the values obtained for $MnCO_3$ and $CsMnF_3$
from microwave measurements (19).

     There is a gap equal to 23.8 GHz in the spectrum of y-magnons
as shown in Fig. 5. As a result for k = 2 x 5 10^5 cm^{-1} the
frequency of y-magnons becomes 29.3 GHz even in zero magnetic
field. As a result it was impossible to observe the parametric
excitation of y-magnons using the two microwave generators we had
(ν_{1p}= 35.4 GHz and ν_{2p}= 51 GHz).

     As follows from the data of Section 2, the minimum frequency
of x-magnons with k = 2.5 10^5 cm^{-1} (for H = 0) is 18.1 GHz.
Accordingly, in experiments with a pumping frequency of 35.4 GHz
no satellites would be expected to appear at half this frequency
in scattering by quasiparticles. Nevertheless at large microwave
power such satellites were observed. Their distinguishing feature
was that they were observed over a fairly broad range of weak
magnetic fields (0-500 Oe), as shown in Fig. 11. The polarization
of light in these peaks differed from the preceding cases: the
scattered light contained components both parallel and
perpendicular to the wave vector of the incident light. In
addition, as is evident from Fig. 11, the frequency of the peaks
was equal to half of the pump frequency and was independent of the
magnetic field. Thus, the appearance of these peaks is not a
consequence of light scattering from parametrically excited
magnons. To explain these observations, it must be assumed that
microwave power input leads, as a result of magnetoelastic
interactions, to parametric pumping of phonons with frequency
close to half the microwave frequency. The frequency of phonons
propagating along the x axis at T ~ 100K is 1 GHz lower (see the
dashed line Fig. 11) than the frequency of the observed peaks.
The phonon frequency may rise by this amount when the temperature

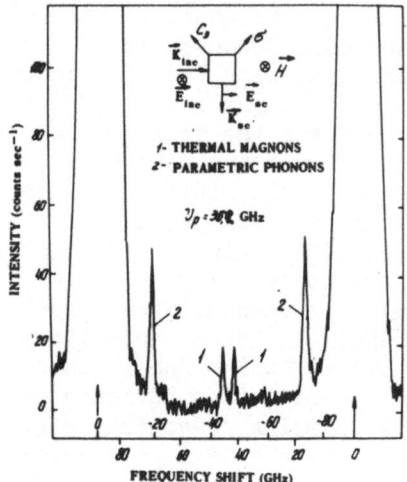

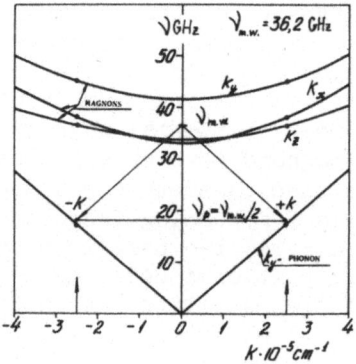

Figure 12. The spectrum of y-particles when AFMR is excited at ν_p = 36.2 GHz.

Figure 13. Parametric excitation of phonons at AFMR.

is reduced from 100K to 2K. Unfortunately, at helium temperatures the thermal phonons cannot be observed directly. It might be thought that the most intense phonon excitation should occur at the point of intersection of the magnon and phonon branches, i.e., at H = 0. However, the sample then divides into domains, and the magnitude of the real field in the domain depends on its size and shape. In any case, when the field is increased above 500 Oe, the magnetoelastic interaction rapidly diminishes and no excitation of phonons occurs. It should be emphasized that for microwave pumping at frequency ν_p= 51 GHz, the peaks at half the frequency with k parallel to the x axis were observed only in a narrow range (±10 Oe) of magnetic fields which corresponded to magnons.

New results were observed by investigation of BMS from a sample in which AFMR was excited at frequency ν = 36.2 GHz. Fig. 12 shows the spectrum of light scattered by y-particles when AFMR is excited at a specially fixed value of magnetic field (21). The spectrum contains, besides the peaks of thermal y-magnons with corresponding frequency ν = 44.4 GHz, additional peaks at the frequency ν = ν (AFMR)/2 = 18.1 GHz. The polarization conditions for these peaks correspond to those for transverse phonons. Their frequency is close to the frequency of one of the transverse phonons, which at room temperature was 16.9 GHz. To explain all these facts, it is reasonable to assume that the additional peaks correspond to phonons which appear through a parametric decay of an AFMR magnon excited with k = 0, to two phonons with half of the magnon frequency and k = ±2.5x10^5 cm^{-1}. Such decay is

schematically shown on Fig. 13. Similar results were observed by
Jantz and Wettling in $FeBO_3$ (22) for phonons propagating along
the z axis. The small intensity of the peaks we have observed in
comparison with those extremely strong peaks in (22) can be
explained if we assume that in our experiments not all the
conditions shown on Fig. 14 are strictly fulfilled.

REFERENCES
1. P.A. Fleury, S.P.S. Porto, L.E. Cheesman, H.J. Guggenheim,
   Phys. Rev. Lett., 17, 84, 1966.
2. P.A. Fleury, S.P.S. Porto, R. Loudon, Phys. Rev. Lett., 48,
   658, 1967.
3. P.A. Fleury, R. Louden, Phys. Rev. 166, 514, 1968.
4. R.J. Elliot, A.J. Smith, J. Phys. Paris 32, Suppl. C1, 585.
5. R. Louden, J. Phys. C. Solid St. Phys., 3, 872, 1970.
6. J.R. Sandercock, W. Wettling, Solid State Comm., 13, 1729,
   1973.
7. J.R. Sandercock, Solid State Comm. 15, 1715, 1974.
8. J.R. Sandercock, Proc. 2nd Int. Conf. on Light Scattering in
   Solids, Ed. by Balkanski (Paris, Flammarion) p. 1-12, 1971.
9. W. Jantz, J.R. Sandercock, W. Wettling, J. Phys. C. 9, 2229,
   1976.
10. W. Jantz, W. Wettling, Appl. Phys. 15, 399, 1978.
11. A.S. Borovik-Romanov, V.G. Jotikov, N.M. Kreines, A.A. Pankov,
    JETP Lett., 24, 207, 1976. Physica 86-88B, 1275, 1977.
12. A.S. Borovik-Romanov, V.G. Jotikov, N.M. Kreines, Sov. Phys.
    JETP 47, 1188, 1978.
13. V.I. Ozhogin, Sov. Phys. JETP 21, 874, 1965.
14. V.G. Barjachtar, M.A. Savchenko, V.V. Tarasenko, Sov. Phys.
    JETP 22, 1115, 1965.
15. T.M. Holden, E.C. Svenson, P. Martel, Can. J. Phys., 50, 687,
    1972.
16. V.V. Eremenko, A.P. Mokhir, Yu. A. Popkov, N.A. Sergienko,
    V.I. Fomin, Sov. Phys. JETP 46, 1231, 1977.
17. A.S. Borovik-Romanov, V.G. Jotikov, N.M. Kreines, A.A. Pankov,
    JETP Lett., 23, 649, 1976.
18. A.S. Borovik-Romanov, V.G. Jotikov, N.M. Kreines, A.A. Pankov,
    Sov. Phys. JETP 43, 1002, 1976.
19. A.S. Borovik-Romanov, L.A. Prozorova, Contemp. Phys., 19, 311,
    1978.
20. V.G. Jotikov, N.M. Kreines, JETP Lett., 26, 360, 1977.
21. V.G. Jotikov, N.M. Kreines, Abstracts of XX-th LT Conference
    (USSR) Vol. 2, p. 36, Chernogolovka, 1978.
22. W. Wettling, W. Jantz, C.E. Patton, Preprint 1979.

OBSERVATION OF PURE SPIN DIFFUSION <u>WITHOUT CHARGE TRANSPORT</u>

BY SPIN FLIP RAMAN SCATTERING

S. Geschwind, R. Romestain[*+], G. Devlin and
R. Feigenblatt[++]
Bell Laboratories, Murray Hill, N.J., [*]CNRS, Grenoble,
France, [†]M.I.T., Cambridge, Massachusetts

INTRODUCTION

In spin flip Raman scattering (SFRS) incident light of wave-vector $\vec{k}_i$ and frequency ω_i interacts via spin orbit coupling with spins in an external magnetic field H_0, producing a spin flip which scatters the light to wavevector $\vec{k}_s$ and frequency ω_s, where $\omega_s - \omega_i = \pm g\mu_\beta H_0/\hbar$. The ± signs refer respectively to the Stokes and anti-Stokes components and correspond to oppositely directed spin reversals. This process was first suggested by Yafet[1] following a treatment by Wolff[2] of Raman scattering from Landau levels in a semiconductor. SFRS was first observed for conduction electrons in InSb[3] where the very large g-value (~ 50) led to tuneable spin flip Raman lasers.[4] It was first observed for bound donors in CdS by Thomas and Hopfield.[5] SFRS has also been applied in a number of experiments to the study of velocities[6] and diffusional motion of donor electrons in semiconductors.[7-9] The diffusional motion appears as a contribution Dq^2 to the SFRS linewidth. In these previous studies the Dq^2 term was related to the diffusion of charge, while in this paper we report the observation of spin diffusion arising from the exchange interaction between bound donors, <u>without any charge transport</u>.

REVIEW OF SFRS FOR CHARGE DIFFUSION

Wolff et al[10] have shown that in SFRS one essentially measures the (q,ω) component of the transverse spin susceptibility, $\chi^+(q,\omega)$. Using the Bloch equations with a spin diffusion term $-D_s\nabla^2 M^+$, where

+ Work performed at Bell Laboratories.

M^+ is the transverse spin magnetization, it was shown that the damping term or linewidth for SFRS is given by[10]

$$\Delta\omega = \frac{1}{T_2} + D_s q^2 \tag{1}$$

where T_2 is the spin lifetime and $D_s q^2$ is the contribution to the linewidth from the diffusive motion. The subscript "s" attached to D_S is to emphasize that it is a spin diffusion constant that one measures. For mobile charges in a semiconductor, D_S will of course be the same as the diffusion constant for electric charge D_C, if electron correlation and many body effects are neglected. In this case, D_C may be related to the electron mobility, μ, by the Einstein relation, $D = kT\mu/e$ for a classical electron gas and for a degenerate electron gas by $D = 2E_F\mu/3e$ where E_F is the Fermi energy. Wolff et al[10] analyzed the temperature dependence of the SFRS linewidth measured by Scott, Damen and Fleury[8] in a sample of CdS with a stated donor concentration of 5×10^{17} and attempted to relate the observed linewidth to the measured temperature variation of μ in thi range. The expression for D for a degenerate electron gas may be equivalently written[9] in terms of the resistivity, ρ, and the donor concentration, N, so that the diffusive part of the SFRS linewidth may be written as

$$Dq^2 = \frac{c}{\rho N^{1/3}} q^2 \tag{2}$$

where $c = \hbar^2 \cdot (3\pi^2)^{2/3}/[3m^*e^2]$. Eq. 2 has been verified for small values of q by Geschwind, Devlin and Romestain[9] for a range of concentrations, N, which spanned the region from the onset to well above the insulator to metal transition in n-CdS.

THE LINEAR k-TERM

Before proceeding to a description of the observation of spin diffusion for bound donors, we briefly review the effects of the linear k-term in SFRS in the metallic samples.

In polar noncentrosymmetric crystals, such as CdS, an additional term linear in the electron momentum operator, $\vec{p}$, appears in the Hamiltonian for conduction band electrons, i.e.,

$$H = \frac{\vec{p}^2}{2m^*} + \lambda\vec{p}\cdot(\vec{c}\times\vec{s}) \tag{3}$$

where $\vec{c}$ is a unit vector along the c-axis and $\vec{s}$ is the spin operator. It was demonstrated[11] that the effect of such a term upon the SFRS linewidth in the presence of diffusive motion of the electrons in the metallic samples, was to yield an asymmetry between the Stokes and anti-Stokes linewidths given by[12]

$$\Delta\omega_{Diff} = D(\vec{q}\pm\vec{q}_0)^2 \qquad (4)$$

with

$$\vec{q}_0 = \frac{\lambda m^*}{\hbar^2}(\vec{c}\times\vec{h}_0) \qquad (5)$$

where $\vec{h}_0$ is a unit vector along the direction of the external magnetic field, $\vec{H}_0$, and the + and - refer to opposite directions of spin flip. The asymmetry in linewidths is maximum when $\vec{c}$, $\vec{H}_0$ and $\vec{q}$ are all perpendicular to each other. The asymmetry in widths reverses with a reversal of the direction of the magnetic field as seen by Eqs. 4 and 5. This behavior was observed[11] in a number of metallic samples of CdS with different donor concentrations and D's, but all yielded the same value $\vec{q}$.

SPIN DIFFUSION WITHOUT CHARGE TRANSPORT

It was quite surprising at first to find the effects of the linear k-term and diffusion described above for metallic samples in a sample whose donor concentration was well below the critical concentration $N_c \sim 10^{18}$, for the insulator to metal transition, i.e., one in which electrons are bound in donor 1s states at low temperatures. The SFRS spectrum for such a sample with a Hall concentration

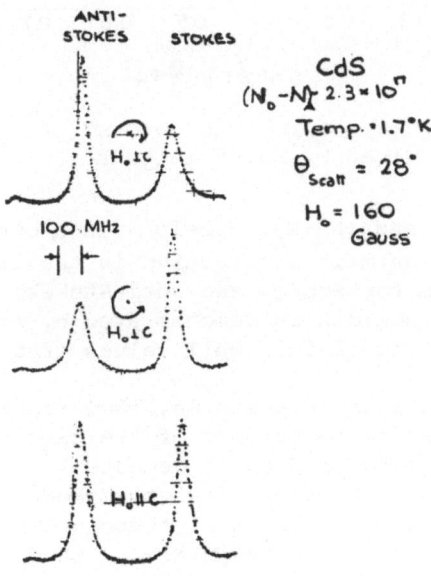

Fig. 1. Linear k-effect for bound donors

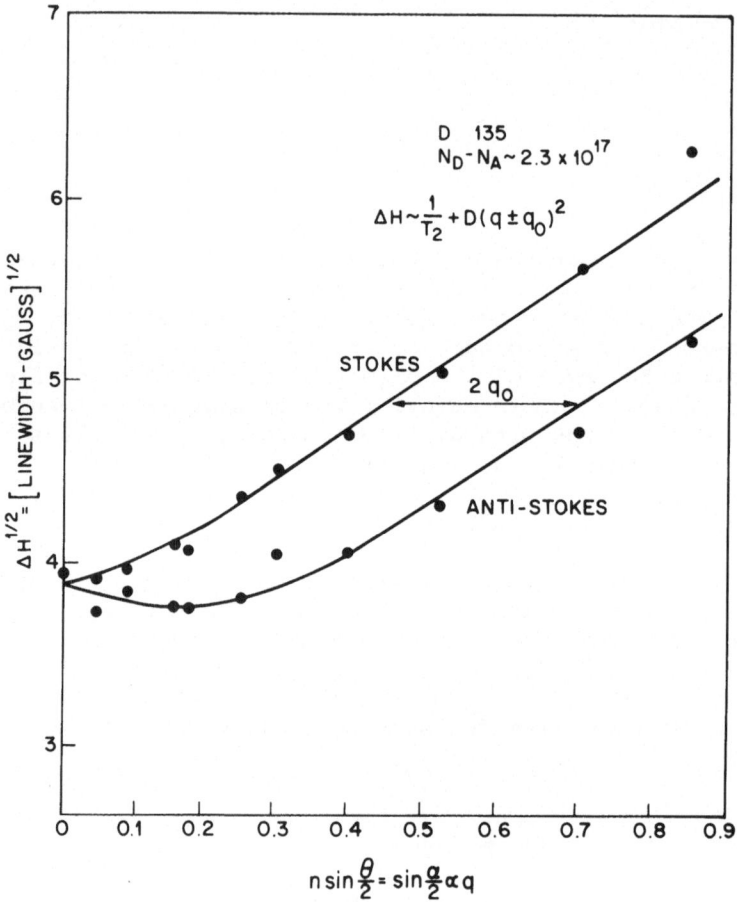

Fig. 2. $\sqrt{\Delta\omega}$ vs $\sin\theta/2 \sim q$ for bound donors. $\Delta\omega \sim 1/T_2 + D(\vec{q}\pm\vec{q}_0)^2$
where q_0 is given by Eq. 5 in text.

of $1.7\ 10^{17}$ at $295°K$ and $(N_D-N_A)\sim 2.3\times 10^{17}$ is shown in Fig. 1. For
$H_0 \parallel \vec{c}$ the linear k-term is ineffective in the linewidth and one
observes equal widths for Stokes and anti-Stokes. The diffusive
nature of the SFRS linewidth is demonstrated by its $(q\pm q_0)^2$
dependence, shown in Fig. 2 for small values of q (up to $\theta \sim 30°$).

The transport data on this sample, shown in Fig. 3, clearly
indicates low temperature freeze-out of the carriers. As a matter
of fact ρ changes by more than three decades from $10°K$ to $1.6°K$,
as shown in Fig. 3, whereas the linewidths shown in Fig. 2 remain
practically constant as a function of temperature in this range of
q. The Hall and resistivity data in Fig. 3 thus completely rule
out charge transport as the mechanism for the diffusion. Moreover,
we further ruled out the possibility that the SFRS was due to
carriers created by the light, as the intensity of the SFRS spectrum

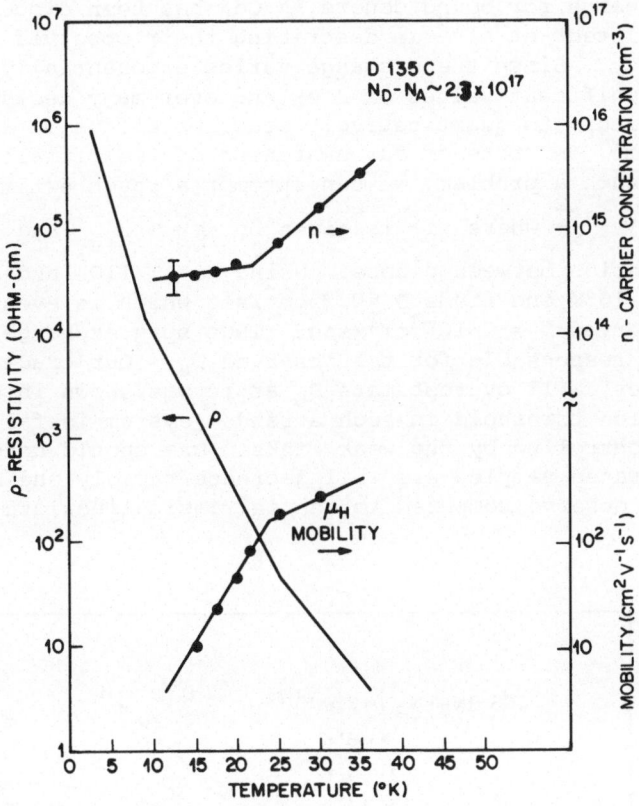

Fig. 3.   Transport data on sample used in Figs. 1 and 2 with room
temperature carrier concentration N = 2.3×10¹⁷, showing
low temperature freeze-out of carriers.

varied linearly with light intensity and gave the same linewidths
at 4965 Å and 4880 Å excitation. It is therefore suggested that the
observed spin diffusion arises from the transverse part, $J_{ij}s_i^+s_j^-$,
of the hydrogenic-like spin exchange between bound donors.

The value of $\vec{q}_0$ determined from Fig. 2 is the same as that
observed in the metallic samples although the value of D is almost
a factor of 100 smaller. Thus experimentally, except for the much
smaller value of D, the diffusive behavior for the bound donors,
including the k-linear term, appears to be similar to that observed
in the metallic samples. It will be shown elsewhere that to a good
approximation this is indeed to be expected. Physically, this is
related to the fact that for one spin direction the localized donor
wavefunction is expanded in plane waves about a conduction band
minimum weighted towards $+\vec{q}_0$, while for the opposite spin direction
it is expanded about $-\vec{q}_0$.

    The exchange for bound donors in CdS has been discussed in
detail by Walstedt et al[13] in describing their observed magnetic
susceptibility. Since the exchange varies exponentially with
distance, significant values of J extend over many decades and it
is very difficult to quantitatively describe diffusion in such a
random system. In spite of our awareness of the pitfalls of using
averages in such a problem, we can attempt a rough estimate of D
by saying $D \sim \dfrac{<a>^2}{<\tau>}$ where $<\tau>$ is given by $<\tau> \sim J_{<a>}$ and $<a>$ is a
median separation between donors. Using $<a> \sim 110\overset{\circ}{A}$ and a median
value $J_{<a>} \sim 10°K$ one finds $D \sim 0.25 cm^2/sec$ which is even larger
than the observed $D = 2\times10^{-3} cm^2/sec$. Thus spin exchange is large
enough to be responsible for the observed D_S. Our crude estimate,
using averages, will overestimate D_S as it was shown in Ref. 13 that
the percolation threshold in such a random system is far below $J_{<a>}$,
i.e. it is controlled by the weak links. One should note that in
less concentrated samples J_{ij} will decrease rapidly and Dq^2 may be
too small to observe compared to the intrinsic linewidth.

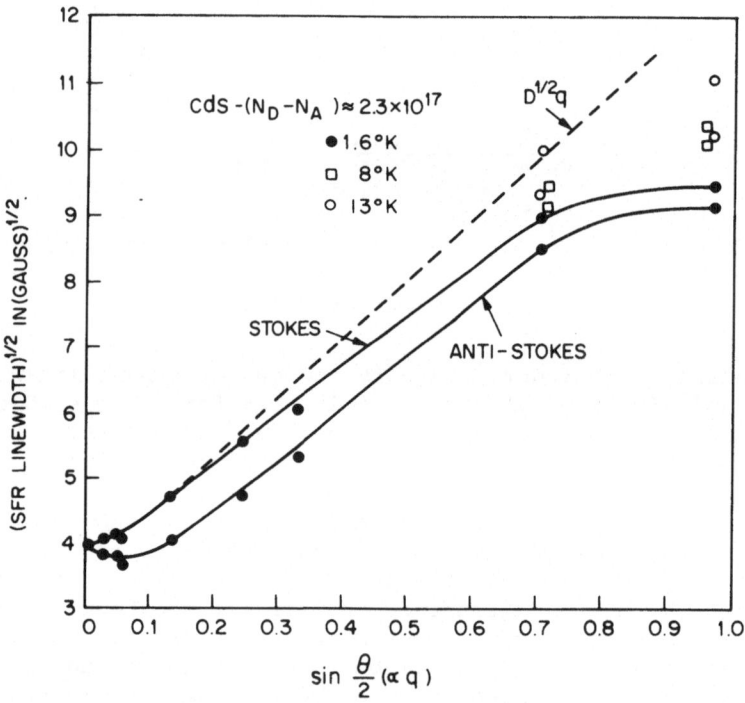

Fig. 4. Breakdown of $\Delta\omega \sim q^2$, i.e. of diffusive behavior at
         large q and recovery with increasing temperature.

CLUSTER MODEL OF SPIN DIFFUSION

Figure 4 extends the data of Fig. 2 to <u>larger</u> values of q where it is seen that $\Delta\omega$ no longer varies as q^2 and becomes almost independent of q at 1.6°K above $\theta \sim 90°$. This corresponds to a breakdown of diffusive behavior at distances $\lesssim 1000$ Å. (The wavelength of light at 4880 Å with n=3 in CdS is 1600 Å.) We suggest that this is connected with formation of spin clusters of roughly this size at this temperature.

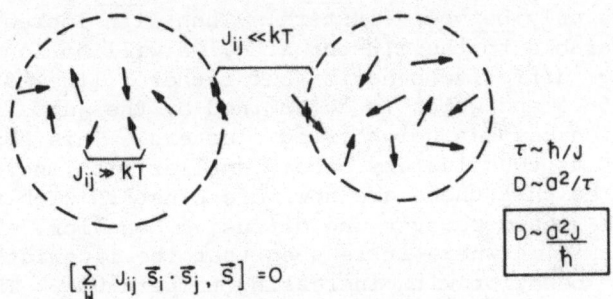

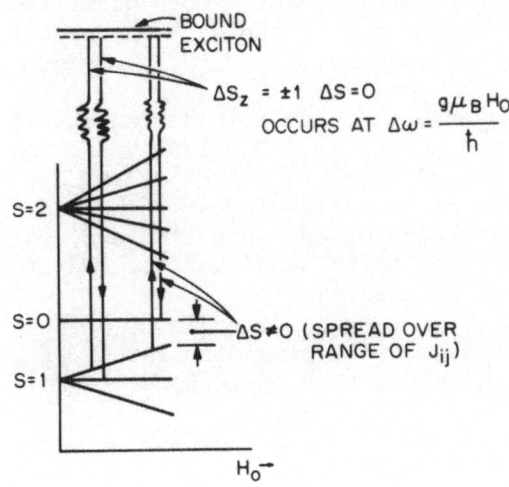

Fig. 5. Cluster model of spin diffusion in amorphous antiferromagnet. Cluster sizes decrease with increasing temperature.

The sample that Walstedt et al studied had a concentration $N \sim 8 \times 10^{16}$. For the sample discussed here with $N \sim 2.3 \times 10^{17}$ one would expect even larger values of J_{ij}, many of which are greater than kT. Thus we expect the spins with $J_{ij} > kT$ to be tightly coupled in clusters separated from other clusters by exchange links $J_{ij} < kT$ as shown in Fig. 5. To the extent that the clusters may be thus decoupled, each cluster will have a spectrum of eigenvalues with total cluster S and S_z as good quantum numbers. In SFRS within a cluster, two types of transitions will be observed, those at the Zeeman frequency $\Delta\omega = g\mu_\beta H_0/\hbar$ ($\Delta S_z = \pm 1$, $\Delta S = 0$) and those for which $\Delta S \neq 0$, which will be spread over the broad spectrum of J_{ij}'s and therefore difficult to see. The spin flip between the eigenstates of a cluster must be regarded as creating or destroying cluster excitations which are <u>not</u> diffusive <u>within</u> a cluster, but are diffusive only <u>between</u> clusters. Thus with increasing q as 1/q becomes comparable to the cluster size, we will no longer expect to see long range diffusive behavior but rather a linewidth that becomes independent of q and which is determined by the jump time of the spin excitation between neighboring clusters. This hypothesis suggests that as the clusters become smaller with increasing temperature, so that there are now more clusters within the 1/q distance, one should recover the diffusive behavior. This trend is displayed in Fig. 4 where it is seen that the linewidth at large q approaches q^2 behavior with increasing temperature. The probe wavelength (value of 1/q) at which diffuse behavior breaks down at 1.6°K is approximately 800 Å. A cluster of this size would contain approximately 50 spins which is consistent with Walstedt's computer calculations[12] on a more dilute sample. Computer calculations are in progress on this sample to study the cluster size distribution as a function of temperature and attempt a more quantitative connection with the observed recovery of the q^2 behavior with increasing temperature for large q.

In summary, we have observed pure spin diffusion without charge transport for bound donors in CdS. This diffusion is governed by spin exchange and seems to correspond to diffusion between clusters of spins of an intra-cluster spin flip excitation at the Zeeman energy. The size of the spin clusters, while not precisely defined, is related to kT such that all couplings J_{ij} within a cluster are greater than kT.

We wish to acknowledge many helpful discussions with P. W. Anderson, D. Fisher, E. O. Kane, L. R. Walker, R. E. Walstedt, P. A. Wolff and Y. Yafet.

REFERENCES

1. Y. Yafet, Phys. Rev. <u>152</u>, 855 (1966).
2. P. A. Wolff, Phys. Rev. Lett. <u>16</u>, 225 (1966).

3. R. E. Slusher, C. K. N. Patel and P. Fleury, Phys. Rev. Lett. 18, 77 (1967).

4. See Review by C. K. N. Patel, Laser Spectroscopy edited by R. G. Brewer and A. Mooradian (Plenum Publishing Co.). See also M. J. Colles and C. R. Pidgeon, Rep. Prog. Phys. 38, No. 3, 329 (1975).

5. D. G. Thomas and J. J. Hopfield, Phys. Rev. 175, 1021 (1968).

6. A. Mooradian, Phys. Rev. Lett. 20, 1102 (1968); see also D. C. Hamilton and A. L. McWhorter in Light Scattering in Solids, edited by G. B. Wright, Springer Verlag, N.Y. (1969).

7. S. R. J. Brueck, A. Mooradian and F. A. Blum, Phys. Rev. B 7, 5253 (197).

8. J. F. Scott, T. C. Damen, and P. A. Fleury, Phys Rev. 6, 3856 (1972).

9. S. Geschwind, R. Romestain and G. Devlin, Proceedings of 14th Intl. Conf. on The Physics of Semiconductors, Edinburgh, Scotland, Sept. 1978, Inst. of Physics (London) 1979, p. 1013.

10. P. A. Wolff, J. G. Ramos and S. Yuen in Theory of Light Scattering in Condensed Matter, edited by Bendow, Berman and Agranovich, Plenum Press, 1976.

11. R. Romestain, S. Geschwind and G. E. Devlin, Phys. Rev. Lett. 39, 1583 (1977). Also see this reference for further references on the linear k-term.

12. Eqs. 4 and 5 hold when collisions are sufficiently rapid so that the internal spin-orbit field associated with the linear k-term, which is along $\vec{k} \times \vec{c}$, is motionally averaged, resulting in a spin quantization which is along $\vec{H}_0$.

13. R. E. Walstedt, R. B. Kummer, S. Geschwind, V. Narayanamurti and G. E. Devlin, J. Appl. Phys. 50, 1700 (1979).

SPIN-FLIP SCATTERING FROM PHOTOEXCITED EXCITONS IN SiC

J. F. Scott and D. J. Toms

Department of Physics, University of Colorado

Boulder, Colorado 80309

and

W. J. Choyke

Westinghouse Research Laboratories

Pittsburgh, Pennsylvania 15235

INTRODUCTION:

Spin flip scattering of laser light from electrons and holes in semiconductors has been examined in a variety of III-V and II-VI compounds, both for application to the development of tunable infra-red lasers and for the study of semiconductor magneto-optics per se.[1] In the latter regard a surprising amount of information has been discovered concerning linewidths and lineshapes,[2] spin diffusion,[3] and phase matched processes.[4] Unexpected scattering processes have also been revealed, including multiple spin flip[5] and spin flip plus LO phonon emission.[6,7] The most recent investigations have emphasized gyromagnetic ratio (g-value) determination in p-type ZnTe. EPR techniques have met with limited success in p-type cubic semiconductors because of inhomogeneous strain, which broadens and splits valence band spin transitions. In p-ZnTe a variety of spin states have been measured via laser spin flip scattering. These include the free heavy hole states,[8] with characteristic g-value 0.92±0.15;[9] photoexcited conduction electron[10] (or related shallow donor) states with g-values 0.39±0.05; and several shallow acceptor levels, (P, As, Li, Na), with bound hole g-values around[11] 0.65.

A surprise afforded by spin flip measurements on p-type ZnTe is illustrated in Fig. 1 below: The spectrum illustrates both bound hole spin flip with g = 0.64 and electron spin flip with g = 0.39. In view of the fact that ZnTe is almost impossible to prepare n-type, the strong electron spin flip was wholly unexpected. Note that the hole spin flip transitions are completely thermalized (no antiStokes spectra), whereas the electron transitions are not. The hot electron spin temperature in this series of spectra was about 100 K at a lattice temperature of 1.6-1.8K. Similar effects were initially reported[12] by Thomas and Hopfield for CdS, where both thermalized hole spin flip and unthermalized electron spin flip processes were observed in n-type specimens. This indicates that the thermalization rates have little to do with the characteristics associated with minority carriers, per se, but instead arise primarily from the strong spin-orbit coupling in the valence bands, and the concommitant thermalization of holes.

A second indication of surprisingly large carrier densities of photoexcited states in ZnTe excited below bandgap with low power cw illumination is afforded by the shallow acceptor electronic transitions. Two strong transitions are observed[13] in ZnTe:As at 169 and 259cm^{-1} which do not arise from the 1S ground state in the hydrogenic series characterizing holes at arsenic acceptors. The field dependences of these transitions have been measured up to H = 14 Tesla and allow the identification of the transitions to be

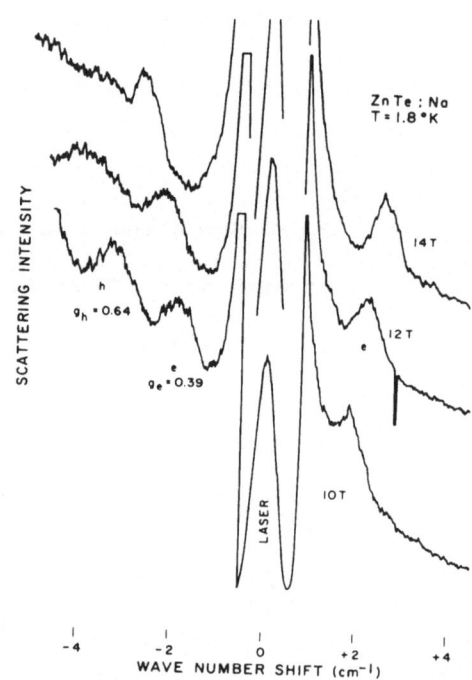

Fig. 1 Spin-flip spectra[10] of ZnTe:Na, showing transitions of photoexcited electrons (e) and of holes bound to As acceptors (h).

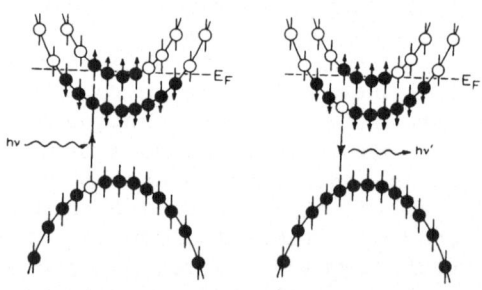

Fig. 2 Electron spin-flip process (schematic).[9]

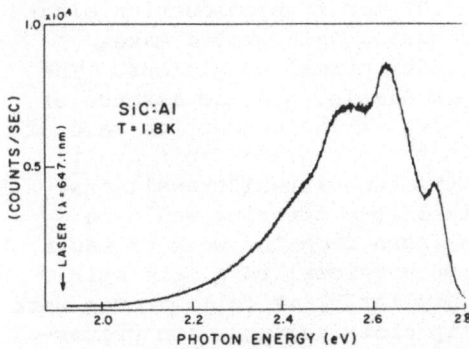

Fig. 3 SiC:Al luminescence with 647.1nm excitation.

made as 1S + LO → 2S and 2P (Γ_7-) → 2S + LO. These transitions are quite remarkable because they originate on levels which are as much as 500 cm^{-1} above the 1S ground state at lattice temperatures of 1.6 to 1.8 K. The thermal population of such states is negligible. And the transitions may be observed strongly with wave lengths of 632.8nm (He-Ne) or 647.1 nm (Kr). Perhaps a hint is given by the fact that one of the photoexcited states is a 1S + LO vibronic level; the role of hot LO phonons is well documented in other systems, such as GaAs.[14]

The studies summarized above suffice to show that very large, nonthermal populations of both electrons and holes may be achieved with relatively low (< 1 W) power cw lasers operating 1 eV or more below the bandgaps of semiconductors. The exact photoexcitation mechanisms are unknown and relatively unimportant for the present study, but presumably involve sequential excitation processes in which rather deep traps play a key role.

SILICON CARBIDE EXPERIMENTS:

We have examined a variety of n-type SiC over the past five years with no success. Our samples were generally nitrogen doped and of excellent optical quality, but their spin flip spectra were so weak as to be obscured by the few counts per second of noise in the detection system, even for carrier concentrations as large as 10^{19} cm^{-3}. We believe that these negative results are reasonable, but important. As in the Sherlock Holmes' mystery, the clue is that the dog didn't bark. Or in our case, the extreme weakness of the spin flip from electrons in SiC contains some useful information: Electron spin flip cross sections vary as $(2 - g)^2$. For SiC the

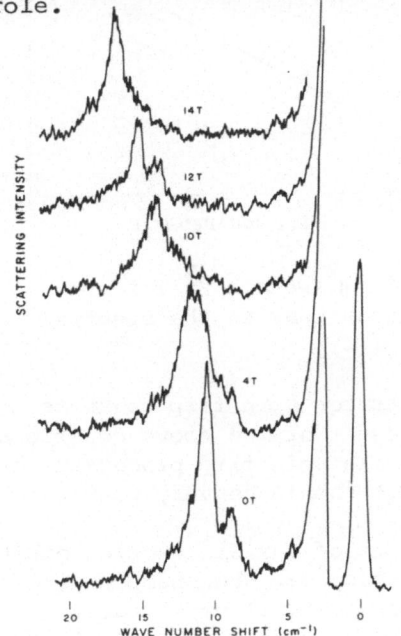

Fig. 4 Exciton spin flip spectra in 6H SiC:Al. 647.1nm excitation. H = 0, 4, 10, 12, and 14 Tesla.

electron g-value varies from 1.97 to 2.01 for free conduction electrons and a wide variety of shallow donors. This should make
electron spin flip unmeasurably weak under normal conditions. The
physical origin of the $(2 - g)^2$ term is simple. In the absence of
spin-orbit coupling, the electron g-value in the conduction band of
any crystal would be 2. Similarly, in the absence of spin orbit
coupling in the valence band, one of the two virtual transitions
diagrammed in Fig. 2 for electron spin flip scattering would be
$\Delta S = 1$ and strictly forbidden. It is known from the work of Laura
Roth[15] that $(2 - g)$ is approximately proportional to Δ, the spin
orbit splitting in the valence band; similarly, it follows from **work**
of Thomas and Hopfield[12] that spin flip cross sections are proportional to Δ^2. Thus, I (spin flip) is proportional to $(2 - g)^2$. For
$\Delta \lesssim E_G$, the intensity I is given by

$$I \cong I_0 \left[\frac{2E_G^3 \Delta}{(E_G^2 - \hbar^2 \omega_0^2)^2} \right]^2 \cong I_0 \ (2-g)^2 (\frac{m_e}{M^*} - 1)^{-2} (1 - \hbar^2 \omega_0^2 / E_G^2)^{-2} \tag{1}$$

where E_G is the band gap; ω_0, the
laser frequency; m_e and M^* the free
and effective masses. It is also
known from the work of Thomas and
Hopfield[12] that hole spin-flip is
very weak and varies roughly as H^2.
This occurs because for hole spin
flip the important intermediate
states are those in the S-like
conduction band, and non-zero cross
section arises from differences in
resonance denominators of form
$(E_g - \hbar \omega_0 + \mu gH/2)$ and
$(E_g - \hbar \omega_0 - \mu gh/2)$; the resulting
matrix elements vary as
$\mu gH/(E_g - \hbar \omega_0)^2$. This is not true
of electron spin flip because of
spin orbit splitting Δ in the
valence band; Δ plays the same role

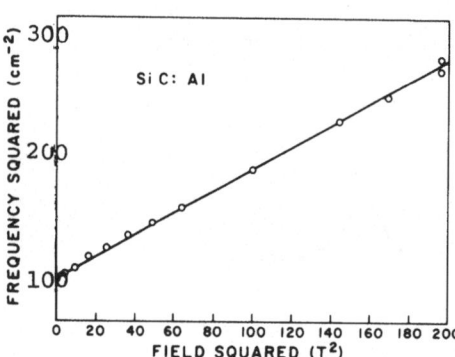

Fig. 5 ω^2 versus H^2 for the
strongest peak in the spectra
of Fig. 4.

in electron spin flip cross sections as H does for hole spin flip.
The result of the above considerations is that both free electron and
free hole spin flip processes (and those of shallow donor or acceptor
states) should be very weak in SiC.

One of our SiC samples exhibits abnormally intense antiStokes
luminescence: When pumped with below bandgap light (for wavelengths
as long as 647.1 nm) it emits very intense luminescence in the
violet and blue-violet. This sample is Al-doped to a few times
10^{18} cm^{-3}. It has polytype 6H and also is thought to contain a
density of nitrogen donors only slightly less than its aluminum concentration. Hopfield calculated in 1964 the criteria for ionized

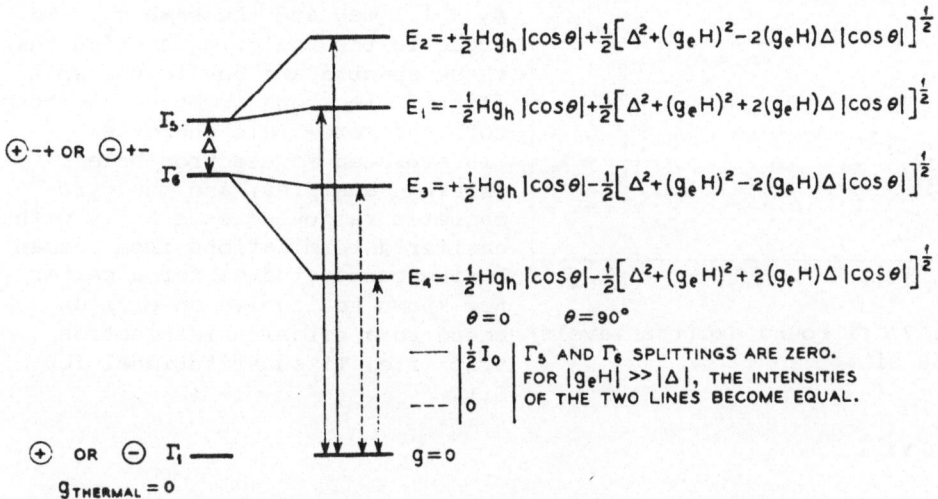

Fig. 6 Exciton levels and field dependences for ionized traps (donors, acceptors, or isoelectronic impurities) at C_{3v} sites.[16]

donors and acceptors to bind excitons stably; based upon his work and subsequent experiments, it is an accepted fact that excitons can bind to neither Al acceptors nor N donors, if these impurities are charged. The exciton spin flip scattering to be described in the present work shows by its field dependence that it does not arise from excitons bound to neutral impurities. Thus, neither Al nor N provides a binding site for the excitons studied here. It is probable, however, that both are necessary for the as-yet to be un-ravelled photoexcitation mechanism that gives rise to such intense up-conversion of red light to violet.

A typical luminescence spectrum is illustrated in Fig. 3. The highest energy transition corresponds to the 3 eV band gap minus the known Al acceptor binding energy of 0.27 eV. The lower energy peaks are thought to be due entirely to N-Al donor-acceptor pair recombination, involving primarily distant pairs (the energy spacing of the main peaks in this luminescence spectrum are very close to the LO energy at the Brillouin zone center, however, a fact that may not be entirely fortuitous).

Raman scattering in this sample was very strong. Data were obtained at seven wavelengths from 647.1 nm to 496.5 nm. Representative spectra are shown in Fig. 4 at fields of 0, 4, 10, 12, and 14 Tesla. Fig. 5 shows that the intense feature in each trace satisfies the equation given below:

$$\hbar\omega = \left[\Delta_0^2 + (\mu_B g_e H)^2 \right]^{\frac{1}{2}} \tag{2}$$

with $\Delta_0 = 1.3$ meV and $g_e = 1.97 \pm 0.02$. The weaker feature yields

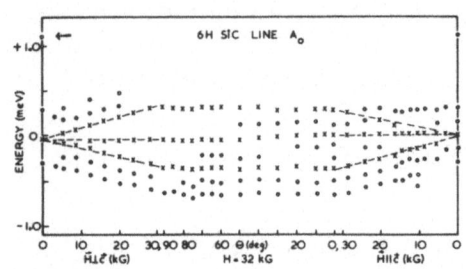

Fig. 7 Ti bound exciton levels[17] in 6H SiC.

$\Delta_0 = 1.1$ meV and the same g_e. We argue in the following section that these spectra are due to the spin flip of electrons in bound excitons. Both the zero-field energies (interpreted as electron-hole exchange energies) and the gyromagnetic ratios agree exactly with earlier determinations from Zeeman luminescence studies for a center now known to involve an exciton bound to a titanium isoelectron trap (i.e. Ti substitutional for Si).

THEORY:

The requisite exciton theory for excitons bound to isoelectronic traps in 6H polytype SiC is exactly the same as that developed by Thomas and Hopfield[16] for CdS, since the impurity sites are of C_{3v} point group in each case. For 6H SiC the only additional complication is that there exist three crystallographically inequivalent sites for the impurities. This is not a serious complication, because our data indicate that only one of these three sites has trapped measurably large numbers of photoexcited excitons.

The exciton levels and transitions corresponding to isoelectronic traps at C_{3v} sites are shown in Fig. 6 below, taken directly from Thomas and Hopfield. For arbitrary geometries the dependences of these levels upon magnetic field are nonlinear and quite complicated. In zero field the two exciton levels are labelled as Γ_5 and Γ_6 and are split by an electron-hole exchange energy Δ_0. For the Ti trap this exchange energy has been measured in the luminescence study by Dean and Hartman[17] and is exactly 1.1 meV, as shown in Fig. 7. For finite values of magnetic field H the exciton levels split in general into a quartet, and increase in separation in a highly non-linear way. However, the geometry in our studies was constrained by the fact that our specimen was a very thin (0.1 mm) platelet, with faces perpendicular to the C_6 axis. In order to collect the scattered light efficiently we were forced to use a geometry with the C_6 axis exactly perpendicular to the applied magnetic field (i.e. light was collected from the large faces of the platelet). For this geometry the formulas in Fig.6 simplify considerably; there is no splitting produced by the field, and the separation of the Γ_5 and Γ_6 levels is given by Eq.(2) above.

Thus, from the theory of Thomas and Hopfield, a single Raman transition from Γ_5 to Γ_6 would be expected, with the field dependence given in Eq. (2). This does not agree perfectly with

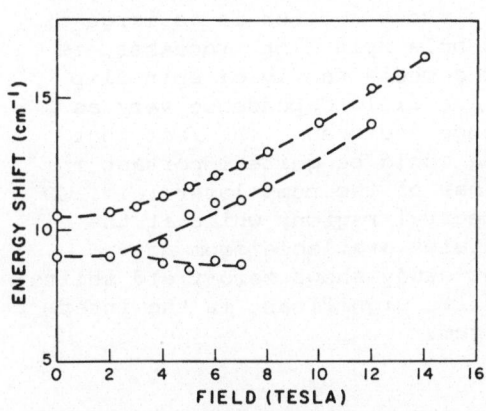

Fig. 8 ω (H) for exciton spin
flip transitions in SiC.

our data, however, since as shown
in Fig. 8, two transitions are
observed, one of which splits into
a doublet with applied H. This
result follows from the theory of
Patrick,[18] who pointed out that
the spin-orbit splitting in SiC
is extremely small in comparison
with that in CdS. The spin-orbit
splitting of traps such as Ti is
only about 0.1 meV, according to
Patrick, and is substantially
reduced from the already small
value of 5 meV known for the
valence band states. In Patrick's
analysis the shallowest of the
three Ti traps will yield two
transitions in zero field, and one
of these two levels will split into
a doublet for H ≠ 0. This is precisely what is shown in Fig. 8.
The splitting of 0.2 meV between our two zero-field transitions
agrees very well with Patrick's estimate, and agrees almost
perfectly with the unidentified "satellite" spectral splitting
measured by Dean and Hartman for this SiC:Ti state (they measure
0.3 meV in zero field).

These considerations show that our spectra are unquestionably
due to electron spin flip from excitons bound to isoelectronic traps
in SiC, and in addition, that the trap is very probably the well
studied Ti substitutional for Si.

FURTHER WORK:

There are some obvious and important implications from our work
to date. Firstly, one can and should study this exciton spin flip
process as a function of excitation power. Changes in the spin
flip spectra should occur as biexcitons are formed. SiC is one of
only a few semiconductors in which electron-hole droplets have been
reported.[19] Thus, spin flip scattering may provide a new probe of
such phenomena, and a probe which is far more sensitive than other
kinds of Raman spectroscopy. Secondly, all of these states --
exciton, biexciton, and electron-hole droplets -- should be
amenable to time-resolved studies. We hope to employ pulsed
nitrogen lasers as the excitation source and to probe the exciton
states thereby photoexcited with weak Kr cw sources. The temporal
kinetics of photoexcited excitons should be measurable via this
technique.

Finally, although serendipity has played an important and
readily admitted role in our work, there is no reason why exciton

spin flip scattering should not be observable in a wide variety of
semiconductors. Similarly, since its cross section is as large or
larger than that of bare electron or hole spin flip processes, it
may well provide a new mechanism for a whole family of spin flip
lasers, in which both cross section and field dependence vary as
functions of the electron-hole exchange energies. The fact that
the zero-field energy shift is finite could be quite important in
shifting the output frequency from that of the pump laser (e.g. CO
or CO_2, HF or DF) to an important spectral region, while at the
same time employing small magnetic fields available from non-
superconducting magnets. The present study shows zero-field shifts
of 10.5 cm^{-1}; this small number is quite significant in the intermediate infrared region of the spectrum.

REFERENCES:

1. J. F. Scott, Physics of Quantum Electronics, edited by
   S. F. Jacobs, M. O. Scully, M. Sargent, and J. F. Scott,
   Addison-Wesley, Reading, Mass. Vol. 2, p 123 (1975), Vol. 4,
   p 325 (1976).
2. J. F. Scott, T. C. Damen, and P. A. Fleury, Phys. Rev. B6,
   3856 (1972).
3. S. Geschwind, this conference, preceding article.
4. R. Romestain, S. Geschwind, G. E. Devlin and P. A. Wolff,
   Phys. Rev. Lett. 33, 10 (1974).
5. J. F. Scott and T. C. Damen, Phys. Rev. Lett. 29, 107 (1972)
6. R. L. Hollis, J. F. Ryan, and J. F. Scott, Phys. Rev. Lett.
   33, 209 (1975).
7. J. F. Scott, Phys. Rev. B (in press, 1979).
8. R. L. Hollis and J. F. Scott, Phys. Rev. B15, 942 (1977).
9. R. L. Hollis, Phys. Rev. B15, 932 (1977); Ph.D. thesis,
   University of Colorado 1975.
10. D. J. Toms, J. F. Scott, and S. Nakashima, Phys. Rev.
    B19, 928 (1979).
11. D. J. Toms, C. A. Helms, J. F. Scott, and S. Nakashima,
    Phys. Rev. B18, 871 (1978).
12. D. G. Thomas and J. J. Hopfield, Phys. Rev. 175, 1021
    (1968).
13. J. F. Scott, F. Habbal, J. H. Nicola, D. J. Toms, and S.
    Nakashima, Phys. Rev. B (in press, March 1979).
14. J. Shah, R.C.C. Leite and J. F. Scott, Sol. St. Comm. 8,
    1089 (1970).
15. L. M. Roth, B. Lax and S. Zwerdling, Phys. Rev. 114, 90
    (1959).
16. D. G. Thomas and J. J. Hopfield, Phys. Rev. 128, 2135 (1962).
17. P. J. Dean and R. L. Hartman, Phys. Rev. B5, 4911 (1972).
18. Lyle Patrick, Phys. Rev. B7, 1719 (1973).
19. D. Bimberg, M. S. Skolnick and W. J. Choyke, Phys. Rev.
    Lett. 40, 56 (1978).

THE SCATTERING OF LIGHT BY SPIN WAVES ON FERROMAGNETIC SURFACES

R. E. Camley and D. L. Mills

Department of Physics
University of California
Irvine, California 92717

ABSTRACT

This paper reviews results of recent experimental studies of Brillouin scattering of light by bulk and surface spin waves on the surface of ferromagnets, and compares features in the data with the results of our theoretical analysis of this phenomenon.

INTRODUCTION

There is by now a rather extensive theoretical literature[1] on the influence of a surface on the properties of Heisenberg magnets. At the same time, the experimental data available is both sketchy and difficult to interpret unambiguously. The data does hint at rather intriguing magnetic anomalies on the surface, however.[2]

Quite recently, there has appeared a sequence of very beautiful experimental studies of the Brillouin scattering of light by spin waves on ferromagnetic surfaces.[2-5] The experiments are done in a backscattering geometry, so the light samples only the spin fluctuations within the optical skin depth δ. The spectra reported to date, carried out on thin films[5] as well as bulk samples[3,4], show features from scattering off bulk spin waves and also from a certain intriguing surface spin wave frequently referred to as the Damon-Eshbach surface spin wave mode.

This experimental method offers our first detailed glimpse at spin dynamics near the surface, and has the potential of addressing issues raised in the earlier theoretical literature. While light is not a microscopic probe of the surface, for the ferromagnetic metals the optical skin depth is roughly 150 Å. Thus, in this case, the backscattered light contains information about the outermost fifty atomic layers.

Also, in recent work, the light scattering method has been applied to the study of standing spin waves in thin films.[5] In essence, through the Brillouin technique one is performing a ferromagnetic resonance experiment, with light rather than microwaves as the exciting source. A comparison between the light scattering studies and conventional ferromagnetic resonance probes[6] of thin films shows the Brillouin technique has the potential of providing far more detailed information than can be obtained by the older method. We discuss this below.

This paper, which is necessarily brief, discusses the principal concepts required to understand qualitative features in the data, and reviews the experimental data available at this writing. We refer the reader to our full papers on this topic,[7] where the theory of backscattering of light from spin waves is developed in quantitative detail.

GENERAL CONCEPTS AND THE EXPERIMENTS

In what follows, we consider a semi-infinite ferromagnet with magnetization $\vec{M}_s$ parallel to the surface. This is the standard configuration realized in a slab geometry, unless an external field re-orients $\vec{M}_s$. We suppose $\vec{M}_s$ parallel to $\hat{z}$, the surface lies in the x z plane with the sample in the half space y < o.

One has the following picture at long wavelengths, where the influence of exchange coupling between the spins can be ignored and the spin wave frequencies are influenced only by the external magnetic field H_o along with demagnetizing fields set up by the precessing spins. If γ is the gyromagnetic ratio and $B = H_o + 4\pi M_s$, the frequency Ω_B of a bulk spin wave can vary from $\Omega_m = \gamma H_o$ to $\Omega_M = \gamma(H_o B)^{1/2}$ depending on the angle $\theta_{\vec{k}}$ between its wave vector $\vec{k}$ and $\vec{M}_s$. If $\theta_{\vec{k}} = 0$, then $\Omega_B = \gamma H_o$, while $\Omega_B = \gamma(H_o B)^{1/2}$ for $\theta_{\vec{k}} = \pi/2$. The situation is quite analogous to optical phonons in a polar crystal with symmetry lower than cubic. At long wavelengths, the frequency of an infrared active optical mode is independent of the magnitude of the wave vector, but does depend on its direction relative to the principal axes. In our case, the magnetization does select out the $\hat{z}$ direction as special, so the magnetic response bears a similarity to lattice dynamics in a uniaxial crystal.

In the long wavelength limit, an intriguing surface spin wave (the Damon-Eshbach (DE) wave) can propagate on the surface.[8] At long wavelengths, its frequency is again independent of the magnitude of its wave vector, but dependent on propagation direction. Let θ be the angle between the wave vector $\vec{k}_\parallel$ of the DE spin wave and the postive x axis, which one should recall is perpendicular to the magnetization $\vec{M}_s$. The Damon-Eshbach mode can then propagate only for a limited range of angles $-\theta_c \leq \theta \leq +\theta_c$, where $\theta_c = \cos^{-1}(H_o/B)^{1/2}$. Thus, we have a wave that can propagate from left to right across the magnetization, but not from right to left! For a given angle θ within the allowed range, the frequency $\Omega_s(\vec{k}_\parallel)$ of the mode is

$$\Omega_s(\vec{k}_\parallel) = \frac{1}{2}\gamma(\frac{H_o}{\cos\theta} + B\cos\theta) \quad . \tag{1}$$

Thus for $\theta = 0$ (propagation perpendicular to $\vec{M}_s$) one has $\Omega_s = \gamma(H_o + B)/2 > \gamma(H_o B)^{1/2}$ so the surface spin wave has frequency <u>higher</u> than any bulk spin wave. As θ approaches the critical angle θ_c, then $\Omega_s(\vec{k}_\parallel)$ drops, to merge with the frequency $\gamma(H_o B)^{1/2}$, which is the maximum frequency allowed for bulk waves.

When the influence of exchange is included in the theory, the frequency of the bulk spin waves is given by the dispersion relation $\Omega_B(\vec{k}) = \gamma(H_o + Dk^2)^{1/2} (H_o + 4\pi M_s \sin^2\theta_k + Dk^2)^{1/2}$. In effect, the Zeeman field H_o is augmented by the exchange contribution Dk^2, where D is the exchange stiffness parameter of the material. Now with the exchange contribution present, bulk spin waves can become degenerate with the DE surface spin wave. In this circumstance, the DE wave necessarily becomes a "leaky" surface wave, with finite lifetime because energy density localized in spin precession near the surface at the DE wave frequency may be radiated into the bulk of the material, with bulk spin waves carrying off the energy. The situation is quite analogous to that encountered in the theory of surface exciton-polaritons, where Maradudin and one of the present authors pointed out that in the presence of spatial dispersion effects, the surface polariton becomes a "leaky" surface wave in precisely the same fashion.[9]

The remarks above provide background for a description of the spectra reported to date. We begin with general comments, then turn to comments on specific studies.

Each spectrum consists of a feature from the DE spin wave, and bulk spin wave lines. In a backscattering geometry, the wave vector $\vec{k}_\parallel$ of the DE wave has the value $(\omega_o/c)[\sin\theta_I - \sin\theta_s] = 2(\omega_o/c)\sin\theta_I$, for the geometries used to date, which examine photons backscattered along the direction of the incident photons. Here θ_I and θ_s are the angles between the respective photons, and the normal to the crystal surface and ω_o is the frequency of the incident light. The most striking feature in the spectra has its origin in the highly nonreciprocal character of the DE wave dispersion relation, with propagation allowed for a single direction across the magnetization. If for a particular geometry, a Stokes event is allowed, then the anti-Stokes event is <u>forbidden</u>, simply because the anti-Stokes wave has wave vector opposite to the Stokes wave. Conversely, if anti-Stokes scattering is allowed, Stokes scattering is forbidden. Thus, the characteristic signature of light scattering from ferromagnetic surfaces is highly asymmetric spectra, with the DE wave missing totally from one side of the laser line.

The structure observed for scattering from bulk spin waves depends on the optical properties of the substrate, most particularly the skin depth δ. With δ finite, as discussed a number of years ago,[10] components of wave vector normal to the surface are not conserved in the scattering process, but the uncertainty δk_z in wave vector is the order of the inverse δ^{-1} of the skin depth. With this in mind, consider light scattering for the simple case where the change in wave vector $\vec{\Delta}_{\parallel}$ of the light parallel to the surface is <u>perpendicular</u> to the magnetization. This is the geometry used in all published spectra to date. Then conservation of wave vector parallel to the surface requires the wave vector of any bulk spin wave created in the scattering process to lie in the plane perpendicular to $\vec{M}_s$. Now if the product $D\delta^{-2} \ll 4\pi M_s$, as will be the case if the skin depth is large, the bulk spin waves give a <u>line</u> spectrum, with the center of the line at $\gamma(H_oB)^{1/2}$, the conventional (uniform-mode) resonance frequency of a thin film with magnetization parallel to the surface.[11] When $D\delta^{-2}$ is not small, this line becomes blurred out into an asymmetric feature that begins a $\gamma(H_oB)^{1/2}$, and has a tail that extends to high frequency. The origin of this tail is coupling to spin waves with $\vec{k}_{\parallel} = \vec{\Delta}_{\parallel}$, but with wave vector normal to surface large and comparable to δ^{-1}.

While the bulk spin waves appear on both the Stokes and anti-Stokes side of the laser line, in contrast to the DE wave, there is one striking feature of the spectra that owes its existence ultimately to the breakdown of time reversal symmetry by the presence of $\vec{M}_s$. Even though one has the inequality $\hbar\Omega \ll k_BT$ satisfied comfortably, the Stokes/anti-Stokes ratio can deviate from unity very considerably. The origin of the Stokes/anti-Stokes asymmetry, as pointed out first by Sandercock and Wetting,[12] is the following.

The light scatters from fluctuations $\delta\varepsilon_{\mu\nu}(\vec{x},t)$ in the dielectric tensor. If $S_\lambda(\vec{x},t)$ is a Cartesian component of spin density, then through terms quadratic in spin density we have

$$\delta\varepsilon_{\mu\nu}(\vec{x},t) = \sum_\lambda K_{\mu\nu\lambda}S_\lambda(\vec{x},t) + \sum_{\lambda\delta} G_{\mu\nu\lambda\delta}S_\lambda(\vec{x},t)S_\delta(\vec{x},t). \quad (2)$$

The quadratic terms contribute to the one magnon cross section, through terms where λ or δ refer to the $\hat{z}$ direction; then $S_\lambda(\vec{x},t)$ or $S_\delta(\vec{x},t)$ may be replaced by S, to leave terms which contribute to the one magnon cross section. The bulk spin wave Stokes/anti-Stokes asymmetry has its origin in the interference between the two terms in Eq. (2).

We refer the reader to our recent papers[7] for a more detailed discussion of the above points, and for a series of numerical calculations that illustrate them. We next turn to the experiments reported to date, which provide illustrations of the points made above. The materials examined include:

(a)   EuO:[3] The Curie temperature of this material is not high ($\sim$ 70K), and the skin depth δ is roughly 1500Å. One has $4\pi M_s \gg D\delta^{-2}$, and the bulk spin wave features in the spectra (taken with $\vec{\Delta}_\parallel$ perpendicular to M_s) are lines centered on $\gamma(H_o B)^{1/2}$ with very large Stokes/anti-Stokes asymmetry.

Grünberg and Metawe[3] report a surprisingly large temperature variation in the frequency of the DE wave. Since thermal fluctuations are known to be enhanced near the surface of a Heisenberg ferromagnet,[13] one may inquire if this may produce a large temperature variation of the DE wave frequency from magnon-magnon interactions. A recent paper[14] explores this question to find the effect of magnon-magnon interactions too small to explain the EuO data, though in other materials one might see the effect, most particularily if $\vec{\Delta}_\parallel$ lies close in direction to the critical angle θ_c.

(b)   Fe and Ni:[4] Here the skin depth δ is only around 150Å at the Ag^+ laser frequency, while the exchange constant D is large by virtue of the high Curie temperatures and itinerant character of the magnetism in these materials. The bulk spin wave features are highly asymmetric, and one sees very clearly that the DE wave sits on top of the high frequency "exchange tail" responsible for the leaky character of the DE wave. In the case of Fe, in early spectra both the DE wave frequency and the bulk spin wave frequency could not be reconciled with the values computed from the bulk magnetization M_s of Fe,[4] although they could be fit by reducing M_s once the influence of exchange on the bulk spin waves was included in the fit.[7] Spectra on samples with surfaces free of oxide give mode positions consistent with the bulk value of M_s, however.[12] Thus, we see here that the spectra are sensitive to the magnetic environment of the surface region. In Ni, while the shape of the spectra are similar to Fe, the peak positions are not in accord with that expected. The reason for this is unclear at this time.

Our theory predicts[7] that if the direction of the wave
vector $\vec{\Delta}_\parallel$ of the DE wave approaches the critical angle
θ_c, then the lifetime of the DE wave should shorten,
since the "radiative leak" produced by coupling to bulk
spin wave increases. In the light scattering spectra
from Fe, we calculate an appreciable increase in linewidth
as θ_c is approached.[7] Evidently, recent data confirms
this expectation,[15] though it remains to place the
observed linewidths in quantitative contact with the
theory.

(c) Amorphous $Fe_{1-x}B_x$ and $Co_{1-x}B_x$ Films: Quite recently, detailed studies on amorphous films of Fe_xB_{1-x} and Co_xB_{1-x}
explore the effect of finite film thickness of the spectra
of spin waves excited by the Brillouin method.[5] The data
is rich in structure; one sees the DE surface wave, while
the broad asymmetric feature produced by scattering off
bulk spin waves breaks up into a sequence of peaks, with
each peak a standing spin wave excitation of the thin
film. In essence, one is performing an analogue of
ferromagnetic resonance here, with the spin waves excited
by light rather than microwaves. The light scattering
method has great flexibility. Among its virtues are:
(i) for each magnetic field, one samples the entire frequency spectrum of spin waves associated with a particular
wave vector $\vec{k}_\parallel$ parallel to the surface while the resonance
method examines power absorbed by one mode of a cavity
resonator. Thus, in the resonance method one must detect
the modes by sweeping them through the resonance frequency
of the cavity by varying the external magnetic field.
One obtains information on the magnetic response of the
material at only a single frequency. (ii) the direction
and magnitude of the wave vector of the spin waves excited may be readily altered by changing the scattering
geometry and (iii) phonons are excited along with spin
waves, so a single technique allows determination of the
elastic constants of the sample as well as its magnetic
parameters.

This concludes our brief summary of the physics that can be
explored through the scattering of light from spin waves on the
surface of a ferromagnet, and the experiments presently completed
on several systems. These are sufficient to illustrate the diverse
possibilities the method offers.

CONCLUDING REMARKS

The experiments to date offer us our first glimpse at spin
fluctuations near ferromagnetic surfaces, under conditions where

the geometry is semi-infinite. While the properties of the DE wave have been known in the literature of magnetism for many years now, we are unaware of any previous studies that explore the behavior of this mode in the simple semi-infinite geometry. It would be of great interest to see measurements on surfaces of single crystals prepared in ultra-high vacuum, and on such surfaces that have been oxidized or upon which adsorbates have been placed in a controlled manner. Also, the temperature and angular variation of the frequency of the DE wave may allow one to probe the spatial variation of the magnetization profile near the surface, since the penetration depth of the DE wave has a strong angle dependence, particularly near θ_c.[14]

The study of the spin wave spectra of thin films is also of great interest since, as remarked earlier, the light scattering method produces a large volume of detailed information on the magnetic response of the film not accessible by the more conventional microwave resonance methods.

We expect a lively future for this area of light scattering, once a larger number of the sophisticated multi-pass spectrometers required for this work become operational.

REFERENCES

1. See, for example, D. L. Mills and A. A. Maradudin, J. Phys. Chem. Solids 28, 1855 (1967), D. L. Mills, Phys. Rev. B 1, 264 (1970), K. Binder and P. C. Hohenberg, Phys. Rev. B9, 2194 (1974).

2. S. E. Trullinger and D. L. Mills, Solid State Communications 12, 819 (1973), C. Demangeat and D. L. Mills, Phys. Rev. B14, 4997 (1976), C. Demangeat, D. L. Mills and S. E. Trullinger Phys. Rev. 16, 52 (1977), D. Castiel, Surface Science 60, 24 (1976).

3. P. Grünberg and F. Metawe, Phys. Rev. Letters 39, 1561 (1977).

4. J. Sandercock and W. Wettling, I. E. E. E. Trans. Magn. 14, 442 (1978).

5. A. P. Malozemoff, M. Grimsditch, J. Aboaf and A. Brunsch (to be published), and M. Grimsditch, A. Malozemoff and A. Brunsch (to be published).

6. See, for example, J. T. Yu, R. A. Turk and P. E. Wigen, Phys. Rev. B11, 420 (1975).

7. R. E. Camley and D. L. Mills, Phys. Rev. B18, 4821 (1978) and Solid State Communications 28, 321 (1979).

8. R. W. Damon and J. R. Eshbach, J. Phys. Chem. Solids 19, 308 (1960).

9. A. A. Maradudin and D. L. Mills, B7, 2787 (1973).

10. D. L. Mills, A. A. Maradudin and E. Burstein, Annals of Physics (N. Y.) 56, 504 (1970).

11. See the discussion in Chapter 17 of C. Kittel, <u>Introduction</u>
      <u>to</u> <u>Solid</u> <u>State</u> <u>Physics</u> (4th Edition) (Wiley, New York, 1971).
12. J. Sandercock (private communication).
13. See the paper by Mills and Maradudin, and the paper by Binder
      and Hohenberg in Reference (1).
14. Talat S. Rahman and D. L. Mills, Phys. Rev. B (to be
      published).
15. This data has recently been reported by J. Sandercock, in paper
      DD4, Bull. Am. Phys. Soc. <u>24</u>, 296 (1979).

RELAXATION OF ENERGY AND POLARIZATION IN THE RESONANT

SECONDARY EMISSION SPECTRA OF SEMICONDUCTORS

A. Klochikhin, Ya. Morozenko, V. Travnikov,
and S. Permogorov

A. F. Ioffe Physical-Technical Institute
Leningrad, USSR

I. INTRODUCTION

Resonant secondary emissions of semiconducting crystals in the exciton energy region often show simultaneously properties which were formerly believed to be characteristic either of Raman scattering or of luminescence. Due to this fact much attention was given to the distinction between the different components of resonant secondary emissions, namely between Raman scattering and hot luminescence.[1-3]

On the other hand there exists a more general approach[4] in which all the properties of secondary emissions —— the spectrum, temporal behaviour, polarization degree, etc., —— can be explicitly calculated with the use of a model which properly describes the relaxation of the excited crystal between the absorption and emission of light. In principle, all the possible intermediate states and scattering processes should be taken into account. Nevertheless, in some cases the experimental results can be adequately described by a simpler model.

The nature of the intermediate states, virtual or real, cannot serve as a criterion for the distinction of Raman scattering from luminescence, since both types of secondary emission include these two kinds of intermediate states. It has been shown that the multiphonon sidebands of exciton luminescence can be described as a successive scattering through the virtual states[5] and in this paper we shall give an example of Raman scattering involving real intermediate states. The only possible distinction between the different components of secondary emission is by the amount of relaxation which takes place during the life-time of the optical

excitation.[4] A quantitative measure of this relaxation is the
number of phonon interactions which occur before the photon
emission. In some cases this relaxation can be described as a
succession of elementary processes,[1,3] and in some it can not be.[2,6]

In this paper we shall report experimental studies of the
secondary emission processes with different amount of relaxation
in some II-VI semiconductors. Special attention will be given to
the multiphonon LO-emission in the region of fundamental absorption,
which was first observed by Leite et al.[7] and by Klein and Porto[8]
in CdS and later in CdSe,[9] ZnSe,[10] ZnO,[10] ZnTe[11] and other polar
semiconductors.

II. CLASSIFICATION OF RESONANT SECONDARY EMISSIONS

The energy $\hbar\omega_s$ and the wave vector k_s of the emitted photons
are related to the energy and wave vector of exciting photon $\hbar\omega_i$,
k_i by the conservation laws

$$\hbar\omega_s = \hbar\omega_i - \sum_{n}^{N} \hbar\Omega_n(k_n),$$

$$k_s = k_i - \sum_{n}^{N} k_n, \tag{1}$$

where N is the number of phonon interactions before the emission
and $\hbar\Omega_n$ is the energy of phonon with wave vector k_n. Since the
energy conservation holds only between the initial ($\hbar\omega_i$) and the
final ($\hbar\omega_s$ + N phonons) states, the energy and the broadening of
the intermediate states do not enter the shape of secondary
emission spectrum directly.

If only a few phonons with energy practically independent of
k_n (e.g., LO phonons) take part in the relaxation, the secondary
emission will consist of narrow lines whose positions are strictly
correlated with the exciting line. Due to the spectral shape and
the small amount of relaxation, these lines should be considered
as Raman scattering regardless of the virtual or real character
of intermediate states involved. Calculation of intensities for
such lines is straightforward.[5,11]

If real intermediate states with life-time long compared
to the phonon scattering times are involved in the course of
relaxation, a quasi-equilibrium distribution of photo-excited
excitons will be established after sufficiently large number of
phonon interactions N. In this case we observe thermalized
luminescence from these real states, and the shape of the emission
spectrum does not depend on the excitation energy.

If phonons with strong dispersion $\hbar\Omega_n(k_n)$ (e.g., acoustical phonons) are involved in the relaxation process, the exact connection between $\hbar\omega_s$ and $\hbar\omega_i$ will be lost after the emission of a few phonons, and we cannot specify the number of phonon interactions for a given $\hbar\omega_s$. At the same time the distribution of the excited excitons will be in nonequilibrium and secondary emission spectrum will strongly depend on the excitation frequency. This intermediate case which is most difficult for theoretical description represents "hot luminescence".

Now we shall give a clear example of simultaneous manifestation of the three above-mentioned components in the secondary emission spectrum. Figure 1 shows the emission spectra of CdS samples with different free exciton life-times taken at 2°K with 4765 Å laser excitation.[12] Free exciton life-times for the samples studied were measured with the aid of Hanle effect (section 4). Indirect LO-assisted absorption of laser light creates the real intermediate state — hot $A, n = 1$ exciton with energy E_i. Kinetic energy of this exciton (10.4 meV) is well below the exciton ionisation energy (28 meV) and less than A-B band splitting (16 meV).

Further relaxation of hot excitons by acoustic phonons leads to the population of $A, n = 1$ exciton band and results in a stationary distribution of exciton kinetic energies, which is strongly dependent on the free exciton nonradiative life-time τ_n. This distribution can be visualized in the emission spectrum through the 2LO-assisted exciton annihilation, the probability of which is independent of the exciton kinetic energy.

The spectral range studied here spans from E_A-2LO up to E_i-2LO. For all samples we can see a narrow line with the energy $\hbar\omega_s = E_i - 2LO = \hbar\omega_i - 3LO$ which corresponds to the secondary emission process with participation of LO phonons only. Despite the real nature of intermediate state E_i this emission should be considered as resonant Raman scattering. Integrated intensity of this line I_{3LO} is nearly the same in all samples and its width is well accounted by the directional dispersion of LO phonons in CdS.[13]

In the samples with the shortest exciton life-time the Raman line dominates the emission spectrum. With the increase of exciton life-time the population of $A, n = 1$ band builds up, and leads to the appearance of first a hot luminescence band, and then a thermalized exciton luminescence band with the threshold E_A-2LO. Characteristic changes in the emission spectral shape are accompanied by the growth of luminescence intensity in comparison to that of Raman scattering.

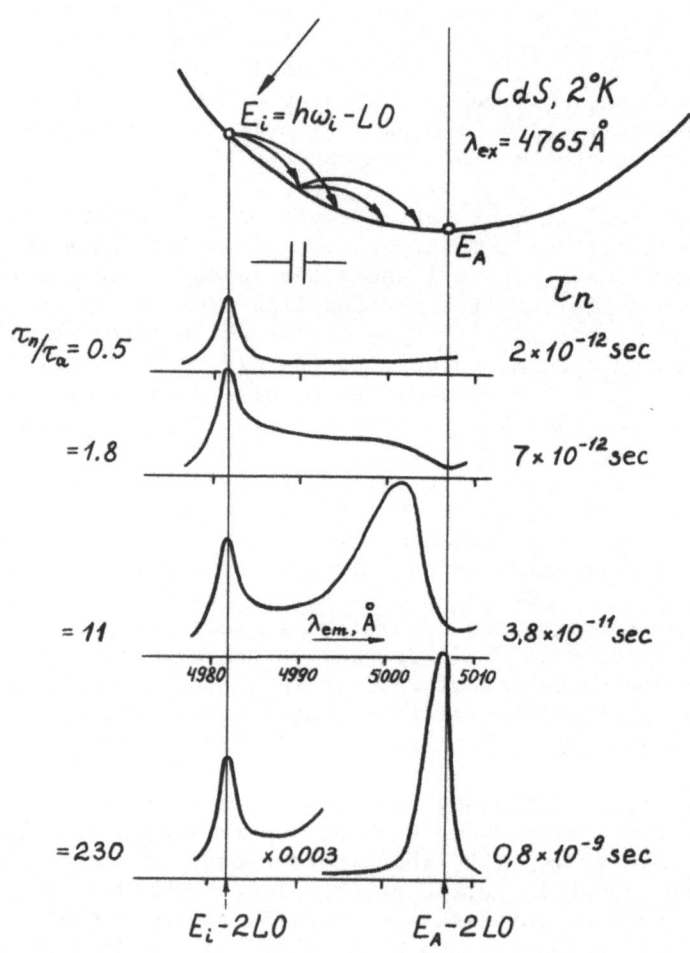

Fig. 1.   Secondary emission spectra of CdS samples with different
          free exciton life-time τ_n. In the right side are
          displayed the values of τ_n measured with the help of
          Hanle-effect (see text). Figures on the left represent
          average number of acoustic phonon interactions estimated
          as the ratio of τ_n to the acoustic relaxation time τ_a
          for point E_i.

With the help of a simple model which assumes that the exciton life-time is limited by nonradiative processes we can express the integrated intensity of the measured portion of emission spectrum I_Σ as

$$I_\Sigma = A\tau_n, \tag{2}$$

where τ_n is the exciton nonradiative life-time. In this case the intensity of Raman line I_{3LO} will be given by

$$I_{3LO} = A\tau_1 = A\,\frac{\tau_n\tau_a}{\tau_n + \tau_a}\ , \tag{3}$$

$$1/\tau_1 = 1/\tau_n + 1/\tau_a\ , \tag{4}$$

where τ_1 is the life-time of the intermediate state E_1 and τ_a is the reciprocal of exciton scattering probability out of state E_1 by acoustical phonons. Then we get

$$R = I_\Sigma/I_{3LO} = \tau_n/\tau_a + 1 \tag{5}$$

and

$$N = \tau_n/\tau_a = R - 1\ . \tag{6}$$

So we can obtain from the relative intensity measurements the quantity τ_n/τ_a which is a good estimate of the number of acoustic phonon interactions preceding the emission. This value is listed on the left of Fig.1. With these results we can determine how many acoustic phonons are necessary to establish thermal equilibrium in the exciton band.

III. MULTIPHONON LO-SCATTERING IN THE FUNDAMENTAL ABSORPTION OF SEMICONDUCTORS

Now we shall discuss multiphonon LO-scattering which takes place in excitation of polar semiconductors in the region of fundamental absorption.[7-11] The secondary emission spectrum in this case contains a series of narrow lines with comparable intensities shifted by nLO from the exciting line. The lines situated in the region of exciton resonance have maximum intensity.

Since only LO phonons are involved in relaxation, we should consider this process as Raman scattering. However, the open question is the nature of intermediate states for this scattering. In an early work[9] we have proposed discrete exciton bands as the intermediate states. On the other hand, Martin[14] and Zeyher[15]

discussed Raman scattering by means of electron-hole pairs.
However, the discrimination between two possibilities was difficult
on the basis of experimental data available. For this reason we
have carefully measured the LO-scattering spectra of ZnTe crystals
under Ar$^+$ laser excitation. Position of the lowest exciton
resonance (5207 Å at 2°K and 5234 Å at 77°K)[16] gives the possibility
of using several Ar$^+$ lines for excitation. Shift of the laser
lines from the exciton resonance in units of the LO energy in ZnTe
is shown in Table I. As an example a LO-scattering spectrum
is shown in Fig.2.

Table I. Shift of Ar$^+$ laser lines from n = 1 exciton
resonance of ZnTe crystals (Δ) in units of
LO energy (210 cm^{-1}).

λ_1, Å	Δ(T = 77°K)	Δ(T = 2°K)
4765	8.96	8.47
4880	6.59	6.10
4965	4.92	4.43
5017	3.93	3.44

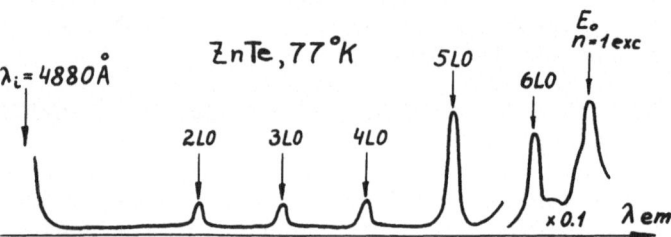

Fig. 2. LO-scattering lines in the secondary emission spectrum
of ZnTe. Position of n = 1 exciton resonance is marked
by arrow.

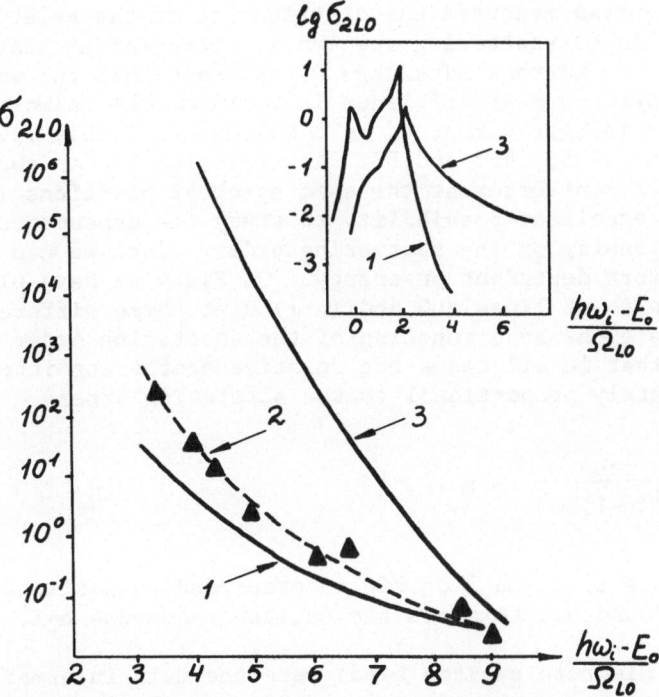

Fig. 3. Experimental dependence of 2LO scattering cross section
on the excitation frequency as measured in ZnTe crystals
with different Ar$^+$ lines at 2 and 77°K (black triangles).
Solid lines 1 and 3 represent theoretical results for
n = 1S and free electron–hole pairs intermediate states,
respectively.

We have measured the dependence of 2LO scattering cross section
on the excitation energy (Fig.3) and compared it with theory. It
can be seen that the measured dependence is closer to that expected
for the pair states[17] than to theoretical prediction for discrete
exciton bands.[5] Insert on Fig.3 compares in the same units the
contribution of discrete excitons and pair states into 2LO scattering, calculated for the parameters of ZnTe crystals.[11] We can
conclude that with an increase of excitation energy scattering
through the pair states (or more precisely through continuum
exciton states) predominates over the scattering through discrete
exciton states. The exact relation of the two processes will
depend on energy and crystal parameters. For example, in the case
of CdS crystals with larger exciton binding energy and split
exciton band edge, the exciton contribution to LO scattering can
extend to much higher energies.

We have also measured the distribution of the relative
intensities in LO scattering spectra at different excitation
energies.   We took advantage of the fact that the energy
difference between some Ar^+ lines is occasionally   almost exactly
equal to the integer number of ZnTe LO phonons (Table I). As a
result, under 4765, 4965 and 5017 Å excitation we can observe LO
lines of different order at the same spectral positions (Fig.4).
It gives an excellent possibility to study the dependence of
relative intensity on the scattering order, since we can exclude
all the factors dependent on energy. On Fig.4 we have plotted the
intensity ratio of lines nLO and (n-1)LO at three different
spectral positions as a function of the scattering order n. We
have found that in all cases the relative scattering intensity S_n
is approximately proportional to the scattering order

$$S_n = \frac{I_{nLO}}{I_{(n-1)LO}} \sim n \; . \qquad\qquad (7)$$

Straight lines 1, 2 and 3 on Fig.4 correspond to position of nLO
line at 2, 1 and 0.4 LO above the exciton resonance n=1.

If the discrete exciton bands were the main intermediate states
of LO scattering, we should expect that the relative intensity S_n
would be independent of the scattering order n. In this case the
scattering cross section can be written as a product of scattering
probabilities through the real intermediate states[5,9] and relative
intensity will depend only on secondary emission frequency $\hbar\omega_s$.
In the case of scattering through continuum states the virtual
intermediate states are important and the cross section of the n-th
order will be proportional to the total number of topologically
nonequivalent diagrams, contributing to the scattering (n!). So
an   additional dependence on n will enter the relative intensity.

IV. POLARIZATION OF RESONANT SECONDARY EMISSION

Excitation of cubic ZnTe crystals with linearly polarized
light results in a high degree of linear polarization of LO lines.
This polarization decreases with increase in scattering order n.
So the relaxation of polarization degree goes in parallel with
the energy relaxation.

Linear polarization of secondary emission can be observed only
if the intermediate states are discrete excitons or electron-hole
pairs which do not dissociate during the relaxation, since phase
correlation between the angular momenta of carriers is necessary
for preserving the linear polarization.[6] On this basis it will be
more correct to consider the electron-hole pairs involved in LO
scattering as continuum exciton states, rather than as free carriers.

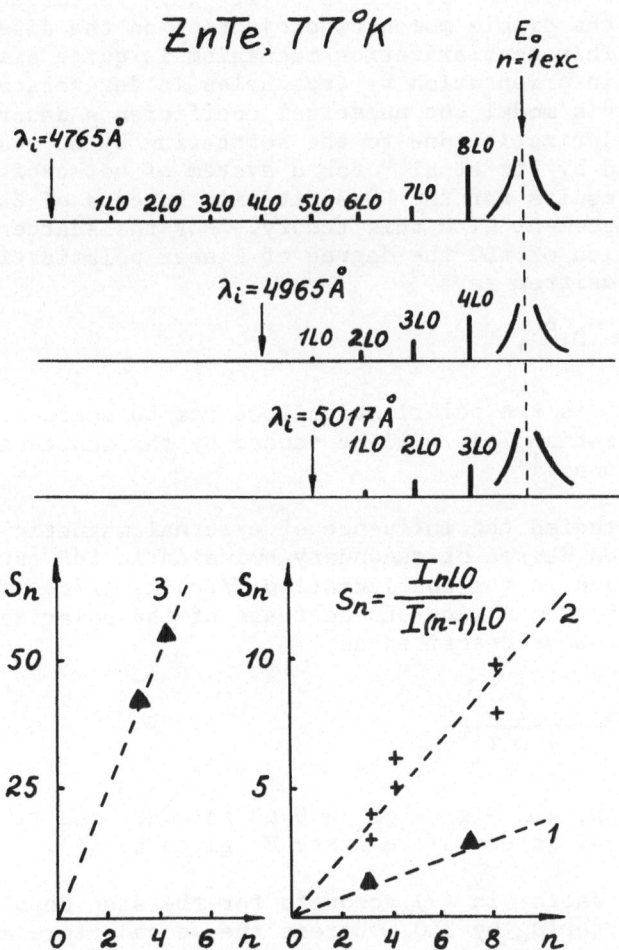

Fig. 4.   Relative position and relative intensity distribution
           S_n of LO-scattering lines in secondary emission of
           ZnTe crystals at different excitation conditions.

Linearly polarized excitation leads to the optical alignment of intermediate (real or virtual) states, i.e., to the linear orientation of dipole moments of these states. Due to the non-degeneracy of the valence bands and the longitudinal-transverse splitting of the exciton states, every scattering in the intermediate state changes the direction of its wave vector and leads to a decrease in the dipole momentum projection on the direction of orientation. This depolarization mechanism is quite similar to the loss of spin orientation by free holes in degenerate valence bands. With this model the numerical coefficients describing the step-like depolarization due to the scattering by LO phonons have been calculated by Bir et al.[6] for a system of hot excitons. Our experimental results for the LO-scattering spectra of ZnTe and CdS[18] are in good agreement with this theory. For the scattering process with the emission of nLO the degree of linear polarization can be approximately written as

$$\rho_n = D_n \approx D_1{}^n \, , \tag{8}$$

where $D_1 \approx 0.85$ is the polarization loss per LO scattering. Similar polarization loss will be caused by the scattering on acoustical phonons.[12]

We have studied the influence of external magnetic field on the polarization degree of secondary emission in CdS crystals under 4765 Å excitation in the configuration $H//c$, $k_{i,s}//c$. We have observed for the Raman line 3LO decrease of the polarization degree (Fig.5) which can be described as

$$\rho_{3LO} = D_1{}^3 \frac{1}{1 + \omega^2\tau_i{}^2} \, , \tag{9}$$

where $\omega = \mu_B g_{ex} H$, $g_{ex} = g_e - g_{h_\parallel} = 0.63$ for CdS, and τ_i is the life-time of real intermediate state E_i given by (4).

The first factor in (9) accounts for the step depolarization due to the scattering by 3LO, whereas the second represents destroying of exciton allignment by the magnetic field during the life-time in the real state E_i (Hanle-effect). From the dependence of polarization degree on magnetic field we have measured τ_i for the samples of Fig.1 and with the help of (4)-(6) calculated the nonradiative life-times τ_n and acoustic relaxation time τ_a. These results are presented in Table II.

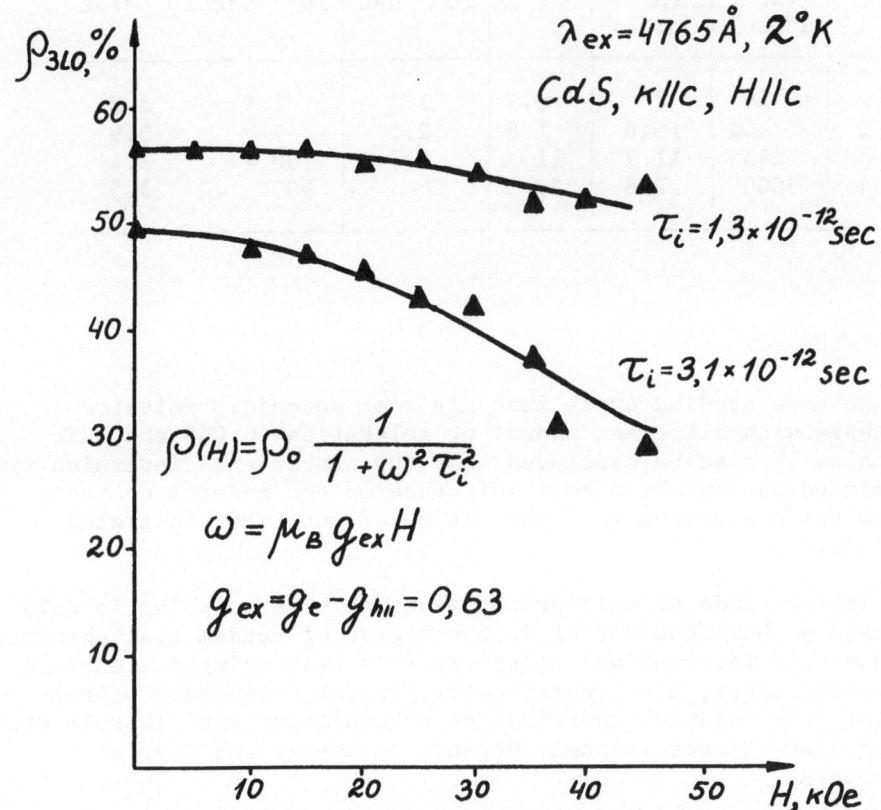

Fig. 5. Dependence of linear polarization degree of the Raman line 3LO (ρ_{3LO}) on external magnetic field H in CdS samples with different exciton life-times.

Table II.   Relative intensity of secondary emission and
            characteristic relaxation times of hot excitons
            in CdS samples at 2°K.

NO.	I_Σ (arb. units)	I_{3LO} (arb. units)	τ_n/τ_a	τ_i 10^{-12}sec	τ_n 10^{-12}sec	τ_a 10^{-12}sec
1	10	7	0.5	1.3	1.9	3.8
2	30	10.8	1.8	2.5	7	3.9
3	145	11.7	11	3.1	38.4	3.4
4	3000	12.8	230	–	800	3.5

V. CONCLUSIONS

We have studied experimentally some secondary emission
processes with different amount of relaxation in CdS and ZnTe
crystals. It can be concluded that the number and dispersion type
of emitted phonons have more influence on the general character of
the emission spectrum than the nature of intermediate states
involved.

In the study of multiphonon resonant LO scattering in ZnTe
crystals we have found that in the region of fundamental absorption
the dominant intermediate states in this scattering are exciton
continuum states, i.e., practically free electron—hole pairs.
However, the relative contribution of continuum and discrete exciton
intermediate states strongly depends on energy and crystal
parameters.

Under the polarized excitation one can observe the relaxation
of secondary emission polarization which takes place in parallel
with the energy relaxation. The main mechanism of polarization
relaxation for the discrete and continuum exciton intermediate
states is connected with the band—splitting. The linear polarization of LO scattering in the region of fundamental absorption
favours consideration of the continuum intermediate states involved
as hot excitons rather than free carriers.

In this work we have been able for the first time to observe
the Hanle effect for the Raman scattering using the magnetic fields
of moderate strength (less than 50 kOe). The observed

depolarization is connected with finite life-time of the real exciton intermediate state.   So the resonant Raman scattering in some cases can have observable life-time.

By measuring the Hanle-effect and relative intensities in the secondary emission spectra of CdS we have obtained information about the characteristic times of exciton relaxation.   From our results free exciton life-time τ_n varies in CdS samples from some picoseconds up to nanoseconds.   The time of acoustical relaxation, measured for excitons with kinetic energy of 10.4 meV, is approximately the same in all samples ($\approx 4 \times 10^{-12}$ sec).   As a result the amount of acoustic phonon relaxation essentially differs in different samples, which is clearly reflected in the shape of secondary emission spectra.

REFERENCES

1.   M. V. Klein, Phys. Rev. B **8**, 919 (1973).
2.   Y. R. Shen, Phys. Rev. B **9**, 622 (1974).
3.   J. R. Solin and H. Merkelo, Phys. Rev. B **12**, 624 (1975).
4.   K. Rebane and P. Saari, J. of Luminescence **16**, 223 (1978).
5.   A. Klochikhin, S. Permogorov, and A. Reznitsky, JETP **71**, 2230 (1976).
6.   G. L. Bir, E. L. Ivchenko, and G. E. Pikus, Isv. Acad. Sci. USSR (phys.) **40**, 1866 (1976).
7.   R. C. C. Leite, J. F. Scott, and T. C. Damen, Phys. Rev. Lett. **22**, 780 (1969); Phys. Rev. **188**, 1285 (1969).
8.   M. V. Klein and S. P. S. Porto, Phys. Rev. Lett. **22**, 782 (1969).
9.   E. Gross, S. Permogorov, Ya. Morozenko, and B. Kharlamov, Phys. Stat. Sol. **59(b)**, 551 (1973).
10.   J. F. Scott, T. C. Damen, W. T. Silfvast, R. C. C. Leite, and L. E. Cheesman, Opt. Commun. **1**, 397 (1970).
11.   A. Klochikhin, Ya. Morozenko, and S. Permogorov, Fiz. Tverd. Tela. **20**, 3557 (1978).
12.   S. Permogorov and V. Travnikov, Sol. St. Commun. **29**, 615 (1979).
13.   S. Permogorov and A. Reznitsky, Sol. St. Commun. **18**, 781 (1976).
14.   R. M. Martin, Phys. Rev. B **10**, 2620 (1974).
15.   R. Zeyher, Phys. Rev. B **9**, 4439 (1974).
16.   R. E. Nahory and H. Y. Fan, Phys. Rev. **156**, 825 (1967).
17.   A. A. Abdumalikov and A. A. Klochikhin, Phys. Stat. Sol. **80(b)**, 43 (1976).
18.   S. Permogorov, Ya. Morozenko, and B. Kazennov, Fiz. Tverd. Tela. **17**, 2970 (1975).

RESONANT RAMAN SCATTERING FROM STRESS-SPLIT FORBIDDEN EXCITONS IN Cu_2O

R. G. Waters,* F. H. Pollak, H. Z. Cummins, R. H. Bruce†
and J. Wicksted
City College and Brooklyn College
City University of New York

Excitons in Cu_2O have been studied extensively since their discovery nearly 30 years ago.[1] Excitons of the fundamental(yellow) series are both very sharp and difficult to observe as a result of the direct forbidden bandgap. Yellow P states are weakly dipole allowed, and 8 or more members of the P series can be observed in good crystals in absorption or luminescence. Yellow S states, however, are strictly dipole-forbidden.

The 1S yellow state was first observed in weak electric quadrupole absorption by Gross and Kaplyanskii in 1960.[2] Quadrupole transitions to higher S states are masked by phonon-assisted absorption to the 1S state, and are usually studied by applying a symmetry-breaking perturbation to the crystal. The currently accepted level assignments for excited yellow S and D states are derived from electroabsorption and electroreflection measurements, largely due to Nikitine and coworkers.[3,4]

Uniaxial stress measurements have played an important role in the study of excitons in Cu_2O. Gross and Kaplyanskii's observations in 1960[5] of the polarization of the stress-split components of the 1S yellow state established its quadrupole character and provided the experimental basis for Elliott's 1961 band assignments.[6] The applied stress in these experiments did not exceed 1 Kbar and the splittings were observed to be linear in the stress. Although stress splitting of some weak absorption features was observed by Agekyan, Gross and Kaplyanskii in 1965,[7] the weakness

*Present address: Optical Information Systems, Elmsford, N.Y. 10523
†Present address: Perkin Elmer Corporation, Norwalk, Conn. 06856

of the absorption precluded detailed analysis. In 1974, Agekyan and Stepanov again investigated the stress splitting of excited S states, but with the addition of a static electric field.[8] The observed dependence of the stress splitting on principal quantum number led them to question the accepted energy level assignments and to propose a new classification scheme for the yellow exciton states. However, they were unable to establish the symmetries of the stress-split excited states because the static electric field mixes P states with S states.

We have previously reported the observation of resonant Raman scattering from odd-parity phonons in Cu_2O when the laser is tuned into resonance with quadrupole exciton states.[9] The first observation of this effect in 1973 by Compaan and Cummins is illustrated in Fig. 1. When the dye laser was tuned 10 cm^{-1} above the 1S yellow exciton, the only Raman feature observed below 250 cm^{-1} was the 220 cm^{-1} two-phonon line which is dipole allowed (a). But with the laser tuned to exact resonance with the 1S yellow exciton, new Raman lines appeared at frequencies corresponding to odd parity phonons which are normally Raman forbidden (b). This effect results from: (1) a quadrupole optical transition from the ground state to the 1S exciton state; (2) an interband electron transition mediated by the electron-phonon interaction; (3) a dipole optical transition back to the ground state. Since the initial optical transition has (quadrupole) symmetry $\hat{E}_o k_o$ and the final optical transition has (dipole) symmetry $\hat{E}_S$, the phonon symmetry must be $\hat{E}_o k_o \hat{E}_S$ which is the odd-parity product of three vectors. Thus quadrupole-dipole Raman scattering is mediated by odd parity phonons in distinction to the usual dipole-dipole Raman scattering which must be mediated by even-parity phonons which transform like the product $\hat{E}_o \hat{E}_S$.

We subsequently recognized that the quadrupole-dipole resonant Raman scattering phenomenon can be exploited as a "quadrupole spectrometer" since resonance enhancement occurs only when the incident laser frequency is tuned to resonance with a quadrupole electronic state, and requires no additional symmetry breaking perturbations. It does not suffer from the presence of phonon assisted absorption which makes optical absorption measurements difficult. By scanning the dye laser through the frequency range of the yellow exciton series we were able to detect excited S and D quadrupole exciton states by observing the resonant enhancement of the 109 cm^{-1} Raman line due to the Γ_{12}^- phonon.[10]

We recently initiated a study of this effect in the presence of uniaxial stress. We have investigated the stress dependent splitting of several quadrupole yellow exciton states and also established the symmetries of the stress-split sublevels. A report of the experimental observations is given below. The theoretical analysis is included in a longer article which will appear elsewhere.[11]

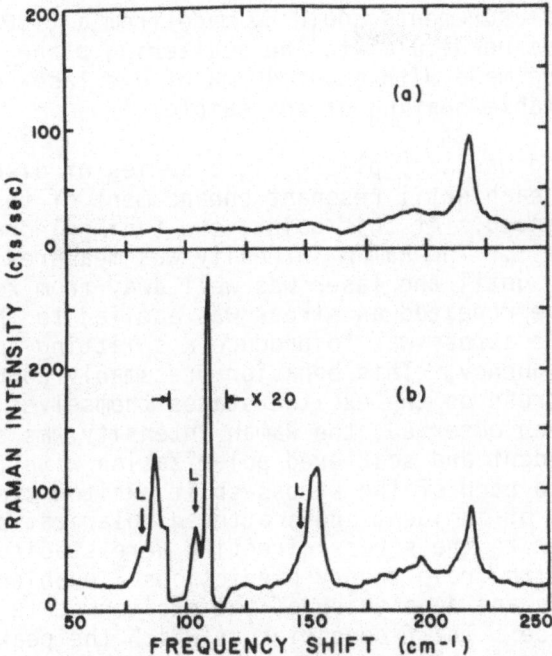

Fig. 1.   Raman spectra of Cu_2O at 4°K with 12 mW incident laser
          power. Incident polarization, [001]; scattered polariza-
          tion, [001] + [110]. Instrumental resolution, 2 cm^{-1}.
          (a) Laser frequency 10 cm^{-1} above the 1S yellow exciton
          frequency. (b) Laser in resonance with the 1S yellow
          exciton. The features labeled L are due to phonon-assisted
          luminescence (from Compaan and Cummins, Ref. 9)

       Experiments were performed at ~4°K on a Cu_2O crystal cut from a
large boule grown by Brower and Parker using the floating zone technique.[12] The crystal was mounted in a uniaxial stress apparatus
(which is described elsewhere[13]) and placed in a Janis "supervari-
temp" dewar. All spectra were obtained with a Coherent Radiation
model 590 dye laser pumped by a Spectra Physics model 165 argon ion
laser. Scattered light was analyzed using a Spex 1401 double-
grating spectrometer and photon counting electronics and was recorded on a strip chart recorder. The dye laser output was passed
through a Spex minimate spectrometer before reaching the sample to
reduce the broadband dye fluorescence. Stepping motors were connected to the tuning drives of the dye laser and the minimate spectrometer which allowed these instruments to be electronically frequency scanned.

       The Cu_2O sample was cut as a parallelepiped (2mm x 2mm x 10mm)
with the [100] directions as principle axes. The surfaces of the
crystal were mechanically polished and etched with concentrated HNO_3.

Backscattering measurements could be made from a (100) surface with stress applied perpendicular to the scattering plane along [001]. Measurements were made with about 40 mW of dye laser power which caused no detectable heating of the sample.

Laser tuning was accomplished in a series of discrete steps of about 3 cm^{-1} each until resonant enhancement of the Γ_3^- (109 cm^{-1}) phonon had commenced. At this point, the tuning increments were reduced to ~1 cm^{-1}. The Raman intensity was measured at each incident frequency until the laser was well away from resonance. The measurements were repeated as stress was applied to the crystal. The effect of the stress was to produce a splitting and/or shift in the resonant frequency. This behavior presumably corresponds to the effect of stress on the exciton states themselves. Once a splitting had been observed, the Raman intensity was studied as a function of incident and scattered polarization with the laser tuned to resonance with each of the stress-split exciton states in turn. All combinations of incident and scattered polarizations parallel and perpendicular to the stress direction were studied. Comparison of the results with group theory predictions[13] enabled us to determine the symmetry and degeneracy of the exciton state responsible for each resonance. The frequencies at which the peak of the resonance enhancement occurs versus applied stress are shown in Fig. 2. The degeneracy of the exciton states is also indicated in the figure.

Since the $^2\Gamma_7^+$ valence and $^2\Gamma_6^+$ conduction bands forming the yellow exciton are both Kramers doublets, the observed initial linear splitting with stress cannot arise from simple deformation splitting of the bands. Rather, it is a second-order effect involving the simultaneous effects of exchange and strain, as first noted by Elliott.[6] The theory of exchange-strain splitting has been discussed by a number of authors,[14] including Kiselev and Zhilich[15] who have worked out much of the theory for Cu_2O.

We have carried out a calculation of the exchange-strain splitting in Cu_2O, following the approaches of Kiselev and Zhilich, of Langer, Euwema, Era and Koda[16] and of Cho[17] based on the effective Hamiltonian formalism of Pikus.[18] The details of the calculation will appear elsewhere.[11]

The results, up to second order perturbation theory, show that a $^3\Gamma_5^+$ exciton state in the unstressed cubic O_h crystal will be split by stress along [001] into a $^1\Gamma_4^+$ and a $^2\Gamma_5^+$ in the stressed D_{4h} crystal with energies

$$E(^2\Gamma_5^+) = h \ X - (4eJ/3\Delta') \ X - (2e^2/\Delta') \ X^2$$

$$E(^1\Gamma_4^+) = h \ X + (8eJ/3\Delta') \ X - (2e^2/\Delta') \ X^2$$

(1)

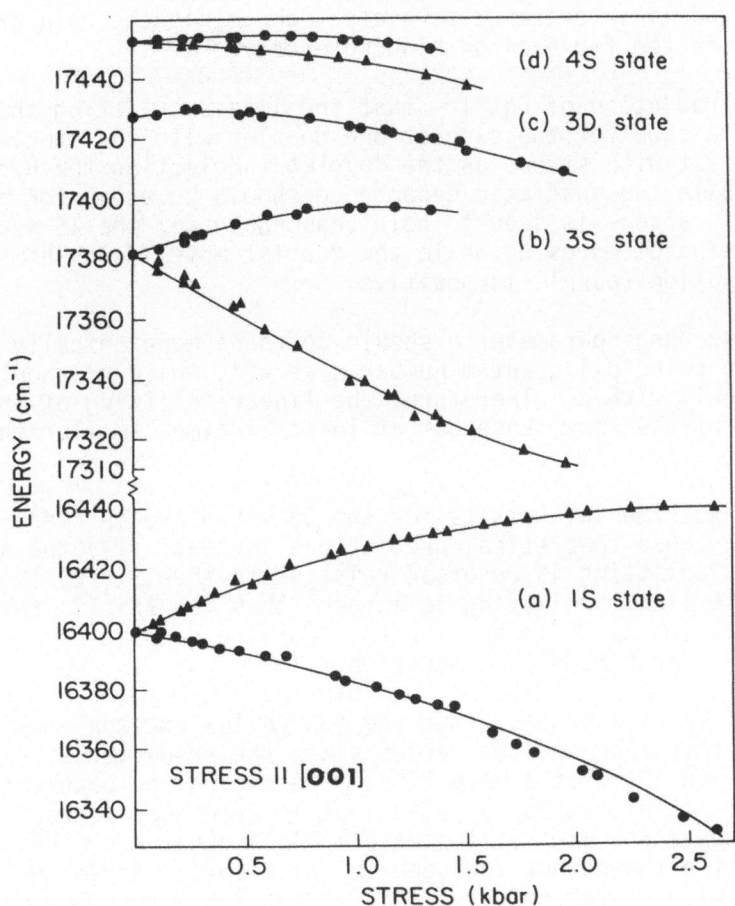

Fig. 2. Energy of resonant Raman scattering maxima vs applied
stress in Cu_2O for tetragonal stress. The data points give
the energy of quadrupole-allowed exciton states. Polarization selection rules allow identification of the doubly-
degenerate ($^2\Gamma_5^+$) and nondegenerate ($^1\Gamma_4^+$) levels into
which the states split under stress.[13] Where a splitting
occurs, these are indicated by circles and triangles respectively. No splitting is resolved for the 3D state.

where X is the applied stress, h is a deformation parameter representing the effect of the hydrostatic component of the stress on the bandgap, e is a deformation parameter representing the shear strain splitting of the $^4\Gamma_8{}^+$ valence band (which is the hole state for green excitons), J is an exchange parameter, and Δ' is the spin orbit energy corrected for exciton binding energy.

The prediction of Eq. 1 - that the linear splitting should be of opposite sign for the singlet and doublet with the singlet moving twice as fast with stress as the doublet (neglecting the hydrostatic effect) while the quadratic dependence should be equal for both and negative in sign - is seen to hold reasonably for the 1S state. That the singlet moves up while the doublet moves down shows that the combination (eJ/Δ') is positive.

The exchange parameter J should decrease monotonically with increasing principal quantum number n as n^{-3}, while Δ' should increase slowly with n. Therefore, the linear splitting of the 3S should be of the same sense but at least 27 times smaller than the 1S.

The experimental results for the 3S state are in clear contradiction to these theoretical predictions in that: (1) the sense of the linear splitting is reversed relative to the 1S; (2) the magnitude of the linear splitting is larger[8] than the 1S; (3) the quadratic splitting is different for the two stress split components of the 3S in both magnitude and sign.

These results suggest that the "3S yellow exciton" may actually belong to the green exciton series since the predicted stress splitting of green $^3\Gamma_5$ states more closely resembles the observed stress dependence of this state, or else that interactions between yellow and green excitons with different principal quantum numbers may be significant although not included in the theory. Additional experiments with different directions of applied stress are in progress which should help to elucidate the surprising observations reported above.

References

1. M. Hayashi and K. Katsuki, J. Phys. Soc. Jpn. 5, 381 (1950).
2. E. F. Gross and A. A. Kaplyanskii, Fiz. Tverd. Tela 2, 379 (1960) [Sov. Phys. - Solid State 2, 353 (1960)].
3. J. L. Diess and A. Daunois, Surf. Sci. 37, 804 (1973); S. Nikitine, in: Optical Properties of Solids edited by S. Nudelman and S. S. Mitra, Plenum Press, N.Y. (1969) p.214 ff.
4. A. Daunois, J. L. Diess, J. C. Merle, C. Wecker and S. Nikitine, in: Eleventh International Conference on the Physics of Semiconductors, Warsaw, (1972) [State Publishing House, Warsaw (1973)] p. 1402.

5. E. F. Gross and A. A. Kaplyanskii, Fiz. Tverd. Tela $\underline{2}$, 2968
    (1960) [Sov. Phys. - Solid State $\underline{2}$, 2637 (1961)].
6. R. J. Elliott, Phys. Rev. $\underline{124}$, 340 (1961).
7. V. T. Agekyan, E. F. Gross and A. A. Kaplyanskii, Fiz. Tverd.
    Tela $\underline{7}$, 781 (1965) [Sov. Phys. - Solid State $\underline{7}$, 623 (1965)].
8. V. T. Agekyan and Yu. A. Stepanov, Fiz. Tverd. Tela. $\underline{17}$, 1592
    (1975) [Sov. Phys. - Solid State $\underline{17}$, 1041 (1975)].
9. A. Compaan and H. Z. Cummins, Phys. Rev. Letters $\underline{31}$, 41 (1973).
10. M. A. Washington, A. Z. Genack, H. Z. Cummins, R. H. Bruce,
    A. Compaan and R. A. Forman, Phys. Rev. B$\underline{15}$, 2145 (1977).
11. R. G. Waters et al (submitted to Phys. Rev. B)
    R. G. Waters, Ph.D. Dissertation, City University of New
    York (1979).
12. W. S. Brower, Jr. and H. S. Parker, J. Cryst. Growth $\underline{8}$, 227
    (1971).
13. B. Berenson, unpublished.
14. F. H. Pollak, Surf. Sci. $\underline{37}$, 863 (1973) and references therein.
15. V. A. Kiselev and A. G. Zhilich, Fiz. Tverd. Tela. $\underline{13}$, 2398
    (1971) [Sov. Phys. - Solid State $\underline{13}$, 2008 (1972)].
16. D. W. Langer, R. N. Euwema, K. Era and T. Koda, Phys. Rev. B$\underline{2}$,
    4005 (1970).
17. K. Cho, Phys. Rev. B$\underline{14}$, 4463 (1976)
18. G. E. Pikus, Fiz. Tverd. Tela $\underline{6}$, 324 (1964) [Sov. Phys. - Solid
    State $\underline{6}$, 261 (1964)].

LOW FREQUENCY EXCITON AND RAMAN SCATTERING SPECTRA OF $CoCO_3$

Yu.A. Popkov, V.V. Eremenko and N.A. Sergienko

Physico-Technical Institute of Low Temperatures
Ukrainian SSR Academy of Sciences
Kharkov, USSR

INTRODUCTION

At present Raman spectroscopy seems to have become one of the principal methods for studying excitons and magnons in magnetically ordered crystals. The interaction of light with the spin system is rather effective in antiferrogmagnets allowing for simultaneous observation of the one- and two-magnon scattering.[1] The two mechanisms are quite independent (in the first case the dominant contribution is due to spin-orbit coupling while in the second it is due to exchange interaction). The first order spectrum provides information on the spin wave (exciton) energies at the center of the Brillouin zone whereas the second order spectrum yields magnon energies at the zone boundary. Thus Raman spectroscopy provides the possibility of a virtually complete reconstruction of the spin-wave spectrum in antiferromagnets from the data of a single experiment.

The one-magnon (and exciton) scattering is most effective in such crystals for which the ground state of the magnetic ion is characterized by a non-zero orbital moment, i.e. where the spin-orbit coupling is high. This is the case with divalent cobalt, and therefore Co-based compounds are the materials most frequently employed for such investigations. Exciton and magnon scattering were observed in CoF_2[2], $KCoF_3$[3], $RbCoF_3$[4], $TlCoF_3$[5], K_2CoF_4[6], Rb_2CoF_4[7], $CoCl_2$[8] and CoO[9] crystals.

In this paper we wish to present results of experimental and theoretical investigations of the Raman spectrum in $CoCO_3$[10,11].

Both one- and two-magnon scattering was observed. Exciton states have been studied in the paramagnetic and antiferromagnetic phases and the energy spectrum and the exciton band intensities have been calculated. The $CoCO_3$ crystal belongs to the rhombohedral calcite type. At $T_N = 18.1$ K it goes over to an antiferromagnetic state with weak ferromagnetism [12]. According to the most recent studies, its sublattice magnetic moments lie in the base plane [13]. Such a structure gives rise to two branches in the spin wave spectra, namely a ferromagnetic (acoustic) and an antiferromagnetic (optical branch). The first branch was the subject of rather detailed earlier investigations by resonance techniques [14] and more recently by the Mandelstam-Brillouin scattering method [15]. The high-frequency branch was revealed in our investigations of Raman scattering [11] and far infrared absorption [16].

Consider the structure of electronic levels of Co^{2+} in the $CoCO_3$ crystal as deduced from experimental results and calculated theoretically.

The ground state of Co^{2+} in $CoCO_3$, i.e. $^4T_{1g}(^4F)$, splits under the influence of the trigonal component of the crystal field and the spin-orbit coupling, into six Kramers doublets with an energy separation between the extreme components of about 1000 cm^{-1}. A group theoretical calculation shows that transitions between all the states are allowed and they are all actually observed in the Raman spectrum [10]. Figure 1 shows schematically the energy levels of Co^{2+}, and the electronic Raman spectrum in the paramagnetic region, with the corresponding frequencies. These data permit the trigonal field constant δ and the spin-orbit coupling constant λ to be calculated. The best agreement between calculated and observed frequencies is obtained with $\delta = 544$ cm^{-1} and $\lambda = 136$ cm^{-1}.

When the sample becomes antiferromagnetic, the Raman spectrum undergoes considerable change [10-11]; namely the exciton bands increase in number, shift in position and change their half-widths. We will discuss in more detail the two lowest doublets for $T \ll T_N$. In this region the experiment shows four bands lying at 35, 57, 178 and 207 cm^{-1} (Fig. 2; the 222 cm^{-1} band corresponds to the E_g-phonon). Analysis of the experimental results allows us to associate the first band with one-magnon scattering on the high-frequency branch of the spin-wave spectrum, the second one with two-magnon scattering and the rest of them with exciton excitations. Note that the magnon bands are almost two orders of magnitude less intense than the exciton bands.

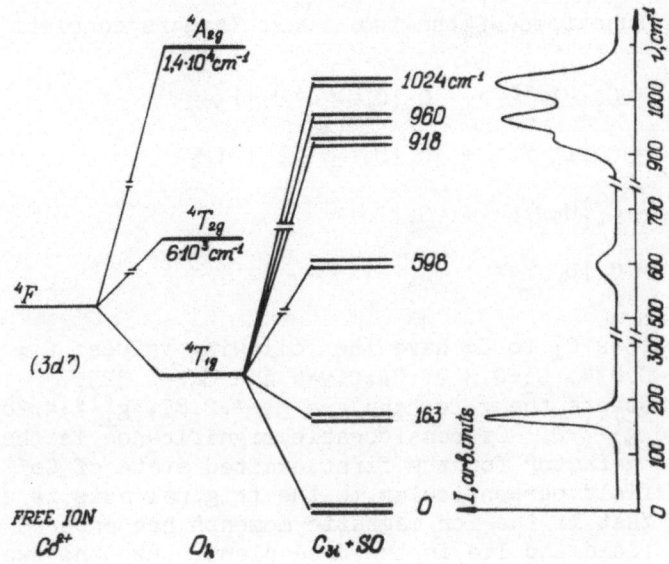

Figure 1. Schematic representation of the energy levels of the
Co^{2+} ion and the measured electronic Raman spectrum in the
paramagnetic phase of a $CoCO_3$ crystal.

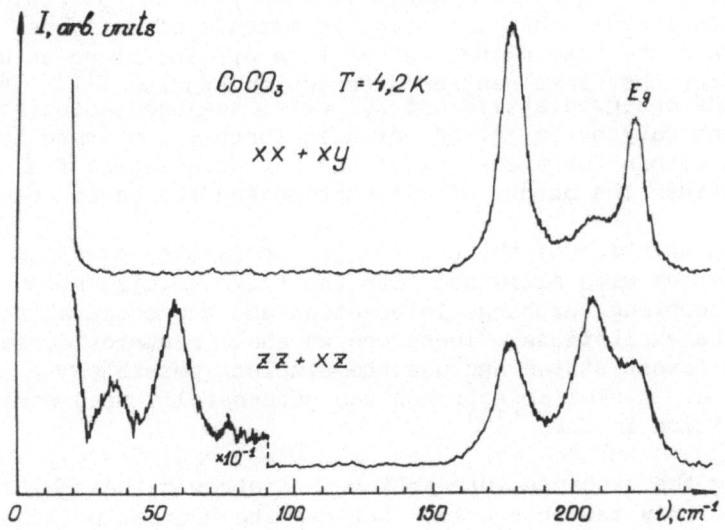

Figure 2. The Raman spectrum of a $CoCO_2$ crystral at T =4.2 K.
The intensities of the (xx+xy) spectra have been reduced by a
factor of 5 compared with the (zz+xz) spectra.

The wave functions of the two lowest Kramers doublets are

$$\phi_1 = C_1 |-1, 3/2 > + \quad C_2 |0, \tfrac{1}{2}> + \quad C_3 |1, -\tfrac{1}{2}>$$

$$\phi_1' = C_1 |1, 3/2> + \quad C_2 |0, -\tfrac{1}{2}> + \quad C_3 |-1, \tfrac{1}{2}>$$

$$\phi_2 = C_4 |0, 3/2> + \quad C_5 |1, \tfrac{1}{2}> \tag{1}$$

$$\phi_2' = C_4 |0, 3/2> + \quad C_5 |-1, -\tfrac{1}{2}>$$

where the factors C_1 to C_5 have the following values: $C_1 = 0.428$, $C_2 = -0.834$, $C_3 = 0.349$, $C_4 = 0.946$ and $C_5 = -0.323$. With such constants the g-factors are $g_\parallel^{(1)} = 2.87$, $g_\perp^{(1)} = 4.28$, $g_\parallel^{(2)} = 5.34$ and $g_\perp^{(2)} = 0$. Of considerable significance is the fact that the g-factor for the first excited state of Co^{2+} with the magnetic field perpendicular to the trigonal axis is zero. This implies that if the ion magnetic moments are ordered due to the exchange field and lie in the base plane, then the two bands observed at 178 and 207 cm^{-1} have nothing to do with removal of the Kramers degeneracy. An external magnetic field $\vec{H} \perp C_3$ of strength up to 50 KOe does not affect the spectral position of those bands, while at $\vec{H} \parallel C_3$ the first band splits into a doublet linearly in the magnetic field with its g-factor equal to 3.7 (the second band is characterized by a much larger width hence it is difficult to draw conclusions about its behavior in the field). These results suggest that the magnetic moments of Co^{2+} are ordered within the base plane, rather than off the plane as was reported after the first neutron diffraction studies [17]. As for the bands observed at 178 and 207 cm^{-1}, they are associated with resonant Davydov splitting which is further confirmed by a calculation within the framework of the self-consistent field model. Consider the method of calculation and the basic results.

An exact solution of the problem is impossible, even for a single $^4T_{1g}$ term with allowance for the trigonal distortion, spin-orbit coupling, exchange interaction and the cooperative nature of the excitations. Therefore we shall restrict ourselves to the four lowest states and use the simplest possible approximation. A similar approach was successfully used earlier for the Co^{2+} ion in CoF_2 [18].

Assuming the exchange interaction is isotropic, we shall take into account only the interaction between the nearest neighbors from different sublattices. The initial Hamiltonian is of form

$$H = \sum_j [-3/2\lambda \vec{L}_j \quad \vec{S}_j - (L_{zj}^2 - \tfrac{2}{3})] + J \sum_{j \neq \ell} (\vec{S}_j \vec{S}_\ell) \tag{2}$$

where J is the exchange interaction constant. The molecular field lifts the Kramers degeneracy hence the two doublets of eq. (1) yield four different states:

$$\psi_0 = a_0 \left[\phi_1 + \phi_1' + \frac{b}{\varepsilon - E_0'} (\phi_2 + \phi_2') \right]$$

$$\psi_1 = a_1 \left[\phi_1 - \phi_1' + \frac{b}{\varepsilon - E_1'} (\phi_2 - \phi_2') \right]$$

$$\psi_2 = a_2 \left[\phi_1 + \phi_1' + \frac{b}{\varepsilon - E_2'} (\phi_2 + \phi_2') \right]$$

$$\psi_3 = a_3 \left[\phi_1 - \phi_1' + \frac{b}{\varepsilon - E_3'} (\phi_2 - \phi_2') \right]$$

(3)

where $E_m' = E_m/zJS$, $a_m = \left[\frac{\frac{1}{2}}{1 + 6^2(\varepsilon - E_m')^2} \right]^{\frac{1}{2}}$, $m = 0,1,2,3$, $\varepsilon = \varepsilon'/zJS$, $\varepsilon' = 164$ cm^{-1} is the energy of the second doublet at $T = T_N$, $z = 6$ is the number of the nearest neighbors, and

$$E_0 = 6JS\left\{ \frac{\varepsilon - a}{2} - \left[\left(\frac{\varepsilon + a}{2}\right)^2 + b^2 \right]^{\frac{1}{2}} \right\}$$

$$E_1 = 6JS\left\{ \frac{\varepsilon + a}{2} - \left[\left(\frac{\varepsilon - a}{2}\right)^2 + b^2 \right]^{\frac{1}{2}} \right\}$$

$$E_2 = 6JS\left\{ \frac{\varepsilon - a}{2} + \left[\left(\frac{\varepsilon + a}{2}\right)^2 + b^2 \right]^{\frac{1}{2}} \right\}$$

$$E_3 = 6JS\left\{ \frac{\varepsilon + a}{2} + \left[\left(\frac{\varepsilon - a}{2}\right)^2 + b^2 \right]^{\frac{1}{2}} \right\}$$

(4)

The factors a and b are: $a = \sqrt{3} \, c_1 c_3 + c_2^2 = 0.954$ and $b = (\sqrt{3}/2) c_2 c_4 + c_3 c_5 = -0.796$.

From the condition of self-consistency we obtain an equation for determining the average moment in the ground state, relating the unknown parameters S and J, viz.

$$S = \frac{a}{2}\left[\left(\frac{\varepsilon + a}{2}\right)^2 + b^2 \right] - \varepsilon(\varepsilon + a)/\left\{ 4\left(\frac{\varepsilon + a}{2}\right)^2 + b^2 \right] \right\}^{\frac{1}{2}}$$

(5)

A second equation for estimating the average spin and the exchange integral can be obtained if we pass to the exciton representation and take some experimental value for the exciton (magnon) energy. Calculations similar to (11) lead to the following expression for exciton energies due to the first, second and third single-ion excited levels:

$$\hbar\omega_m^{\pm} = (\varepsilon_m^2 + \beta_m^2 - \alpha_m^2 \pm 2\varepsilon_m \beta_m)^{\frac{1}{2}}$$

(6)

Here m = 1,2 and 3

$$\alpha_{1(3)} = J\gamma_q \left\{ (S_{01(03)}^z) - (S_{01(03)}^y)^2 \right\} \qquad \alpha_2, \beta_2 = -J\gamma_q (S_{02}^x)^2,$$

(7)

$$\beta_{1(3)} = J\gamma_q \left\{ (S_{01(03)}^z)^2 + (S_{01(03)}^y)^2 \right\};$$

$$\gamma_q = \exp(iq_z\frac{d}{2}) \, [\, \exp(iq_y c) + \exp(-iq_y\frac{c}{2}) 2\cos\frac{c\sqrt{3}}{2} q_x] + \qquad (8)$$

$$+ \exp(-iq_z\frac{d}{2}) \, [\, \exp(-iq_y c) + \exp(iq_y\frac{c}{2}) 2\cos\frac{c\sqrt{3}}{2} q_x].$$

where the coordinate axes have been chosen in the following way: z is along the trigonal crystal axis and x along the antiferro-magnetic vector: $\vec{q}$ is the exciton (magnon) wave vector: c and d are projections of the basic vector of the $CoCO_3$ unit call on the plane perpendicular to the trigonal axis and on the axis itself: finally, S_{ik}^{α} are matrix elements of spin operators using the functions given in eq. (3).

Taking for the magnon frequency at q=0 an experimental value of 35.5 cm^{-1} we can find from (5) and (6) J =2.57 cm^{-1} and S =1.06. (In Ref. 11 there were minor errors in the expressions for α_i (7) which resulted in different values of S and J, and accordingly, of the exciton energies). Now we are able to calculate energies of the exciton levels at any point of the Brillouin zone. The results for the points Γ (0,0,0) and Z (0,0 π/d) are summarized in Table 1 which also contains a comparison with the experimental data.

Table 1. Calculated and observed exciton and magnon frequencies (cm^{-1}) in $CoCO_3$ for T << T_N.

Calculation			Experiment
Γ	Z	The Brillouin zone center	Boundary
35.5	31.1	35.5	29
173.2		178	
173.7	181.5		193
189.5	181.7		
189.4		207	

As can be seen from the Table, the calculated magnitude of the Kramers splitting for the lowest excited levels of Co^{2+} is negligible in the exchange field while at the boundary (point Z), the Davydov components are almost completely degenerate too. For other boundary points the calculations show the splitting to be about several cm^{-1}, both for the magnon and for excitons. This allows us to estimate the magnon and exciton frequencies at the boundary of the Brillouin zone, using experimental results. Since the 178 and 207 cm^{-1} bands are Davydov components, their energy at the boundary can be estimated as 193 cm^{-1}. The magnon energy at the boundary can be found from the frequency value (57 cm^{-1}) of the two-magnon scattering. Assuming that the contribution of the magnon-magnon interaction to the frequency is small and the magnon density of states is a maximum at the boundary, we obtain for the magnon frequency at the boundary the value of 29 cm^{-1}. Both estimates support the previously suggested mechanism [10,19] for the anomalous behavior near T_N of both the frequency and the band half-width of the 222 cm^{-1} E_g -phonon. This is associated with the additional phonon relaxation channel below T_N involving decay of the phonon into an exciton and a magnon (Fig. 3). The resonance coincidence of energies is favorable for the proposed mechanism.

Thus, comparison of experimental and theoretical results provides rich information on the energy spectrum of CoCO$_3$, even including dependence on dispersion. This is the first time that characteristics of an antiferromagnet have been studied so completely on the basis of Raman spectrum data alone.

Consider now the exciton and magnon band intensities using the theory of spin and orbital dependent electric polarizability developed by Moriya [18,20]. In the case of Co^{2+} ions the spin dependent part of the polarizability that appears only in the 3rd order of perturbation theory can be neglected compared with the orbital-dependent part contributing already in the 2nd order .

The form of the polarizability tensor of the ion is determined by the symmetry of its local environment. The local symmetry of Co^{2+} in CoCO$_3$ is nearly octahedral and may be well approximated by a spherically symmetric tensor expressed in terms of the ground-state L components of the j-th ion:

$$\alpha_j = \begin{vmatrix} AL_x^2 & -iBL_z+A\{L_xL_y\} & iBL_y+A\{L_zL_x\} \\ iBL_z+A\{L_xL_y\} & AL_y^2 & -iBL_x+A\{L_yL_z\} \\ -iBL_y+A\{L_zL_x\} & iBL_x+A\{L_yL_z\} & AL_z^2 \end{vmatrix} \qquad (9)$$

For the Co^{2+} ion we have A $=-(10^{-24}- 10^{-25})$ and B $\sim \dfrac{\hbar\omega_o}{1.1 \times 10^5}$ A. where ω_o is the incident photon frequency. Then the crystal polarizability is obtained by summation over all the ions in the two sublattices.

To determine intensities of the exciton bands, the polarizability tensor components should be expressed in terms of exciton operators, retaining only linear terms of the operators, as these are responsible for first order light scattering.

Calculations result in the following expressions for the $CoCO_3$
crystal excitations:

For the 35 cm^{-1} magnon

$$\sqrt{N} \begin{vmatrix} 0 & 0 & (-0.456B+0.028A)a_1^+ \\ 0 & 0 & 0 \\ (0.456B+0.028A)a_1^+ & 0 & 0 \end{vmatrix} \quad (10)$$

For the 173.2 and 173.7 cm^{-1} excitons

$$\sqrt{N} \begin{vmatrix} 0.088Aa_2^+ & i(0.018B-0.164A)a_3^+ & 0 \\ -i(0.018B+0.164A)a_3^+ & 0.068Aa_2^+ & 0 \\ 0 & 0 & 0.021Aa_2^+ \end{vmatrix} \quad (11)$$

For the 189.5 and 189.4 cm^{-1} excitons

$$\sqrt{N} \begin{vmatrix} 0 & 0 & (0.616-0.138A)\bar{a}_3^+ \\ 0 & 0 & i(-0.646B+0.132A)\bar{a}_2^+ \\ -i(0.616B+0.138A\bar{a}_3^+ & i(0.646B+0.132A)\bar{a}_2^+ & 0 \end{vmatrix} \quad (12)$$

Here a_1^+ and a_j^+ are creation operators of the excitations
specified.

The differential extinction coefficient, determining the
scattered intensity, for the $\mu\nu$ component (where μ and ν represent
the incident and the scattered light polarizations, respectively
can be written as

$$\frac{d^2h}{d\Omega d\omega} = \frac{\omega_0\omega^3}{2\pi c^4} \int e^{i(\omega-\omega_0)t} <\alpha_{\nu\mu}(0)\alpha_{\mu\nu}(t)> dt, \quad (13)$$

where ω_0 and ω are the incident and the scattered frequencies, Ω denotes the solid angle, and the angular brackets <..> stand for statistical averaging. Numerical values of the exciton band intensities and exciton polarizations are listed in Table 2 (for $\omega_0 = 1.58 \ 10^4 \ cm^{-1}$).

Table 2. Magnitudes of $dh/d\Omega$ for single exciton scattering of light in CoCO$_3$ (10^{-11}).

M o d e	a_1^+	a_2^+	a_3^+	$\bar{a}_2^+$	$\bar{a}_3^+$
Frequency (cm^{-1})	35	173.2;	173.7	189.5;	189.4
xx	–		29		–
yy	–		17		–
zz	–		1.5		–
xy	–		102		–
yx	–		100		–
xz	7		–		96
zx	0.4		–		60
yz	–		–		90
zy	–		–		44

As is easy to see, there is good agreement between the calculated and observed intensities for the exciton and magnon scattering (see Fig. 2). Indeed, the exciton bands are much more intense than those of one-magnon scattering. The theory also provides a rather good description of the polarization properties because: single-magnon scattering is observed only in the xz(yz) polarization (in our experiment we did not distinguish between the x and y axes because this would have required a weak magnetic field applied in the base plane); the 178 cm^{-1} band is most intense in the xx+xy spectra while the 207 cm^{-1} band is predominantly observed in the xz(yz) spectra. The good qualitative agreement between theory and experiment supports our assumption about the nature of the bands discussed and suggests that the energy parameters of the CoCO$_3$ crystal as determined here are correct.

In conclusion, the authors would like to express their sincere gratitude to A.P. Mokhir and V.I. Fomin for their assistance in the experimental work.

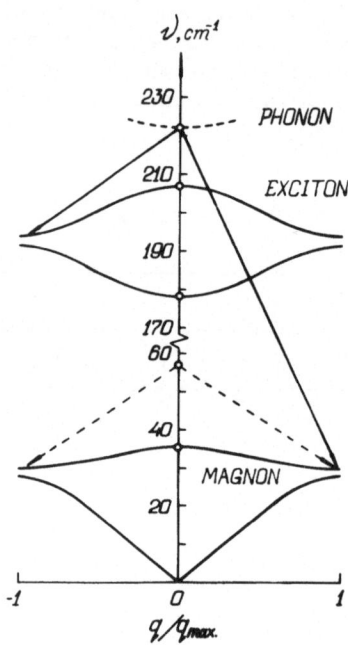

Figure 3. Schematic view of the lowest excitation spectrum of the CoCO$_3$ crystal (antiferromagnetic phase). The dots represent experimental frequencies. The solid lines with arrows illustrate the mechanism of phonon decay into an exciton and a magnon, the dashed line shows two-magnon excitation.

REFERENCES

1.   P.A. Fleury and R. Loudon, Phys. Rev. 166, 514 (1968).
2.   R.M. Macfarlane, Phys. Rev. Lett. 25, 1454 (1970).
3.   P. Moch and C. Dugautier, Phys. Letters 43A, 169 (1973).
4.   J. Nouet, D.J. Toms and J.F. Scott, Phys. Rev. B7, 4874 (1973).
5.   D.J. Toma, J.F. Ryan, J.F. Scott and J. Nouet, Phys. Letters 44A, 187 (1973).
6.   J.Y. Gesland, M. Quilichini and J.F. Scott, Solid State Commun. 18, 1243 (1976).
7.   D.J. Lockwood, I.W. Johnstone, C. Mischler and P. Carrata, Solid State Commun. 25, 565 (1978).
8.   G. Mischler, M.C. Schmidt, D.J. Lockwood and A. Zwick, J. Phys. Chem. Solids 27, 1141 (1978).
9.   H-h Chou and H.Y. Fan, Phys. Rev. B13, 3924 (1976).
10. Yu.A. Popkov, A.P. Mokhir and N.A. Sergienko, Fiz.Tverd. Tela (Leningrad) 18, 2053 (1976) Sov. Phys. Solid State 18, 1194 (1976)

11. V.V. Eremenko, A.P. Mokhir, Yu.A. Popkov, N.A. Sergienko and
     V.I. Fomin, Zh. Eksp. Teor. Fiz. 73, 2352 (1977) Sov. Phys.
     JETP 46, 1231 (1977) .

12. A.S. Borovik-Romanov and M.P. Orlova, Zh. Eksp. Teor. Fiz.
     31, 579 (1956) Sov. Phys. JETP 4, 531 (1957) .

13. B.J. Brown, P.L. Welford and J.B. Forsyth, J. Phys. C6, 1405
     (1973).

14. E.C. Rudashevski, Zh. Eksp. Teor. Fiz. 46, 134 (1964) Sov.
     Phys. JETP 19, 96 (1964).

15. A.S. Borovik-Romanov, V.G. Jotikov and N.M. Kreines. Pis'ma
     Zh. Eksp. Teor. Fiz. 24, 233 (1976); Zh. Eksp. Teor. Fiz.
     74, 2286 (1978).

16. V.M. Naumenko, V.V. Eremenko, A.I. Maslennikov and A.V.
     Kovalenko, Pis'ma Zh. Eksp. Teor. Fiz. 27, 20 (1978).

17. R.A. Alikhanov, Zh. Eksp. Teor. Fiz. 36, 1690 (1959) Sov.
     Phys. JETP 9, 1204 (1959) .

18. A. Ishikava and T. Moriya, J. Phys. Soc. Japan 30, 117 (1971).

19. Yu.A. Popkov, V.V. Eremenko, V.I. Fomin and A.P. Mokhir, in
     "Theory of Light Scatt. in Condensed Matter", ed. by B.
     Bendow, J.L. Birman and V.M. Agranovich, Plenum Press,
     N.Y.-London, 1976, p. 485.

20. T. Moriya, J. Phys. Soc. Japan 23, 490 (1967).

THEORETICAL AND EXPERIMENTAL DETERMINATIONS OF RAMAN SCATTERING CROSS SECTIONS IN SIMPLE SOLIDS

Manuel Cardona, M.H. Grimsditch, and D. Olego

Max-Planck-Institut für Festkörperforschung
Heisenbergstrasse 1
7000 Stuttgart 80, F.R.G.

The scattering "power" of a solid for a given first order Raman phonon is usually represented by the <u>scattering efficiency</u> S per unit solid angle and unit length (in cm^{-1} $ster^{-1}$ or $Å^{-1}$ $ster^{-1}$). It can also be characterized by the cross section σ (in cm^2 $ster^{-1}$ or $Å^2$ $ster^{-1}$) per atom or per primitive cell. The latter are related to S by

$$\sigma = S/N, \tag{1}$$

where N is the number of atoms (or primitive cells) per unit volume. The scattering efficiency S (and also σ) depends explicitly on temperature through the Bose Einstein statistical factors. It also depends on the <u>polarization</u> of the incident and scattered radiation (we assume <u>allowed</u> scattering, independent of the scattering vector). It is therefore convenient to express the scattering "power" in terms of the Raman tensors $\overleftrightarrow{R}_j$ instead of S or σ. The Raman tensors represent the change in the polarizability of a primitive cell produced by an atomic displacement corresponding to a suitably normalized phonon normal mode. For the Raman active phonons of crystals with diamond (and also zincblende) structure there are three $\overleftrightarrow{R}_j$'s corresponding to the triple degeneracy of these modes (T_{2g} or $\Gamma_{25'}$). In the case of diamond the three

R_j's have only one independent component usually designated[1] as a (sometimes the designation $\underline{P}$ is used[2]). We then have for the Stokes scattering:

$$S = \frac{2\hbar \, \omega_s^4 \, N^2}{\rho \, c^4 \, \Omega} \, (n_o + 1) \sum_j |\hat{e}_s \cdot \overleftrightarrow{R}_j \cdot \hat{e}_i|^2 \qquad (2)$$

where ω_s is the frequency of the scattered light, N the number of primitive cells per unit volume, n_o the Bose-Einstein factor, ρ the density, c the velocity of light and Ω the phonon frequency. The vectors $\hat{e}_i$ and $\hat{e}_s$ represent the polarization of the incident and scattered light and $\overleftrightarrow{R}_j$ is the Raman tensor, the subindex j allowing for phonon degeneracy. The three components of the tensor $\overleftrightarrow{R}_j$ are:

$$R_3 = \begin{pmatrix} o & a & o \\ a & o & o \\ o & o & o \end{pmatrix} \quad R_2 = \begin{pmatrix} o & o & a \\ o & o & o \\ a & o & o \end{pmatrix} \text{ and } R_1 = \begin{pmatrix} o & o & o \\ o & o & a \\ o & a & o \end{pmatrix} \qquad (3)$$

Because of the LO-TO splitting in zincblende-type materials, two independent components are required to represent the tensor $\overleftrightarrow{R}_j$. The additional "allowed" component is related to the first order electro-optic coefficients[3] and produces a difference between the LO and the TO scattering powers not represented by Eqs. (2,3). This effect is small for III-V compounds[3] and will not be discussed here. In the definition of a in Eqs. (2) and (3) the normalization of the lattice distortion has been chosen such that R_1 represents the change in the polarizability of a primitive cell for atomic displacements equal to unity and of opposite directions for each one of the two basis atoms.

The determination of the absolute value of the scattering efficiency leads to an experimental value for the magnitude (not the sign) of the "Raman polarizability" a which, in turn, can be obtained theoretically from the band structure of the material[2,4,5]. The sign of a can be determined experimentally from quantum mechanical interference effects when two scattering mechanisms are present[6]. The comparison of the experimental magnitudes and signs of the polariza-bilities a with the results of band structure calcula-tions is of interest as a test of our microscopic understanding of the mechanism of Raman scattering.

The experimental determination of the absolute
scattering cross section is, however, not easy as a
number of geometrical factors must be taken into account.
It has become customary to determine a by comparison
with a "standard" substance. The absolute a of the
standard substance must be determined by careful
consideration of all geometrical factors involved. An
alternative method developped recently[1] is to compare
the Raman with the Brillouin scattering power: the
latter is given by expressions similar to Eq. (2) with
a replaced by a suitable linear combination of
elastooptic constants which can be obtained from a
conventional piezobirefringence experiment at least in
the region of transparency[8]. In this manner a value
of $|a| = 4.4$ Å^2 ster^{-1} has been determined for diamond.
This value of a is independent of frequency throughout
the whole region of transparency[7]. According to the
resonance Raman data of Ref. 7 the sign of a should be
negative. This sign agrees with the result of
calculations based on the pseudopotential band structure[2]
(a =-1.82). Although the experimentally determined a is
larger than the calculated one, we believe the value
of a = -4.4 Å determined as discussed above for diamond
should be quite accurate and can be used as a standard
of scattering power. It agrees well with the value
determined from the electric field induced infrared
absorption[9].

Diamond is particularly suitable as a standard for
the determination of the scattering power of materials
transparent to the scattered radiation. When opaque
materials are measured, one must correct for the finite
penetration depth of the radiation. In order to avoid
errors due to the depth of focus of the collecting
lens-spectrometer system a thin wafer of diamond must
be used. In this manner crystalline silicon has been
measured[10]. The results are represented by the black
squares in Fig. 1. The results of previous
measurements[11-13] are also shown in this figure. The
crosses represent the results of a comparison with the
intensity of the Brillouin line, whereby one must remark
that the elastooptic constants are not known very
accurately in the opaque region under consideration[8].
These points agree reasonably well with those obtained
from the comparison with diamond for photon energies
up to 2.5 eV. For higher photon energies the crosses
(Brillouin technique) lie much higher than the black

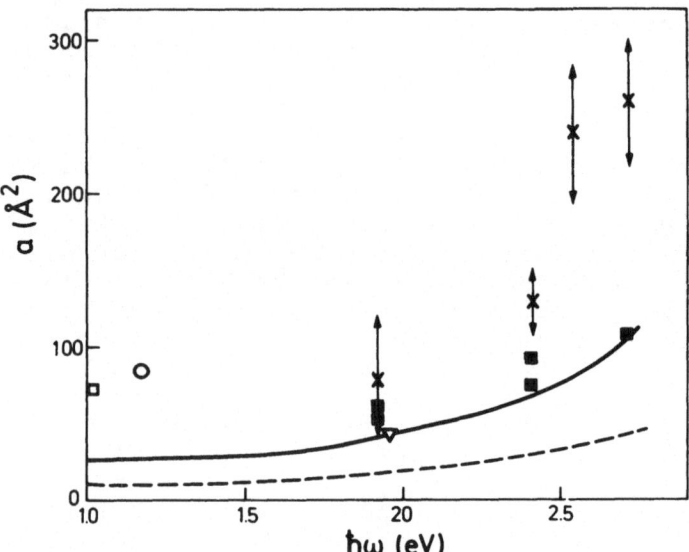

Figure 1. Absolute cross-section of first order Raman
          scattering of silicon. The full and dashed
          lines are theoretical calculations[1],[2],
          the X and ■ are values obtained in Ref. 10,
          and ∇, O and ▢ are previously determined
          values in the literature[11],[12],[13].

squares (obtained by comparison with diamond). The
reason for the discrepancy is not known.

     The dashed line in Fig. 1 displays the result of
a numerical calculation based on the pseudo-potential
band structure[2]. Except for a factor of ∿2 these
calculations represent well the dispersion of the black
experimental points. A similar discrepancy of ∿ a factor
of two was found between the experimental and calculated
values of a for diamond, as mentioned earlier. The
possibility of an error in the numerical calculations
appears, especially in view of the fact that more recent
numerical calculations[5] yield for a values in agreement
with the experimental ones.

It is possible, in some cases, to get around the numerical calculations and thus obtain an analytic expression for the scattering cross sections. This is done by using so-called "two band" models[14] in which the bands in the neighborhood of a critical point are approximated by parabolic bands extending to infinity. In this manner the Raman polarizability a is obtained quite accurately in the dispersive region near the critical point in terms of one or more deformation potential constants. The procedure introduces additive constants <u>away</u> from the critical point. The dispersion shown in Fig. 1 for the a of Si is due to the so-called E_1 critical points (direct gap along the {111} directions). The contribution of these E_1 critical points to a is given by (in atomic units):[10]

$$a(\omega) = \frac{1}{4\sqrt{6}} \, d_{3,0}^5 \, [\frac{\omega}{\omega_g} \frac{d\chi(\omega)}{d\omega} + \frac{\chi(\omega)}{\omega_g}] + \text{const.}, \qquad (4)$$

where ω_g is the energy of the E_1 gap, χ the frequency dependent susceptibility, and $d_{3,0}^5$ a deformation potential constant known from band structure calculations to equal $\simeq +27$ eV[15]. Using for $\chi(\omega)$ experimental data obtained from reflectivity measurements[16] and assuming that the additive constant in Eq. (4) is zero we obtain with Eq. (4) the solid line in Fig. 1 which fits the experimental data obtained by comparison with diamond (black points) quite well. The sign of a in Si has also been determined[6] to be <u>positive</u> below the E_1 gap. This agrees with the predictions of Eq. (4) and also with the complete numerical calculations[2,5].

While Eq. (4) applies to the contribution of the E_1 gap to a, it is also possible to write a similar equation for the corresponding contribution of the $\Gamma_{25'} \rightarrow \Gamma_{2'}$ (or $\Gamma_{15} \rightarrow \Gamma_1$ in zincblende) gap (the so-called $E_0 - E_0 + \Delta_0$ gap). This contribution must be included in materials like germanium or GaAs. It is given by:

$$a = \frac{C_0'' \, a_0^2}{\pi \, E_0} \frac{\sqrt{3}}{128} \, d_0 \{-g(x_0) + \frac{4E_0}{\Delta_0}[f(x_0) - (\frac{E_0}{E_0 + \Delta_0})^{3/2} \, f(x_{os})]\} \quad (5)$$

where C_0'' is approximately related to the real part of the dielectric constant through

$$\epsilon_r = C_0''[f(x_0) + 0.435 \, f(x_{os})] + \text{constant}. \qquad (6)$$

a_o is the lattice constant, d_o the phonon deformation potential for the Γ_8 valence state, E_o the energy gap, Δ_o the spin orbit splitting, $x_o = \hbar\omega/E_o$, $x_{os} = \hbar\omega/(E_o + \Delta_o)$ and $g(x)$ and $f(x)$ are given by

$$g(x) = x^{-2}[2-(1+x)^{-1/2} - (1-x)^{-1/2}] \qquad (7)$$

$$f(x) = x^{-2}[2-(1+x)^{1/2} - (1-x)^{1/2}]. \qquad (8)$$

The full line of Fig. 2 represents the best fit to the measured a of GaAs according to[17]

$$a = \left| A_1 \cdot \left\{ -g(x_o) + \frac{4E_o}{\Delta_o}\left[f(x_o) - \left(\frac{E_o}{E_o+\Delta_o}\right)^{3/2} f(x_{os}) \right] \right\} + A_2 \left[\frac{1}{1-x_1^2} + \left(\frac{E_1}{E_1+\Delta_1}\right)^2 \frac{1}{1-x_{1s}^2} \right] + A_3 \right| \qquad (9)$$

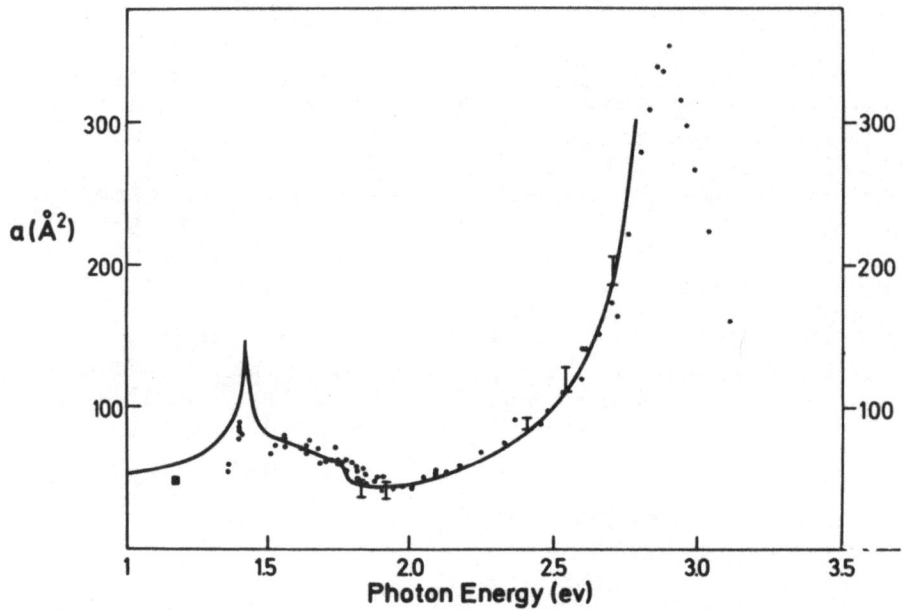

Figure 2. Raman tensor a of GaAs as a function of photon energy at room temperature. The length of the vertical bars indicates our estimated error. The full square is the result reported in Ref. 3. The solid line is the fit with Eq. 9.

where Δ_1 is the spin-orbit splitting of the E_1 gap and A_1, A_2, and A_3 are adjustable parameters. The A_2 contribution is an analytic expression which corresponds to Eq. (4). Using for GaAs E_o = 1.43 eV, Δ_o = 0.34 eV, E_1 = 2.89 eV, and Δ_1 = 0.23 eV, we obtain the excellent fit to the experimental points shown in Fig. 2. From this fit we obtain A_1 = 7, A_2 = 18, and A_3 = -5 in units of $\overset{o}{A}$. Using the value of C'' = 1.53 as determined from the piezobirefringence induced by a [111] stress (same symmetry as the TO phonon) we obtain d_o = 48 eV and $d_{3,0}^5$ = 37 eV. These results are in satisfactory agreement with theoretical calculations of Zeyher[18], i.e. d_o = 31.5 eV, $d_{3,0}^5$ = 41 eV.

A somewhat similar analysis of the contribution of the E_o edge to the first order Raman polarizabilities has been performed for the wurtzite type material ZnO. In this case, however, because of the lowered symmetry, five different deformation potential constants are obtained.

REFERENCES

1 M. GRIMSDITCH and A.K. RAMDAS, Phys. Rev. B 11: 3139 (1975)
2 L.R. SWANSON and A.A. MARADUDIN, Solid State Commun. 8: 859 (1970)
3 D. JOHNSTON and I.P. KAMINOW, Phys. Rev. 128: 1209 (1969)
4 M. CARDONA, Solid State Commun. 9:819 (1971)
5 H. WENDEL, Solid State Commun., in press
6 M. CARDONA, F. CERDEIRA, and T.A. FJELDLY, Phys. Rev. B10:3433 (1974)
7 J.M. CALLEJA, J. KUHL, and M. CARDONA, Phys. Rev. B17:876 (1978)
8 M. CHANDRASEKHAR, M.H. GRIMSDITCH, and M. CARDONA, Phys. Rev. B18:4301 (1978)
9 E. ANASTASSAKIS and E. BURSTEIN, Phys. Rev. B2: 1952 (1970)
10 M. GRIMSDITCH and M. CARDONA, in "Physics of Semiconductors 1978", B. Wilson ed., The Institute of Physics, London, 1978, p. 639
11 J.P. RUSSELL, Appl. Phys. Lett. 6:223 (1965)
12 J.M. RALSTON, R.L. WALSACK, and R.K. CHANG, Phys. Rev. Lett. 25:814 (1970)

13 A. MORADIAN in laser Handbook, ed. by T. ARECCHI
   and E.O. SCHULZ-DUBOIS (North Holland Publ. Co.,
   Amsterdam (1972), p. 1409
14 M. CARDONA, in "Atomic Structure and Properties of
   Solids", E. Burstein ed., Academic Press, N.Y.,
   1972, p. 514
15 J.B. RENUCCI, R.N. TYTE, and M. CARDONA, Phys. Rev.
   B11:3885 (1975)
16 H.R. PHILIPP and E. TAFT, Phys. Rev. 120:37 (1960)
17 M.H. GRIMSDITCH, D. OLEGO, and M. CARDONA, Phys.
   Rev., to be published
18 R. ZEYHER, private communication.

RESONANT SECONDARY EMISSION BY IMPURITIES IN CRYSTALS

K.K. Rebane

Institute of Physics of the Estonian SSR
Academy of Sciences
Tartu 202400, USSR

1. INTRODUCTION

Problems of resonant Raman scattering have been among the
essential ones in light scattering studies for quite a long time.
New horizons, opened by tunable lasers, stimulated remarkable
growth of activities in this field (see, e.g. (1)). In the case
under consideration resonance actually means that the exciting
light is absorbed to some extent by the matter under study. The
photons emitted by matter, after they had been captured (really
absorbed) for a certain period of time, should be different from
those which were not really absorbed but only scattered. These
differences should be checked and studied in order to get a better
understanding of the physical situation and to get more
information out of experimental data. So, in comparison with
nonresonant scattering the situation turns out to be more
complicated and, at the same time, more informative. When matter
consists of non-interacting atoms or molecules and only a pair of
resonant levels is to be taken into account we have the well-known
case examined by Heitler (2). The relation between the spectral
widths of the excited level and that of the excitation line
is decisive. If the latter is much smaller than the former,
i.e. $\Gamma \gg \Delta\omega$, resonant secondary emission (RSE) (light emitted by
an atom) performs as scattered light; in an opposite case,
RSE should be, according to Heitler, interpreted as resonance
fluorescence*. In intermediate cases it is proper to speak about

---

*The other possibility: to the extent that there is no transverse
relaxation present, all the RSE should be interpreted as
scattering.

RSE. Note that the same classification may be formulated when comparing the duration of the excitation pulse and the lifetime of the resonant level.

Note also, that fluorescence- and scattering-like emissions appear together, if the excitation intensity becomes high enough (see, e.g. (3)).

The situation becomes still more complicated if the light-transforming material system is a piece of condensed matter, or consists of molecules with various inner relaxation processes taking place within characteristic times much shorter than the lifetime of the excited electronic state. When a relaxation process is a thousand times faster than the rate of the optical transitions, which is the usual case for an overwhelming majority of luminescence centers in solids and liquids, the intensity of luminescence is about a thousand times stronger than that of the scattering (only if strong quenching is not taking place). That is why reasonable theories of resonant Raman scattering in condensed matter should take account of relaxation processes and not ignore luminescence (see, e.g. (4)). Moreover, one should not always expect an enhancement of scattering in resonance, because the energy denominator becomes small for the whole RSE cross section, the main part of which, as was stressed already, reduces to luminescence.

Resonant excitation is especially important when small amounts of impurities in condensed matter are under study. That was understood and made use of in luminescence studies decades ago: even very small amounts of impurities excited selectively could give strong luminescence under proper conditions. Investigations via luminescence at selective resonant excitation are among the most sensitive methods in spectral analysis. The light scattering is here completely neglected. Due to the very high intensity of luminescence compared to the scattered light this approach is always reasonable provided the quantum yield of the luminescence is not too low and we are not looking for scattering (e.g. in that part of the spectrum where there is no luminescence at all). Nevertheless the problem of distinction of luminescence from the scattered light on one hand, and from the ordinary thermal radiation on the other hand, had been carefully examined by the founders of the modern science of luminescence - P. Pringsheim and by S. Vavilov (5) (see also (6)). The spectral characteristics and properties of time dependence were taken into account in their discussions. B. Stepanov and P. Apanasevich pointed out the decisive role of the intermediate processes in the excited state in transforming scattering into luminescence (7).

As an important extension of Raman scattering studies resonant excitation was applied to liquids and liquid mixtures by P.P.

Shorygin more than twenty years ago (8) (see also (10)). It is a
pity that up to now, when this method has become very well known,
especially in connection with the Raman scattering studies in
chemistry, very little attention has been paid to the question of
what really happens to the other possible components of the RSE -
luminescence and hot luminescence. The interpretation may become
especially complicated when the liquid under study behaves like an
inhomogeneous host matrix (9).

The present status could be characterized by a gradually
widening variety of Raman studies both linear and non-linear,
involving resonance effects, and a rapid growth of a number of
relevant experimental as well as theoretical (see, e.g.
(10-12)) publications.

The first part of this paper will give a brief review of the
published studies about RSE of condensed matter and related
problems. The list of references is quite a long one but far from
being complete. The aim of the second part is to demonstrate via
theoretical time-dependent spectra of an impurity centre in a
crystal how all the three components of RSE come into being, how
the luminescence lines develop and the pure electronic line
gradually turns into the well-known very narrow Lorentzian-shaped
line - the optical analog of the Mossbauer line (13,14).

2. RESONANT SECONDARY EMISSION: LUMINESCENCE, HOT LUMINESCENCE, LIGHT SCATTERING

In the case of rapidly relaxing centers of luminescence there
is a clear physical distinction between resonant scattering (RS),
ordinary (OL) and hot luminescence (HL). Slightly different
versions of the theory of RSE based upon two-photon processes
including all the three components mentioned above were worked out
in (4,14-18). The criteria of distinction may be formulated via
transverse and longitudinal relaxations. The RSE components can
be determined as follows (18):
     1. OL is the emission from the thermal equilibrium vibrational
state, i.e. from the state where transverse as well as
longitudinal vibrational relaxations are finished.
     2. HL is the emission from a state where the phase memory of
the excitation has failed already, the vibrational distribution,
however, has not yet reached equilibrium. In other words, HL
occurs after the phase (transverse) relaxation is over but before
the energy (longitudinal) relaxation is finished.
     3. Resonance scattering is the emission from a state which
still has a memory of excitation, i.e. before the phase as well as
energy vibrational relaxations take place. In other words, no
actually absorption nor "real" transition into the excited
electronic state take place.

The conclusions of the theory are in reasonable agreement with experiments (see an earlier review (19). HL of crystals was first observed on a sample of $KCl-NO_2$ in (20). Afterwards, in the laser-excited RSE of molecular anions all three components - OL, HL and RRS are clearly demonstrated: on $KI-Se_2$ by L. Rebane and T. Haldre (21), on $KCl-NO_2$ by P. Saari (23,24). Recently full RSE spectra were obtained and investigated in the case of mixed (24) and pure molecular crystals (25); particularly, a rich RSE spectrum in the strong exciton absorption region of anthracene shows pecularities caused by polariton effects (26).

The HL studies have provided information on different energy relaxation pathways and the corresponding characteristic times of picosecond duration (see review papers (19,25)). HL data combined with the studies of homogeneous linewidths enable to get estimates for the transverse relaxation times to be obtained as well (27).

3. TIME-DEPENDENT (TRANSIENT) SPECTRA OF RSE

There is quite a number of general problems about time-dependent spectra such as mathematical definitions of what is a time-dependent spectrum, and how to take into account the role of the spectral apparatus when real physical spectra are concerned (28-31)). Recent success in pico- and subpicosecond pulse experiments requires a corresponding development of theory, and recently a number of papers on time-dependent RSE spectra of luminescence centers in crystals has been published (30-33). Naturally, time-dependent spectra display very clearly how all three RSE components - scattered light, HL and OL - come into being after a short-pulse excitation, how the intensities and shapes of the lines of luminescence develop with the time of collecting photons and how they depend on the choice of the collection time interval. The models used in (30,32,34) and especially in (35-37) by V. Hizhnyakov and I. Rebane are quite complete for a proper discussion of the problem (see also (38,39)).

I shall review some recent results of the theory mentioned above. The details of the models under study and the corresponding formulae may be found in (32,34,35,37). Let us give here the list of the notations and the main features of the models.

The emission center is characterized by a usual potential energy diagram with two parabolic curves of different curvatures representing the local vibration in ground and excited electronic states (Fig. 1). As we know, it is most important to take into account the vibrational relaxation, without which we cannot get the correct picture of RSE. Here it is supposed that the n-th level of the oscillator decays exponentially with the characteristic time $\tau_{||,n} = \tau_{||,1}/n = (2\Gamma_{||}n)^{-1}$ (model 1)

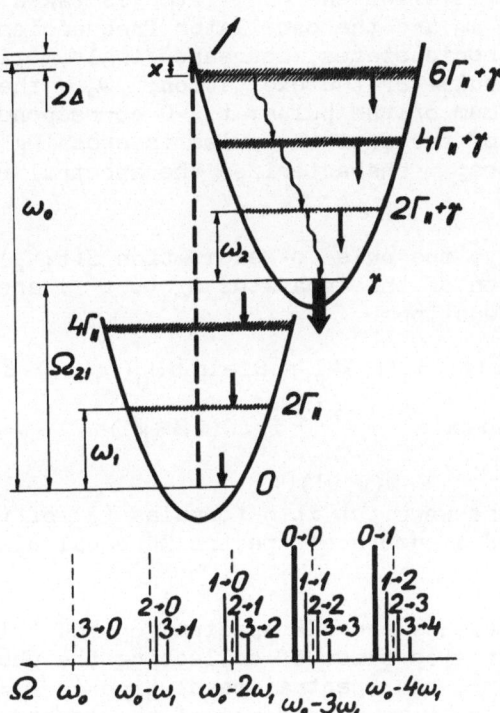

Fig. 1. The diagram of the potential energy curves and the scheme of the vibronic resonant secondary emissions lines of an emission center in a crystal. The transitions and lines of ordinary luminescence (broad lines), hot luminescence (narrow lines), and scattered light (dashed lines) are shown. Because of the different frequencies of vibration in the ground and excited electronic states the luminescence lines corresponding to different transitions are separated; owing to the excess of the excitation energy x over the vibronic level the lines of scattering are shifted from the luminescence ones (35,37).

(34). Model 2 takes account of the transverse relaxation as well, i.e. the same $\tau_\perp$ for all levels of the excited electronic state is introduced; so the width of the n-th level is $\sigma_n = 2\Gamma_{||}n + 2\Gamma + \gamma$, where γ stands for the optical linewidth.

Everywhere in the Figures the relaxational halfwidth $\Gamma_{||}$ of the first excited level of the oscillator is taken as the unit for frequency; ω_1 and ω_2 are the oscillator frequencies in ground and excited electronic states, correspondingly; Δ is the spectral halfwidth of the pulse of the excitation; ω_0 - the frequency at the spectral maximum of the pulse; $t = 0$ corresponds to the moment when the maximum of the exciting pulse is crossing the luminescence centre; η characterizes the spectral resolution of the apparatus.

More precisely, the pulse of excitation $S(t_1 t_2)$ and the spectral resolution of the apparatus $C_\Omega(\nu,\nu')$ are represented via correlation functions

$$S(t_1 t_2) \equiv \langle S_1^*(t_1)S_2(t_2)\rangle_R = S_0 \exp\{i\omega_0(t_1-t_2)-\Delta|t_1|-\Delta|t_2|\}, \quad (1)$$

$$C_\Omega(\nu,\nu') \equiv \iint dx dx' e^{ix\nu-ix'\nu'} f_\Omega(\Omega+x, \Omega+x') =$$

$$= C_0 e^{-\eta|\nu|} e^{-\eta|\nu'|} \equiv C(\nu,\nu'). \quad (2)$$

These functions were put into formulas (3) of (32) (see also (28-30,35,37)) and a number of spectra were calculated on a computer.

Figure 2 represents the RSE spectra for model 1 corresponding to three different stop times of collecting the photons: $t = 0$, $0.5\ \Gamma^{-1}$ and $1.0\Gamma^{-1}$, the spectral resolution is $\eta = \Gamma$. As follows from (2) the time dependence of the spectrum is actually rather complicated. It depends not only on the time t but on η as well. The sensitivity decreases with increase of η.

The spectrum at $t = 0$ represents the situation when the first half of the excitation pulse has crossed the system. We have mainly scattered light. Because of the absence of transverse relaxation ($\tau_\perp=0$) in model 1 the lines at $3 \to 1$ and $3 \to 2$ have to be interpreted as the Raman ones. We do not have here any phase memory losses except the energy relaxation and consequently in this model we do not have HL lines starting from level 3. Because in this case the half pulse crossing time is about the same as the vibrational relaxation time of the third level, some relaxation takes place and the HL lines begin to form.

The $t = 0.5\ \Gamma^{-1}$ the spectrum has well-pronounced HL lines already.

At $t = 1.0\ \Gamma^{-1}$ the hot lines starting from levels 2, 1 are

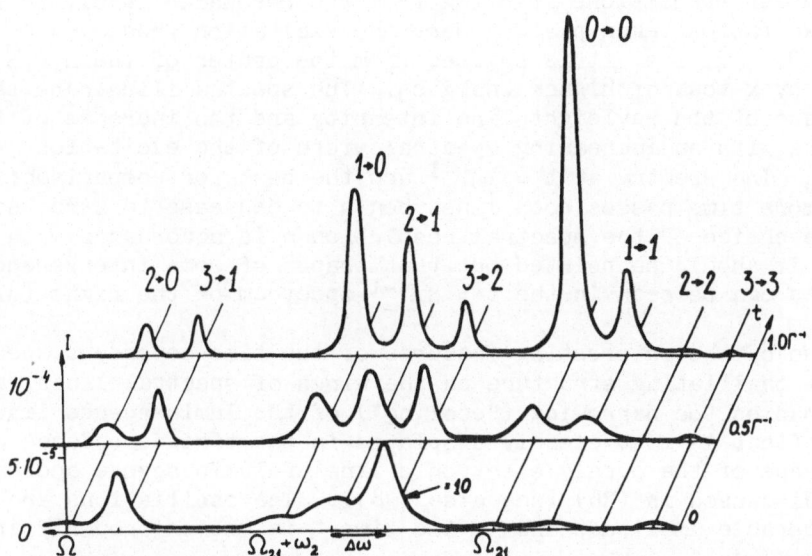

Fig. 2. Time dependence of the computer-calculated theoretical RSE
spectrum at the very beginning of the formulation of the spectrum
of luminescence. All lines of luminescence are hot. Because of
the absence of transverse relaxation in model 1 lines 3   1,2,3
are to be interpreted as scattering. Here $\Gamma \equiv \Gamma_{||}$, $\omega_0 = \Omega_{21} +$
$3\omega_2$, $\omega_1 = 150\Gamma$, $\omega_2 = 120\Gamma$, $\Delta = 5\Gamma$, $\eta = \Gamma$, $\gamma = 0.002\Gamma$, Stokes'
shift $S = 0.5$.

considerably stronger. The $0 \to 0$ line - the only one in the
picture which in the future becomes an OL line - already shows a
trend to continue its growth far beyond the vibrational relaxation
time. The RS lines $3 \to 1,2,3$ have not grown in comparison with
their intensities at $t = 0.5\,\Gamma^{-1}$. All lines are quite broad,
including the $0 \to 0$ line, which in this case is a hot line too,
but a slight narrowing trend is visible already. In other words,
all lines are transient in a transient spectrum. That is why it
seems to be preferable to use the terms "light scattering" and
"hot luminescence" and not to call "transient" any lines in a
conventional (stationary) spectrum.

The results of the study and computer calculations of the
transient RSE spectra for model 2 with two relaxation times by I.

Rebane, A. Tuul and V. Hizhnyakov (37) are presented in Fig. 3.

The main difference in comparison with model 1 ($\Gamma_\perp = 0$) occurs for transitions starting from the resonance level, to which the excitation takes place. Here the excitation frequency $\omega_o = \Omega_{12} + 3\omega + x$, i.e. it is shifted from the center of the n = 3 level by x towards higher energies. The spectra illustrate the decrease of the Rayleight line intensity and the increase of the HL line with an increasing spectral width of the excitation pulse. The spectra at $t = 1\,\Gamma^{-1}$ are the best for comparison; when more time passes both lines begin to decrease to zero because of the choice of the spectral resolution η in accordance with (2). It should be pointed out that traces of some interference effects can be seen in the $t = 3\,\Gamma^{-1}$ spectrum on the right (37).

One of the interesting features of the time-dependent spectra is the oscillating structure on the wings of spectral lines, which accompanies the narrowing ("cooling") of the luminescence lines. This effect is most clearly displayed in the time dependence of the shape of the purely electronic line of luminescence and has been discussed in (36) (see also (38)). The oscillations are of considerable amplitude and if the time t is chosen properly are

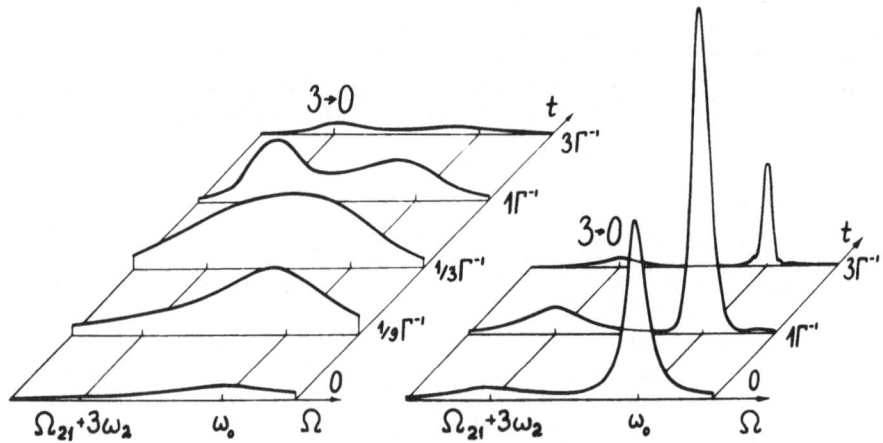

Fig. 3. The Rayleigh line $\Omega = \omega_o = \Omega_{12} + 3\omega_2 + x$ and hot luminescence line $3 \rightarrow 0$ $\Omega = \Omega_{12} + 3\omega_2$ (the latter exists due to the transverse relaxation $\Gamma_\perp \neq 0$) are separated because the excitation frequency is shifted by x to higher energies from the n = 3 level. Here $\Gamma_\perp = \Gamma$, $\eta = 0{,}5\Gamma$; x = 20Γ; the spectral width of the excitation pulse $\Delta = \Gamma$ for the spectra on the right and $\Delta = 9\Gamma$ on the left of the figure (37).

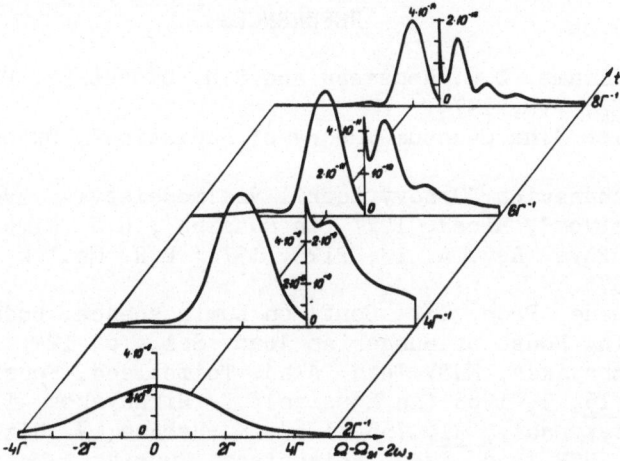

Fig. 4. Oscillating sidebands of the hot luminescence line 3 ∿→ 2 → 0.

well separable in the frequency domain. But the inhomogeneous broadening of optical spectral lines in crystals makes the experimental study rather difficult. Up to now the oscillating structure on the wings and line narrowing, including narrowing beyond the natural linewidth, has been studied only for the case of Mössbauer lines (40,41) (see also (38).

Fig. 4 represents the oscillating structure on the sidebands of the HL line 3 ∿ 2 → 0. In the case of the purely electronic line (in the absence of inhomogeneous broadening) this structure is always present, if only the photon collecting time t is properly fixed. With HL lines the situation is more complicated: a very special choice of the resolution η and spectral width Δ of the excitation pulse is needed to get such a relatively distinct structure as displayed in Fig. 4. It should be pointed out that these choices are analogous to those leading to remarkable and special narrowing effects of time-dependent HL lines, including narrowing far beyond the widths determined by the lifetimes of the levels involved in the transition (37).

ACKNOWLEDGEMENTS
    The author is grateful to P. Saari, V. Hizhnyakov, I. Rebane for discussions and for kind permission to review their very recent results. Thanks are also due to L. Pedosar and E. Vaik for their valuable help in preparing this publication.

REFERENCES

1. P.F. Williams, D.L. Rousseau and S.H. Dworetsky, Phys. Rev. Lett., 32, 196, 1974.

2. W. Heitler "The Quantum Theory of Radiation", Oxford, 1954, p. 196.

3. P.A. Apanasevich "Osnovy teorii vzaimodeistviya sveta a veshchestvom", Minsk, 1977 (in Russian); H.J. Kimble and L. Mandel, Phys. Rev. A, 13, 2123, 1976; B.R. Mollow, ibid., 12, 1919, 1975.

4. K.K. Rebane, Proc. Int. Conf. on Luminescence, Budapest, 1966, Publishing House of Hungarian Acad. Sci., p. 124; I.J. Tehver, V.V. Hizhnyakov, ENSV Tead. Akad. Toimetised, Fuusika * Matem., 15, 9, 1966 (in Russian); V. Hizhnyakov, I. Tehver, phys. stat. sol., 21, 755, 1967; K. Rebane, V. Hizhnyakov, I. Tehver, ENSV Tead. Akad. Toimetised, Fuusika * Matem., 16, 207, 1967.

5. S.I. Vavilov, Complete works, V. 2, Publishing House of USSR Acad. Sci., 1952, p. 188 (in Russian).

6. A.E. Adirovich "Teorija luminestsentsii kristallophosphorov", Moscow, 1951 (in Russian).

7. B.I. Stepanov, P.A. Apanasevich, Izv. Akad. Nauk SSSR, ser. fiz., 22, 1380, 1958 (in Russian).

8. P.P. Shorygin, Dokl. Akad. Nauk SSSR, 87, 201, 1952 (in Russian); Sov. Phys. Uspekhi, 16, 99, 1973.

9. V. Hizhnyakov and I. Tehver, J. Luminescence, 18/19, 673, 1979; K.K. Rebane, R.A. Avarmaa, A.A. Gorokhovski, Izv. Akad. Nauk. SSSR, ser. fiz., 39, 1793, 1975 (in Russian).

10. Proc. of the Sixth Int. Conference on Raman Spectroscopy, Bangalore, India, Sept. 4-9, 1978. Eds. E.D. Schmidt, R.S. Krishan, W. Kiefer, H.W. Schrotter. Heyden, London - Philadelphia - Rheine, 1978; see also J. Brandmuller and W. Kiefer, in: The Spex Speaker, 1978, Raman Anniversary Issue.

11. J. Luminescence, 18/19, 1979, Proc. of Intern. Conf. on Luminescence, Paris, July, 1978.

12. Light Scattering in Solids, Third Int. Conf., Campinas Brazil, July 1975. Eds. M. Balkanski, R.C.C. Leite, S.P.S. Porto, Flammarion, Paris, 1978.

13. K.K. Rebane, Impurity Spectra of Solids, Plenum Press, N.Y.-London, 1970.

14. K. Peuker, E.D. Trifonov, phys. stat. sol., 30, 479, 1968; Fizika Tverdogo Tela, 10, 1705, 1968 (in Russian).

15. Y. Toyozawa, J. Phys. Soc. Japan, 41, 400, 1976; A. Kotani, Y. Toyozawa, J. Phys Soc. Japan, 41, 1699, 1976.

16. S. Mukamel, J. Jortner, J. Chem. Phys., 62, 3609, 1975; A. Nitzan, J. Jortner, J. Chem. Phys., 57, 2870, 1972; S. Mukamel, A. Nitzan, J. Chem. Phys., 66, 2462, 1977.

17. J. Friedman, R.M. Hochstrasser, Chem. Phys. Lett., 32, 414, 1975; R.M. Hochstrasser, F.A. Novak, Chem. Phys. Lett., 41, 407, 1976; 48, 1, 1977; 53, 3, 1978.

18. Rebane, K.K., Tehver I.Y., Hizhnyakov V.V., in: "Theory of Light Scattering in Condensed Matter". Eds. B. Bendow, J. Birman, V. Agranovich, Plenum Press, N.Y. and London, 1976, p. 393; see also Comment by Y.R. Shen, p. 407.

19. K. Rebane and P. Saari, J. Luminescence, 16, 223, 1978.

20. P. Saari and K. Rebane, Solid State Commun., 7, 887, 1969.

21. L.A. Rebane, T.J. Haldre, Pis'ma JETP, 26, 674, 1977 (in Russian).

22. Ultrafast Relaxation and Secondary Emission, Proc. of the Int. Symposium "Ultrafast Phenomena in Spectroscopy", Tallinn, Sept. 27 - Oct. 1, 1978, published by Estonian Acad. Sci.

23. See in (22) p. 14.

24. P. Saari, in the present Proceedings.

25. P. Saari, in (22), p. 142.

26. J. Aaviksoo, P. Saari, T. Tamm, Pis'ma JETP, 29, 388, 1979.

27. L.A. Rebane, in (22), p. 89.

28. J.H. Eberly, K. Wodkiewicz, J. Opt. Soc. Am., 67, 1252, 1977.

29. E. Courtens, A. Szoke, Phys. Rev., A15, 1588, 1977.

30. V. Hizhnyakov, Tech. Rep. of Inst. Solid State Phys., Univ. Tokyo, Ser. A, No. 860, 1977.

31. P. Saari, ENSV Tead. Akad. Toimetised, Fuusika * Matem, 27, 109, 1978 (in English); see also: P. Saari, in the present issue.

32. V.V. Hizhnyakov, I.K. Rebane, Zh. Eks. Teor. Fiz. (JETP), 74, 885, 1978 (in Russian).

33. T. Takagahara, E. Hanamura and R. Kubo, J. Phys. Soc. Japan, 43, 802, 1977; 43, 311, 1977; 43, 1522, 1977; 44, 728, 1978.

34. Inna Rebane, ENSV Tead. Akad. Toimetised, Fuusika * Matem., 27, 192, 1978.

35. Inna Rebane, ENSV Tead. Akad. Toimetised, Fuusika * Matem., 27, 459, 1978.

36. I. Rebane, A. Tuul, ENSV Tead. Akad. Toimetised, Fuusika * Matem., 27, 463, 1978 (in Russian).

37. I.K. Rebane, V.V. Hizhnyakov, A.L. Tuul, Preprint F-10, Institute of Physics, Estonian SSR Acade. Sci., 1979; I.K. Rebane, A.L. Tuul, V.V. Hizhnyakov, JETP, to be published.

38. K.K. Rebane, J. Luminescence, 18/19, 693, 1979.

39. K.K. Rebane, in (22), p. 7.

40. T. Kobayashi, S. Shimizum Phys. Lett., 54A, 311, 1975.

41. E. Realo, R. Koch, in: Magnetic Resonance and Rel. Phen. (XX Congr. AMPERE, 1978), Tallinn, 1979, p. 246.

THE RAMAN SCATTERING AND HOT LUMINESCENCE OF

SELF-TRAPPING EXCITONS

V. Hizhnyakov

Institute of Physics
Estonian SSR Academy of Sciences
Tartu, USSR

SUMMARY

A theoretical consideration of light scattering by an excitation of self-trapping excitons in the absorption band is presented. The whole secondary radiation spectrum, including ordinary and hot luminescence (OL and HL), is studied. Polariton effects are disregarded.

Two different cases of exciton self-trapping are treated as follows: (1) the strong interaction of a localized exciton packet as well as a free exciton with phonons; and (2) the strong interaction of a localized exciton packet with phonons, but weak interaction of a free exciton with phonons. In the first case the secondary radiation spectrum is shown to be analogous to the corresponding spectrum of an impurity centre in a crystal. The main differences are the validity of the quasi-momentum conservation law and a certain motion of an exciton in the hot vibrational state. The former significantly affects only the Raman scattering cross sections of some first orders. The latter leads to an asymmetry in scattering excitation profiles and to an enhancement of the blue part of HL spectrum.

The main feature of the second case is the coexistence of free and self-trapped excitons, which are separated by a potential barrier. It is shown that in this case the resonance scattering of small order can be described by formulas analogous to those for the case of weak exciton-phonon coupling. The multiphonon resonance scattering, however, differs sharply in that it is primarily conditioned by transitions involving a self-trapped state as an intermediate one. Formulas are presented describing two kinds of such resonance scattering: (1) tunnel HL created by the penetration of an exciton through the self-trapping barrier and the subsequent relaxation and (2) HL from the top of the barrier.

In this paper, the resonance light scattering of a crystal with excitation in the absorption band of self-trapping excitons is discussed. The resonance scattering of light is defined in a very general sense, i.e., a process in a substance transforming primary photons into secondary ones. Consequently, both ordinary and hot luminescence belong to the resonance scattering of light.

Here we assumed that the interaction of light with excitons is much weaker than the interaction of excitons with phonons, allowing us to neglect the polariton effects. In such a case an important characteristic of the light transformation process in a crystal is the light scattering cross section described as

$$I(\Omega_1, \Omega_2) = \frac{1}{2\pi} \int_{-\infty}^{\infty} dt\, e^{-i(\Omega_1 - \Omega_2)t} < P^+(t)P(0) > . \qquad (1)$$

Here Ω_1 and Ω_2 are the frequencies of the primary and secondary photons, $P(t) = e^{itH}Pe^{-itH}$, H is the Hamiltonian of the substance, and P is the polarizability operator, $h = 1$. When the excited light and the exciton band are in resonance (within a multiplicative constant), the polarizability operator is determined to be

$$P = i \int_{0}^{\infty} d\tau\, e^{i\Omega_1 \tau} a_{k_2} e^{-i\tau H - \hat{\gamma}\tau} a^+_{k_1} e^{i\tau H - \hat{\gamma}\tau} , \qquad (2)$$

where k_1 and k_2 denote the wave vectors of the exciting and scattered light ($k_1 = \Omega_1/c$, $k_2 = \Omega_2/c$), $a^+_{k_1}$ and a_{k_2} are the creation and annihilation operators of the excitons with wave vectors k_1 and k_2, respectively, and $\hat{\gamma}$ is the radiation damping operator.

Let us assume that only one resonance Frenkel exciton band exists. In such a case, the Hamiltonian H may be written in a conventional form: $H = H_e + H_L + H_{eL}$, where

$$H_e = \sum_k E_k a^+_k a_k , \qquad (3)$$

$$H_L = \sum_{qr} \omega_{qr} b^+_{qr} b_{qr} , \qquad (4)$$

$$H_{eL} = \sum_{kqr} (V_{qr} a_k^+ a_{k+q} b_{qr}^+ + h.c.) \tag{5}$$

are the Hamiltonians of excitons, phonons and their coupling, E_k is the energy of an exciton with the wave vector k, and b_{qr}^+, b_{qr} are the creation and annihilation operators of a phonon with the wave vector q of the branch r.

In this study, the main assumption concerning the exciton-phonon interaction is the fulfillment of the conditions

$$S = \sum_{qr} |V_{qr}/\omega_{qr}|^2 \sim (\sigma/\bar{\omega})^2 \gg 1 , \quad S > B\sqrt{\omega} , \tag{6}$$

where B is the exciton bandwidth, $\sigma^2 = \sum_{qr} |V_{qr}|^2 (2n_{qr}+1) \sim S\bar{\omega}^2$, σ is the width of the exciton absorption band caused by dispersion, $\bar{\omega}$ is the mean frequency of phonons, and n_{qr} is the number of phonons with the frequency ω_{qr} at the temperature T. The condition $S \gg 1$ means that an exciton trapped at one site interacts strongly with phonons, and the condition $S > B\sqrt{\omega}$ indicates the occurrence of exciton self-trapping.

An important parameter of the model is

$$S = (\sigma/B)^2 \sim S(\bar{\omega}/B)^2 , \tag{7}$$

which determined the interaction strength of the band exciton with phonons. When conditions (6) are fulfilled, this parameter may be either more or less than unity.

If $S \gg 1$, the resonance absorption band is approximately a Gaussian shape [1] with a halfwidth $\sigma 2\sqrt{2} \cdot ln2$ and a maximum at the frequency $E = \Omega_0 + B/2$, where Ω_0 is the frequency of the exciton band minimum. For $s \gg 1$, the photo-induced excitons localized at one of the lattice sites lose coherence during the time $\sim \sigma^{-1}$, which is short compared with the delay time on a site, as well as with the time of full vibrational relaxation. This localization is analogous to the Andersen localization because of exciton energy fluctuations in a rigid lattice with "frozen" vibrational modes. Inclusion of the vibrations leads to an incoherent hopping of the exciton from one site to another with a probability $\tau_0^{-1} \sim B^2/\sigma d$ [2] (d is the number of nearest neighbours). However, the transfer is probable only when the exciton energy remains within the interval $E \pm \sigma$. During the time of vibrational relaxation $t_0 \sim \bar{\omega}^{-1}[(2\bar{n}+1)/S]^{\frac{1}{4}} \ll \bar{\omega}^{-1}$ [2] the exciton leaves this energy interval. If $t_0/\tau_0 \ll 1$, the transfer probability is small and $(t_0/\tau_0)^{\frac{1}{2}}$ may be taken as a small parameter of the theory.

In a zeroth-order approximation, a model of an oriented gas may be used, which takes into account the periodic structure of the crystal and the exciton-phonon interaction but neglects the exciton motion. In that case, using the site representation, we obtain the following formula for the scattering cross section for one lattice site:

$$I^{(0)}(\Omega_1,\Omega_2) = 2\mathrm{Re}\left\{ \int_0^\infty \int dt\,d\tau\,d\tau'\, e^{i\Omega_2 t - i\Omega_1(t+\tau'-\tau) - \gamma(\tau+\tau')} \right.$$

$$\left. \times N^{-2} \sum_{n_1 n_2} e^{i(k_1-k_2)(n_1-n_2)} A^{(0)}_{n_1 n_2} \right\} , \tag{8}$$

$$A^{(0)}_{n_1 n_2} = \langle e^{iH_{n_2}\tau'}\, e^{iH_L t}\, e^{-iH_{n_1}\tau}\, e^{-iH_L(t+\tau'-\tau)} \rangle ,$$

where γ is the radiation damping constant, n_1 and n_2 denote the lattice sites, whose number is N,

$$H_n = e^{\nabla_n} H_L e^{-\nabla_n} + \omega_0 \tag{9}$$

is the vibrational Hamiltonian of a crystal with an exciton at the n-th site, e^{∇_n} is the unitary displacement operator,

$$\nabla_n = \sum_{qr} \left(\frac{V_{qr}^* e^{iqn}}{\omega_{qr}} b_{qr} - \text{h. c.} \right),$$

$$\omega_0 = E - \sum_{qr} |V_{qr}|^2 / \omega_{qr} .$$

In an harmonic approximation

$$A^{(0)}_{n_1 n_2} = \langle e^{\nabla_{n_2}}\, e^{-\nabla_{n_2}(\tau')}\, e^{\nabla_{n_1}(t+\tau')}\, e^{-\nabla_{n_1}(t+\tau'-\tau)} \rangle$$

$$= \exp\{ i\omega_0(\tau'-\tau) + g(\tau') + g(-\tau) + K_{n_1 n_2}(t,\tau,\tau') \} , \tag{10}$$

where $g(\tau) \equiv g_{nn}(\tau)$, $K_{n_1 n_2}(t,\tau,\tau') = g_{n_1 n_2}(t) + g_{n_1 n_2}(t+\tau'-\tau)$

$-g_{n_1 n_2}(t+\tau') - g_{n_1 n_2}(t-\tau)$, with

$$g_{n_1 n_2}(t) = <\nabla_{n_2}\nabla_{n_1}> - <\nabla_{n_2}\nabla_{n_1}(t)>$$

$$= \sum_{qr} e^{iq(n_1-n_2)}\left|\frac{V_{qr}}{\omega_{qr}}\right|^2 [(e^{i\omega_{qr}t} - 1)(1 + n_{qr})$$

$$+ (e^{-i\omega_{qr}t} - 1)n_{qr}]. \tag{11}$$

Let us expand correlator (10) in powers of $K_{n_1 n_2}$. After an integration over t, τ and τ', we get

$$I^{(0)}(\Omega_1,\Omega_2) = \sum_n I_n^{(0)}(\Omega_1,\Omega_2), \tag{12}$$

where $I_n^{(0)}(\Omega_1,\Omega_2)$ describes the resonance Raman scattering of the n-th order. At $T = 0$, it is described by a formula[*)]

$$I_n^{(0)}(\Omega_1,\Omega_2) = \frac{\pi^2}{n!\sigma^2} \sum_{q_1 r_1} \cdots \sum_{q_n r_n} \left|\frac{V_{q_1 r_1}\cdots V_{q_n r_n}}{\omega_{q_1 r_1}\cdots\omega_{q_n r_n}}\right|^2$$

$$\times \left|\sum_{p_1\cdots p_n=0}^{1} (-1)^{p_1+\cdots p_n} \Phi(\Omega_2+p_1\omega_{q_1 r_1} + \cdots p_n\omega_{q_n r_n})\right|^2$$

$$\times \delta(\Omega_2+\omega_{q_1 r_1} +\cdots\omega_{q_n r_n} -\Omega_1)\delta(k_2-k_1+q_1+\cdots q_n), \tag{13}$$

where
$$\Phi(\Omega) = i \sqrt{\frac{2}{\pi}} \sigma \int_0^\infty d\tau e^{i(\Omega-\omega_0)\tau-\gamma\tau-g(\tau)}$$

$$= i \sqrt{2\pi} \sigma \kappa(\Omega) + \sqrt{\frac{2}{\pi}} \sigma \int \frac{\kappa(x)dx}{\Omega - x} \tag{14}$$

[*)] It is quite easy to generalize formula (13) for the case $T \neq 0$; however, the corresponding expression is quite cumbersome.

is the dispersion of the complex refractive index in the exciton absorption region. In particular,

$$I_0^{(0)}(\Omega_1,\Omega_2) = |\Phi(\Omega_1)|^2 \delta(\Omega_1-\Omega_2)\delta(k_1-k_2), \tag{15}$$

$$I_1^{(0)}(\Omega_1,\Omega_2) = \sum_{qr} \left|\frac{V_{qr}}{\omega_{qr}}\right|^2 \left|\Phi(\Omega_1) - \Phi(\Omega_1-\omega_{qr})\right|^2$$

$$\times \delta(\Omega_1-\Omega_2-\omega_{qr})\delta(k_2-k_1+q), \tag{16}$$

$$I_2^{(0)}(\Omega_1,\Omega_2) = \frac{1}{2} \sum_{q_1r_1} \sum_{q_2r_2} \left|\frac{V_{q_1r_1}V_{q_2r_2}}{\omega_{q_1r_1}\omega_{q_2r_2}}\right|^2$$

$$\times \left|\Phi(\Omega_1)+\Phi(\Omega_2)-\Phi(\Omega_1-\omega_{q_1r_1})-\Phi(\Omega_2-\omega_{q_2r_2})\right|^2$$

$$\times \delta(\Omega_1-\Omega_2-\omega_{q_1r_1}-\omega_{q_2r_2})\delta(k_2-k_1+q_1+q_2). \tag{17}$$

Here $I_0^{(0)}$ describes the contribution of the exciton transition to the forward elastic scattering, whereas $I_1^{(0)}$ and $I_2^{(0)}$ are the resonance Raman scattering of the first and second orders, respectively. In the case under consideration, $\sigma \gg \bar{\omega}$ for $n < s^{\frac{1}{2}}$:

$$I_n^{(0)}(\Omega_1,\Omega_2) = \frac{\pi^2}{n!\sigma^{2(n+1)}} |\Phi_n(z)|^2 \sum_{q_1r_1} \cdots \sum_{q_nr_n}$$

$$\times |V_{q_1r_1}\cdots V_{q_nr_n}|^2 \delta(\Omega_1-\Omega_2-\omega_{q_1r_1}-\cdots\omega_{q_nr_n})$$

$$\times \delta(k_1-k_2-q_1\cdots-q_n), \tag{18}$$

where

$$\Phi(z) = ie^{-z^2} + \frac{2}{\sqrt{\pi}} w(z),$$

$z = (\Omega_1 - E)/\sigma\sqrt{2}$, $w(z) = e^{-z^2}\int_0^z e^{t^2}dt$ is the Dawson function,
$\Phi_n(z) = d^n\Phi(z)/dz^n$.

It follows from (18) that, in the case considered, the exciting profiles of RRS are determined by the same function $|\Phi_n(z)|^2$ as in the case of impurity centres with large Stokes shifts [4]. However, the relation of the scattering cross sections of various orders differs greatly from that in the case of an impurity centre as a result of quasimomentum conservation in an ideal crystal. In particular, in the case under consideration, because of the smallness of k_1 and k_2, the first-order scattering cross section may turn out to be considerably less than the second-order one. This is the situation with the scattering from acoustic and LO phonons for which $V_{qr} \to 0$, $q \to 0$.

For multi-phonon processes, quasimomentum need not be conserved. Therefore, the δ function for wave vectors in (13) may be replaced for multi-phonon processes by const $= N^{-1}$. Such a replacement is equivalent to a formal introduction of the sum over k_1, normalized to unity. However, as $N^{-1}\sum_{k_1} \exp[ik_1(n_1-n_2)] = \delta_{n_1 n_2}$, when describing multi-phonon transitions, it is possible to take in (8) $A^{(0)}_{n_1 n_2} = \delta_{n_1 n_2} A^{(0)}$, where

$$A^{(0)} = <e^{iH_n\tau'} e^{iH_L t} e^{-iH_n\tau} e^{-iH_L(t+\tau'-\tau)}>$$

$$= \exp\{i\omega_0(\tau-\tau') + g(t) + g(t+\tau'-\tau) + g(-\tau)$$

$$+ g(\tau') - g(t+\tau') - g(t-\tau)\} \ . \tag{19}$$

As a result, the problem of the multi-phonon scattering of light which is in resonance with the absorption band of self-trapped excitons reduces to an analogous problem of resonant scattering of light from an impurity centre investigated earlier [4]. The most intense part of the spectrum is the ordinary luminescence (OL) of the self-trapped exciton, and the rest is hot luminescence(HL) emitted during the self-trapping process. The shape of the OL spectrum is Gaussian-like whereas the HL spectrum has a complicated shape, which contains maxima either with a form $\sim |\omega_i - \Omega_2|^{-\frac{1}{2}}$ (to the blue from OL) or $|\Omega_2 - \omega_i'|^{-\frac{1}{2}}$ (to the red from OL) plus parts with a square root growth towards OL [2,5]. The radiation frequencies ω_i and ω_i' are from the classical turning points of the relaxing configurational coordinate. The maxima result from HL emitted during the rapid vibrational relaxation stage, and the square root growth is induced by HL emitted during the slow relaxation stage.

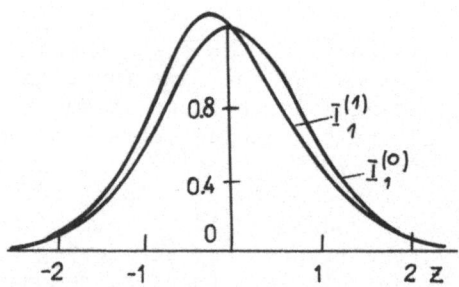

Fig. 1. The excitation profile of the first-order Raman
         scattering.

        Let us now consider the effects due to the exciton motion.
If $t_0/\tau_0 \ll 1$, slight asymmetries appear in the excitation
profiles of resonance Raman scattering [3]. In particular, the
correction for the first-order scattering cross section becomes
(see Fig.1):

$$I_1^{(1)} - I_1^{(0)} = 2^{3/2} \frac{I_1^{(0)}}{\sigma} [E(k_1) + E(k_2) - 2E]w(z). \qquad (20)$$

As $E(k_1) \simeq E(k_2) \simeq \Omega_0$, the sign of the correction depends on
whether or not the minimum of the exciton band coincides with the
point Γ.

        If $t_0/\tau_0 > 1$ (but $\sigma > B/d$), a new phenomenon, hot migration,
arises [2] (i.e., an incoherent exciton transfer from one site to
another in a non-relaxed vibrational state). The migration not
only leads to a possibility of transporting the exciton over the
crystal, which is important, especially at low temperatures when
the self-trapped exciton is immobile, but also affects the rate
of the exciton vibrational relaxation. The time for the vibrational
relaxation of the "dangerous" energy interval $E \pm \sigma$ increases by
$N^{1/2}$ times [2]:

$$t_{HM} \sim t_0 N^{1/2} \sim B^2/\bar{\omega}^3 \, Sd, \qquad (21)$$

causing the corresponding enhancement of the blue part of the HL
spectrum. Here $N \sim B^4/d^2 S\sigma\bar{\omega}^3$ is the number of hot migration
jumps. Formula (21) holds if $S^{1/2}(2\pi+1)^{3/2} d^2 > N > 1$.

Now we consider the conditions S >> 1 and s << 1, where the
interaction of the phonon with the small-sized exciton wave packets
is strong and that with the large-sized exciton wave packets is
weak. It follows from (5) and (7) that these conditions may be
realized if the exciton bandwidth is large, $B >> \bar{\omega}$. Here the
adiabatic approximation may be used to describe the exciton-phonon
system for both self-trapped and band excitons. Rashba [6],
Sumi and Toyozawa [7] have shown that in such a case band and
self-trapped excitons coexist and are separated by a potential
barrier on the adiabatic surface. These theoretical inferences
have been experimentally confirmed for many ionic and rare gas
crystals [8,9].

The height of the self-trapped barrier and the exciton radius
in the barrier region depend to a considerable degree on the ratio
of B to $S\bar{\omega}$. In particular, if $B \sim S\bar{\omega}$, the radius of this state
is small and the height of the barrier [10] is:

$$\lambda = B^2/36 \, S\bar{\omega} \; . \tag{22}$$

The presence of the self-trapping barrier is the most
peculiar aspect of this case. Further, we consider some evidence
of this barrier in optical spectra. The absorption spectrum has
a form analogous to that for the case of weak exciton-phonon
interaction and the absence of self-trapping (S << 1). Specifi-

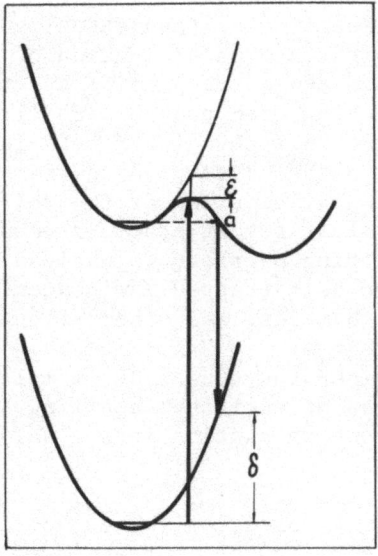

Fig. 2. Adiabatic surface of the self-trapping exciton.

cally, if the exciton band minimum coincides with the point Γ,
the spectrum reveals a narrow resonance no-phonon line and a
relatively weak but broad wing of indirect transitions involving
phonons. There is, however, a difference in that at low temperatures a peculiarity can be observed at wavelengths toward the blue
of the resonance line, which is connected with the direct creation
of the self-trapped exciton in the barrier region. Indeed, if
$T = 0$, the part of the absorption spectrum associated with a direct
creation of a self-trapped exciton (see Fig.2) may be described in
a δ-function approximation which is well known in the theory of
the absorption spectra of photodissociative molecules [11]:

$$\kappa'(\Omega_1) \sim \int_{-\infty}^{\infty} \rho(x)\delta[U(x) - \Omega_1]dx \sim \rho(x_0)|U'(x_0)|^{-1} , \qquad (23)$$

where x is the self-trapping configurational coordinate, $\rho(x)$ is
the distribution function of x on the zeroth vibrational level,
$U(x)$ is the self-trapping adiabatic surface counted from the zero-
point vibrational energy of the crystal, $U'(x) = dU(x)/dx$, and
x_0 is the root of the equation $U(x) - \Omega_1 = 0$.[*)] The maximum of
the spectrum $\kappa'(\Omega_1)$ lies in the region $\Omega_1 \sim U_{max}$. Actually, in
accordance with (23), the spectrum differs from zero only in the
region $\Omega_1 \ll U_{max}$. On the other hand, with the decrease of
$U_{max} - \Omega_1$, the spectrum increases at the expense of both $\rho(x)$
and the densities of the final states $(U'(x))^{-1}$ when x approaches
the top of the barrier.

In our case, the spectrum of resonance light scattering,
including that of OL with thte participation of a small number of
phonons, is analogous to the corresponding spectra in the case of
excitons weakly interacting with phonons $(S \ll 1)$. However, the
multi-phonon spectra are entirely different. In the present case,
the main contribution to the corresponding transitions is made by
the processes with the participation of the self-trapped state,
which is lacking at $S \ll 1$. At that state, a considerable role is
played by the self-trapping barrier, which leads not only to the
coexistence of free and self-trapped luminescence but also to
tunnel HL, HL from the barrier, and other effects.

First let us consider the tunnel HL, a multi-phonon radiation
emitted in the processes of exciton tunnelling through the barrier
and during the consequent relaxation into a self-trapped state.

[*)]The approximation can be used for the transitions to the right
from the point a (Fig.2) in the nontransparent part of the
barrier on the distance $>(\lambda/\alpha\omega)^{1/2} > \bar{\omega}$ from its top (α is the
curvature of the adiabatic surface at the barrier).

In the light of Bardeen's theory [12], taking the expression

$$T_{if} = \int \psi_i^*(H - E_i)\psi_f dx \sim \lambda \int \psi_i^*\psi_f dx \tag{24}$$

as the matrix element of tunnelling from the state ψ_i to ψ_f, we get for the spectrum of tunnel radiation

$$I_{tun}(\Omega_2) \sim \lambda^2 Re \int\int_0^\infty\int dt d\tau d\tau' e^{i(\Omega_2-\Omega_1)t-i\Omega_0(t+\tau'-\tau)-\gamma(\tau+\tau')} A^{(0)},$$

$$\tag{25}$$

where $\Omega_0 = E - B/2$ is the frequency of the resonance radiation line of band excitons. In simple physical terms, expression (25) describes the process of a free exciton transformation into light in second-order perturbation theory, in which the self-trapped exciton state is treated as an intermediate state.

Let us find the envelope of the spectrum in (25). Expanding $\ln A^{(0)}$ in a power series of t and taking into account the term linear in t, we obtain

$$I_{tun}(\Omega_2) \sim \lambda^2 \int\int_0^\infty d\tau d\tau' e^{g(\tau'-\tau)-\gamma(\tau+\tau')}$$

$$\times \delta(\Omega_2-\Omega_0+g_1(\tau')+g_1(-\tau)-g_1(\tau'-\tau)), \tag{26}$$

where

$$g_1(\tau) = \sum_{qr} \frac{|V_{qr}|^2}{\omega_{qr}} [(e^{i\omega_{qr}\tau}-1)(1+n_{qr})+(e^{-i\omega_{qr}\tau}-1)n_{qr}]. \tag{27}$$

Integral (26) may be separately investigated in the region of short and long times [$(\tau+\tau')/2$ compared with $\bar\omega^{-1}$]. For short times, in the region describing the radiation in the course of tunnelling and immediately after it, the functions g and g_1 may be expanded in a power series of τ and τ'. Keeping terms up to the second power, we then get:

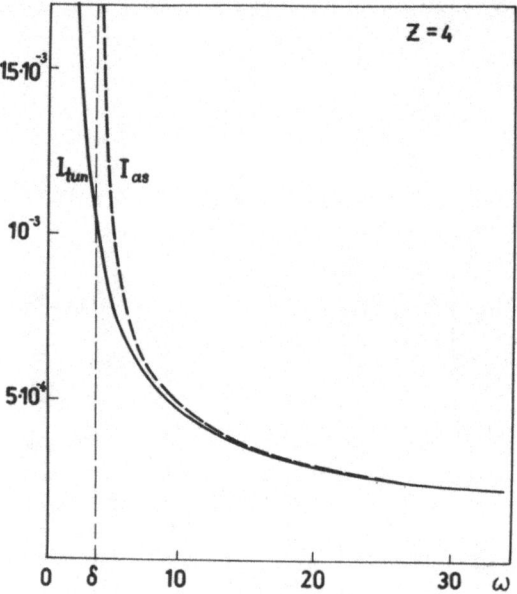

Fig. 3.   The spectrum of tunnel HL.

$$I_{tun}(\Omega_2) \sim \lambda^2 \left| \int_0^\infty d\tau d\tau' \exp\{\frac{iB}{2}(\tau-\tau') - \frac{\sigma^2}{2}(\tau-\tau')^2 - \gamma(\tau+\tau')\} \right.$$

$$\times \; \delta(\Omega_2 - \Omega_1 + m\tau\tau')$$

$$= \frac{2\lambda^2}{m} \int_0^\infty dx (x^2+1)^{-1/2} \; \exp\{-2x^2\omega - z^2/2(x^2+1)\}, \qquad (28)$$

where $m = \sum\limits_{qr}\omega_{qr}|V_{qr}|^2 \sim S\bar{\omega}^3$, $\omega = (\Omega_0 - \Omega_2)\sigma^2/m \leq 0$, $z = B/2\sigma \gg 1$. The
spectrum is presented in Fig.3. At small ω, the spectrum (28)
diverges logarithmically. However, this divergence has no
physical meaning since $I_{tun}(\Omega_2) \sim -\ln(\Omega_0 - \Omega_2)$ only in the region
$0 < \Omega_0 - \Omega_2 \ll \bar{\omega}$, where the envelope does not characterize the real
spectrum. With $\Omega_0 - \Omega_2$ increasing in the region $1 \ll |\omega| \ll z^2/4$, the
spectrum (28) decreases nearly exponentially, whereas in the region
of $\omega - z^2/4 \gg 1$ it asymptotically tends to the form characteristic
of HL:

$$I_{as} \sim \lambda^2 \sqrt{\frac{\pi}{2}} \frac{\exp(-z^2/4)}{(\omega - \frac{z^2}{4})^{1/2}} \ . \tag{29}$$

At T=0, $z^2/4 \sim B^2/16S\bar{\omega}^2 = \delta\sqrt{\omega}$ (see Fig.2). This is understandable since at T=0, after tunnelling, the system is at point a (see Fig.2) and the HL, in accordance with the Franck-Condon Law, has the frequency $\Omega_2 = \Omega_0 - \delta$. The temperature dependence of spectrum (29) in the region of $\omega - z^2/4 \gg 1$ is also peculiar to the tunnel radiation. At high frequencies, $I(\Omega_2) \sim \exp(-\lambda/kT)$.

When the relaxation in a self-trapped state proceeds slowly, calculation of the spectrum of the tunnel HL requires that integral (26) in the region $(\tau+\tau')/2 > \bar{\omega}^{-1} \gg \sigma^{-1}$ also is considered. The spectrum obtained is totally analogous with the HL of impurity centres from the successive turning points of the configurational coordinate in the course of its relaxation [2,5].

Recently, the tunnel HL of self-trapped excitons of this type was observed by Kink, Lōhmus and Selg on a Xe crystal at 10 K[*] (see Fig.4).

In conclusion, let us consider the HL from the barrier. Such HL is observable as a maximum in the envelope of resonance light scattering with a frequency $\Omega_1 \sim \Omega_b < \Omega_0 - \varepsilon$ (see Fig.2) in the classical interval of high temperatures $T \gg \bar{\omega}/k$. At $\Omega_b - \Omega_1 \gg (kT\lambda/\alpha)^{1/2}$, this maximum is asymmetrical. If $\Omega_b - \Omega_2 \gg (kT\lambda/\alpha)^{1/2}$, then $I(\Omega_2) \sim (\Omega_b - \Omega_2)^{-1}$. However, if $\Omega_2 > \Omega_b$ then $I(\Omega_2) \simeq 0$. At $|\Omega_b - \Omega_1| \ll (kT\lambda/\alpha)^{1/2}$, the HL maximum is approximately symmetrical and in the region $|\Omega_b - \Omega_2| \ll (\lambda kT/\alpha)^{1/2}$ can be described by the formula $I(\Omega_2) \sim \frac{1}{2}|\Omega_b - \Omega_2|^{-1}$, which differs from the inverse square root lineshape because of the transparency of the barrier in the region, determined by the parameter $(\lambda kT/\alpha)^{1/2}$. Note that the HL from the barrier is also observable at low temperatures when the excitation occurs in the region of $\Omega_1 = \Omega_0 + \lambda$, where the transitions creating an exciton on the barrier are essential.

The author is grateful to R. Kink, A. Lōhmus and M. Selg for the kind permission to report their recent results.

---

[*] The radiation observed is connected with the exciton self-trapped near the vacancy.

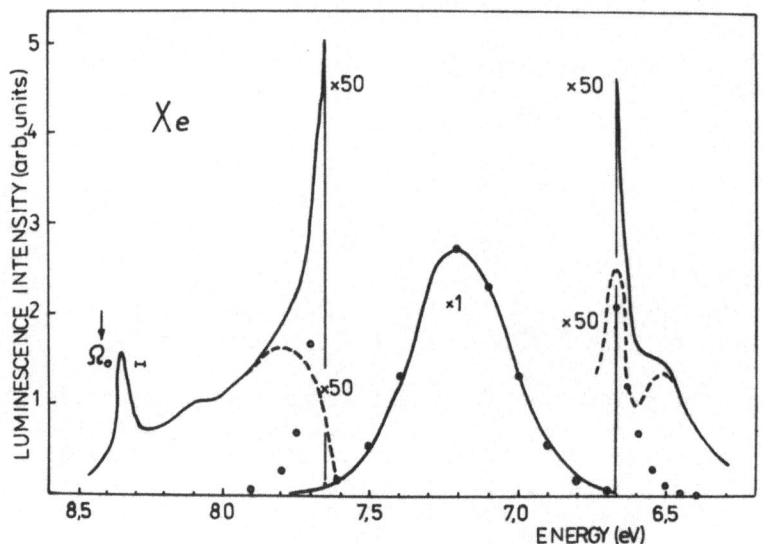

Fig. 4.   The emission spectra of an exciton in Xe, self-trapping
           at a vacancy. Dashed curve shows the difference between
           the total emission spectrum and the ordinary luminescence
           spectrum which has a Gaussian shape.

REFERENCES

1. Y. Toyozawa, Progr. Theor. Phys. (Kyoto) 27, 89 (1962).
2. I. J. Tehver and V. V. Hizhnyakov, JETP 69, 599 (1975);
    V. V. Hizhnyakov, Phys. Stat. Sol. (b) 76, K69 (1976).
3. V. V. Hizhnyakov and A. V. Sherman, Phys. Stat. Sol. (b) 92,
    177 (1979).
4. V. V. Hizhnyakov and I. J. Tehver, Phys. Stat. Sol. 21, 755
    (1967).
5. V. V. Hizhnyakov and I. J. Tehver, J. Luminescence 18/19,
    673 (1979).
6. E. I. Rashba, Optika i Spektroskopiya 2, 75, 88 (1957).
7. H. Sumi and Y. Toyozawa, J. Phys. Soc. Japan 31, 342 (1971).
8. V. V. Hizhnyakov and A. V. Sherman, Trudy Inst. Fiz. AN ESSR,
    No. 46, 120 (1976).
9. I. L. Kuusman, P. H. Liblik and Ch. B. Luschik, Pisma v JETP
    21, 161 (1975).
10. I. Ya. Fugol', Advan. Phys. 27, 1 (1978).
11. G. Herzberg, "Molecular Spectra and Molecular Structure I,
    Spectra of Diatomic Molecules" 2nd ed. (Van Nostrand,
    Princeton, N. J., 1950), p. 392.
12. J. Bardeen, Phys. Rev. Lett. 6, 57 (1961).

PHENOMENOLOGICAL DESCRIPTION OF LIGHT SCATTERING AND THERMAL RADIATION

D.N. Klyshko

Chair of Optics, Moscow State University, Moscow USSR

SUMMARY

The connections between induced and spontaneous observable effects in a quasiequilibrium sample are discussed. A generalized Kirchoff's law (GKL), expresssing the statisics of thermal radiation (TR) in terms of the sample's elastic scattering matrix (SM) and the temperature, is obtained in the one-photon approximation. A similar expression describes the statistics of Raman-scattered light (including the Stokes-polariton and Stokes-antistokes correlation) in terms of the SM of the pumped sample in the parametric approximation. In the case of weak two-photon nonlinearity the 2rd, 3d and 4th moments of the TR are determined by the quadratic and cubic SM-s. It is shown that the TR of noncentrosymmetric sample should contain a stationary term $\langle E^3 \rangle$ cubic in the electric field. It is suggested that the relation between spontaneous and induced phenomena and peculiarities of two-photon fields may lead to metrological applications.

1. INTRODUCTION

There is a very convenient and realistic model in optics: a quasiequilibrium sample at a constant (in space and time) temperature, emitting thermal radiation (TR). The spectral brightness B of the TR is connected with the sample's absorptivity A by the famous Kirchoff's law, which for one polarization may be written in the form:

$$N(\vec{k},\vec{H}_o) = A(-\vec{k},-\vec{H}_o) \, N, \tag{1}$$

where

$$N \equiv \frac{B}{B_{vac}}, \quad B_{vac} \equiv \frac{\hbar\omega^3}{8\pi^3 c^2}, \quad N \equiv \frac{1}{e^{\beta\omega}-1} \tag{2}$$

As is well known, Eq. (1) does not take into account diffraction and elastic scattering and it says nothing about the spatial correlation of TR. Furthermore, it neglects the optical nonlinearity of matter. It is natural to ask if at least some of these shortcomings may be remedied. Is it possible to derive more general and systematic relations between spontaneous and induced properties of a quasiequilibrium sample, including multiphoton effects?

2. GENERALIZED KIRCHOFF'S LAW (GKL) IN ONE-PHOTON APPROXIMATION

Ritov and Levin (1) have used the fluctuation-dissipation theorem (FDT) to obtain an expression for the second moments of TR in terms of the linear susceptibility $\hat{\chi}$ and the Green's function, $\hat{G}$, for macroscopic Maxwell equations. Their result (1),

$$\langle \vec{E}^+ \vec{E} \rangle_{\omega r} \sim \hat{G} \cdot (\hat{\chi} - \hat{\chi}^+) \cdot \hat{G}^+ \tag{3}$$

may be considered as a generalization of (1). Eq. (3) fully determines the Gaussian statistics of one-photon TR, but in contrast to (1) it contains quantities, which are defined only within a certain macroscopic model, and are not measurable in practice.

It is very simple to derive a linear GKL (2, 3). In terms of measurable quantities, let us divide the transverse field into three parts: a probe "in-field" and two "out-fields" - spontaneous and induced. In the one-photon approximation the out-field is linearly related to the in-field by

$$\vec{a}_{out} = \hat{U} \cdot \vec{a}_{in} + \vec{a}_{sp} \, ,$$

$$\vec{a} \equiv \{a_k, a_{k'}, \ldots\}, U_{kk'} \sim \delta(\omega_k - \omega_{k'}), \tag{4}$$

Here a_k (a_k^+) is an annihilation (creation) operator. The
SM of the sample $U_{kk'}$ can be measured by interferometric methods.

Using (4) we form the second-moment matrix of the observed
out-field:

$$\hat{N}_{out} \equiv \langle \vec{a}^+\vec{a} \rangle_{out} = \hat{U}\cdot\hat{N}_{in}\cdot\hat{U}^+ + \hat{N}_{sp} \qquad (5)$$

Let us for a moment place the sample inside a large cavity, filled
with TR of the same temperature. Then $\hat{N}_{out}=\hat{N}_{in}=\hat{N}_{sp}=\hat{N}$
(where $N_{kk'}\equiv N\delta_{kk'}$) and we get the linear GKL:

$$\hat{N}_{sp} = (\hat{I} - \hat{U}\cdot\hat{U}^+)\cdot\hat{N} \equiv \hat{A}\cdot\hat{N}, \qquad (6)$$

where we defined an absorption matrix $\hat{A}$, which measures the
"nonunitarity" of the sample's transformation of the in-field into
out-field.

A diagonal component of (5) gives the brightness of the
out-field in "vacuum" units (see Eq. (2)):

$$N_k^{out} = \sum_{k'} G_{kk'} N_{k'}^{in} + A_k N, \qquad (7)$$

$$N_k \equiv N_{kk}, \; G_{kk'} \equiv |U_{kk'}|^2, \; A_k \equiv 1 - \sum_{k'} G_{kk'} \qquad (8)$$

(we suppose here, that the probe light is noncoherent). Eq. (7)
is the scattering-corrected Kirchoff's law and (8) gives an
operational definition of the absorptivity. Another definition of
A_k follows from the Onsager symmetry $U_{kk'}(\vec{H}_o) = U_{k'k}(-\vec{H}_o)$

3. GKL FOR RAMAN SCATTERING

The next step is to take into account multiphoton processes.
Let us first consider two-photon transitions with a classical or
coherent pump. If we neglect the change of pump statistics the
transformation of a weak probe field is again linear. Instead of
Nyquist or Langevin methods used in the derivation of (6) one can
use a kinetic equation method with an effective energy interaction

$$V_{ef} = \sum_{jk} (C_{jk_1} a_{k_1}^+ + C_{jk_2} a_{k_2})\sigma_j^+ + h.c., \qquad (9)$$

where $C_{jk_1} \sim E_L e^{-i\omega_L t}$, $C_{jk_2} \sim$ const or $E_L^* e^{i\omega_L t}$,

k_1 refers to Stokes modes and σ_j^+ is an operator which transfers the j-th molecule to its excited states. Eq. (9) describes two processes. In the first k_2 is the polariton's wave vector and (9) describes interaction between molecules, Stokes photons and polaritons. In the second k_2 refers to antiStokes waves and we are looking at Stokes-anti-Stokes interaction.

By using (9) in a Markovian kinetic equation for the density operator (or characteristic function) of light, we can calculate the time evolution of light statistics. This was done by Shen (4) and others for one or two modes. In a multimode model we can obtain the "space evolution" by setting the interaction time equal to infinity. The result has the following structure:

$$<\vec{a}_i^{out}> = \hat{U}_{ii}^* \cdot <\vec{a}_i^{in}> + \hat{U}_{ij}^* \cdot <a_j^{in}>^* \quad (i,j = 1,2) \tag{10}$$

$$\hat{N}_i^{out} = \hat{U}_{ii}\hat{N}_i^{in} \cdot \hat{U}_{ii}^+ + \hat{U}_{ij}(\tilde{N}_j^{in}+\hat{I}) \cdot \hat{U}_{ij}^+ + \hat{A}_1 (N_o + \delta_{ii}) \tag{11}$$

$$<\vec{a}_1 \vec{a}_2>_{sp}^* = \hat{U}_{21} \cdot \tilde{U}_{11} + (\hat{U}_{21} \cdot \tilde{U}_{11} - \hat{U}_{22} \cdot \tilde{U}_{12}) N_o \tag{12}$$

$$\hat{A}_1 \equiv \hat{U}_{11} \cdot \hat{U}_{11}^+ - \hat{U}_{12} \cdot \hat{U}_{12}^+ - \hat{I}, \ \hat{A}_2 \equiv \hat{I} - \hat{U}_{22} \cdot \hat{U}_{22}^+ - \hat{U}_{21} \cdot \hat{U}_{21}^+ \tag{13}$$

Eq. (10) and (13) give the definitions of Raman SM and absorption matrixes which can be measured by means of "inverse" Raman experiments. Eq. (11) and (12) may be considered as the GKL for Raman scatterings: they determine the statistics of spontaneously scattered light in terms of SM. The kinetic equation also determines the full statistics of the out-field. In the case of Gaussian in-field it is "quasigaussian" with correlated amplitudes of different frequencies (see Eq. (12)).

4. COHERENT RAMAN SCATTERING

In case $\omega_1 + \omega_2 = 2\omega_L$ the induced part of (11) corresponds to the so called "inverse" and "coherent antistokes" Raman scattering (CARS).

In the absence of in-fields Eq. (11) describes spontaneous inelastic one-phonon (or one-polariton) scattering to all orders of the pump. The scattered field contains two parts: one depends on the sample's temperature directly through the Planck function N_o and the other is due to zero-point fluctuations of the in-field and molecules (2).

In the case of a macroscopic sample the SM component $U_{12} = U_{sa}$ has a noticeable value only for phase-matched modes ($\vec{k}_s + \vec{k}_a = 2 \vec{k}_L$). Till now this effect of "spontaneous CARS" has been observed only in case of powerful pump pulses, when it depends on the pump exponentially. In the lowest order in the pump power Eq. (11) describes phenomenologically the usual linear Raman scattering and the so far unobserved quadratic coherent effects, which may be considered as resonant 4-photon parametric (or light by light) scattering.

5. STOKES-ANTISTOKES CORRELATION

Thus it follows from the phenomenological considerations, that Stokes (k_1) and antiStokes (k_2) fields in the phase matching directions are correlated (2). In the case of a pump with definite phase the 4th moments are expressible in terms of the 2nd moments.

$$\langle a_1^+ a_2^+ a_1 a_2 \rangle = N_1 N_2 + |\langle a_1 a_2 \rangle|^2 \tag{14}$$

Of course, the existence of such correlation follows also from the elementary 4-photon interaction picture, in which two pump photons convert into a pair of s- and a-photons. On the other hand, it can be easily understood with the help of classical Mandelshtam modulation model of the Raman scattering, in which the s- and a-sidebands are produced by modulation of the pump field by a quasi-monochromatic noise $Q(t) e^{-i\omega_o t}$.

These quantum and classical pictures correspond to two basic types of light correlation, which may be called correlation of photons and of intensities. The latter effect was first observed by Brown and Twiss and it gives an "accidental" coincidence rate (the first term in Eq. (14)) of the same order or less than the "true" coincidence rate. On the other hand, photon correlation gives no accidental coincidences. It follows from Eq. (14), that in the limit of weak pump and low temperature the a-photons are emitted only in pairs with s-photons.

A numerical estimate of this effect gives several coincidences per second in the case of a 1-Watt pump, focused into a liquid nitrogen sample. It should be much easier to observe classical correlation by means of quasielastic scattering by macroparticles. For the purpose of dividing the correlated fields in space a two beam pump may be used, e.g., a standing wave pump. Here the beams scattered in opposite directions should fluctuate simultaneously (in the absence of multiple scattering). This experimental arrangement could be useful for investigation of the multiplicity of scattering.

6. SCATTERED LIGHT AS A BRIGHTNESS STANDARD

Let us consider now the stokes-polariton interaction ($\omega_1 + \omega_2 = \omega_L$). We define as "parametric" the scattering by a transparent sample which has zero absorptivity matrices A_i in Eq. (11), so that the sample temperature does not influence the scattering directly:

$$\hat{N}_1^{out} = \hat{U}_{11} \cdot \hat{N}_1^{in} \cdot \hat{U}_{11}^+ + \hat{U}_{12} \cdot (\tilde{N}_2^{in} + \hat{I}) \cdot \hat{U}_{12}^+ , \tag{15}$$

$$\hat{N}_1^{sp} = \hat{U}_{12} \cdot \hat{U}_{12}^+ = \hat{U}_{11} \cdot \hat{U}_{11}^+ - \hat{I} , \tag{16}$$

$$<\vec{a}_1^+ \vec{a}_2^+>^{sp} = \hat{U}_{21} \cdot \tilde{\hat{U}}_{21} = \hat{U}_{22} \cdot \tilde{\hat{U}}_{12} . \tag{17}$$

We see, we may think of parametric scattering as being due to vacuum fluctuations of the in-field (one photon per a couple of interacting modes). Thus the GKL for parametric scattering provides a means for measuring the brightness of the in-field (5), (6). In contrast to the usual Planck standards, here we do not need the temperature and absorptivity calibrations.

7. A STANDARD PHOTON GENERATOR

Another possible metrological application of parametric scattering stems from the absence of single photons (5). In the case of a sufficiently weak pump $U_{ii} \approx I \gg U_{ij}$ and with $N_{in} = 0$ it follows from Eq. (15) - (17) that $|<a_1 a_2>|^2 \gg N_1 N_2$. This means that every signal photon is accompanied by an idler one with frequency, direction and polarization determined by energy and momentum conservation laws. We can use this knowledge to measure the absolute quantum efficiency of photo detectors: $n_c/n_1 = \eta_1 \eta_2 n / \eta_1 n = \eta_2$ Here n is the (unoberved) number of pairs, emitted during the sampling time and belonging to the frequency and angle acceptance intervals of the signal detector: n_1, n_2, and n_c are observed numbers of signal, idler and coincidence counts; and η is the probability of counting the emitted photon.

One can independently measure or calculate all optical losses
and thus determine the detector efficiency alone.   This procedure
was carried out by Burnham and Weinberg (7) in a reversed sense –
to demonstrate the two-photon character of the parametric
scattering.   Alternately, we can set in the place of the idler
detector an optical gate, monitored by the signal detector and
make a standard photon generator with a priori known number and
departure times of photons (5).   It should be mentioned that in
parametric scattering the pecularities of indirect (correlation)
quantum measurements can be simply demonstrated.   If we measure
the frequency ω_1 (or momentum $\vec{k}_1$) of a signal photon, then we
know that the corresponding idler photon acquires the frequency
$\omega_L - \omega_1$ (or momentum $\vec{k}_L - \vec{k}_1$).   So if we use classical
language we are forced to believe that some long range force
influences the idle photon.

8. TWO-PHOTON KIRCHOFF'S LAW

There are two types of two-photon decay – cascade with a real
intermediate level, and direct decay of a metastable level.   In
both cases the photons go off in pairs and therefore the
statistics of the TR should differ from the usual predictions of
Kirchoff's law of the FDT.

It is not difficult to correct the Kirchoff law for two-photon
effects in the case of direct transitions.   Let us proceed from
the effective interaction energy

$$V_{ef} = \sum_{jkk'} C_{jkk'} \; \sigma_j^+ \; a_k a_{k'} + h.c. \tag{18}$$

(for simplicity we neglect the antistokes term).   The
corresponding kinetic equation for one or two modes was considered
by Shen (4) and others.   We are interested here in the multimode
case, which in the first approximation gives the following
increments for the field moments (3):

$$\Delta \langle a_1 \rangle = N_o \sum_2 u_{12} \langle a_2 \rangle^{in} - \sum_{234} u_{(12)34} \langle a_2^+ a_3 a_4 \rangle^{in}, \tag{19}$$

$$\Delta \langle a_1^+ a_2 \rangle = \sum_3 u^*_{(13)(32)} [(N_o (N_2^{in} + N_3^{in} + 1) - N_2^{in} N_3^{in}] + (1;2)^* \qquad (20)$$

$$\langle a_1^+ a_2^+ a_1 a_2 \rangle^{sp} = A_{12} N_o, \quad A_{12} \equiv 2Re \ u_{(12)(12)} \qquad (21)$$

Here $u_{12} \equiv \sum u_{(13)(32)}$; $u_{(12)34} \equiv u_{1234} + u_{2134}$ and $u_{(12)34}$ should be considered as a phenomenological cubic SM, measurable by frequency conversion experiments $\omega_4 + \omega_3 - \omega_2 \rightarrow \omega_1$. Eq. 20-21) give an approximate two-photon GKL. The first term in Eq. (19) describes amplitude enhancement in a sample with no population inversion.

9. ODD MOMENTS OF TR

Let us consider the TR of non-centrosymmetrical molecules. Now we should take into account one- and two-photon transitions simultaneously. The kinetic equation method gives the following additional terms (3):

$$\Delta \langle a_1 \rangle = -\sum (u_{123} \langle a_2 a_3 \rangle^{in} + v_{(12)3} \langle a_2^+ a_3 \rangle^{in}) \qquad (22)$$

$$\langle a_1^+ a_2^+ a_3 \rangle^{sp} = N_o (u_{3(21)} + v^*_{(12)3}). \qquad (23)$$

Eq. (22) defines the quadratic SM of the sample and from Eq. (23) it follows, that there is a non-zero third moment of the electric field in the TR, expressible through the temperature and SM.

Of course, this result also follows from microtheory. Let us consider now the three-level case. It is clear, that three harmonics, emitted by an excited molecule, should have definite phase correlations which give a stationary component $\langle E^3 \rangle \neq 0$. One can demonstrate this effect by the opposite process of mixed one-two-photon excitation. The counting rate should depend on relative phase-shifts in the optical path between source and detector.

If there is phase matching inside the sample, then the contributions from single molecules are summed.

CONCLUSION

My aim has been to show the usefulness of the phenomenological description of TR in terms of operationally defined quantitites. There remains the problem of finding general connections between arbitrary moments of TR and nonlinear SM-S (8).

REFERENCES

1. M.L. Levin, S.M. Ritov, J. Exp. Theor. Phys. 65, 1382, 1973.
2. D.N. Klyshko, J. Exp. Theor. Phys., 64, 1160, 1973; Quant. Electr. 4, 1341, 1977.
3. D.N. Klyshko, Dokl. Acad. Nauk USSR, 244, 563, 1979.
4. Y.R. Shen, Phys. Rev., 155, 921, 1967.
5. D.N. Klyshko, Quant. Electr., 4, 1056, 1977.
6. H.A. Kitayeva, A.N. Penin, V.V. Phadeyev, Yu. A. Yanite, Dokl. Acad. Nauk USSR, in print (1979).
7. D.C. Burnham, D.L. Weinberg, Phys. Rev. Lett., 25, 84, 1970.
8. R.L. Stratonovitch, Dokl. Acad. Nauk USSR, in print (1979).

RAMAN SCATTERING FROM PLASMON-PHONON COUPLED MODES IN GaP

J. E. Kardontchik[a] and E. Cohen[b]

[a]Department of Physics, Technion, Haifa, Israel

[b]Bell Telephone Laboratories, Murray Hill, NJ   07974
On leave of absence from Technion, Haifa, Israel

INTRODUCTION

Raman scattering from phonon-plasmon coupled modes has been observed in several doped semiconductors.[1] The experimental observations were limited to scattering from a single component plasma (electrons in a single-valley conduction band) with negligible damping. This case has been fully analyzed within the framework of the total dielectric function.[2,3] In the present report we extend the use of Raman scattering to the case of a photoexcited electron-hole plasma (EHP) in the indirect gap semiconductor GaP. Apart from the interest in the scattering itself, this method provides an independent estimate of the electron-hole pair density and verifies its variation with excitation intensity as observed by photoluminescence studies.

EXPERIMENTAL

Single crystals of undoped GaP were excited by a pulsed dye laser pumped with nitrogen laser. The peak power of the dye laser was 5-10 kW, its pulse width was 2 nsec in the green and 4 nsec in the blue. The highest photon energy used was 2.80 eV. Luminescence and Raman scattering spectra were obtained with a triple spectrometer and analyzed using a boxcar integrator with a gate of 1 nsec. Most of the studies were done with the crystals immersed in liquid helium. Some data were taken at higher temperatures up to 300 K. The low temperature luminescence spectrum of all the crystals studied under pulse excitation (taken with a 1 nsec gate and delay of 0-20 nsec after excitation) was that characteristic of the EHP in GaP.[4,5] In the scattering experiments the same excitation pulse is used to create the EHP and to scatter from it. Figure 1

presents typical Raman spectra obtained in a nearly back-scattering
configuration from a <100> natural face. The directions refer to
<100> axes with the x-axis normal to the crystal face. Under these
conditions the TO phonon should ideally be missing and the LO phonon
should be observed only in the ZY polarization. The appearance of
the TO phonon in the spectra is due to a slight crystal misalignment.
When a low power cw Ar$^+$ laser excitation was used, only the phonon
lines were observed. Under pulse excitation two additional lines
are observed. The frequency of these lines (denoted ω_- and ω_+ in
Fig. 1) increases upon increasing the laser excitation energy E_{exc}
from 2.37 eV up to 2.80 eV. However, when a fixed laser energy
was used and its intensity attenuated by a factor of 20, the ω_- and
ω_+ lines did not shift. Similar scattering experiments performed
on GaP:N crystals (bulk N doping level of $2 \times 10^{18} cm^3$) yielded
identical results (same ω_- and ω_+). Also, the scattering persisted
up to room temperature indicating that the same entity which
scattered the light existed at high temperatures.

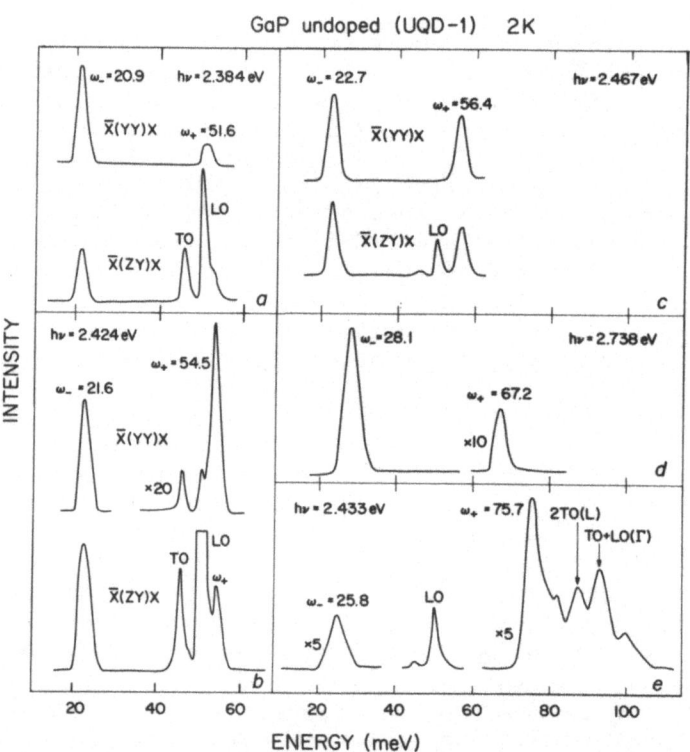

Fig. 1: Pulsed-laser Raman scattering spectra of undoped GaP at
2 K for various photon excitation energies. All spectra were
taken in back-scattering configuration from a <100> face.

ANALYSIS

We interpret the ω_- and ω_+ lines as due to coupled plasmon-phonon modes of the photoexcited plasma. Within the framework of the random phase approximation for a two component plasma, the plasma frequency ω_P is given by[3] $\omega_P^2 = 4\pi e^2 n / \varepsilon_\infty m^*$, where the effective mass m^* is given by

$$\frac{1}{m^*} = \frac{2}{3}\frac{1}{m_\perp} + \frac{1}{3}\frac{1}{m_\parallel} + \frac{1}{m_{lh}}\frac{1}{1 + \left[\frac{m_{hh}}{m_{lh}}\right]^{3/2}} + \frac{1}{m_{hh}}\frac{1}{1 + \left[\frac{m_{lh}}{m_{hh}}\right]^{3/2}} \tag{1}$$

where $m_\perp$ and $m_\parallel$ are the transverse and longitudinal electron masses and m_{hh} and m_{lh} are the heavy and light hole masses. This gives for GaP $m^* = 0.18\, m_0$, where m_0 is the free electron mass. For e-h pair densities between $(1-4)\times10^{18}\,cm^{-3}$ the plasmon energy would lie between 30-60 meV. As the optical phonons in GaP lie also in this range, we need consider explicitly the lattice contribution to the dielectric function. The plasmon energy is then given by the solutions of

$$\varepsilon(\omega) = \varepsilon_\infty\left(1 + \frac{\omega_L^2 - \omega_T^2}{\omega_T^2 - \omega^2}\right) - \frac{4\pi e^2 n}{m^* \omega^2} = 0 \tag{2}$$

where ω_L and ω_T are the longitudinal and transverse optical phonon frequencies. We expect thus two branches

$$\omega_\pm^2 = \frac{1}{2}\left[\omega_L^2 + \omega_P^2\right] \pm \frac{1}{2}\left[\left(\omega_L^2 + \omega_P^2\right)^2 - 4\omega_T^2\omega_P^2\right]^{1/2} \tag{3}$$

The full-line curves in Fig. 2 give ω_- and ω_+ as a function of ω_P as obtained by Eq. (3). An experimental determination of ω_- or ω_+ would give us then directly ω_P, i.e., the density n of electrons and holes in the plasma.

We now consider the data obtained at 2 K. As can be seen in Fig. 1, the general behavior of coupled plasmon-phonon modes predicted by the model described above is actually observed. The energy variations of the ω_+ and ω_- lines indicate that the electron-hole pair density in the plasma depends on excitation conditions. It is observed that the pair density is affected by changing the dye laser energy E_{exc}. The origin of this effect is not yet understood. E_{exc} has been changed from about 50 meV above the indirect gap up to 200 meV below the direct gap. The linear absorption coefficient varies between 20-1000 cm^{-1} in this range.[6] Therefore, if the effect were due simply to changes in penetration depth with

E$_{exc}$ (and hence in excitation density), a similar change could have been achieved by varying the laser intensity. This has not been observed (for the limited range of attenuation which still allowed observation of Raman scattering).

As discussed below the density of electron-hole pairs can be estimated using Eq. (3) and it varies between $(0.5-5) \times 10^{18} cm^{-3}$. The same density range has been observed in EHP luminescence experiments in GaP:N.[7] We note that the threshold density $(5 \times 10^{17} cm^{-3})$ is about an order of magnitude larger than expected for the Mott transition (using the known exciton parameters for GaP).

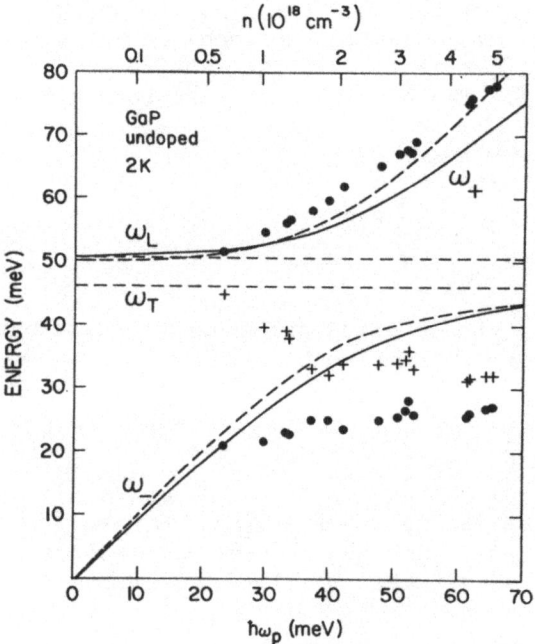

Fig. 2: Plasmon-phonon energies as a function of ω_p. Full circles – experimental points. Full lines – plasmon-phonon branches given by Eq. (3), crosses – effective ω_T. Broken lines – plasmon – phonon branches calculated with the $\varepsilon_{inter}(\omega)$ term.

In attempting to extract the density n from Eq. (3) using the known values of ω_T and ω_L, we found that for every pair of experimental values (ω_-, ω_+) the plasma frequency ω_p obtained using ω_- was different from that obtained using ω_+ . Thus the simple model represented by Eq. (3) does not apply here. For the sake of simple and compact presentation of the experimental results the set of (ω_-, ω_+) were plotted against the abscisae ω_p by calculating ω_p through the relation $\omega_p^2 = \omega_-^2 + \omega_+^2 - \omega_L^2$, which follows from Eq. (3). This procedure still leaves an additional free parameter. It can be taken as either ω_T $(=\omega_+\omega_-/\omega_p)$ or ε_0, the static dielectric constant which is obtained by the Lyddane–Sachs–Teller relation.

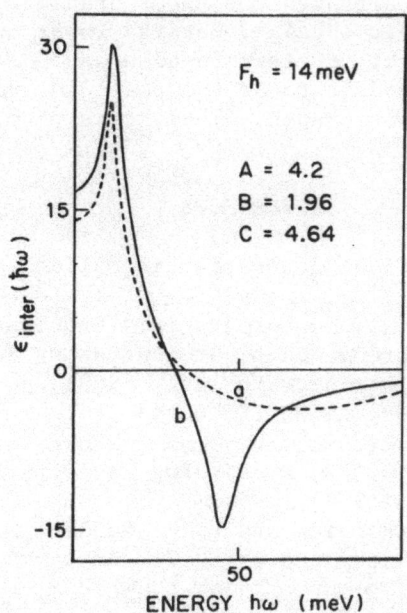

Fig. 3: Calculated interband hole contribution to the dielectric function. The valence band parameters are given in the figure. A hole Fermi energy of 14 meV is taken as an example. Broken curve cubic approximation, full curve – exact.

We have plotted in Fig. 2 the ω_T obtained for each pair (ω_-,ω_+) as a measure of the deviation of the experimental results from the simple model given by Eq. (3). The fact that ω_T obtained in this way is different for each ω_p shows the inapplicability of this model.

We tried an improved model by adding the contributions to the dielectric function of the interband transitions between heavy and light holes[8] as well as transitions to the split-off band (which is removed by 80 meV from the top band). The calculated $\varepsilon_{inter}(\omega)$ is shown in Fig. 3. The broken curve corresponds to the contributions of the light to heavy holes only. The full curve is an exact calculation including the split-off band. The valence band parameters and the hole Fermi energy are given in the figure. Adding the calculated $\varepsilon_{inter}(\omega)$ to Eq. (2) and solving for the roots of $\varepsilon(\omega)$ we obtain the broken curves shown in Fig. 2 for the plasmon-phonon coupled modes. Although the upper branch (ω_+) is better fitted, no improvement is obtained for the lower branch. Tzoar[9] has calculated the effect of electron-hole and hole-hole correlations on the plasmons and found it to be too small in order to account for the behavior of the lower branch. Thus, this problem is still unresolved.

REFERENCES

1. M. V. Klein, in: "Light Scattering in Solids", M. Cardona, ed., Springer Verlag, (1975).
2. A. Mooradian and A. L. McWhorter, in: "Proceedings of the International Conference on Light Scattering Spectra of Solids", G. B. Wright, ed., Springer Verlag, (1969).
3. P. M. Platzman and P. A. Wolff, Solid State Physics, Supplement 13, Academic Press, (1973).
4. J. Shah, R. F. Leheny, W. R. Harding and D. R. Wight, Phys. Rev. Lett. 38:1164 (1977).
5. D. Bimberg, M. S. Skolnick and L. M. Sander, Phys. Rev. (1979).
6. P. J. Dean, G. Kaminsky and R. B. Zetterstorm, J. Appl. Phys. 38:3551 (1967).
7. J. E. Kardontchik and E. Cohen, Phys. Rev. (1979).
8. M. Combescot and P. Nozieres, Solid State Comm. 10:301 (1972).
9. N. Tzoar, private communication.

INTERACTION BETWEEN LOCALIZED CARRIERS IN THE ACCUMULATION LAYER AND EXTENDED BULK LO PHONONS IN InSb AND GaSb: RAMAN INTERFERENCE LINESHAPES

Ralf Dornhaus, Roger L. Farrow,[*] Richard K. Chang

Department of Engineering and Applied Science
Yale University
New Haven, CT 06520

and

Richard M. Martin

Xerox Palo Alto Research Center
Palo Alto, CA 94304

Presented are Raman observations (see Fig. 1) of the interference between forbidden LO phonon scattering (discrete) and a broad emission which occurs when the incident photon energy $\hbar\omega_i$ is nearly resonant with the E_1 or $E_1 + \Delta_1$ gaps of InSb and GaSb. While the resonance enhancement of the forbidden LO phonons (with energy $\hbar\omega_{LO}$ and wavevector q) induced by the Fröhlich mechanism[1] is well documented,[2,3] the physical mechanism causing the broad emission which can interfere with the phonon Raman process is less understood.

Even though the broad bands remain centered at E_1 and $E_1 + \Delta_1$ for all $\hbar\omega_i \geq E_1 + \Delta_1$, we exclude them as luminescence for two reasons: 1) they are polarized along the polarization of $\hbar\omega_i$; and 2) they exhibit a coherent interference with the forbidden LO phonon Raman process [see Fig. 2(b)]. We interpret the continuum as intraband electronic Raman scattering of carriers near Γ which are coupled to photo-induced carriers at the E_1 or $E_1 + \Delta_1$ gaps--that is, $\hbar\omega_i$ photons are absorbed, creating electron-hole pairs along the [111] axes, and these pairs then transfer some energy and momentum via the Coulomb interaction to carriers near Γ before recombining and emitting the broad emission (see Fig. 3). These photo-induced electron-hole pairs can also transfer some energy and momentum via the Fröhlich mechanism[1] to

---
*Present address: Sandia Laboratories, Livermore, CA 94550.

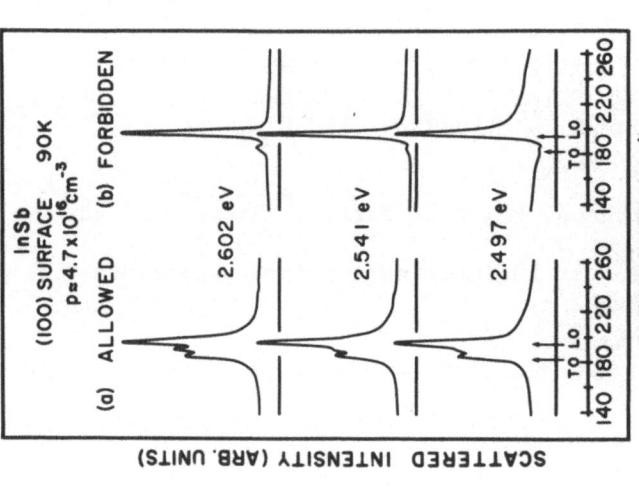

Fig. 2. Comparison of (a) allowed LO phonon $(\hat{e}_i \| \hat{e}_s)$ and (b) forbidden LO phonon $(\hat{e}_i \| \hat{e}_s)$ Raman spectra for InSb with different $\hbar\omega_i$ near $E_1 + \Delta_1$.

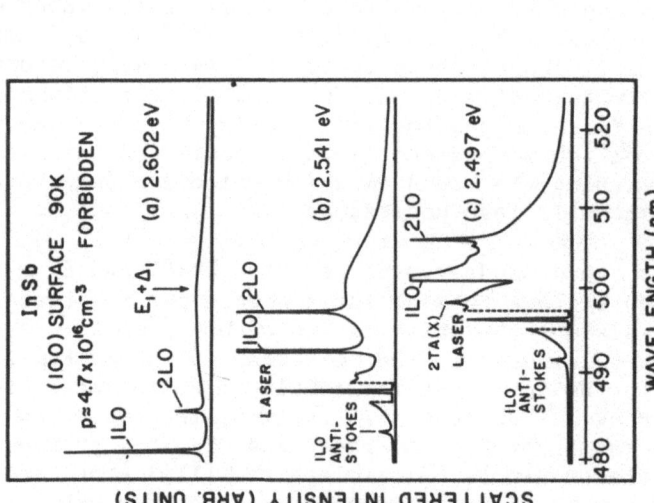

Fig. 1. Raman spectra $(\hat{e}_i \| \hat{e}_s)$ for InSb with different $\hbar\omega_i$ near $E_1 + \Delta_1$. All spectra have the same initial and final wavelength.

LO phonons, which interact with the longitudinal polarization field of the carriers near Γ (see Fig. 3). Consequently, both scattering processes have the same final state and temporal coherence, leading to a Fano-type lineshape.[4] This type of electronic Raman scattering, which has only been predicted earlier,[5] differs from the previously observed charge density scattering caused by $\vec{p} \cdot \vec{A}$ and $|\vec{A}|^2$ processes and can be resonantly enhanced at all energy gaps regardless of the location of carrier populated states in the Brillouin zone.[6]

Discrete continuum interference has been observed from InSb and GaSb with (100) and (110) surfaces in the standard backscattering geometry. Raman scattering from TO phonons is not allowed for (100) surfaces but is observable due to a slight sample misorientation and certain resonance conditions.[2] For a (100) surface, forbidden LO phonon scattering is observed with the polarization vectors of the incident and scattered radiation ($\hat{e}_i$ and $\hat{e}_s$) parallel to the [010] crystal axes, while allowed LO phonon scattering is observed with $\hat{e}_i$ parallel to [010] and $\hat{e}_s$ parallel to [001]. The following observations are common to undoped and heavily doped p-InSb and p-GaSb: 1) broad emissions centered at E_1 and $E_1 + \Delta_1$ are present for $\hat{e}_i || \hat{e}_s$ and absent for $\hat{e}_i \perp \hat{e}_s$; 2) when the forbidden LO phonon scattering is near the maximum of the broad bands, a Fano-type lineshape is developed with a minimum centered at ω_{TO} and a peak at ω_{LO}. 3) no in-interference occurs between the 2LO phonon Raman peaks and the broad bands, consistent with the fact that there is no direct Coulomb coupling between two phonons and the intraband single-particle transitions; and 4) for $\hat{e}_i \perp \hat{e}_s$, the allowed LO phonons are coupled to the plasmons and become $\omega_{\pm}(q)$ modes.

The free surface of InSb and GaSb has the Fermi level pinned close to the valence band.[7,8] For heavily doped crystals (p $\simeq$ 4 x 10^{18} cm^{-3}), the surface accumulation layer thickness $d_a \simeq 20$ Å is considerably less than the optical penetration depth $d_{op} \simeq 300$ – 600 Å and, therefore, the observed scattering is mainly from the bulk.[9] For undoped crystals (p $\simeq$ 4 x 10^{16} cm^{-3}), $d_a \simeq 150$ Å, which is a sizable fraction of d_{op} (see Fig. 4 for GaSb). Thus, the scattered radiation is from an inhomogeneous system where the carriers are localized in a two-dimensional accumulation layer of a semiconductor and the LO phonons are extended throughout the bulk.

The localized carriers can be coupled to the extended LO phonons in two different ways. 1) With $\hat{e}_i \perp \hat{e}_s$, the plasmons localized in the accumulation layer are coupled to the extended LO phonons [with wavevectors in the range $0 \rightarrow 2\pi(1/\lambda + 1/d_a)$]. The scattered spectrum [see Figs. 2(a) and 4] consists of a tail extending from ω_{LO} to below ω_{TO}, which we have ascribed to a distribution of coupled LO phonon-plasmon modes. Lineshape calculation of these modes is difficult because knowledge is needed of the charge distribution and dielectric constant $\varepsilon(q,\omega)$ as a function of distance away from the surface. Qualitatively, we expect a superposition of $\omega_-(q)$ modes starting with ω_{LO} and

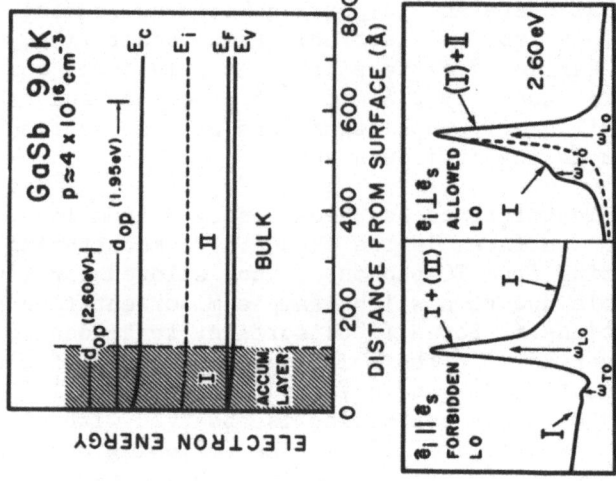

Fig. 4.   Carriers localized in the surface accumulation layer (region
I) and phonons extended into
the bulk (region II). Regions
in which typical allowed and
forbidden LO phonon spectra of
GaSb originate are shown.

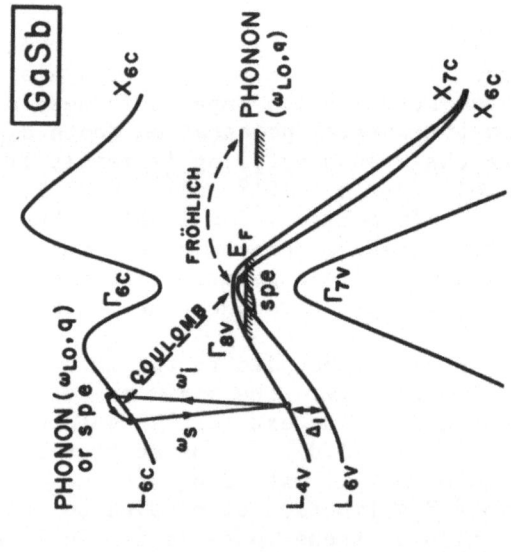

Fig. 3.   Illustration of the two interactions: 1) photo-induced
electron-hole pairs near E1
with carriers near Γ; 2) forbidden LO phonon with carriers
near Γ.

extending to below ω_{TO}. The peak at ω_{LO} [see Fig. 2(a)] is associated with the uncoupled extended LO phonons. 2) With $\hat{e}_i || \hat{e}_s$, the single-particle excitations (spe) localized in the accumulation layer exhibit interference when coupled to the extended forbidden LO phonons (having large wavevectors because of resonance conditions[1]). The extended forbidden LO phonons with no interference give rise to a symmetric peak at ω_{LO} with a linewidth γ.

The scattering spectrum consists of two contributions: one representing the interference of the localized excited states and the extended phonon states and the second having the characteristic of only the extended phonon states. The resultant lineshape for Coulomb coupling between the surface and bulk modes can be expressed as follows:

$$I(\omega) = \text{Im}\left[C\left(\frac{\lambda}{\lambda[\varepsilon_{ph}(\omega) - 1] + 1} - 1\right) + \frac{1}{\varepsilon_{ph}(\omega)}\right] , \qquad (1)$$

where

$$\varepsilon_{ph}(\omega) = 1 - \frac{(\omega_{LO}^2 - \omega_{TO}^2)}{(\omega^2 - \omega_{TO}^2 + 2i\omega_{TO}\gamma)} \qquad (2)$$

and

$$\lambda = 1 + VG_{11}^o = 1 + \Delta. \qquad (3)$$

The effective Coulomb coupling constant for the surface and bulk modes is V, and the unperturbed surface electronic response function is G_{11}^o which can be approximated as the flat continuum for the spe case. The ratio of coupled to uncoupled contribution is C. The dielectric constant $\varepsilon_{ph}(\omega) \to \infty$ at $\omega = \omega_{TO}$ for well defined bulk phonons, assuming a small γ. Consequently, the scattering has a minimum[5] at ω_{TO} independent of other coupling mechanisms of the electrons, phonons, and photons [see Eq. (1)]. We have fitted (see Fig. 5) our experimental spectra at different $\hbar\omega_i$ with Eq. (1) with two adjustable parameters $\text{Re}(\Delta)$ and $\text{Im}(\Delta)$ while keeping C fixed. We also fitted the same spectra to the usual Fano lineshape with parameters for the asymmetry Q and the discrete continuum coupling Γ. At this time, it is not clear why these parameters [$\text{Re}(\Delta)$ and $\text{Im}(\Delta)$ or Q and Γ] vary with $\hbar\omega_i$. Both the Coulomb and Fano lineshape expressions fit the observed spectra with reasonable success. Therefore, it is not possible to state which description is better. However, the Coulomb lineshape expression always predicts the scattered intensity to have a minimum near ω_{TO}, while the minimum in the Fano description depends on Q.

The intensity dependence of the forbidden LO, allowed LO, and the spe scattering as a function of $\hbar\omega_i$ is shown in Fig. 6 with CaF_2 as the reference crystal.

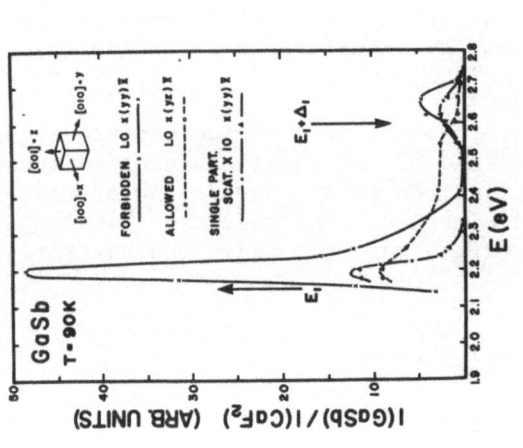

Fig. 6. Intensity dependence of the forbidden LO, allowed LO and spe
from GaSb [$p \approx 4 \times 10^{16}$ cm^{-3},
90 K, (100) surface, $\hat{e}_i \| \hat{e}_s$] as
a function of $\hbar\omega_i$.

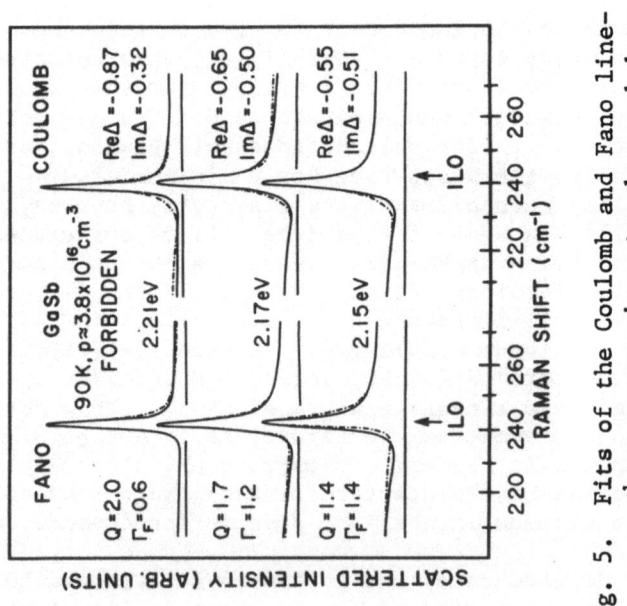

Fig. 5. Fits of the Coulomb and Fano lineshape expressions to observed data
for GaSb [$p \approx 4 \times 10^{16}$ cm^{-3}, 90 K,
(110) surface, $\hat{e}_i \| \hat{e}_s$] for different
$\hbar\omega_i$ near E_1.

We thank the Office of Naval Research for partial support of this work under Contract No. N00014-76-C-0643.

REFERENCES

1. Richard M. Martin, Resonance Raman scattering near critical points, Phys. Rev. B 10:2620 (1974).
2. W. Dreybrodt, W. Richter, F. Cerdeira, and M. Cardona, Orientation-dependent resonant Raman scattering in InSb and GaSb at the E_1--$E_1 + \Delta_1$ region. Phys. Stat. Sol (b) 60:145 (1973).
3. Peter Y. Yu and Y. R. Shen, Resonance Raman scattering in InSb near the E_1 transition, Phys. Rev. Lett. 29:468 (1972).
4. U. Fano, Effects of configuration interaction on intensities and phase shifts, Phys. Rev. 124:1866 (1961).
5. Miles V. Klein, B. N. Ganguly, and Priscilla J. Colwell, Theoretical and experimental study of Raman scattering from coupled LO-phonon-plasmon modes in silicon carbide, Phys. Rev. B 6:2380 (1972).
6. E. Burstein, A. Pinczuk, and S. Buchner, 1979, Resonance inelastic light scattering by charge carriers at semiconductor surfaces, in: "Physics of Semiconductors 1978," B. L. H. Wilson, ed., The Institute of Physics, Bristol.
7. C. A. Mead and W. G. Spitzer, Fermi level position at metal-semiconductor interfaces, Phys. Rev. 134:A713 (1964).
8. I. Lindau, P. W. Chye, C. M. Garner, P. Pianetta, C. Y. Su, and W. E. Spicer, New phenomena in Schottky barrier formation on III-V compounds, J. Vac. Sci. Technol. 15:1332 (1978).
9. Ralf Dornhaus, Roger L. Farrow, Richard K. Chang, and Richard M. Martin, Interaction of single-particle excitations with LO phonons in the bulk and surface layer of p-GaSb near E_1 and $E_1 + \Delta_1$ resonances (to be published).

INELASTIC LIGHT SCATTERING BY THE TWO DIMENSIONAL ELECTRONS IN SEMICONDUCTOR HETEROJUNCTION SUPERLATTICES

A. Pinczuk, H. L. Störmer, R. Dingle, J. M. Worlock,
W. Wiegmann and A. C. Gossard

Bell Laboratories
Holmdel and Murray Hill, NJ

Physical systems of reduced dimensionality have been lately much studied, both experimentally and theoretically. A very popular subcategory is the two dimensional electron gas (2DEG) which is confined in a narrow space charge region at a semiconductor surface or interface.

Burstein, Pinczuk and Buchner[1] proposed resonant inelastic light scattering as a new tool to study the electronic excitations of these semiconductor 2DEG's. This proposal and the production of high mobility 2DEG's at selectively doped GaAs-AlGaAs heterojunctions grown by molecular beam epitaxy[2,3,4], has led recently to two successful light scattering experiments. Abstreiter and Ploog[5] found evidence of one intersubband excitation in the spectra of the 2DEG at a single GaAs-AlGaAs interface. The present authors[6] observed several intersubband excitations of the multilayer 2DEG in a modulation doped GaAs-AlGaAs heterojunction superlattice. In this paper we wish to describe our experiments, concentrating on those aspects which elucidate the light scattering mechanisms involved.

The electronic states in <u>undoped</u> GaAs-AlGaAs heterojunction superlattices consist of a set of subbands constructed of discrete levels for motion perpendicular to the layers and two dimensional bands for motion in the plane of the layers. This picture has been verified by the spectroscopy of optical transitions between valence and conduction subbands[7,8].

<u>In modulation doped</u> superlattices, the charge can be confined in the GaAs layers as illustrated in Fig. 1. The potential wells and therefore the subband spacings are then strongly affected by

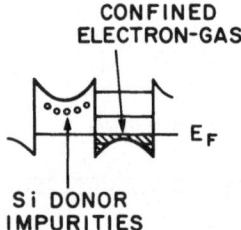

Fig. 1. Model of the conduction band edge of modulation doped
         heterojunction superlattices.

the charge distributions in the layers. Light scattering experiments are the first to measure the spacings between the subbands.

We have published elsewhere[6] the comparison between our measured and calculated subband levels. We wish, in this paper, to concentrate on the resonant enhancement of the scattering, which supports the mechanism proposed by Burstein et al.[1]

Two superlattice samples grown on (001) substrates were investigated in detail. The thicknesses of the GaAs and AlGaAs layers, d_1 and d_2, and their Al concentrations x are given in Table I. The table also shows the carrier concentrations, n, and Fermi energies E_F, obtained from Shubnikov-deHaas measurements[3], as well as the Hall mobilities μ.

Pinczuk et al.[9] found that light scattering by both single particle and collective excitations of the 3DEG in bulk n-type GaAs is strongly enhanced near the $E_0 + \Delta_0$ energy gap, while the luminescence is relatively weak. For this reason we also measured inelastic light scattering spectra with incident photon energies (1.89-2.02eV)

Table 1. Sample Parameters

Sample	d_1 (Å)	d_2 (Å)	x	n 10^{12} cm^{-2}	E_F (meV)	μ (cm^2/sec volt)
1	400	415	0.28	1.1	29	19,000
2	221	218	0.24	3.1	59	6,300

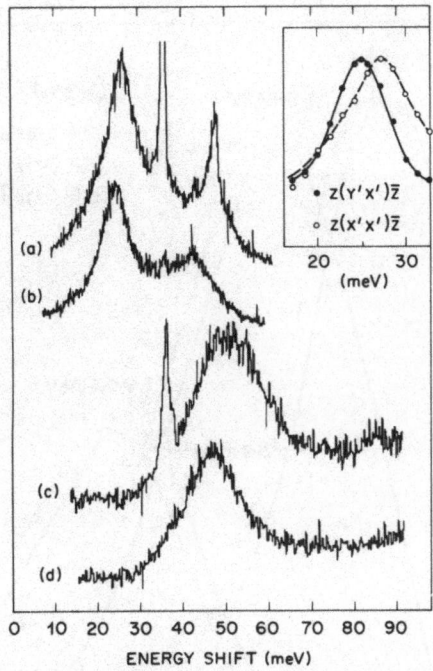

Fig. 2.   Light Scattering spectra from the doped superlattices. (a)
          and (b) are z(x'x')z̄ and z(y'x')z̄ spectra of sample 1; (c)
          and (d) are z(x'x')z̄ and z(y'x')z̄ spectra of sample 2. The
          insert shows the lowest bands of spectra (a) and (b).

near the $E_O+\Delta_O$ resonance. We obtained backscattering spectra with
the samples immersed in superfluid He, in z(x'x')z̄ and z(y'x')z̄ configurations. z is the (001) direction normal to the layers. x'
and y' are (110) and (1$\bar{1}$0) directions, in the plane of the layers.
The angle of the scattering wavevector $\vec{k}$ with the $\hat{z}$-direction
varied between 2^0 and 8^0.

     Figure 2 shows spectra from both samples. The <u>narrow</u> bands at
35.3meV and 46.6meV in the z(x'x')z̄ spectra (curves a and c) occur
at the LO_1 and LO_2 phonon energies of AlGaAs[10]. The existence of
these bands, which do not show the effects of coupling to charge carrier excitations[11], has led us to conclude that the electrons are
indeed confined as a multilayer 2DEG in the GaAs layers.

     The broad bands in both z(x'x')z̄ and z(y'x')z̄ spectra we have
interpreted as arising from intersubband excitations of the multilayer 2DEG. This interpretation was supported by: (a) the fact that
these bands are not found in the resonant Raman spectra of undoped
superlattices[12], (b) the qualitative observation that the intersubband energies are higher for sample 2 with the thinner layers,

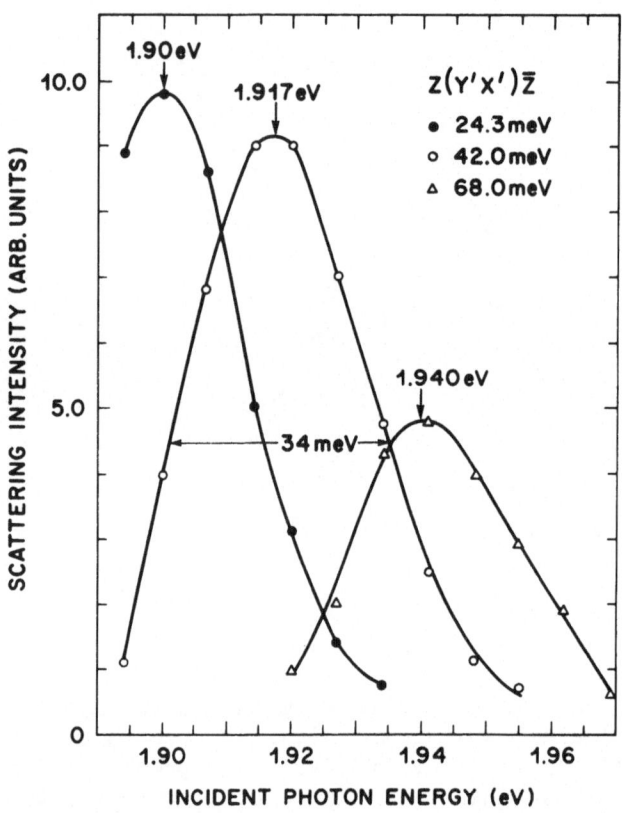

Fig. 3. Normalized scattering intensity as function of incident
         photon energy for the three lowest bands in the z(y'x')z̄
         spectra of sample 1.

(c) the quantitative agreement of the positions of these bands with
our calculation of the subband energies; and (d) the different
resonant behavior of the intensities of the bands as the incident
photon energy was varied. We wish to discuss here more fully only
the last of these reasons, which is related to the light scattering
mechanism by the intersubband excitations.

     Figure 2 shows two of the broad bands observed in spectra from
sample 1 and only one from sample 2. By changing the incident photon
energy, we were able to observe a number of these intersubband excitations. Each spectral band had a different resonant behavior. Figure 3 shows resonant enhancement curves of light scattering by the
intersubband excitations. Remarkable features of these curves are:
(1) the photon energies at the maxima depend on the energies of the
spectral bands, (2) the separation between peaks is approximately
equal to the differences in energy between the corresponding bands;

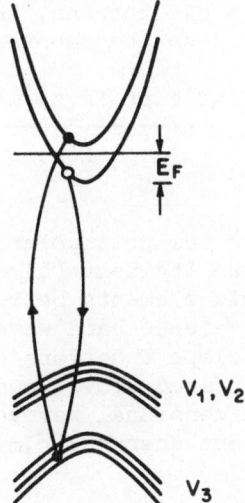

Fig. 4.   Schematic diagram of conduction and valence subbands, showing
          the optical transitions contributing to light scattering
          near the $E_o + \Delta_o$ resonance.

and (3) the enhancement peaks are narrow, with a width (~ 34meV) only
slightly larger than the Fermi energy.

     We wish to compare these results with the implications of the
mechanism proposed by Burstein et al.[1]. The optical transitions
involved are shown in Fig. 4. For photons with wavevectors normal
to the layers the transitions are vertical. The incident photon
promotes an electron in an <u>intermediate</u> valence subband to an
unoccupied conduction subband, the <u>final single particle state</u>. The
creation of the scattered photon is associated with the recombination
of an electron in an occupied conduction subband, the <u>initial single
particle state</u>, with the hole in the valence subband. Such processes
exhibit resonances at all optical energy gaps that involve electron
states occupied by the carriers such as the E_O and $E_O + \Delta_O$ gaps of
n-type GaAs[1,9].

     We examined the light scattering matrix element[13,14] for this
process in the effective mass approximation, and specialized to a
two dimensional layered electron gas in which the subband envelope
functions are localized in the GaAs layers. We found that the
polarization selection rules are similar to those of light scattering
by the excitations of the 3DEG of n-type GaAs[11,14] in that spin-flip
excitations (intersubband spin-density fluctuations) are predicted
in the $z(y'x')\bar{z}$ spectra, with non-spin-slip excitations (intersubband
charge-density fluctuations) in the $z(x'x')\bar{z}$ spectra.

In the case of spin-flip excitations, the matrix element for the light scattering process described above can be written as:

$$M_{if}(\vec{\kappa}) \cong \hat{e}_1 \times \hat{e}_2 |P_{cv}|^2 \sum_m \frac{<f|\exp(-ik_1 z)|m><m|\exp(ik_2 z)|i>}{E(\vec{\kappa},m) - \hbar\omega_1} \qquad (1)$$

where $\hat{e}_{1,2}$, $\vec{k}_{1,2}$ and $\omega_{1,2}$ are the unit polarization vectors, wave-vectors and frequencies of the incident (1) and scattered (2) photons. P_{cv} are momentum matrix elements between the cell-periodic parts of the conduction and valence band wavefunctions. The states $|i>$, $|m>$ and $|f>$ are the envelope functions of the initial, inter-mediate and final states described above, and the sum m is over the valence subbands. The two dimensional wavevector $\vec{\kappa}$ is common to all these states. The resonant energy $E(\vec{\kappa},m)$ is given by

$$E(\vec{\kappa},m) = E_G + E_f + E_m + \frac{\hbar^2 \kappa^2}{2\mu} \qquad (2)$$

where E_G is the $E_o+\Delta_o$ energy gap of GaAs, E_f and E_m are the $\vec{\kappa}=0$ energies of subbands $|f>$ and $|m>$. They are measured respectively from the bottom of the conduction band and the top of the valence band of GaAs. $\mu^{-1} = m_e^{-1} + m_h^{-1}$, where m_e and m_h are the effective masses of the conduction and V_3 valence bands.

Although all processes $i \rightarrow f$ described by the matrix element are degenerate in energy, each $\vec{\kappa}$ value leads to a unique final state. As a result, the scattering cross section, at T=0, is obtained by squaring $M_{if}(\vec{\kappa})$ and then summing over all $|\kappa| < \kappa_F$, i.e. over all occupied states in subband i. κ_F is the Fermi wavevector. We consider first the contribution of a single intermediate valence subband m. We find that there is a symmetrical resonance enhancement peak at an energy $E_R(m,f)$ given by:

$$E_R(m,f) = E_G + E_f + E_m + \frac{1}{2}\left(\frac{\hbar^2 \kappa_F^2}{2\mu}\right) \qquad (3)$$

The width of the resonance is given by

$$W_R(m,f) = \frac{\hbar^2 \kappa_F^2}{2\mu} + 2\gamma \qquad (4)$$

where γ is a phenomenological damping assigned to the optical transi-tions ($\gamma \lesssim 10$ meV).

The absence of structure in the experimental resonant enhance-ment curves of Fig. 3 implies that only closely spaced valence

subbands are contributing to $M_{if}(\vec{k})$. The fact that the experimental
widths are comparable to the Fermi energy, as in Eq. 4, implies that
those which do contribute are nearly degenerate (within $\sim$ 5meV). Furthermore, since the spacing between the experimental resonant enhancement peaks is approximately equal to the energy differences between
the intersubband excitations, and making use of our earlier result[6]
that all these transitions originate from two almost degenerate initial states, we conclude from Eq. 3 that in all cases it is the same
group of valence subbands that makes the dominant contribution to
the $M_{if}(\vec{k})$.

The approximate energy of these subbands can be obtained from
Eq. 3 using the values of E_R given in Fig. 3, the calculated values
of E_f[6] and E_G=1.86eV[15]. We find that these states are at $\sim$ 20meV
below the top of the valence band of GaAs measured at the center of
the GaAs layer. Because the band bending in the GaAs layers is
$\sim$ 100meV[6], these states are completely confined in the GaAs layers.
At the present time we do not know the reason why only these valence
states contribute to $M_{if}(\vec{k})$. However, we would like to point out the
fact that, due to band bending, the conduction and valence subband
states tend to localize in different regions of the GaAs layers. The
relevant valence subbands have the apparently unique property of
having a considerable overlap with both the initial and final conduction subbands.

Experimentally, the resonant enhancement curves for the non-
spin-flip excitations in the $z(x'x')\bar{z}$ spectra are very similar to
those shown in Fig. 3. However, because these transitions are associated with charge flow, they are subjected to depolarization electric fields due to Coulomb interactions, giving rise to additional
terms in $M_{if}(\vec{k})$[1,14]. Evidence of effects associated with the depolarization fields are indeed found in the light scattering spectra. The
insert to Fig. 2 shows that the lowest band in the $z(x'x')\bar{z}$ spectra
is shifted to a slightly higher energy. Further work on the theory of
these intersubband excitations, as well as experiments on different
scattering geometries will doubtless help to clarify these effects.

In summary, we have discussed our recent work on resonant inelastic light scattering by the multilayer 2DEG in modulation doped GaAs-
AlGaAs heterojunction superlattices. The resonant enhancement of spectra assigned to spin-flip intersubband excitations have been compared
with calculations based on the mechanism proposed by Burstein et al.[1].
Good agreement between measured and calculated enhancements is found,
but only by assuming that, at the $E_0+\Delta_0$ energy gap, only a closely
spaced group of valence subbands contributes to the resonance enhancement. Resonant inelastic light scattering is a unique and
powerful method for studying the two dimensional electron gases at
semiconductor surfaces and interfaces, including the energies
of the intersubband excitations and their Coulomb interactions.

ACKNOWLEDGMENTS

We gratefully acknowledge illuminateing discussions with
E. Burstein, P. A. Wolff and D. L. Mills.

REFERENCES

1. E. Burstein, A. Pinczuk and S. Buchner, in "Physics of Semicon-
    ductors 1978", B. L. H. Wilson ed., The Institute of Physics,
    London (1979), p. 1231.
2. R. Dingle, H. L. Störmer, A. C. Gossard and W. Wiegmann, Appl.
    Phys. Lett. 33:665 (1978).
3. H. L. Störmer, R. Dingle, A. C. Gossard, W. Wiegmann and R. A.
    Logan, in Ref. 1, p. 557.
4. H. L. Störmer, R. Dingle, A. C. Gossard, W. Wiegmann and M. D.
    Sturge, Solid State Commun. 29:705 (1979).
5. G. Abstreiter and K. Ploog, Phys. Rev. Lett. 42:1308 (1979).
6. A. Pinczuk, H. L. Störmer, R. Dingle, J. M. Worlock, W. Wiegmann
    and A. C. Gossard, submitted for publication.
7. L. Esaki and R. Tsu, IBM J. Res. Develop. 14:61 (1970); L. L.
    Chang, L. Esaki and R. Tsu, Appl. Phys. Lett. 24:593 (1974).
8. R. Dingle, W. Wiegmann and C. H. Henry, Phys. Rev. Lett. 33:827
    (1974); R. Dingle, A. C. Gossard and W. Wiegmann, Phys. Rev.
    Lett. 34:1327 (1975).
9. A. Pinczuk, G. Abstreiter, R. Trommer and M. Cardona, Solid
    State Commun. 30:429 (1979).
10. R. Tsu, H. Kawamura and L. Esaki, in "Proceedings of the 11th
    International Conference on the Physics of Semiconductors",
    Polish Scientific Publications, Warsaw (1972), p. 1135.
11. M. V. Klein, in "Light Scattering in Solids", M. Cardona ed.,
    Springer-Verlag, Berlin-Heidelberg (1975), p. 147, and refer-
    ences therein.
12. G. A. Sai-Halasz, A. Pinczuk, P. Y. Yu and L. Esaki, Solid State
    Commun. 25:381 (1978).
13. P. A. Wolff, Phys. Rev. Lett. 16:225 (1966).
14. D. C. Hamilton and A. L. McWhorter, in "Light Scattering Spectra
    of Solids", G. B. Wright ed., Springer-Verlag, Berlin (1969),
    p. 309.
15. D. E. Aspnes and A. A. Studna, Phys. Rev. B7:4605 (1973).

ON THE DISTINCTION BETWEEN RESONANT SCATTERING AND HOT

LUMINESCENCE: APPLICATION OF THEORY TO EXPERIMENT

P. Saari

Institute of Physics
Estonian SSR Academy of Sciences
Tartu, USSR

SUMMARY

Referring to experiment, some particular problems
of the distinction are discussed in this paper.

1. If the vibronic spectra of an impurity crystal
exhibits clear-cut quasi-line structure scattering,
hot and ordinary luminescence can be definitely distinguished with steady state monochromatic excitation in
both theory and experiment.

2. Resonant scattering does not dominate nonresonant
scattering for a condensed system with rapid relaxational processes. Even though the excitation line is
within the absorption band of a given electronic transition, the scattering spectrum may have significant
contributions from other nonresonant electronic terms.
Thus for a proper interpretation, polarization and
other additional measurements are required.

3. Although invoking T_2-processes generally results
in conversion of a part of the scattering into lumi-
nescence-like emission, it will not suffice to introduce (frequency-independent) relaxation constants to
describe properly the resonant secondary emission of a
crystal, but rather a detailed dynamic model is
required. Using the former approach, the decomposition of the entire flux of resonant emission into
components may turn out to be impossible since the
contributions are of a mixed nature, even though the
relaxation criterion generally works well (especially
in nonlinear spectroscopy).

4. While its applicability to the study of more
complicated systems is beyond doubt, the experimentally
realizable time resolution does not provide any

additional possibilities of distinguishing scattering and fluorescence in the case of the two-level
system, the test model for the classification problem.

The classification of the components of resonant secondary
emission (RSE), widely discussed since the early studies by Vavilov
[1], provides a fundamental problem for both theoretical and
practical spectroscopy. Some progress in the understanding of the
problem has been achieved since 1960, after a unified theoretical
approach to the RSE components of impurity crystals was demonstrated [2, 3]. A weak emission, when first observed [4], was
called hot luminescence (HL) and described by the theory as an
intermediate component of RSE in addition to scattering and luminescence. Later the concept of HL was applied to RSE in semiconductors [5, 6] where controversies existed about the origin of
inelastic emission [7, 8]. The experiment on time-resolved RSE
from molecular iodine [9] triggered considerable theoretical
activity in the field of using time criteria to distinguish between
scattering and fluorescence (see refs. in [10]). Thus, the RSE
classification problem remains a pressing one since laser spectroscopy is developing and its applications are being extended to
various investigations which require a proper interpretation of
new results.

We are going to discuss here the following key questions about
the classification of experimentally observed results.
1. Do systems exist for which scattering and HL are definitely
    separable both in theory and experiment?
2. What complications may arise in interpreting a real RSE spectrum
    with theoretical models based on "the resonance excitation
    case"?
3. Will the introduction of phenomenological constants of longi-
    tudinal and transverse relaxations suffice to describe the RSE
    of condensed matter adequately?
4. What is to be expected from the time resolution of RSE?
5. What complications may be encountered when investigating real
    inhomogeneous systems?

We are going to deal mainly with molecular centres in host
crystals at low-temperature which, apart from their practical
importance, are good model systems for the problems under study.
As we are certainly not able to give a comprehensive answer to all
these questions, our treatment is restricted to certain topics only.

I. ON THE DISTINCTION BETWEEN SCATTERING AND HL DUE TO LOCALIZED
MODES

Let us consider the model of a luminescence centre with
localized modes, which change their equilibrium position and

frequency upon electronic transition. Assuming not very rapid exponential decay of modes and low temperature,

$$\gamma_r << \gamma_k << |\omega_k^* - \omega_k| << |\omega_k - \omega_{k\pm1}| >> kT/h,$$

where γ_r, γ_k are the decay constants of the electronic and vibrational states, respectively, and ω_k is the energy of the k-th vibrational level. An asterisk denotes a quantity in the excited electronic state. An expression for RSE intensity is (detailed formula for the case of single mode has been obtained in [11]; see also [12]):

$$I(\omega_1,\omega_2) = BK(\omega_1,\Omega_g^e)\left[W_{OL}(\omega_2 - \Omega_g^e) + W_{HL}(\omega_2 - \Omega_g^e) + W_S(\omega_2 - \omega_1)\right],$$

(1)

where

$$W_{OL}(\omega_2 - \Omega_g^e) = \frac{1}{\gamma_r} \sum_{f=0}^{\infty} F_{0f}^2(\gamma_f + \gamma_r)/\left[(\omega_2 - \Omega_g^e + \omega_f)^2 + (\gamma_f + \gamma_r)^2\right],$$

$$W_{HL}(\omega_2 - \Omega_g^e) = \sum_{n=1}^{i-1} \frac{1}{\gamma_n^*} \eta_{in} \sum_{f=0}^{\infty} F_{nf}^2(\gamma_n^* + \gamma_f)$$

$$/\left[(\omega_2 - \Omega_g^e + \omega_f - \omega_n^*)^2 + (\gamma_n^* + \gamma_f)^2\right],$$

$$W_S(\omega_2 - \omega_1) = \frac{1}{\gamma_i^*} \sum_{f=0}^{\infty} F_{if}^2 \gamma_f/\left[(\omega_2 - \omega_1 + \omega_f)^2 + \gamma_f^2\right].$$

Here ω_1, ω_2 and Ω_g^e are the frequencies of the exciting light, secondary emission, and pure-electronic transition, respectively. B depends on the choice of units and has smooth frequency dependences; η_{in} is the probability of relaxational transition from the level i to the level n and F_{if}^2 is the probability of the radiative vibronic transition to the level f in the ground state; and $K(\omega_1,\Omega_g^e) = F_{0i}^2\gamma_i^*/\left[x^2 + (\gamma_i^*)^2\right]$ is the absorption probability. it is assumed that ω_1 is in resonance (detuning $x << |\omega_i^* - \omega_{i\pm1}^*|$) with the absorption line of frequency $\Omega_g^e + \omega_i^*$.

As follows from (1), the scattering spectrum W_S consists of scattering lines whose frequencies are shifted from ω_1 by various vibrational frequency combinations in the ground electronic state only. In contrast, vibrational frequencies and widths of both electronic states appear in the HL spectrum W_{HL} and in the ordinary luminescence spectrum W_{OL}. It is noteworthy that in the given model there is no fluorescence-type emission from the resonant level i as long as the modulational broadening of vibrational levels (T2-process), which is negligible only at low temperatures, is not introduced. Nevertheless, if the exciting line is not monochromatic but has a width $\Delta > \gamma_i^*$, the ω_1 integration of (1) within Δ must be carried out, which results in the replacement of scattering terms by HL-like terms (HL from the level i).

The experiment clearly confirms the conclusions drawn from the model considered. If the absorption and luminescence spectra possess a distinct quasi-line structure, there are no problems in distinguishing HL; if the excitation is tuned onto the broad modes (e.g., phonon side-bands), the same holds for scattering [12-15]. In Fig.1, the RSE spectrum of anthracene molecules in a fluorene matrix excited on the absorption line $\Omega_g^e + 1500$ cm^{-1} is presented.

At wavelengths shorter than the 0-0 line, a series of weak lines can be observed which are interpreted as being due to scattering and HL on the basis of the known frequencies of the system. The lines corresponding to the transitions from the 1500 cm^{-1} level are classified as resonant scattering, keeping in mind that any modulation process in this level is practically absent.

II. THE INCLUSION OF NONRESONANCE SCATTERING

It must be stressed that for rapidly relaxing systems one should not expect   dominance of resonant scattering over nonresonant scattering [16], because the energy denominator of the RSE cross section is small. The main part of RSE, however, reduces to ordinary luminescence. This is obvious from (1) since the ratio of W_S (as well as that of W_{HL}) to W_{OL} is of the order of γ_r/γ_i^*, i.e., $10^{-3} - 10^{-5}$ for condensed matter even in the case of allowed dipole transitions. From this result the following can be deduced: 1) In impurity systems, nonresonant RS from host modes is comparable to resonant RS from impurity modes (see Fig.1). In an organic crystal, where the frequencies of both kinds of modes nearly coincide, fine scanning of ω_1 is needed for a proper interpretation [12, 14]. 2) Even though the excitation line falls within the absorption band of a given electronic transition, the scattering line intensities may have significant contributions from other nonresonant electronic terms. (See RSE spectra of KCl-NO$_2^-$ and pure NaNO$_2$ in [14, 16].) 3) Therefore, polarization and excitation profile measurements are necessary for a proper interpretation of scattering lines.

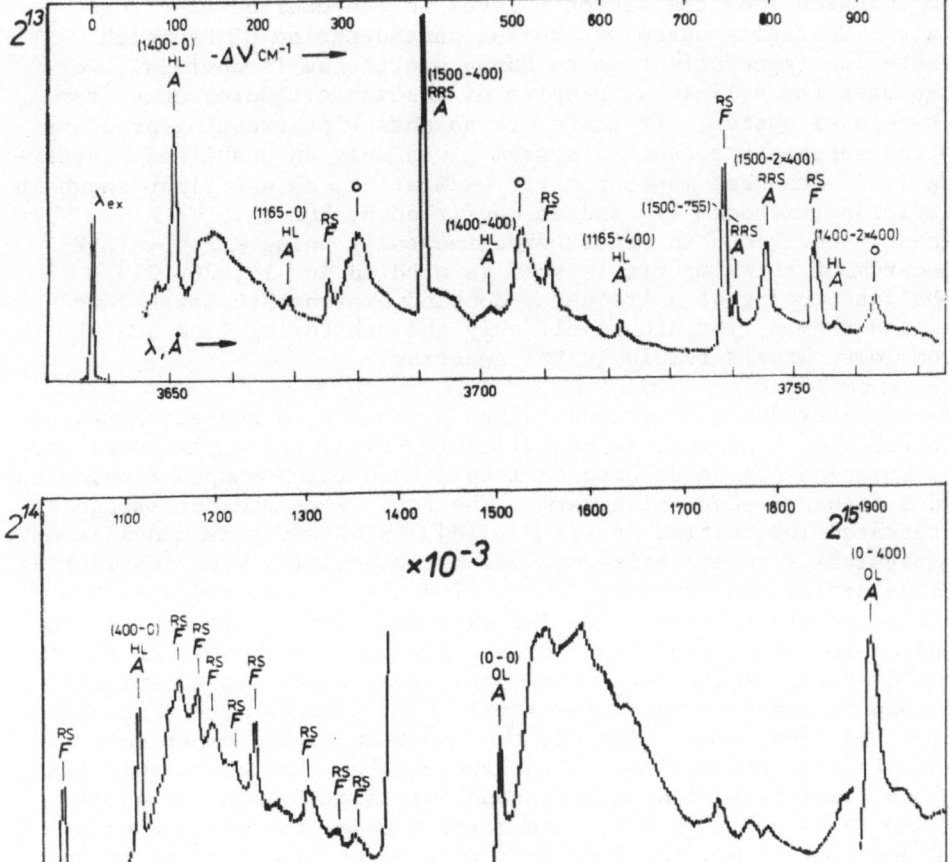

Fig. 1. RSE spectrum of an anthracene/fluorene mixed crystal at
4.2 K. RRS-lines indicate resonance scattering from
anthracene (A). HL and OL-lines are hot and ordinary
luminescence of anthracene, RS/F-lines indicate nonresonant
scattering from the fluorene (F) matrix. 0-bands are a
foreign emission.

III. THE RELATION BETWEEN RS AND HL AT FINITE TEMPERATURES

As was stressed in Sect. 1, for the systems under examination the question of the distinction between RS and HL may arise only for emission from the vibronic level in the excited electronic state. Confining ourselves to the consideration of Rayleigh scattering (generalization to Raman scattering is obvious), we encounter the well-known problem of resonance fluorescence from a two-level system. If there are no phase-interrupting processes in the upper state, such a system gives only an unshifted scattering line with weak monochromatic excitation, as was first shown in Heitler's textbook [17] and as confirmed by Eq.(1). We have recently checked this result experimentally using $KCl-NO_2^-$ (RSE spectrum of this impurity system is studied in [16, 20, 21].) Upon fine tuning of a frequency-doubled oxazine dye laser down shifted from a vibronic level, only the scattering line and HL from lower levels remain in the spectrum.

By introducing fast modulation processes, a phase relaxation contribution Γ appears in the width $(\gamma + \Gamma)$ of the upper level and the emission can be decomposed into a Lorentzian-shaped fluorescence and a δ-shaped scattering term. The ratio R of the corresponding integrated intensities is Γ/γ [3, 18]. However, this model leads to physically unrealistic conclusions, especially when considering solids at low temperatures. R remains the same regardless of how much ω_1 is off resonance.*) For example, although in the case of a pure electronic line Γ may exceed the radiative decay constant by $\sim 10^2$ times even at helium temperatures, there are no suitable phonons to modulate the upper level if ω_1 is much more than kT/h below the resonance. Indeed, the above-mentioned experiment supports this assumption. When studying the pure electronic line region, luminescence disappears and only scattering remains if $\Omega_g^e - \omega_1 > 10$ cm^{-1}. Hence, in a correct model the ratio R must depend on ω_1, i.e., the frequency dependence of Γ must be taken into account. This is possible in a dynamic description of relaxation processes [22-25]. Based on [25] where the RSE spectrum has been considered for $\Gamma \neq$ const., we have carried out computer calculations of $\Gamma(\omega_1 - \Omega_g^e)$ for $\omega_1 - \Omega_g^e < 0$ and a low-temperature case. Out results show that R should indeed rapidly decrease after $\Omega_g^e - \omega_1 > 10$ cm^{-1}. We may conclude that to properly account for phase relaxation in describing the RSE of a crystal at finite temperatures it will not suffice to introduce the phenomenological relaxation constants, but rather a detailed dynamic model is required.

_____

*)Consideration of the problem in the time domain gives an analogous result for the ratio of slow and fast emission components [10].

IV. ON TIME RESOLUTION IN SCATTERING AND LUMINESCENCE

As mentioned previously, a number of investigations have recently been concerned with the time behaviour of RSE. It is interesting to note that while the steady-state spectrum of the two-level system, modulated by a T_2-process, and with exact resonance excitation, is distinctly decomposable into two components (see, Sect. 3),[*] the time dependence of RSE intensity has only the slow component. In other words, time resolution cannot distinguish scattering when ω_1 is at exact resonance even if we consider a model where T_2-relaxation actually decomposes the emission into two components. For a simple model, where only the population damping (T_1-process) is taken into account, the conclusion is the same [26]. Moreover, in the pre-resonance case, the appearance of the fast and slow components can be attributed to two different processes with a high degree of accuracy. This discrepancy between the spectral and time pictures is due to the fact that in the latter case a sufficiently fast excitation switch-off is needed, inevitably leading to the uncertainty in ω_1.

Therefore, it is rather exciting to make clear what can be achieved by applying both time and spectral resolutions. The time-dependent spectrum of the two-level system under study has been calculated in [27]. However, when dealing with transient spectra, a question of the physical meaning of the quantity arises. In fact, although the expression of an emission spectrum through the time derivative of a photon emission probability into the given mode works well in a steady-state case, this derivative does not experimentally give an observable transient spectrum. A rigorous definition of a time-dependent spectrum must take into account the measurement processes [28-30], particularly the uncertainty $T \cdot \delta\omega_2 = 1$, where T and $\delta\omega_2$ are extremal time and spectral resolutions, respectively. This measurement uncertainty was not taken into account in the first calculations of transient spectra [27, 31, 32].

Let us proceed to calculation of the time-dependent spectra of the two-level system.[**] Most instructive is the behaviour of the emission after the excitation is switched off. Therefore, we choose as the exciting field a monochromatic wave at the frequency ω_1 in the form of a rectangular pulse. If the pulse duration is assumed to be much longer than the characteristic times of the system and the pulse termination is at $t=0$, then the general

---

[*] Experimental separation, however, may be unperformable if scattering linewidth is comparable to that of fluorescence.

[**] Preliminary results relative to the separability of scattering and fluorescence have been published in [30].

expression of the measurable time-dependent spectrum obtained in [30] reduces to:

$$I_T(\omega_1,\omega_2,t) = T^{-2} \int_{t-T}^{t} dt_1 \int_{t-T}^{t} dt_1' e^{i\omega_2(t_1'-t_1)} \int_{-\infty}^{t_1} dt_2 \int_{-\infty}^{t_1'} dt_2' K\cdot G_e.$$

$$(2)$$

Here $K\cdot G_e$ is the product of the correlation functions of the material system and exciting field, which for the case under study can be expressed as:

$$K\cdot G_e = Y(-t_2)Y(-t_2')\exp\{g(t_2 + t_2' - t_1 - t_1') + i\Delta_1(t_2 - t_2') +$$

$$+ 2\Gamma[\min(t_1,t_1') - \max(t_2,t_2')]Y(t_1 - t_2')Y(t_1' - t_2)\}, \quad (3)$$

where Y is the Heavyside unit step, $g = \gamma + \Gamma$ is the sum of the damping constant and the modulation broadening of the upper level, and $\Delta_1 \equiv \Delta\omega_1 = \omega_1 - \Omega_g^e$. A statistical averaging has been carried out within the limit of fast modulation of the transition frequency Ω_g^e[18]. After an appropriate partitioning of the time domain, integration of Eq. (2) can be performed analytically to obtain:

$$I_T(\omega_1,\omega_2, t) = \frac{1}{T^2} \frac{1}{\Delta_1^2 + g^2} E_T(\Delta_1,\Delta_2,t),$$

where $\Delta \equiv \Delta\omega_2 = \omega_2 - \Omega_g^e$ and the emission term $E_T(\Delta_1,\Delta_2,t)$

$$= \delta_T(\omega_1 - \omega_2) + 2\Gamma\gamma^{-1}L_T(\Delta_2) \qquad\qquad \text{if } T \leq 0;$$

$$= \delta_{T-t}(\omega_1 - \omega_2) + 2\Gamma\gamma^{-1}L_{T-t}(\Delta_2) + L_t'(\Delta_2) + \ldots, \quad 0<t<T;$$

$$= L_T'(\Delta_2)\exp[-2\gamma(t - T)], \qquad\qquad t \geq T. \qquad (4)$$

The following notation has been used:

$$\delta_\tau(\omega_1 - \omega_2) = \frac{\sin^2(\omega_1 - \omega_2)\tau/2}{[(\omega_1 - \omega_2)/2]^2},$$

$$L_\tau(\Delta_2) = \frac{g\tau}{\Delta_2^2 + g^2} + \frac{(\Delta_2^2 - g^2) - e^{-g\tau}[(\Delta_2^2 - g^2)\cos\Delta_2\tau + 2\Delta_2 g\sin\Delta_2\tau]}{(\Delta_2^2 + g^2)^2},$$

$$L_\tau'(\Delta_2) = \frac{g^2}{\gamma^2(\Delta_2^2 + g^2)} + \frac{(\gamma^2 - \Gamma^2)e^{-2\gamma\tau}}{\gamma^2[\Delta_2^2 + (\gamma - \Gamma)^2]}$$

$$- \frac{2g[(\Delta_2^2 + \gamma^2 - \Gamma^2)\cos\Delta_2\tau + 2\Delta_2\Gamma\sin\Delta_2\tau]e^{-g\tau}}{\gamma(\Delta_2^2 + g^2)[\Delta_2^2 + (\gamma - \Gamma)^2]}.$$

By (...) we denote complicated interference terms, which depend on all arguments. For $t \leq 0$ we have the steady-state regime, and we recognize the well-known expression consisting of two components with the integrated intensity ratio $R = \Gamma/\gamma$. A finite spectral resolution causes the scattering line not to be infinitely narrow and the fluorescence line not to be a pure Lorentzian. After the excitation is switched off, the scattering line begins to broaden and gradually disappears. If the measurement time interval $t \rightarrow T + t$ does not exceed the excitation time duration, i.e., if $t \geq T$, only exponentially decaying fluoresence remains.

To analyse various situations, we have used a computer to calculate the spectra according to (2) - (4). We can see from the resultant plots (see Figs.2-5) that the calculation confirms the expected picture of a time-resolved spectrum for the case of off-resonance excitation, where the scattering and fluorescence lines remain distinctly separated even with a spectral resolution which is less than one width (Fig.2). In fact, this is just the condition which makes possible the observation of the difference in time behaviour of the two lines: 1) scattering decays within the reciprocal of the instrument spectral bandwidth, and 2) fluorescence, after a small increase caused by the excitation switch-off, decays exponentially with the depopulation rate of the upper level $2\gamma = 0.6$.

In a pre-resonance case (Fig.3), the components become less separable in the time/frequency plane. If one chooses a better spectral resolution, the difference in time behaviour is lost and both components decay within the spectrometer response time. In contrast, if we try to improve the time resolution, the two spectral components fuse into one exponentially decaying band.

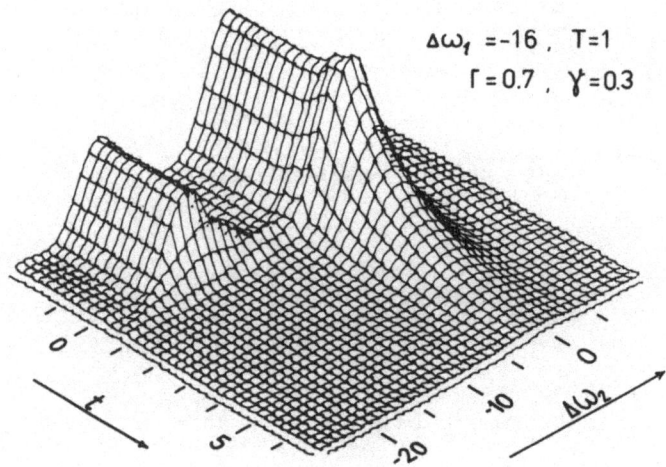

Fig. 2.   Time-dependent spectrum of the two-level system at off-
resonance excitation (detuning is 16 fold width of the
absorption line g = $\Gamma + \gamma$, which has been taken as the
frequency unit, g^{-1} as the time unit).

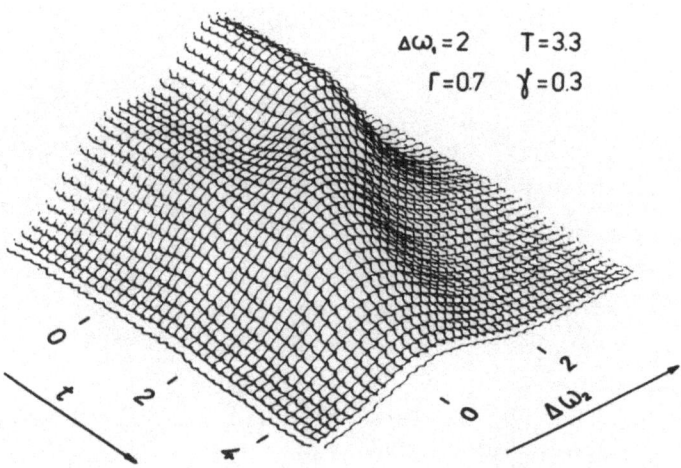

Fig. 3.   Time-dependent spectrum at pre-resonance excitation.

In the case of exact resonance (cf. [27]), it is not possible to observe the fast decay of the scattering line. In order to separate the components, one should choose a high spectral resolution in a steady-state regime. Time resolution does not give any advantage in this case (Fig.4).

Finally, when the phase relaxation is negligible ($\Gamma \ll \gamma$) in the upper state, there is, as we know, only scattering in the steady-state regime (Fig.5). The switched-off excitation causes an emission at the shifted frequency. Thus, the introduction of the time resolution enables us to observe the transformation of scattering into fluorescence afterglow (again, only if the excitation is not too close to resonance).

We conclude that experimentally realizable time resolution does not give any significant additional possibilities of distinguishing scattering and fluorescence, as was shown for the case of the two-level model, a test model for the RSE problem. This conclusion however, by no means reduces the significance and potential of the theoretical and experimental study of transient spectra in more complicated systems in particular, for separating HL and for relaxation studies (see, e.g., [16, 32]).

V. EFFECT OF INHOMOGENEITY ON THE RELATION BETWEEN SCATTERING AND LUMINESCENCE

A wide distribution of $\hbar\Omega_g^e$ for pure electronic transitions causes additional difficulties in the distinction between scattering and luminescence [33], since the luminescence spectrum of the latter is now coupled and shifts with the frequency ω_1. A theory generalizing the model of Sect. 1 to inhomogeneous systems has been developed in [34]. What is useful for experimentalists to note is that the inhomogeneity broadens the luminescence spectrum and does not affect the scattering spectrum, whereas the effect of nonmonochromaticity in the excitation is exactly the opposite. To prove that, one should integrate expression (1) over either Ω_g^e or ω_1. Consequently, for strongly inhomogeneous systems, for which scanning the excitation is non-productive, one may try to use the above-mentioned difference to distinguish scattering from luminescence (see Fig.6).

VI. CONCLUSION

We are of the opinion that it is rather meaningless to discuss the RSE classification problem in general terms since the physical picture seems to be quite clear on the level of principle. If some process causes transverse relaxation of the intermediate state (T_2-process) a part of the scattering is converted into fluorescence-type emission which, in turn, can be decomposed into hot and ordinary luminescence depending on the subsequent

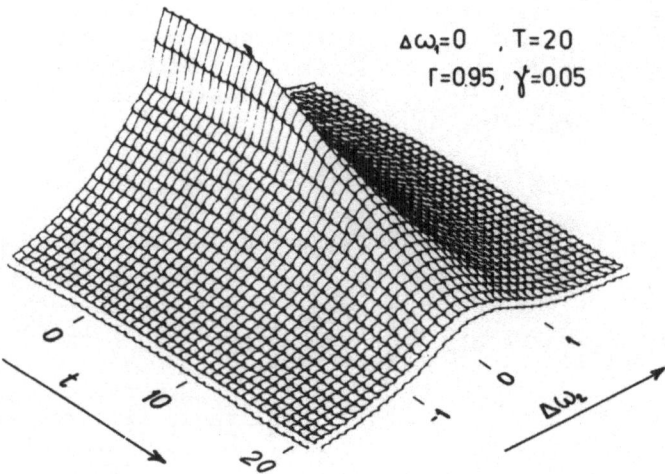

Fig. 4.   Time-dependent spectrum at exact resonance. A high
           spectral resolution (T=20) has been chosen in order to
           separate the δ-shaped scattering line from the
           fluorescence band.

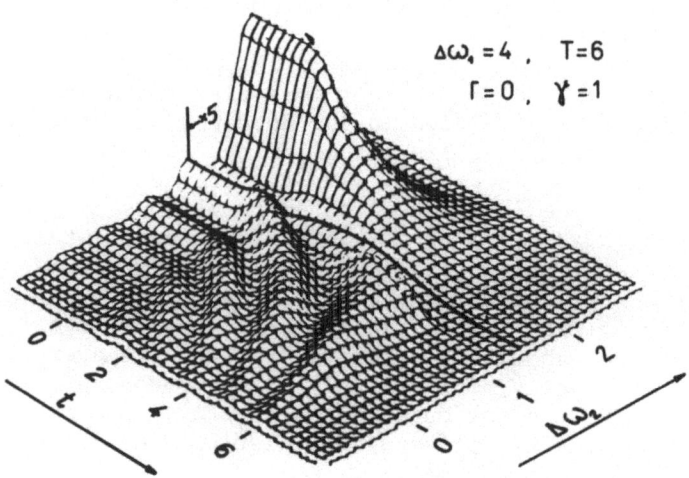

Fig. 5.   Time-dependent spectrum of the two-level system with
           negligible modulational broadening. After excitation
           switch-off at t=0 one can observe a nonmonotonic
           interference behaviour of the emission intensity around
           the shifted frequency, which transforms into exponential
           decay at t=6, i.e., after the spectrometer has "forgotten"
           the initial field.

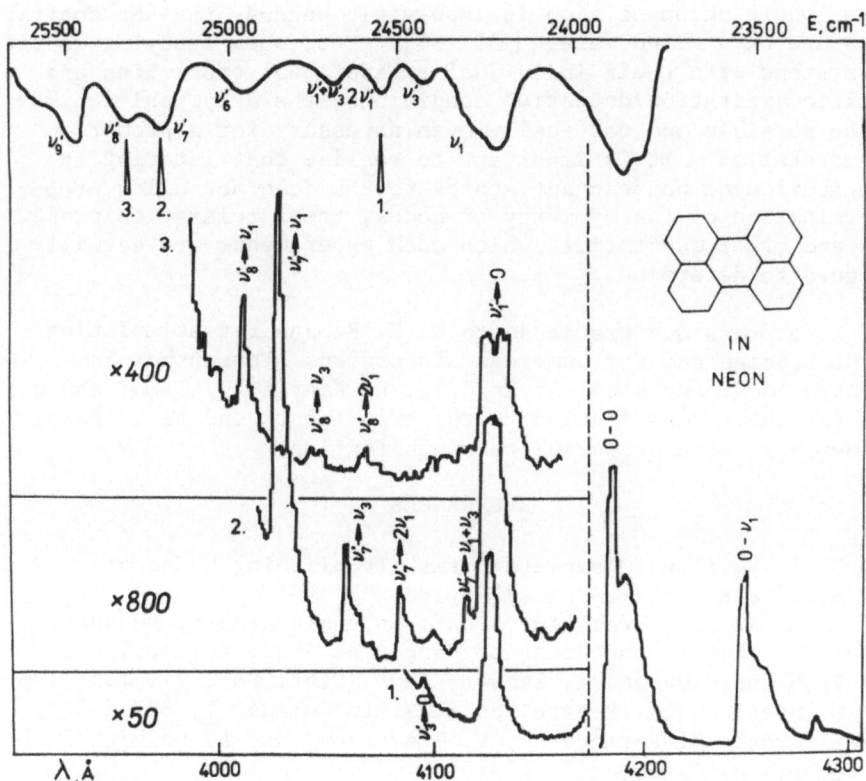

Fig. 6.   HL spectrum with the first bands of the OL spectrum of
          matrix-isolated perylene molecules. The top curve shows
          the corresponding positions of exciting frequencies
          relative to absorption bands. The lines originating from
          the resonant levels (ν_8', ν_7', $2\nu_1'$) do not show phonon side-
          bands enhanced by inhomogeneity and may be interpreted as
          proof of their scattering nature, as can also be inferred
          from the model of Sect. 1.

relaxation processes [35]. Indeed, there are a number of models, real systems and experimental situations (especially in nonlinear spectroscopy) where this relaxational criterion works well. Nevertheless, the decomposion of the entire flux of RSE is artificial, generally speaking, and there are also a number of cases where decomposition is impossible because the RSE contributions are of a mixed nature [35, 36]. Yet, when studying particular systems with their individual relaxational properties and specific excitation/detection conditions, the decomposition may not be possible and not really even necessary for a proper interpretation. It is important to realize that ignoring the contribution of nonresonant states to RSE does not allow proper determination of the symmetry of modes, their relaxation properties and other quantities, which such experiments are actually designed to determine.

We express our gratitude to K. K. Rebane for stimulating the HL studies and for numerous discussion. The author is indebted to J. Aaviksoo, A. Anijalg, R. Kaarli, T. Tamm, and A. Vill for their contribution to the experiments and to L. Pedosar for her help with preparing this publication.

REFERENCES

1. S. I. Vavilov, "Complete Works" (Publishing House of Acad. Sci. USSR, 1952), Vol. 2, p.188.
2. K. K. Rebane, Proc. Int. Conf. on Luminescence, Budapest, 1966. (Publishing House of Hungarian Acad. Sci.)p.124.
3. V. Hizhnyakov and I. Tehver, Phys. Stat. Sol. 21, 755 (1967).
4. P. Saari and K. Rebane, Solid State Commun. 7, 887 (1969).
5. E. Gross, S. Permogorov, V. Travnikov, and A. Selkin, J. Phys. Chem. Solids 31, 2595 (1970).
6. S. Permogorov, Phys. Stat. Sol. (b) 68, 9 (1975).
7. M. V. Klein, Phys. Rev. B 8, 919 (1973).
8. Y. R. Shen, Phys. Rev. B 9, 922 (1974).
9. P. F. Williams, D. L. Rousseau, and S. H. Dworetsky, Phys. Rev. Lett. 32, 196 (1974).
10. S. Mukamel, A. Ben-Reuven, and J. Jortner, Phys. Rev. A12, 947 (1975); J. Chem. Phys. 64, 3971 (1976).
11. V. Hizhnyakov and I. Tehver, Phys. Stat. Sol. 82, K89 (1977).
12. K. Rebane and P. Saari, J. Luminescence 16, 223 (1978).
13. Ultrafast Relaxation and Secondary Emission. Proc. of the Int. Symposium "Ultrafast Phenomena in Spectroscopy", Tallinn, Sept. 27 - Oct. 1, 1978, published by Estonian Acad. Sci.
14. P. Saari, ibid. p.142.
15. J. Aaviksoo, P. Saari, and T. Tamm, Pis'ma v JETP 29, 388 (1979).
16. K. Rebane, J. Luminescence 18/19, 693 (1978); in [13], p.7; and in the present volume.

17. W. Heitler, "The Quantum Theory of Radiation" (Oxford, 1954), p.196.
18. D. L. Huber, Phys. Rev. $\underline{158}$, 843 (1967); $\underline{170}$, 418 (1968); $\underline{178}$, 93 (1969).
19. K. Rebane, "Impurity Spectra of Solids" (Plenum Press, N. Y. -London, 1970).
20. K. K. Rebane and L. A. Rebane, J. Pure Appl. Chem. $\underline{37}$, 161 (1974).
21. R. Avarmaa and P. Saari, Phys. Stat. Sol. $\underline{36}$, K177 (1969).
22. M. A. Krivoglaz, Fiz. Tverd. Tela. $\underline{6}$, 1707 (1964).
23. K. K. Rebane and V. V. Hizhnyakov, "Secondary Emission of Impurity Centre —— Luminescence, Hot Luminescence and Scattering" Preprint FAI-2, Tartu, 1973 (in Russian).
24. A. A. Maradudin, Solid State Phys. $\underline{18}$, 273 (1966).
25. A. Kotani and Y. Toyozawa, J. Phys. Soc. Japan $\underline{41}$, 1699 (1976).
26. S. Mukamel and J. Jortner, J. Chem. Phys. $\underline{62}$, 3609 (1975).
27. T. Takagahara, E. Hanamura, and R. Kubo, J. Phys. Soc. Japan $\underline{43}$, 1522 (1977).
28. J. H. Eberly and K. Wodkiewicz, J. Opt. Soc. Am. $\underline{67}$, 1252 (1977).
29. E. Courtens and A. Szöke, Phys. Rev. A$\underline{15}$, 1588 (1977).
30. P. Saari, Eesti NSV Teaduste Akad. Toimetised, Füüs. Mat. $\underline{27}$, 109 (1978). (in English)
31. Y. Toyozawa, J. Phys. Soc. Japan $\underline{41}$, 400 (1976).
32. V. V. Hizhnyakov and K. K. Rebane, JETP $\underline{74}$, 885 (1978); Eesti NSV Teaduste Akadeemia Toimetised, Füüs. Mat. $\underline{26}$, 260 (1977).
33. K. Rebane, R. Avarmaa, and A. Gorokhovski, Izv. Akad. Nauk SSSR, Ser. Fiz. $\underline{39}$, 1793 (1975).
34. V. Hizhnyakov and I. Tehver, J. Luminescence $\underline{18/19}$, 673 (1979).
35. K. K. Rebane, I. J. Tehver, and V. V. Hizhnyakov, in "Theory of Light Scattering in Condensed Matter", Proc. of the First Joint USA-USSR Symposium, edited by B. Bendow, J. L. Birman, and V. Agranovich (Plenum Press, N. Y., 1976), p.393.
36. Y. Toyozawa, A. Kotani, and A. Sumi, J. Phys. Soc. Japan $\underline{42}$, 1495 (1977).

LIGHT SCATTERING NEAR STRUCTURAL PHASE TRANSITION POINTS

IN PURE CRYSTALS AND IN CRYSTALS CONTAINING DEFECTS

V. L. Ginzburg, P.N. Lebedev Physical Institute, Acad. Sci. USSR,
Moscow, USSR: A.P. Levanyuk, A.V.Shubnikov Crystallography
Institute, Acad. Sci. USSR, Moscow,USSR: A.A. Sobyanin, All-Union
Scientific Research Institute of Metrological Service, Moscow,
117334 USSR: A.S. Sigov, Institute of Radio Engineering
Electronics and Automation, Moscow 123275, USSR.

1. INTRODUCTION

The study of light scattering near structural phase
transitions represents a growing field of research, and a number
of reviews on the subject are available (see for example,
Refs[1-6]). However we believe it is worthwhile to present one
more review aimed at elucidation of the question in terms of
phenomenological theory, which is rather simple, but quite
general. In this case it is especially convenient to use that
phenomenological theory which describes a crystal as a continuous
medium, i.e. from the macroscopic point of view, and so makes it
possible to neglect inessential details of structures and
interactions. Below we use the Landau theory only since the
critical (scaling) region does not seem to have been observed
reliably for structural phase transitions. We do not claim a
comprehensive explanation of experimental data but try to
characterize the state and possibilities of the theory. All
questions touched upon here are discussed in far more detail in
the authors' article[5] at the Fourth International Meeting on
Ferroelectricity (Leningrad, September 1977). The present paper
differs from Ref. 5 mainly by more emphasis on the "central peak
problem."

First we briefly discuss the expected anomalies of integrated
intensity of light scattered by thermal fluctuations (sec. 2) and
then the temperature evolution of the shift of spectral density of
scattered light (sec. 3). We also consider (Sec. 4) the influence
of crystal defects on both integral intensity and spectral density
anomalies near phase transition points.

2. INTEGRATED INTENSITY

As is well known[8] the scattered light integrated intensity is

$$I(\vec{q}) \sim <|\Delta\varepsilon(\vec{q})|^2>,\tag{1}$$

where ε is the relevant combination of components of the dielectric tensor ε_{ij}, $\vec{q}$ is the difference between wave vectors of the incident and scattered light, $< >$ denotes the statistical average, and $\Delta\varepsilon(\vec{q})$ is the spatial Fourier transform of $\Delta\varepsilon(\vec{r},t) \equiv \varepsilon(\vec{r},t) - <\varepsilon>$. The ε - fluctuations are determined by fluctuations of different variables describing the state (configuration) of a system. Near phase transition points it is natural to consider fluctuations of the order parameter η although the coupling with some other variables may also prove to be essential.

The simplest case corresponds to linear coupling[7] between ε_{ij} and η. However such coupling is allowed by symmetry for proper ferroelastics only when we may choose as order parameter a component (or components) of the strain tensor, i.e. a quantity of the same transformation properties as those of ε_{ij}, . The temperature dependence of $I(\vec{q})$ in proper ferroelastics reflects that of $<(\Delta\eta)^2>$ (if the incident and scattered light are polarized properly). This conclusion remains valid even when the coupling of ε_{ij} with other variables is taken into account, and the presence of this coupling affects mainly the amplitude of the $I(\vec{q})$-temperature dependence . The above statement seems to be supported in general by experiments [3,9] on light scattering in $PrAlO_3$ and KH_2PO_4 although further experimental investigations are desirable for the purpose of a more detailed comparison with the theory.

For the more usually encountered case of bilinear coupling [10,11] one has $\varepsilon = \varepsilon_0 + a\eta^2$ and the first order light scattering intensity is given within mean field theory by

$$I_I(\vec{q}) \sim \eta_e^2 <|\eta(\vec{q})|^2>,\tag{2}$$

where η_e is the equilibrium value of η . We see that in this case first-order scattering by η fluctuations takes place in the nonsymmetrical phase only. Within the Landau theory the intensity $I_I(\vec{q})$ as given by formula (2) increases in the vicinity of a second-order phase transition temperature $(T=T_c)$, only if the transition is close to the tricritical point-but even for such transitions the scattered light intensity is predicted to increase in solids by no more than several fold. This is due to the fact that in the nonsymmetrical phase η-fluctuations are coupled linearly with mass density fluctuations which, in turn, are accompanied by shear deformations in solids.[12]

In the symmetrical phase η-fluctuations lead solely to second order scattering which is determined by the value of $\langle|\eta^2(\vec{q})|^2\rangle$, i.e. by the fluctuations of fourth order. The intensity of this scattering may be estimated as

$$I_2(\vec{q}) \sim \frac{r_c}{d} \tilde{I}_2 \ , \qquad\qquad (3)$$

where r_c is the order parameter correlation radius, d is the interatomic spacing and $\tilde{I}_2$ is a constant which is comparable in magnitude to second order light scattering intensity far from the phase transition point. One can estimate $\tilde{I}_2$ as $\tilde{I}_1(T_c/T_a)$ where T_a is the characteristic "atomic" temperature $(T_a \sim 10^4 - 10^5 K)$. It should be stressed that in the region of applicability of the Landau theory $I_2(\tau)$ is less than $I_1(-\tau)$ $(\tau \equiv (T-T_c)/T_c)$ becoming comparable with $I_1(-\tau)$ on the boundary of this region. **Closer to** T_c, **i.e.** in the critical region, the intensities $I_1(-\tau)$ and $I_2(\tau)$ have the same temperature dependence and order of magnitude [13]. Second order light scattering by η - fluctuations seems to have been observed for the first time in Refs.[14,15,45]. Its intensity proves to be two orders less than the intensity of the first order scattering, which indicates that the Landau theory is well applicable to the phase transition **which was** investigated in the experiments, **namely in** Hg_2Cl_2.

Near phase transition points in solids the intensity of light scattering by static inhomogeneities and defects may suffer a great enhancement as evidenced by experimental data for structural phase transitions in quartz[16-18], NH_4Cl[19,20]-$SrTiO_3$[21,22] $Pb_5Ge_3O_{11}$[23,45], KH_2PO_4[25,9], $KH_3(SeO_3)_2$[26]. So, the extraction of the part due to thermal fluctuations from the total scattering intensity is generally a rather hard task especially in the case of transitions which are not proper ferroelastic.

3. SCATTERED LIGHT SPECTRAL DENSITY

When considering an anomaly of the integrated light scattering intensity one may restrict oneself effectively by proper consideration of η- fluctuations alone, the dependence of ε_{ij} on other variables being included implicitly in the general formulae. On the contrary, in the consideration of the scattered light spectral density such an approach often proves to be unsuccessful and one needs to know the power spectra not only of the η-fluctuations but of a number of other variable as well as mixed fluctuation spectra. Nevertheless the basic qualitative aspects of the problem may be elucidated by discussion of the

simplest examples. That is why special attention will be given
below to such examples and only some remarks about possible
generalizations will be presented.

Let us begin with consideration of the first-order scattering
and suppose that ε depends on η only. Then the scattered light
spectral density is expressed through the function $<|\eta(\vec{q}\approx 0),\Omega|^2>$.

To calculate this function one should know the equation of
motion for $\eta(q\approx 0) \equiv \eta'$. The simplest form of the equation (for
displacive phase transitions in which we are interested) is

$$m\ddot{\eta}' + \gamma\dot{\eta}' + \Phi_{\eta\eta}\eta' = h(t) , \qquad\qquad (5)$$

where $\Phi_{\eta\eta}$ is the second derivative of the thermodynamic potential
as given, say, by the Landau theory; h is the generalized force
conjugate to η. Within the Landau theory for second-order phase
transitions

$$\Phi_{\eta\eta} \equiv \frac{\partial^2\Phi}{\partial\eta^2}\bigg|_{\eta=\eta_e} \sim |\tau| \equiv |1 - T/T_c| \qquad\qquad (6)$$

and the coefficients m and γ are considered to be temperature
independent. Equations (5),(6) express the essence of the
so-called soft mode concept[27]; η being the soft mode normal
coordinate. As follows from Eqs. (5),(6) the soft mode
frequency $\Omega_0 = (\Phi_{\eta\eta}/m)^{\frac{1}{2}}$ goes to zero as $\tau \to 0$, the soft mode
sidebands merging into an unshifted line at $\Omega_0 = \gamma/m\sqrt{2}$.[28]

With the use of Eq. (5) it is not difficult to understand the
character of the second-order scattering spectrum as well. The
soft phonon branch dispersion being small, the spectral density
contains two side maxima at the frequences $\Omega\approx\pm 2\Omega_0$ and a central
maximum whose width is $2\gamma/m$. In terms of quantum theory the side
components correspond to phonon scattering with absorption or
emission of two phonons, whilst the central peak corresponds to
scattering with absorption of one and emission of another phonon
with approximately the same frequency.[45]

Equation (5) strictly holds only for small enough frequencies.
For higher frequencies one has generally to add in Eq. (5) terms
with higher order time derivatives of η or, in other words, to
take account of the frequency dispersion of the coefficients $\Phi_{\eta\eta}$
and γ . The physical reason for this dispersion consists
evidently in coupling of the "η -oscillator" with other degrees
of freedom of a system, which form a "thermal bath" for the

oscillator. Of course, such coupling is partly included in Eq.
(5) but only in a very approximate way through the damping
term $\gamma \dot{\eta}$ and the temperature dependence of $\Phi_{\eta\eta}$. Ideally one
should treat this coupling in more detail. As a result, the above
conclusions about the spectral density shape may be invalid. In
particular the soft mode frequency, determined naturally not by a
static but by a high-frequency value of the "stiffness
constant" $\Phi_{\eta\eta}$, may not go to zero as $\tau \to 0$ since the
high-frequency value of this constant need not be zero at $\tau = 0$.

The simplest theory of the frequency dispersion of the
coefficients $\Phi_{\eta\eta}$ and γ is the Mandelstam-Leontovich theory[29]
which was applied to the analysis of the order parameter kinetics
near the phase transition point in Ref. 30 and to the problem of
the scattered light spectrum near phase transition in Ref. 31. We
discuss the theory at greater length as, based on it, one can have
a better insight into more complicated cases too.

In the spirit of Mandelstam -Leontovich theory let us assume
that in the system there is a variable ξ of the same
transformation properties as η but obeying a different
kinetics. Namely, we suppose that the variable ξ has a
relaxation nature. For instance, one can imagine a ferroelectric
crystal of the type KD_2PO_4. Here one can consider
polarization due to heavy ion displacements as the order
parameter η and the degree of ordering in the system of
hydrogen bonds as the relaxation variable ξ . Instead of (5) one
has now:

$$m\ddot{\eta}' + \gamma\dot{\eta}' + \Phi_{\eta\eta}^{\xi} \eta' + \Phi_{\eta\xi} \xi = h(t) , \tag{7}$$

$$\gamma_\xi \dot{\xi} + \Phi_{\eta\xi} \eta' + \Phi_{\xi\xi}^{\eta} \xi = 0 , \tag{8}$$

where the superscript indicates that derivatives are taken at
fixed values of ξ or η , respectively. Expressing the
Fourier-transform of $\xi(t)$ through that of $\eta'(t)$ from (8) and
substituting this into (7) one obtains for $\eta'(\Omega)$ an equation with
frequency-dependent coefficients

$$[-m\Omega^2 - i\gamma(\Omega)\Omega + \Phi_{\eta\eta}(\Omega)]\eta'(\Omega) = h(\Omega) , \tag{9}$$

where

$$\gamma(\Omega) = \gamma + \frac{m\delta^2 \Omega_\xi}{\Omega_\xi^2 + \Omega^2} , \tag{10}$$

$$\Phi_{\eta\eta}(\Omega) = \Phi_{\eta\eta}^{\zeta} - \frac{m\delta^2\Omega_{\xi}^2}{\Omega_{\xi}^2 + \Omega^2} = \Phi_{\eta\eta}^{g} + \frac{m\delta^2\Omega^2}{\Omega_{\xi}^2 + \Omega^2} \quad . \tag{11}$$

Here $\Omega_{\xi} \equiv \Phi_{\xi\xi}^{\eta}/\gamma_{\xi}$ is the relaxation rate of the variable ξ, $\Phi_{\eta\eta}^{g}$ is the generalized static stiffness constant corresponding to η (g is the generalized force conjugate to ξ), $m\delta^2 \equiv \Phi_{\eta\eta}^{\xi} - \Phi_{\eta\eta}^{g} = \Phi_{\eta\xi}^2/\Phi_{\xi\xi}^{\eta}$ is the difference between the high frequency and static values of $\Phi_{\eta\eta}$. As distinct from $\Phi_{\eta\eta}^{\xi}$ the quantity $\Phi_{\eta\eta}^{g}$ goes to zero as $\tau\to 0$. The expression in brackets on the right-hand side of Eq. (9) is the inverse generalized susceptibility $\chi(\Omega)$ corresponding to η. As is known

$$<|\eta(\Omega)|^2> = \frac{T}{\Omega} \text{ Im } \chi(\Omega) \quad . \tag{12}$$

One can see from Eqs. (9)-(11) that if Ω_{ξ} is greater than all the other characteristic frequencies, the picture of the temperature evolution of the spectrum is analogous to that in the simplest case discussed above. It is quite natural since, Ω_{ξ} being great enough, the relaxing nature of ξ does not manifest itself in the η-fluctuation spectrum because ξ has time to follow η in its changes. In the opposite limit when $\Omega_{\xi} \ll \Omega_{\infty} \equiv (\Phi_{\eta\eta}^{\xi}/m)^{\frac{1}{2}}$, ξ cannot follow η. Thus the position and the intensity of the sidebands depend more on the high frequency value of stiffness constant $\Phi_{\eta\eta}^{\xi}$, which remains finite at $\tau=0$. The minimum value of Ω_{∞}, i.e., $\Omega_{\infty}(\tau=0)$, has been designated above by δ. Thus if damping is small enough ($\delta > \gamma/m$), the side components do not merge, up to $\tau=0$. As to the total intensity, in the case of linear coupling between ε_{ij} and η (or for neutron scattering), this intensity $I_{tot} \sim \Phi_{\eta\eta}^{g} = \Phi_{\eta\eta}^{\xi} - \Phi_{\eta}^2/\Phi_{\xi\xi}^{\eta}$ and grows as $\tau \to 0$ at the expense of the central peak, whose fractional intensity tends to unity as $\tau \to 0$ and whose width is proportional to $\Phi_{\xi\xi}^{h} \sim \tau$.

Such a picture of the evolution of the spectrum suggested for the first time in Ref. 31, has recently attracted much attention. In part or in full, it was observed in a number of experiments (see, for example[32-35,45]) it is the explanation of the picture for concrete cases, i.e. the elucidation of the nature of the slow relaxation variables, that constitutes in essence, the so-called "central peak problem".

In the case of bilinear coupling between ε_{ij} and η, within the Landau theory and for second-order phase transitions far from the tricritical point I_{tot} does not rise as $\tau\to 0$. Then the increase of the fractional intensity of the central peak takes place at the expense of decreasing side component intensity ($I_{side} \sim \eta_e^2$).

Now look at the results obtained from a somewhat different point of view. When neglecting $\eta \leftrightarrow \xi$ coupling, the η -fluctuation spectrum contains side components, a central peak being absent, and the ξ-fluctuation spectrum contains a central peak only. As a result of coupling between η and ξ the central peak manifests itself in the spectrum of η -fluctuations too, although in a somewhat modified form: the width of the peak is determined by $\Phi_{\xi\xi}^{\eta}$ and not by $\Phi_{\xi\xi}^{\eta}$.

These simple considerations also provide an insight into the situation when frequency dispersion of the coefficients in Eq. (5) arises due to the nonlinear coupling between η and other variables forming the thermal bath. Treating nonlinear coupling is certainly of major interest, since for many experimentally observed central peaks one fails to find in the system a slow relaxing variable coupled linearly with the order parameter.

Nonlinear coupling effects were discussed in connection with dynamic phenomena near phase transition by many authors (see, for example, Ref. 36-38 and the literature cited in Ref. 38). The main distinct feature of the nonlinear case is that one needs to deal here with an infinite set of extra variables (as $V \rightarrow \infty$, V is the system volume). In the framework of the approach based on the continuum approximation nonlinear interactions between fluctuations are described by terms of the type η^{4}, $\eta^{2}\xi$, $\eta\xi^{2}$, $\eta^{2}\xi^{2}$ etc. in the thermodynanic potential density. The equation of motion for $\eta'(\vec{q})$ will then contain sums of the type

$$\sum_{\vec{k}_1+\vec{k}_2+\vec{k}_3=\vec{q}}\eta(\vec{k}_1)\eta(\vec{k}_2)\eta(\vec{k}_3)\,,\quad \sum_{\vec{k}_1+\vec{k}_2=\vec{q}}\eta(\vec{k}_1)\xi(\vec{k}_2)\,,\quad \sum_{\vec{k}_1+\vec{k}_2=\vec{q}}\xi(\vec{k}_1)\eta(\vec{k}_2)$$

and so on. As a result instead of one frequency Ω_{ξ} characterizing the frequency dispersion of the coefficients $\Phi_{\eta\eta}$ and γ there now appears a whole spectrum of characteristic frequencies, and the dispersion of these coefficients acquires, generally speaking, a complicated character. Below, when considering the effects of nonlinear coupling we shall use perturbation theory. Such an approach termed the "mode-coupling" theory, has been widely used in the study of dynamical critical phenomena. Although **specific** computations even in the framework of perturbation theory often turn out rather complicated, the qualitative character of the results obtained may be understood from the following simple considerations.

We have seen that in the case of linear coupling the characteristic features of the fluctuations of one quantity may manifest themselves in the spectrum of another quantity. Namely, the linear coupling between an oscillation variable $\eta' \equiv \eta(\vec{q})$

and a relaxation variable $\zeta \equiv \zeta(\vec{q})$ leads to the appearance of a
central peak in the fluctuation spectrum of $\eta(\vec{q})$ in
addition to the side components, if the relaxation rate of the
variable $\xi(\vec{q})$ is less than the oscillation frequency of $\eta(\vec{q})$. On the
other hand, a nonlinear coupling, e.g. one described by the
term $\eta\xi^2$ in the thermodynamic potential density, may be
considered as a linear coupling between the variable $\eta(\vec{q})$ and a
set of variables $\xi(\vec{k}) \equiv \xi(\vec{k})\xi(\vec{q}-\vec{k})$. The spectrum of fluctuations of
the $\xi(\vec{k})$ always contains the central maximum which corresponds to
the one in the second order scattering spectrum discussed above.
Due to interactions of $\eta\xi^2$ -type the central maximum will in
addition manifest itself in the η -fluctuation spectrum if the
width of the maximum is less than the η - oscillation frequency.
In fact, the central peak in the ξ^2 -fluctuation spectrum is a
superposition of the central peaks corresponding to various $\vec{k}$ and
hence of a rather complicated shape. Due to the nonlinear
coupling this "complex" central peak will appear in the η
-fluctuation spectrum provided that damping of the   "ξ -branch"
vibrations is small enough.

We now discuss the case when coupling of the $\eta\xi^2$-type is
realized. For one-component variables η and ξ it is allowed only
in the nonsymmetrical phase. The interaction coefficient is then
proportional to η_e. Near a phase transition point the effects
due to this coupling are small. For the two-component
variable $\xi = \{\xi_1, \xi_2\}$ the $\eta\xi^2$-type interaction may already take place
in both phases. In this case, as a detailed treatment shows, the
fractional intensity of the central maximum may be estimated as
the expression

$$\frac{I \text{ central}}{I \text{ side}} \sim \frac{T_c}{T_a} \mid \tau \mid^{-1}. \tag{13}$$

Thus in this case the intensity of the central peak becomes
comparable with that of the side components at $|\tau| \sim T_c/T_a$ only, i.e.,
quite near the phase transition point. Note that in the case of a
linear coupling between η and ξ the small parameter T_c/T_a does
not appear.

When writing the interaction term between η and ξ in the form
$\eta\xi^2$, we neglected, in fact, the dependence of the nonlinear
coupling coefficient on wave vector $\vec{k}$. Due to this dependence
the number of phonon branches, which may interact with the soft
mode via coupling of the $\eta\xi^2$-type, increases considerably, with
the interaction with longitudinal acoustic branch being of special
interest because the acoustic branches are usually the most
underdamped ones in crystal. The interaction with the acoustic

branch is described by the term

$$\sum_{\vec{k},\vec{q}} r(\vec{k}, \vec{k} - \vec{q})\rho(\vec{k})\rho(-\vec{k} - \vec{q})\eta(\vec{q})$$

in the thermodynamic potential. Here ρ is the mass density, $r(\vec{k},\vec{k}-\vec{q})$ is the coupling coefficient. In the expression for the thermodynamic potential density the terms with spatial derivatives correspond to nonzero vectors $\vec{k}$. For example for a ferroelectric which is piezoelectric in the nonpolar phase (e.g. for crystals of KH_2PO_4-type) the thermodynamic potential may contain the following term (the role of η is played here by the polarization vector component P_z):

$$D\int P_z \frac{\partial \rho}{\partial x} \frac{\partial \rho}{\partial y} dV = \sum_{\vec{k},\vec{q}} D\, k_x(k_y - q_y)\rho(\vec{k})\rho(-\vec{k}-\vec{q})\eta(\vec{q}) \quad .$$

Taking into account that the greatest phase volume corresponds to the states with large k and following the same reasoning as above we come to the conclusion that in the η -fluctuation spectrum such an interaction leads to a central maximum whose width is about $\Gamma_\rho(k_{max})$ where Γ_ρ is the acoustic phonon damping constant, and k_{max} is of the order of reciprocal lattice spacing. Of course, the central maximum is observable if $\Gamma_\rho(k_{max}) < \Omega_0$. To estimate the intensity of this maximum one may use Eq. (13).

We would like to stress that although the nonlinear $\eta\rho^2$-type interaction in connection with the central peak problem was considered first in Ref. 37, another type of central peak has been discussed there. It arises when in a region of wave vectors $\vec{k}$ the coefficient r does not depend on k (usually $\Gamma \sim k^2$ at $k < k_{max}$). Such a situation might occur in the so-called second-sound regime when τ_N, the characteristic time of N-processes under which the phonon quasimomentum is conserved, is much less than τ_v, the characteristic time of Umklapp-processes. In this case the coefficient Γ_ρ turns out to be practically a constant for wave vectors $(C_s\tau_u)^{-1} \lesssim k \lesssim (C_s^2 \, \tau_u \, \tau_N)^{-1/2}$. **The width of this** central peak is determined by the lifetime of acoustic phonons in this region of wave vectors and is much less than characteristic phonon frequencies. However the intensity of the central peak is very low due to the smallness of the phase volume corresponding to the vectors $\vec{k}$ for which Γ_ρ is almost independent of k. Thus the observation of such a central peak seems to be hardly possible (recall also the difficulties of fulfilling the condition $\tau_u \gg \tau_N$ in real crystals).

Since the energy U, and entropy S possess the same
transformation properties as the density ρ,
all the aforesaid on the nonlinear interaction of soft and
acoustic modes is valid also for an interaction of soft and
entropy (heat-conductive) modes due to terms with spatial
derivatives of η and S in the thermodynamic potential density.
The width of the central maximum arising here is determined by the
thermal diffusion relaxation rate at values of k also comparable
with reciprocal lattice spacing.

Let us now discuss an interaction of the $\eta^2\xi$ -type which may
occur for one-component order parameter, if ξ transforms
according to unit representation of the symmetry group of the
symmetrical phase. In the spirit of the above consider first the
power spectrum of fluctuations of the quantity $\eta(\vec{k})\xi(-\vec{k}-\vec{q})$. If
both $\eta(\vec{k})$ and $\xi(\vec{k})$ are of oscillary nature, this spectrum may
contain side components at the frequencies $\pm\Omega_\eta(\vec{k})\pm\Omega_\xi(\vec{k})$. Two
inner components may merge into a single central line but this is
possible only in a rather limited temperature interval not very
close to the phase transition temperature and, therefore, is of no
particular interest here. When ξ-fluctuations are of
relaxational character, there is no central peak in the power
spectrum of $\eta(\vec{k})\xi(-\vec{k}-\vec{q})$ fluctuations (of course, this is true
if the soft mode is underdamped). For this reason the
statement[39] that the interaction of the type $\eta^2 S$ **may**
lead to the appearance of a central peak in the power spectrum
of η -fluctuations seems to us to be incorrect. An analogous
conclusion can also be drawn regarding η^4 -type interaction
since the power spectrum of η^3 -fluctuations may contain
maxima near the frequencies $\pm\Omega_\eta$ and $\pm3\Omega_\eta$ **but not at** $\Omega = 0$.

The above examples of nonlinear interactions are most
important. All the remaining types of nonlinear interactions (say
$\eta^2\xi^2$) manifest themselves in the power spectrum of η
-fluctuations much more weakly. At the same time we should like
to stress that the light scattering spectrum is by no means
determined by the form of the correlation function $<|\eta(\Omega)|^2>$
only· Even in the hypothetical case when ε is independent of
all the variables except η , there is second-order scattering
in addition to first-order scattering, the former being determined
by the correlation function $<|\eta^2(\Omega)|^2>$.Nonlinear coupling not only
changes the form of the correlation function, but also leads to
interference of first and second-order scattering spectra so that
they cannot be considered independently[2,40-42].

As a result there may occur a redistribution of intensity
among spectral features of first and second-order scattering even

in the region of applicability of the Landau theory. Thus it may
be said that the second-order scattering whose intensity is
usually very low a factor $(\sim T_a/T \sim 10^5 K/T)$ times less than the
first-order scattering intensity, may be greatly enhanced due to
nonlinear coupling.

Note that although qualitatively the influence of nonlinear
coupling on the scattered light spectrum is rather understandable,
a direct comparison of theory and experiment is quite a
complicated problem. Indeed, the theoretical formulae include
integrals over k-space and it is necessary to know the laws of
dispersion of interacting branches, as well as the dependence of
the corresponding coupling constants upon wave vectors.

One can however argue that in this way it is impossible to
account for very narrow central peaks observed in some cases (with
width 4-5 orders of magnitude less than the characteristic phonon
frequencies). The most plausible origin of such peaks is the
presence of defects in the crystal.[21,23,45,25]

4.   INFLUENCE OF DEFECTS UPON LIGHT SCATTERING NEAR
PHASE TRANSITION POINTS

The possibility of a considerable increase of the intensity of
light scattering by static inhomogeneities, i.e. by crystal
defects near a phase transition became clear after it had been
shown in Ref. 17 that the anomaly of scattering at the $\alpha \leftrightarrow \beta$
-transition in quartz is connected mainly just with such
inhomogeneities.   At the present time analoguous data has also
been obtained for other transitions (see for example Refs. 23-26).

Possible mechanisms are discussed below for the appearance of
a central peak in the power spectrum of the order parameter
fluctuations near phase transition points in systems with frozen
defects. The proposed mechanisms are: (i)the linear coupling of
the order parameter fluctuations with the thermal diffusion mode
due to frozen defects inducing a nonzero value of the order
parameter in their vicinity above T_c,(ii)interaction of the soft
and longitudinal acoustic mode in the vicinity of such defects,
(iii) coupling, due to the defects, of the soft mode and another
relaxational or vibrational phonon mode which is fully symmetrical (iv) coupling near defects of the order parameter fluctuations with fluctuations of the concentration of a component in
solid solution.

As evidenced by experimental data the spectrum of the order
parameter fluctuations has a rather complicated form near points

of a number of phase transitions . In particular one observes
frequently even for displacive transitions a temperature dependent
central peak along with the soft mode sidebands in neutron or
light scattering spectra.   A variety of central peaks has been
observed with different contributions to the total scattering
intensity and with different types of temperature evolution near
the transition point[45,47]. It has been speculated that at least
part of the central peaks investigated are connected with the
presence of defects in the system[46].

An example has been given in Ref. [46] of the situation when
defects provide the linear coupling of the order parameter soft
mode with some other relaxational mode. In this reference defects
were considered which induce in their vicinity a nonzero value η_0
of the (one-compnent) order parameter above the transition **temperature** T_c.   Such defects (say, impurities) may occupy two
equivalent positions in a unit cell differing in sign of η_0 , and
may hop between the two positions. If the value of η averaged
over a region of the crystal is zero then the concentrations of
N_+ and N_- defects in mentioned positions should be equal in
equilibrium in this region. On the other hand, if in a
volume $\langle\eta\rangle > 0$, then in the volume $N_+ > N_-$,since the
impurity position, in which the signs of η_0 ,and $\langle\eta\rangle$ coincide,
is now more advantageous. In other words there exists linear
coupling between η and the variable $n = N_+ - N_-$. The approach
of n to its equilibrium value is, of course, of relaxational
character. The corresponding relaxation rate is determined by the
temperature and the height of the energy barrier separating the
states of the defects with opposite signs of η_0: $\Omega_{R\eta} = \Omega_a \exp(\Delta E / K_B T)$,
where Ω_a is a characteristic attempt frequency of order 10^{13}
sec^{-1} (phonon).   Thus we come to the situation described above
with n playing the role of ξ - **relaxor.**

Now we discuss other mechanisms for the appearance of a
dynamic central peak in the spectrum of η -fluctuations, which
are not connected with hopping defects and which may occur even if
the defects are completely frozen. For simplicity we restrict
ourselves to temperatures $T > T_c$ (symmetrical phase).

One such mechanism consists in the interaction, between the
soft and the heat conductive modes due to defects. As is known,
in the nonsymmetrical phase the η- vibrations are followed by
temperature changes ($\Delta T = T - T_{eq}$), and that gives rise to a heat
conductive (Landau-Placzek) central peak in the spectrum of the
order parameter fluctuations. In the symmetrical phase there is
no linear coupling between η and ΔT as the interaction term has
the form $\eta^2 \Delta T$. However, if the defect induces a nonzero

value η of the order parameter in its vicinity, linear coupling between η and ΔT is induced in the symmetrical phase too. Under these conditions the oscillator $\Delta\eta(\vec{q})$ ($\vec{q}$ is the scattered wave vector) is coupled linearly with many "T-relaxors" since in the Fourier representation the interaction term is proportional to $\sum\limits_{k}\eta_{eq}(-\vec{k}-\vec{q})\Delta T(\vec{k})\Delta\eta(\vec{q})$.   Here $\eta_{eq}(-\vec{k} - \vec{q})$ is the Fourier component of the space equilibrium distribution $\eta_{eq}(\vec{r})$
of the order parameter. As the value of $\eta_{eq}(\vec{k})$ decreases rapidly for $k > r_c^{-1}$ (r_c is the order parameter correlation length[44]), one can define the maximal characteristic rate of "T-relaxors" $\Omega_T \equiv \kappa r_c^{-2}/C_\eta$, where κ is the heat conductivity and C_η is the specific heat of the system at fixed value of η . If $\Omega_{0\eta} > \Omega_T$ then a central peak would be present in the spectrum of η – fluctuations whilst the position and intensity of the soft mode sidebands would be determined by the adiabatic generalized susceptibility corresponding to η , which is the same as for the pure crystal. Indeed, for $\Omega_T < \Omega_{0\eta}$ the "slow" temperature variations have no time to follow the "rapid" η – vibrations, and therefore the η-vibrations are not influenced by defects, as opposed to temperature variations. The total magnitude of η -fluctuations is determined by the isothermal susceptibility so that the fractional intensity of the central peak makes up $\Delta C_\alpha/C$, where ΔC_α is the contribution of defects to the specific heat C of the crystal. According to Ref.[44], the ratio $\Delta C_\alpha/C$ increases as $T \rightarrow T_c$ like $(T-T_c)^{-3/2}$ and may reach about 10^{-1} near T_c even for moderate defect concentrations (about 10^{-18} cm-3)so that the central peak is certainly observable. This peak has a rather unusual shape, the intensity close to $\Omega=0$ being proportional to $1- (\Omega/2\Omega_T)^{1/2}$.

In the symmetrical phase the η vibrations can be coupled linearly, due to defects, not only with the temperature (entropy) variations but with any variable which is invariant under operations of the high-symmetry group of the crystal. The mass density δ or the volume deformation $v=-\Delta\rho/\rho_0$ is another important example of such a variable. The appropriate term of interaction between $\Delta\eta(\vec{q})-$ oscillator and these of $v(\vec{k})$ has the form $\sum\limits_{\vec{k}}\eta_{eq}(-\vec{k}-\vec{q})v(\vec{k})\Delta\eta(\vec{q})$.

The spectrum of $v(\vec{k})$ -fluctuations contains two maxima at frequencies $\Omega = \pm C_s k$ (C_s is the longitudinal sound velocity). In the η -fluctuation spectrum these maxima are superimposed forming a central peak of non-Lorentzian shape[45]. Since $\eta_{eq}(\vec{k})$ falls rapidly for $k > r_c^{-1}$,the characteristic frequency determining the linewidth of this peak is now $\Omega_s = C_s r_c^{-1}$. Of course, it was assumed above that $\Omega_{0\eta} >> \Omega_s$. Notice that within the mean-field

approximation used here the temperature dependences of $\Omega_{0\eta}$
and Ω_s are the same.   Therefore if the condition $\Omega_{0\eta} \gg \Omega_s$ is
fulfilled anywhere near T_c, it remains fulfilled up to the
transition point.   The fractional intensity of this central peak
makes up $\Delta\lambda_d/\lambda$ ($\Delta\lambda_d$ is the contribution of defects to the
hydrostatic compression modulus λ) and increases like
$(T-T_c)^{-3/2}$ as $T \to T_c$ (see Ref.[44]).

In an analogous way one might consider the coupling of the
soft mode with any fully symmetrical phonon mode which can occur
near defects.   If this extra mode is overdamped and its relaxation
rate is less than $\Omega_{0\eta}$, the spectrum of η -fluctuations will
contain a central peak too.   The intensity of this peak is of the
same order as in the cases discussed previously.

In solid solutions the ξ-variable may have the meaning of
the concentration of a component. Due to coupling of η-and ξ-
fluctuations near defects the central peak arising in this case is
completely analogous to that caused by coupling of the soft and
thermal diffusion modes-the linewidth of this peak being
determined by the inverse time of diffusion of the component to
the distance of order r_c.

Thus we see that there may be many reasons for the appearance
of central peaks in real crystals.   In order to identify these
peaks properly one should be able to carry out measurements at low
enough frequencies [5,46,47].   The control of the specimen quality
and defect concentrations is also necessary.

REFERENCES

1. V.L. Ginzburg, Usp. Fiz.Nauk 77, 621 (1962); Sov.Phys.
    Uspekhi 5, 649 (1963).
2. J.F. Scott. Rev.Mod. Phys. 46, 83 (1974).
3. P.A. Fleury in: Theory of Light Scattering in Condensed
    Matter (eds. B. Bendow, J.L. Birman and V.M. Agranovich).
    Plenum Press, N.Y., 1976, p.13.
4. P.A. Fleury in: Light Scattering in Solids (eds. M.
    Balkanski R.C.D. Leite and S.P.S. Porto). Flammarion
    Sciences, Paris, 1976, p. 747.
5. V.L. Ginzburg, A.P. Levanyuk and A.A. Sobyanin. Ferro-
    electrics 20, 3 (1978).
6. W. Hayes and R. Loudon. Scattering of Light by Crystals. J.
    Wiley and Sons, N.Y., 1978.
7. I.L. Fabelinskii. Molekularnoe rasseyanie sveta, Nauka,
    Moskva, 1965; Molecular Scattering of Light. Plenum Press,
    N.Y., 1968.
8. M.A. Krivoglaz and S.A. Rybak, Zh. Eksp. Teor.Fiz. 33,139
    (1957). Sov. Phys. J.E.T.P. 6 107 (1957).

9. E. Courtens. Phys. Rev. Lett. $\underline{41}$, 1171 (1978).

10. V.L. Ginzburg, Dokl. Akad. Nauk SSSR $\underline{105}$, 240 (1955).

11. V.L. Ginzburg and A.P.Levanyuk, J.Phys. Chem. Sol. $\underline{6}$, 51(1958).

12. A.P. Levanyuk, Zh. Eksp.Teor.Fiz.$\underline{66}$,2256 (1974).Sov. Phys. JETP, $\underline{39}$ 1111 (1974).
    V.L. Ginzburg and A.P. Levanyuk, Phys. Lett. $\underline{47A}$, 345 (1974).

13. A.P. Levanyuk.Zh.Eksp.Teor Fiz.$\underline{70}$,1253 (1976); Sov.Phys. JETP $\underline{43}$, 652 (1976).

14. C. Barta, B.S. Zadokhin, A.A. Kaplyanskii and Yu.F. Markov, Pis'ma (in Russian) ZhETF $\underline{26}$, 480 (1977).

15. J.P. Benoit, Cao Xuan An, Y.Luspin, I.P. Chapelle, J.Lefebre. J. Phys.C: Solid State Phys. $\underline{11}$, L721 (1978).

16. I.A.Yakovlev, T.S. Velichkina and L.F. Mikheeva, Krystallo-grafiya $\underline{1}$, 123 (1956) (Sov. Phys.Crystallogr. $\underline{1}$, 91 (1956); I.A. Yakovlev and T.S. Velichkina, Usp.Fiz.Nauk $\underline{63}$, 411 (1957).(In Russian)

17. S.M.Shapiro and H.Z. Cummins, Phys.Rev. Lett. $\underline{21}$, 1578 (1968).

18. G. Dolino and J.P. Bachheimer, Phys. Stat. Sol. (a) $\underline{41}$, 673 (1977).

19. O.A.Shustin, pis'ma Zh.Eksp.Teor.Fiz.$\underline{3}$, 491 (1966), (JETP Lett. $\underline{3}$, 320 (1966)).

20. P.D. Lazay, J.H. Lunacek, N.A. Clark and G.B. Benedek, p. 593 in Light Scattering in Solids (ed. G.B. Wright), Springer-Verlag, N.Y., 1969.

21. E.F. Steigmeier, H. Anderset and G. Harbeke, p. 153 in Solid State Commun. $\underline{12}$, 1077 (1973).

22. J.B. Hastings, S.M. Shapiro and B.C. Fraser, Phys.Rev. Lett., $\underline{40}$, 237 (1978).

23. D.J. Lockwood, J.W. Arthur, W.Tyalor and T.J. Hosea, Solid State Commun. $\underline{20}$, 703 (1976).

24. W.Taylor, D.J. Lockwood and H.Vass. Solid State Commun. $\underline{27}$, 547 (1978).

25. L.N. Durvasula and R.W. Gammon, Phys.Rev. Lett. $\underline{38}$,1081 (1977).

26. T.Yagi, H.Tanaka and I.Tatsuzaki, J.Phys.Soc. Japan $\underline{41}$,717 (1976); Phys.Rev.Lett. $\underline{38}$, 609 (1977).

27. V.L.Ginzburg, Zh.Eksp.Teor.Fiz.$\underline{19}$,36 (1949).(In Russian)
    P.W.Anderson, Fizika dielektrikov,ed.by G. Skanavi (AN SSSR, Moscow, 1960), p. 290.
    W. Cochran, Adv. in Physics $\underline{9}$, 387 (1960); $\underline{10}$,40 (1961).

28. V.L. Ginzburg and A.P. Levanyuk, Zh. Eksp.Teor.Fiz. $\underline{39}$,192 (1960)(In Russian)

29. L.I.Mandelstam and M.A.Leontovich,Dokl.Akad.Nauk SSSR $\underline{3}$,111 (1936);Zh.Eksp.Teor.Fiz.$\underline{7}$,438 (1937).

30. L.D. Landau and I.M. Khalatnikov, Dokl.Akad.Nauk. SSSR, $\underline{96}$, 469 (1954), see also L.D. Landau, Collected Papers (Nauka, Moscow 1969), vol.2, p.218.(Translation ed. D.Ter Haar, Pergammon 1970).

31. A.P. Levanyuk and A.A. Sobyanin,Zh.Eksp.Teor.Fiz. $\underline{53}$, 1024 (1967); Sov.Phys. JETP $\underline{26}$,612 (1968).

32. T.Riste,E.J.Samuelsen, K.Otnes and J.Feder Solid State Commun. $\underline{9}$, 1455 (1971).

33. S.M. Shapiro, J.D. Axe, G.Shirane and T. Riste Phys. Rev.$\underline{B6}$, 4332 (1972).

34. N.Lagakos and H.Z.Cummins. Phys. Rev. $\underline{B10,}$ 1063 (1974).

35. H.G. Unruh, J.Kruger and E. Sailer, Ferroelectrics $\underline{20}$, 3 (1978).

36. A.P. Levanyuk. Zh. Eksp.Teor.Fiz. $\underline{49}$,1304 (1965). Sov.Phys. JETP $\underline{22}$, 901 (1966).

37. R.A Cowley and G.J. Coombs. J. Phys. C. Solid Sta. Phys. $\underline{6}$,143 (1973).

38. P.C. Hohenberg and B.I. Halperin.Rev.Mod.Phys.$\underline{49}$,435(1977).

39. C.P. Eng.Solid State Commun. $\underline{15}$,459 (1974).

40. J.F. Scott. Phys. Rev. Lett. $\underline{21}$, 907 (1968).

41. J. Ruvalds and A.Zawadovski. Phys. Rev. $\underline{B2}$, 1172 (1970).

42. Y.Yacoby, R.A. Cowley,T.J. Hosea, D.J. Lockwood and W.Taylor J. Phys. C: Solid St. Phys. $\underline{12}$, 387 (1979).

43. A.P. Levanyuk, V.V. Osipov and A.A. Sobyanin in: Theory of Light Scattering in Condensed Matter (eds. B. Bendow, J.L. Birman and V.M. Agranovich) Plenum Press, N.Y. 1976, p. 517.

44. A. P. Levanyuk, V.V. Osipov, A.S. Sigov and A.A. Sobyanin Zh. Eksp. Teor. Fiz. $\underline{76}$,345 (1979).(In Russian).

45. K.B. Lyons and P.A. Fleury. Phys. Rev. $\underline{B17}$, 2403 (1978). Solid St. Comm. $\underline{23}$, 477,(1977).

46. B.I. Halperin and C. Varma. Phys. Rev. $\underline{B14}$, 4030 (1976).

47. G. Shirane, Rev. Mod. Phys. $\underline{46}$, 437,(1974).

48. F. Schwabl, Anharmonic Lattices, Structure Transitions and Melting, edited by T. Riste Noordhoff, Leiden,(1974).

RAMAN SCATTERING FROM CHARGE DENSITY WAVES AND SUPERCONDUCTING GAP

EXCITATIONS IN 2H-TaSe$_2$ AND 2H-NbSe$_2$

R. Sooryakumar, D. G. Bruns*, and Miles V. Klein

Department of Physics and Materials Reasearch Laboratory
University of Illinois at Urbana-Champaign
Urbana, Illinois 61801

INTRODUCTION

Many transition-metal dichalcogenides undergo phase transitions of the charge-density-wave (CDW) type.[1] The metal atoms are found at the center of hexagonal layers where they bind covalently and metallically to chalcogen atoms in adjacent planes. Weak interlayer bonding is of the Van der Waals type. In the 2H polytype of 2H-TaSe$_2$ and 2H-NbSe$_2$ there is one electron per metal atom in a narrow half-filled "d$_z$2" conduction band above a filled s-p valence band and below a broad, empty d-like conduction band. This normal high temperature structure is unstable against the formation of a periodic structural distortion (PSD) of the lattice associated with an electronic CDW. The wave vector of the distortion is determined by the geometry of the Fermi surface and is close to one-third the smallest in-plane reciprocal lattice vector $\vec{a}$*.

The PSD primarily involves motions of the metal atoms of the longitudinal type (LA), and precursor effects are seen in the normal phase as a softening of the LA phonon (Kohn anomaly) near $\vec{a}$*/3.[2] The data on TaSe$_2$ show further, but incomplete, softening just above the onset temperature T_O of a transition to an incommensurate CDW (ICDW) where the wave vector q_O is initially about 2% less than a*/3. A central component is seen a few degrees above T_O. At a lower temperature T_d the CDW locks into the lattice, forming a commensurate charge density wave (CCDW) and $q_O = a$*/3. For many samples of 2H-TaSe$_2$ T_d = 90 K, but T_d seems to be 100 K for some grown at the University of Illinois. For 2H-NbSe$_2$ T_O is 33 K. Lock-in is not observed.[2]

---

*Present address: Laser Systems Division, Hughes Aircraft Co., Culver City, CA 90230.

TANTALUM DISELENIDE

The low temperature structure of 2H–TaSe$_2$ has been determined
by neutron diffraction to within an unknown phase of the eigenvector.[2]
The structure has a superposition of three PSD's each with a different equivalent $\vec{q}$ (triple-q state). If inversion symmetry is
assumed, then the structure has the same space group (D^4_{6h}) as the
original lattice. Proof of inversion-symmetry is based on Raman
results.[3] One expects one amplitude and one phase mode derived from
the LA phonon for each of the three q's. (These will be explained
shortly.) With the assumed symmetry these will give one A$_{1g}$ and one
double degenerate E$_{2g}$ mode of each type. Raman data are shown in
Fig. 1.[4] If one assumes that the phase modes occur at lower
frequencies than the amplitude modes, then the 44 cm^{-1} A$_{1g}$ and 46 cm^{-1}
E$_{1g}$ peaks are phase modes, and the 78 cm^{-1} A$_{1g}$ and 63 cm^{-1} E$_{1g}$ peaks
are amplitude modes. The 80 cm^{-1} E$_{2g}$ peak is actually a doublet
that results from folding to the zone center of a pair of zone corner
modes at K.[3] They are rendered Raman-active by the CDW but are otherwise unaffected by it.

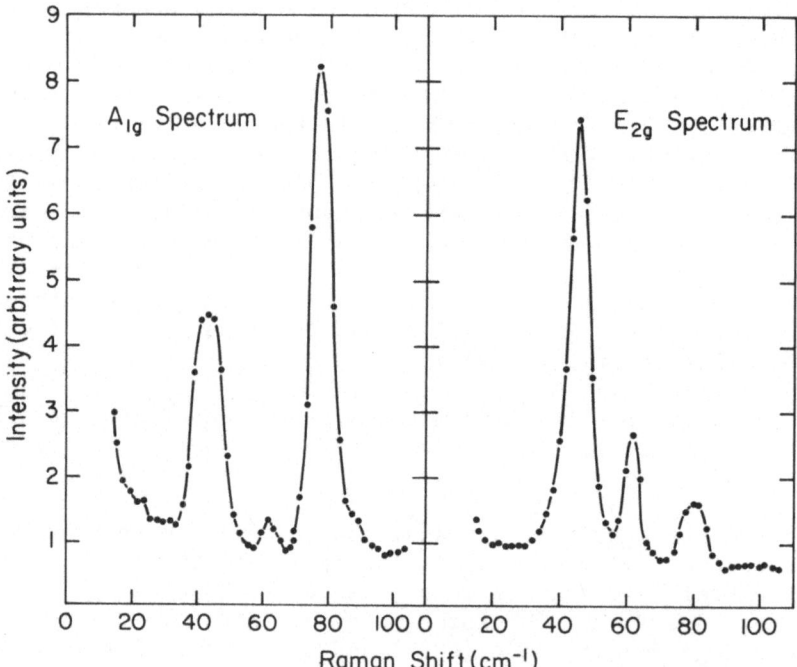

Fig. 1. New Raman-active modes that appear at low temperature (44 K)
in the commensurate charge density wave state of 2H–TaSe$_2$.[3,4]

AMPLITUDE MODES AND PHASE MODES

It is useful to digress to consider a simple picture of what happens to a soft mode such as that of the LA phonon near $\vec{a}*/3$ just below the transition temperature T_O. In the presence of the static distortion phonons with wave vector $\vec{q}_O$ and with wave vector $-\vec{q}_O$ will now be found ("folded") at $\vec{q} = 0$. In lowest order only the modes with wave vectors $\vec{q} + \vec{q}_O$ and $\vec{q} - \vec{q}_O$ will interact to produce new modes of the low temperature structure having wave vector $\vec{q}$. The symmetric combination may be described as a modulation of the amplitude of the static distortion ("amplitude mode"). The anti-symmetric combination may be described as a modulation of the phase of the static distortion ("phase mode"). For $\vec{q}_O$ commensurate with the reciprocal lattice both modes have non-zero frequency as $q \to 0$, but in the incommensurate case the frequency of the phase mode is linear in q. It must vanish at q=0, for it should be possible to slide the phase of the distortion an arbitrary amount without the expenditure of any energy. The expected form of the dispersion curves is sketched in Fig. 2. The presence of impurities is expected to "pin" the phase mode and raise its frequency to a finite value.[5]

When nonlinearities are taken into account, the static distortion will not be given by a single spatial Fourier component. For a short distance span the dependence closely matches that of a commensurate structure with a wave vector $\vec{q}_c$ close to $\vec{q}_O$. The phase changes nearly discontinuously at a series of equally spaced "discommensurations."[5] In three dimensions these entities become "walls" separating "domains" of essentially commensurate material.[6]

The four strongest peaks in Fig. 1 show strong position and shape changes as the temperature is raised close to T_d (110 K) and T_O

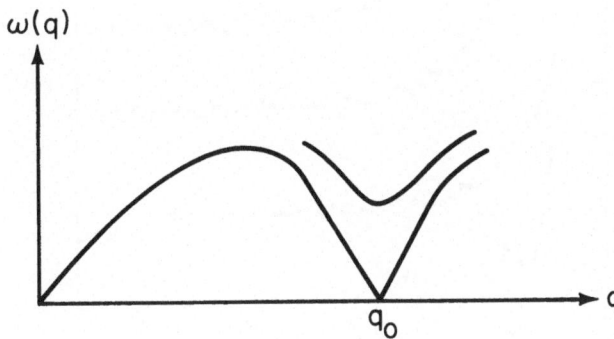

Fig. 2. Behavior of wave vector dependence of the amplitude mode frequency (upper curve) and phase mode frequency (lower curve) for a one-dimensional soft mode at an incommensurate wave vector $\vec{q}_O$.

(123 K). The above mentioned elementary notions of CDW's would
suggest that the phase modes should soften to zero frequency at T_d
and remain either as acoustic-phonon-like modes (see Fig. 2) or
as overdamped central peaks until T_O is reached, where the amplitude
modes should soften to zero frequency. What actually happens is
partially masked by the great growth in damping that reduces peak
intensities, but the simple predictions do not hold. Evolution of
the E_{2g} spectra with rising temperature is shown in Fig. 3.[4]
The 63 cm^{-1} peak from the 44 K data of Fig. 1 has softened and broadened sufficiently by 91 K that it appears as a barely observable
shoulder at 55 cm^{-1} in the top curve. At higher temperatures
the remaining single peak broadens and softens, becoming nearly
overdamped at about 105 K. The data fit well a lineshape predicted

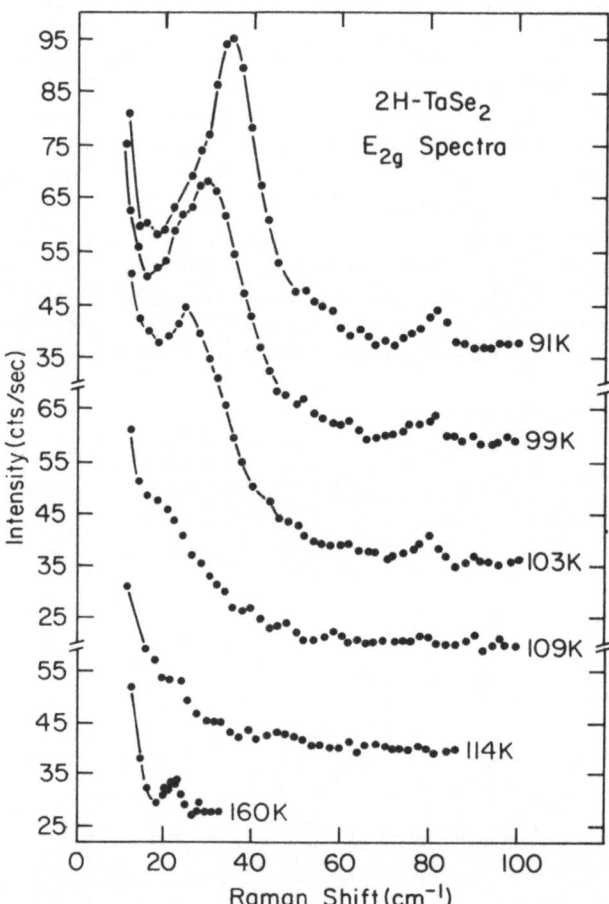

Fig. 3. Temperature dependence on the soft E_{2g} mode(s) in 2H-TaSe$_2$.[4]

from a damped harmonic oscillator.[4,7] The oscillator frequency $\omega_0(T)$
varies with temperature as $(T_o-T)^{-1/3}$ and damping $\Gamma(\Gamma)$ varies as
$(T-T_o)^{-1}$. Recent unpublished work on Illinois-grown samples with
$T_d = 110$ K has verified these dependences below T_d and a saturation of the oscillator damping beyond T_d.[4] It remains nearly overdamped and its frequency stays constant (at about the value seen
by neutron scattering near T_o)[2] until T_o, where it vanishes in
intensity.

This recent study of the E_2g spectrum has also found a central
peak of width less than 1 cm^{-1}. Data at 82 K are shown in Fig. 4.
The height is a maximum at 80 K. This behavior can be modelled by
replacing the response function of a damped harmonic oscillator

$$R(\omega) = [\omega^2 - \omega_0(T)^2 - i\omega\Gamma]^{-1} \tag{1}$$

by one in which the square of the soft-mode frequency $\omega_0(T)^2$ becomes

$$\omega_\infty^2(T) - \frac{\gamma\delta^2}{\gamma-i\omega} \tag{2}$$

One may interpret ω_∞ as the "bare" frequency of the soft mode coupled
via a coupling constant δ to a "relaxation mode" with relaxation rate
γ. A divergent central peak appears when $\omega_\infty^2 = \delta^2$. The physical
origin of this peak is unknown; it could be due to impurities, or it
may be the sign of a second phase transition. Some samples of
$2H-TaSe_2$ grown at Bell Laboratories show evidence of two lower phase
transitions--one at 93 K and the other near 113 K, as revealed by
dilatometric studies.[8]

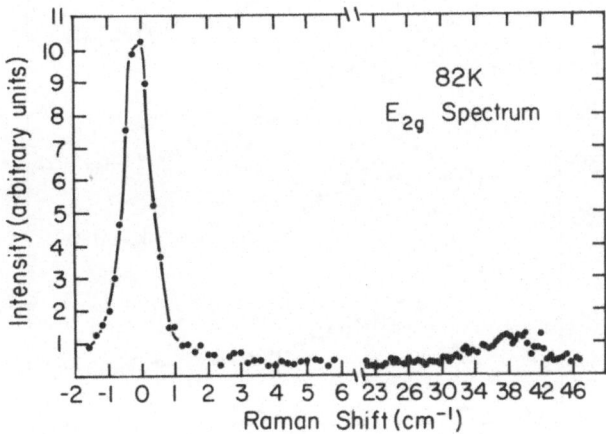

Fig. 4. E_2g Raman spectrum of $2H-TaSe_2$ at 82 K taken with an iodine
cell and a double monochromator.

The evolution with temperature of the $A_{1}g$ spectra is shown
in Fig. 5. The two peaks, originally assigned to phase and amplitude modes, have about equal intensity and merge at 94 K. At still
higher temperatures they soften and broaden together. Oscillator
fits below 100 K give an oscillator frequency that varies as
$(T_O-T)^{-1/3}$.[4]

NIOBIUM DISELENIDE

Because of the narrow temperature range between T_c and T_O
for $2H-TaSe_2$ and because in this range the Raman peaks of the soft
modes are strongly damped, it is difficult to learn much about the
ICDW state via Raman scattering. This is not the case for $2H-NbSe_2$,
which remains incommensurate below $T_O = 33$ K.[2] Raman modes of the

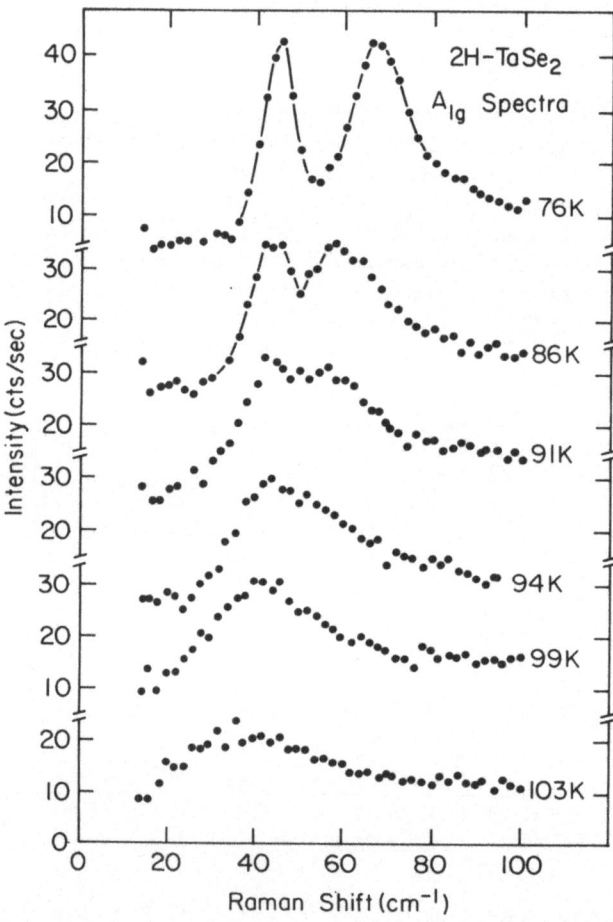

Fig. 5. Temperature dependence of the soft $A_{1}g$ modes in $2H-TaSe_2$.

ICDW were first seen by Tsang et al.[9] We have recently been studying them in more detail.[10] Fig. 6 shows that a distinction exists between the position of A and E peaks, providing evidence of the three-q nature of the ICDW, [11] for a superposition of Raman spectra from many single-q domains would give identical A-like and E-like spectra.

GAP EXCITATIONS IN SUPERCONDUCTING NbSe$_2$

Below T_c = 7 K 2H-NbSe$_2$ becomes superconducting.[12] The energy gap at 1.6 K is 2Δ = 17.2 ± .4 cm^{-1}, as determined by the position of the peak in the infrared transmission.[13] The Raman spectra with the sample immersed in superfluid helium at 2 K are shown in Fig. 7. Two new peaks are seen: one of A$_{1g}$ symmetry at 19 cm^{-1} and an E$_{2g}$ peak at 15.5 cm^{-1}. Their weighted average (16.7 cm^{-1}) agrees with the position of the infrared peak. The E$_{2g}$ Raman peak has been studied in a magnetic field applied parallel to the layers. The peak softens as H rises and is below 10 cm^{-1} at 26 kG, where it is hidden by the elastic tail. This material is a type-II superconductor. Critical fields at 2 K may be estimated from published data[14] and are H_{c1} = 250 G and H_{c2} = 95 kG for H parallel to the layers. The downward shift in the Raman frequency with increasing field provides good evidence that the Raman peaks are associated with excitations of quasi-particles across the superconducting energy gap.

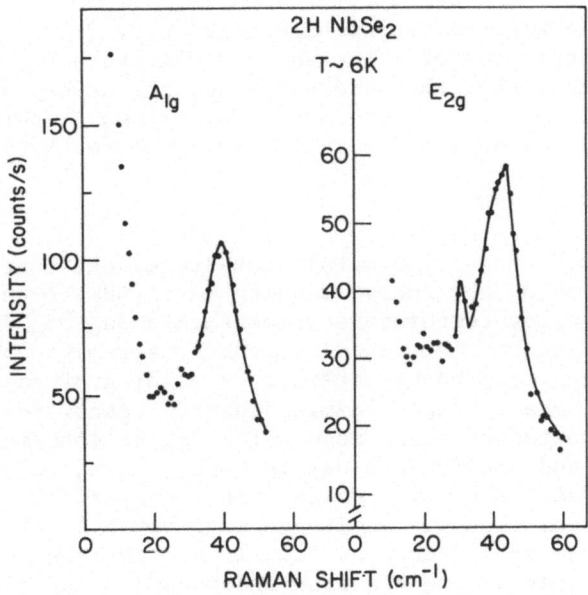

Fig 6. A-like and E-like Raman spectra of the soft modes in the incommensurate charge density wave state of 2H-NbSe$_2$ at about 6 K.[10]

There are a number of features that make a full theory of Raman scattering by excitations of quasi-particles in superconductors formally complex: (1) wave vector non-conservation due to attenuation of optical radiation fields; (2) screening of matrix elements by electron-electron coulomb interactions; (3) resonance enhancement when the laser energy is close to that of interband transitions; (4) effects of anisotropy of the Fermi surface; (5) effects of the superconducting gap on the light scattering response; and (6) interactions among excited quasi-particles. Theories of such effects have been given by several authors.[15-19] All take (1) and (5) into account. None consider (6). Cuden considers (3) and (4) and deals with the screening of some, but not all matrix elements.[19] Abrikosov and Genkin consider (2) and (4) using an effective mass approximation for interband transitions,[18] which, when screening is taken into account, gives a scattering amplitude for an excitation of an electron at $\vec{k}$ on the Fermi surface proportional to

$$\gamma_k = \sum_{ij} e_{si}\left(\frac{\partial^2 E}{\partial k_i \partial k_j} - \overline{\frac{\partial^2 E}{\partial k_i \partial k_j}}\right) e_{\ell j} \, , \tag{6}$$

where $\vec{e}_s$ and $\vec{e}_\ell$ are polarization unit vectors for scattered and laser light, $E(\vec{k})$ is the electron's energy at the Fermi surface, and the average is taken over the Fermi surface. Such a term could give both A-like and E-like Raman spectra for a layered compound with a Fermi surface as complicated as that of 2H-NbSe$_2$.[20] These effects would be analogous to those in many-valley semiconductors.[21]

The data in Figure 7 show sharp peaks near 2Δ. The cited theoretical treatments predict a threshold at 2Δ and a monotonic rise above 2Δ. We feel that the observed lineshape is evidence of interactions either among the excited quasi-particles or between them and the CDW; these will have to be included in future theories.[22]

SUMMARY AND CONCLUSIONS

Some layered transition-metal dichalcogenides undergo phase transitions of the charge-density-wave (CDW) type. An ambiguity in the low temperature commensurate $3a_0 \times 3a_0$ structure of 2H-TaSe$_2$ determined by neutron diffraction is removed by Raman data, which show four strong peaks obtained by folding of the soft LA phonon. Their behavior upon warming towards transition temperatures in complex. Instead of phase mode softening at lock-in, merging of phase modes and amplitude modes is seen. An E$_g$ mode becomes almost overdamped 5 K below a phase transition at 110 K. A central peak is also seen, whose intensity is maximum at 80 K. The complex Raman behavior seen near lock-in for 2H-TaSe$_2$ shows that the CDW's in this material are far from understood. Additional measurements of CDW structures are in order. In 2H-NbSe$_2$ onset occurs at $T_0 = 33$ K to an incommensurate CDW that persists through a

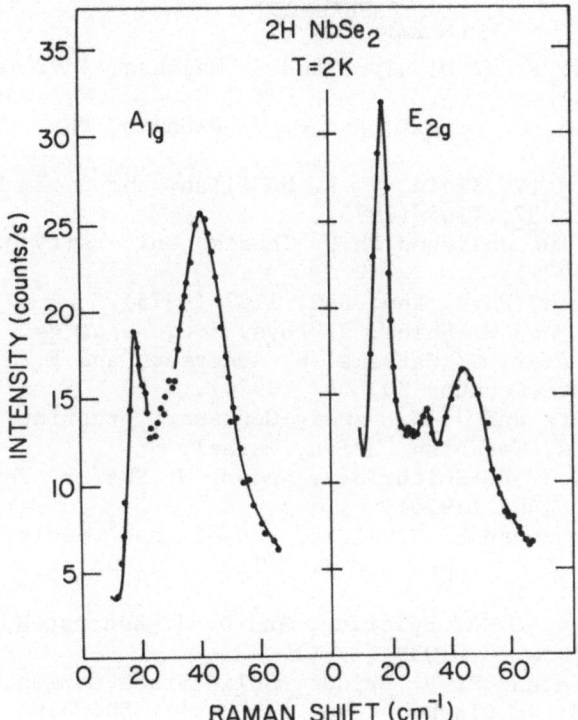

Fig. 7. A-like and E-like Raman spectra for superconducting 2H-NbSe$_2$ immersed in superfluid He at 2 K.[10] The two peaks not present in Fig. 6 are assigned to quasi-particle excitations at $\omega \approx 2\Delta \approx 17$ cm^{-1}.[13]

superconducting transition at $T_c = 7$ K. Between T_c and T_0 Raman-active amplitude modes of the CDW are seen near 40 cm^{-1}. At 2 K new modes are seen near 17 cm^{-1}. Their shift in a magnetic field suggests that they are excitations of quasi-particles across the superconducting gap. This is the first observation of such excitations by Raman scattering.

ACKNOWLEDGEMENTS

The authors thank W. L. McMillan and D. E. Moncton for many helpful discussions and S. E. Meyer and R. D. Coleman for providing the samples of TaSe$_2$ and NbSe$_2$. This work was supported in part by the National Science Foundation under the MRL grant DMR 77-23999.

REFERENCES

1. J. A. Wilson, F. J. DiSalvo, and S. Majahan, Adv. in Phys. 24, 117 (1975).
2. D. E. Moncton, J. D. Axe, and F. J. DiSalvo, Phys. Rev. B16, 801 (1977).
3. J. A. Holy, M. V. Klein, W. L. McMillan, and S. F. Meyer, Phys. Rev. Lett. 37, 1145 (1976).
4. D. G. Bruns, unpublished Ph.D. Thesis, University of Illinois, Urbana (1979).
5. W. L. McMillan, Phys. Rev. B12, 1187 (1975).
6. K. Nakanishi and H. Shiba, J. Phys. Soc. Japan 44, 1465 (1978).
7. E. F. Steigmeier, G. Harbeke, H. Auderset, and F. J. DiSalvo, Solid State Commun. 20, 667 (1976).
8. M. O. Steinitz and J. Grunzweig-Genossar, preprint, Department of Physics, Technion, Haifa, Israel.
9. J. C. Tsang, J. E. Smith, Jr., and M. W. Shafer, Phys. Rev. Lett. 37, 1407 (1976).
10. R. Sooryakumar and M. V. Klein, unpublished results.
11. C. Berthier, D. Jerome, and P. Molinie, J. Phys. C. 11, 797 (1978).
12. E. Revolinsky, G. A. Spiering, and D. J. Beernsten, J. Phys. Chem. Solids 26, 1029 (1965).
13. B. P. Clayman and R. F. Frindt, Solid State Commun, 9, 1881 (1971); B. P. Clayman, Can. J. Physics 50, 3193 (1972).
14. P. de Trey, Suso Gygax, and J.-P, Jan, J. Low Temp. Physics 11, 421 (1973).
15. A. A. Abrikosov and L. A. Falkovskii, Zh. Eksp. Teor, Fiz. 40, 262 (1961) [Trans: Soviet Physics JETP 13, 179 (1961)].
16. S. Y. Tong and A. A. Maradudin, Mat. Res. Bull, 4, 563 (1969).
17. D. R. Tilley, Z. Physik 254, 71 (1972); J. Phys. F. 3, 1417 (1973).
18. A. A. Abrikosov and V. M. Genkin, Zh. Eksp. Teor. Fiz. 65, 842 (1973) [Trans: Soviet Physics JETP 38, 417 (1974)].
19. C. B. Cuden, Phys. Rev. B13, 1993 (1976).
20. J. E. Graebner and M. Robbins, Phys. Rev. Lett. 36, 422 (1976).
21. For a review see M. V. Klein in Light Scattering in Solids, edited by M. Cardona (Springer Verlag, Heidelberg, 1975).
22. W. L. McMillan, private communication.

QUASIELASTIC LIGHT SCATTERING NEAR

STRUCTURAL PHASE TRANSITIONS

K. B. Lyons and P. A. Fleury

Bell Telephone Laboratories

Murray Hill, N. J. 07974

The focus of the this binational symposium lies in the consideration of the interaction of light with matter. One manifestation of this interaction is found in the phenomenon of light scattering. In many situations, in order to understand the light scattering spectrum, it is sufficient to consider first order (ie: one-phonon) scattering processes only. However, the study of higher order processes is also of interest. One kind of system where higher order (multiphonon) scattering may be important is a crystal undergoing a structural phase transition. In the present paper we shall discuss the way in which recent light scattering studies of such systems demonstrate the importance of higher order processes. We will discuss these effects with emphasis on the role of second order or two-phonon scattering.

Our understanding of structural phase transitions has evolved in the last decade through several stages from the simple soft mode and mean field theories[1] (MFT) to the modern coupled-mode, renormalization group[2] and dynamic scaling[3] ideas. Crucial to this evolution has been an increasingly detailed interpretation of the experimental data, particularly those resulting from scattering experiments.

In the simplest theory[1], the order parameter fluctuations are characterized by a soft mode whose frequency goes to zero at the transition temperature T_c as

$$\omega_s^2 \propto |T_c - T|^{2z} \quad ; \quad 2z = 1.0 \tag{1}$$

The initial thrust of phase transition scattering studies[1] was to extract soft mode frequencies and compare these to predictions of MFT.

The dynamic susceptibility associated with the order parameter was assumed to be quasi-harmonic, of the form

$$\chi_s \propto [(\omega^2 - \omega_{so}^2) + i2\Gamma_s\omega]^{-1} \quad . \tag{2}$$

Using this approximation it was possible to extract values for ω_{so} and Γ_s from neutron and light scattering spectra. It was soon found,, however, that the values obtained did not always behave as expected. The value of ω_{so}^2 did not extrapolate to zero at the transition. Moreover, in the same temperature region where this deviation was evident, a new spectral component was observed, centered at zero energy, with a very narrow width. This feature has since been called the "Central Peak". In the early observations of this phenomenon, via neutron scattering[4], the peak width was instrumentally limited. Nevertheless it was possible to parametrize the spectra by adding a term to Γ_s of the form

$$\Sigma(\omega) = \frac{\delta^2 \tau}{1 - i\omega\tau} \qquad (3)$$

representing the coupling of the soft mode to a relaxation process, of unknown origin, characterized by a relaxation time τ and a coupling strength δ. This has the effect, in Eq. (2), of replacing ω_{so}^2 by the quantity $\omega_s^2 = \omega_{s\infty}^2 - \delta^2/(1+\omega^2\tau^2)$, and Γ_s by $\Gamma_s' = \Gamma_s + \delta^2\tau/(1+\omega^2\tau^2)$, where $\omega_{s\infty}^2 = \omega_{so}^2 + \delta^2$. Since the central peak was instrumentally narrow, no information was available on τ, but δ was measurable by extrapolation of the observed value of $\omega_s^2 (\sim\omega_{s\infty}^2)$ to the transition temperature, where $\omega_{so} = 0$ by definition. This treatment also yielded a quantitative description of the central peak intensity.

The obvious question then arose as to the nature of the relaxation process responsible for the self energy in Eq. (3). A number of mechanisms[5-8] were proposed, all of which yielded self energy forms similar to Eq. (3). However, in the absence of experimental information on the value of τ and its dependence on temperature, wavevector, and other experimental parameters, it was not possible to differentiate among these mechanisms.

A natural approach to this problem was to utilize the greater energy resolution capability of light scattering to study the same phenomenon. However, for a number of years, quantitative study of any central component was prevented by the very strong elastic scattering from sample defects. Recently, a technique, developed by the present authors and reported elsewhere[9], has been used to circumvent this problem. It is based upon use of a molecular iodine reabsorption filter in conjunction with appropriate computer data analysis. By this technique, it has been possible to observe in detail the spectral profile at energies as low as 0.002 meV (0.5 GHz), and thus investigate central peaks near structural phase transitions in various materials.

Before considering the results of these investigations, it is important to understand the differences between the light scattering and neutron scattering experiments. These differences lie in the scattering wavevector, the selection rules, and in the properties of the iodine filter. The wavevector q involved in light scattering is typically in the range $1-4\cdot10^5$ cm^{-1}, at least an order of magnitude less than the resolution of a typical neutron scattering experiment. The effect of this difference is to introduce the acoustic modes into the low frequency light scattering spectrum. These modes, in light scattering, lie in the range 0.01-0.25 meV, and thus may interact with the soft mode near T_c. The observed spectrum is that of the coupled modes, which, although considerably different in appearance, contains information similar to that in the neutron spectrum. For a soft *optic* mode, the soft mode frequency is virtually q-independent, while the acoustic mode frequency increases as q. Hence, at large q, for the neutron scattering spectrum, the modes are effectively uncoupled. In the light scattering spectrum, on the other hand, this coupling may be significant. The light scattering spectrum of such a coupled mode system may be written[10]

$$S(\omega) = \sum_{i,j} \omega_i \omega_j F_i F_j \chi_{ij}, \qquad (4)$$

where F_i represent the scattering strengths of the uncoupled modes. In the simple case of two coupled modes χ_{ij} are expressed in terms of χ_i, the uncoupled susceptibilities, as

$$\chi_{ii} = \frac{\chi_i}{1 - A^2 \chi_a \chi_s}, i = a, s; \quad \chi_{as} = \frac{A \chi_a \chi_s}{1 - A^2 \chi_a \chi_s} \quad , \tag{5}$$

where A is the coupling constant, and the subscripts a,s refer to the acoustic and soft modes respectively. This coupling can strongly modify the spectrum for light scattering from that seen in neutron scattering. Although the coupling is of course allowed at larger q, its effect is made negligible by the difference in the characteristic frequencies of the modes.

A second difference lies in a peculiarity of the molecular iodine reabsorption technique. Any feature which lies within the absorption notch of the iodine will not be visible in the spectrum, even after computer analysis[9]. This "blind spot" does not exist for neutron scattering. Thus, any part of the neutron scattering central peak which is static in origin (unshifted) or very narrow (<300 MHz) will be removed from our spectra.

The final and most important difference is due to a selection rule which is operative in light scattering and not in neutron scattering. This results in the neutron scattering cross section being simply proportional to the fourier transform (FT) of the dynamic susceptibility $\chi(r,t) \propto \langle \delta\psi(r,t)\delta\psi(0,0)\rangle$, where we consider a spatially and temporally fluctuating order parameter, $\psi = \psi_o + \delta\psi(r,t)$, deviating from the equilibrium value ψ_o. The light scattering, on the other hand, is related to the autocorrelation function of the refractive index, $\langle \delta n(r,t)\delta n(0,0)\rangle$. We can write $\langle \delta n = a\delta\psi + b(\delta\psi)^2$. If $a \neq 0$ is allowed by symmetry, then the two cross sections are proportional. If not, then the light scattering cross section is given by the autocorrelation function for the *square* of the order parameter:

$$S(q,\omega) \propto FT G_2(r,t) \equiv FT \langle [\psi(r,t)]^2 [\psi(0,0)]^2 \rangle \tag{6}$$

In the latter case the soft mode is Raman inactive above T_c. To see this in a simple way, we use a mean field approach and write (6) using $\psi = \psi_o + \delta\psi$ to obtain

$$S(q,\omega) \propto FT \psi_o^2 \langle \delta\psi(r,t)\delta\psi(0,0)\rangle \quad . \tag{7}$$

Above T_c, the static order parameter ψ_o is zero, so the mode is inactive. This is the case for all the phase transitions under discussion in the present paper. Thus, the large anomaly in the total intensity observed in neutron scattering is removed, for light scattering, by the ψ_o^2 term in (7). We should note, however, that the above factorization of the fourth order correlation function ignores the contribution of other terms, in particular $\langle \delta\psi(r,t)^2\delta\psi(0,0)^2\rangle$. This amounts to ignoring the higher order scattering processes. As we shall see, this approximation fares poorly near T_c, where $\delta\psi$ becomes large.

One system in which this behavior is manifested especially clearly is $Pb_5Ge_3O_{11}$, near its ferroelectric phase transition at 451 K[11]. Well below and above T_c, Raman[12] and neutron scattering[13] have shown that the soft mode behaves in a simple mean-field way. As T_c is approached from below, the soft mode becomes overdamped at $T \sim T_c - 40°$, and continues to narrow, still obeying a simple Landau-type behavior. This behavior persists down to $(T_c - T)/T_c \lesssim 0.01$, whereupon a change is observed. The soft mode stabilizes at an overdamped width of about 2.5 cm^{-1}, and a central peak appears near zero frequency. The soft mode--central peak system interacts with the LA acoustic phonon to produce the unusual spectral shape shown in Fig. 1. The soft mode wing is evident in the figure, in the lower resolution spectra. Using the coupled mode formalism outlined above, it is possible to fit the spectra below T_c, with most of the parameters determined from other experiments[11]. In doing this, we use a pairwise coupling model, in which the soft mode is first coupled to a relaxation process, introducing a relaxing self energy into its response function as given by Eq.'s (2) and (3). The acoustic mode is treated as quasiharmonic. This

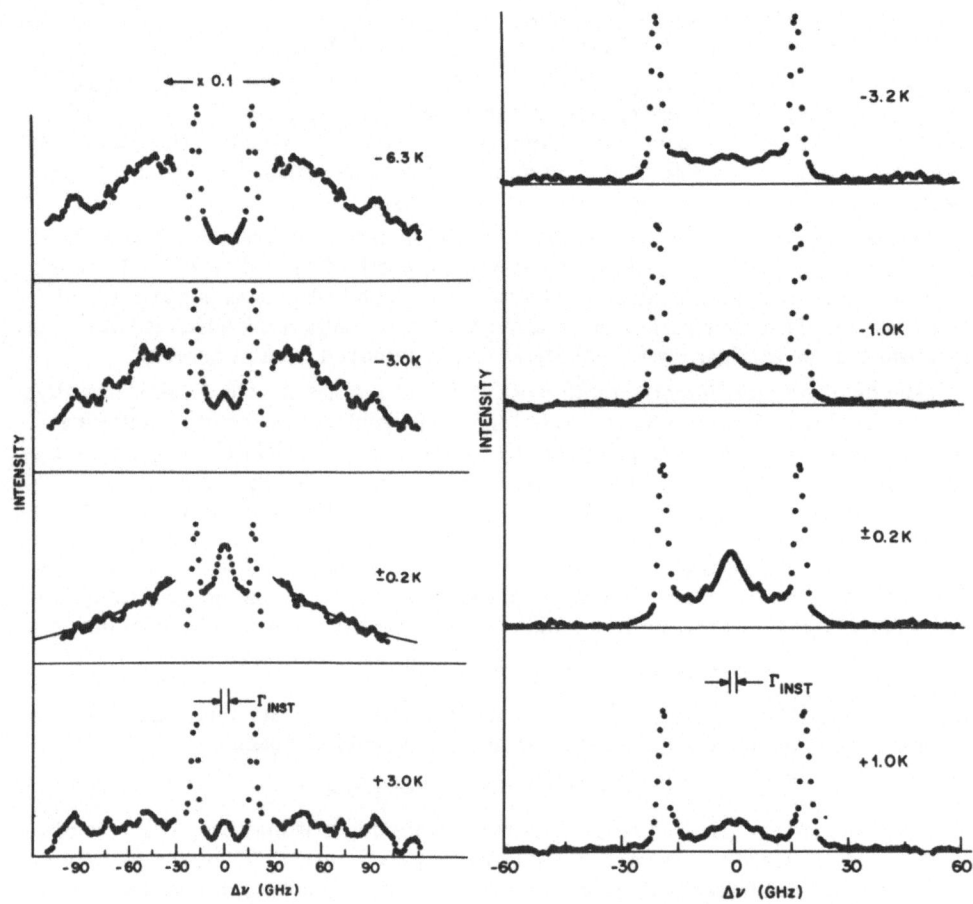

Fig. 1: Light scattering spectra of $Pb_5Ge_3O_{11}$ observed near its ferroelectric phase transition. Temperatures are given as $T_c - T$. The lower resolution spectra, on the left, show clearly the existence of the high-frequency wing of the soft mode at the same time that the narrow central peak, shown most clearly at high resolution on the right, is present.

fit is shown in Fig. 2. From these fits we find the characteristic relaxation frequency to be $\tau^{-1} \sim 29$ GHz. For the data above T_c, also shown in Fig. 2, the theory breaks down, since no scattering is expected. The fits shown above T_c represent simply parametrization of the spectra, with the various parameters adjusted in an *ad hoc* fashion to fit the data. The spectral lineshape observed does *not* depend perceptibly upon the scattering wavevector. However, as is evident in the figure, the central width *is* strongly *temperature dependent* near T_c, demonstrating a strong renormalization of the 29 GHz relaxation process.

In the simple theory outlined above, no scattering is expected above the phase transition. The fact that the central peak persists above T_c argues for the participation of the

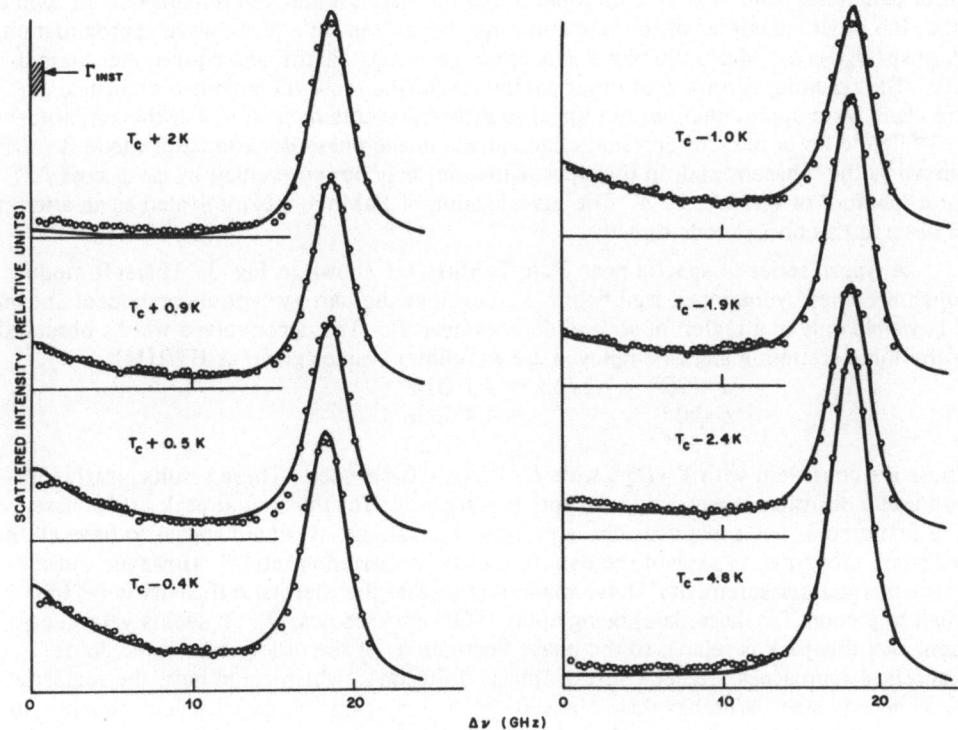

Fig. 2: A least squares fit to the spectra of $Pb_5Ge_3O_{11}$, similar to those shown in Fig. 1, using the coupled mode formalism described in the text. The data are shown by the points, while the fit is given by the solid lines. The value of τ which results from the fits below T_c is 29 GHz.

higher order processes mentioned above. Since such processes would be averaged over the Brillouin zone, the observed lack of dependence upon scattering wavevector would be expected. Another way of viewing this is to say that very close to T_c, the fluctuations of $\delta\psi$ become comparable to, or even larger than, the static order parameter value ψ_o. Hence the term dropped in the derivation of (7) becomes important and, in fact, may dominate the scattering near T_c.

Similar spectral features have been observed in other systems, including $SrTiO_3$ and $BaMnF_4$, representing different symmetry classes of phase transitions. The mode observed in $SrTiO_3$ is similar to that in $Pb_5Ge_3O_{11}$, in that the associated relaxation time is 15 GHz, and the lineshape is not observably q-dependent. The central peak width in $BaMnF_4$, on the other hand, is much smaller (1.6 GHz HWHM for right angle scattering), is temperature insensitive, but is strongly q-dependent.

The phase transition which occurs in $BaMnF_4$ is one of a recently discovered type, which involves a transition to an incommensurate phase. That is, the periodicity of the order parameter is not a simple multiple of the prototypical unit cell dimension. In such a case, the order parameter of the transition may be written, in a plane wave approximation, as $\psi \exp(iq_o r + i\phi)$, where the phase ϕ is arbitrary and q_o^{-1} is the order parameter periodicity. The resulting spectrum of order parameter fluctuations will have two branches. In the plane wave approximation, the upper branch represents fluctuations in the amplitude ψ, while the lower branch represents fluctuations in the phase ϕ. The latter mode is known as the "phason" and, in this approximation, may be represented by $\phi = \phi_o \cos(\vec{q} \cdot \vec{r})$ for a "phason" of wavevector $\vec{q}$. The investigation of BaMnF$_4$ was motivated as an attempt to observe this phase mode directly.

A typical series of spectra near T_c in BaMnF$_4$ are shown in Fig. 3. The soft mode, which becomes overdamped well below T_c, develops the narrow central component shown. It is visible only in a region of several degrees near T_c. The deconvolved widths observed at the three scattering angles employed are as follows (values given as HWHM):

$$\theta = 125° \qquad \Gamma/2\pi = 2.1 \text{ GHz}$$
$$\theta = 90° \qquad \quad = 1.4 \text{ GHz}$$
$$\theta = 55° \qquad \quad \leq 0.8 \text{ GHz}$$

These are consistent with $\Gamma = Dq^2$, with $D = 0.14 \pm 0.02 \text{cm}^2/\text{sec}$. These results clearly indicate that a diffusion process of some sort is responsible for the central peak in this case. The first process one must consider is entropy fluctuations. It would appear to have all the necessary properties to explain the data (intensity, polarization, etc)[14]. However, subsequent thermal measurements[15] have made it clear that the thermal diffusivity is far too small to account for these data, being about 0.005 cm^2/sec near T_c. It seems very likely, then, that this peak is related to the phase fluctuations of the order parameter. Some theoretical approaches predict a sort of "phase diffusion"[16] which could have the requisite $\Gamma \propto q^2$ dependence. Whether this is indeed the source of our observed spectra remains an open question until further corroboration can be obtained.

In any case, it is curious, though, that the observed q-dependence persists above T_c. Furthermore, the scattered intensity depends strongly on the *direction* of $\vec{q}$, dropping off very rapidly as $\vec{q}$ moves away from the plane perpendicular to $\vec{q}_o$. If the mechanism for the scattering above T_c were simply a higher-order process involving phonons averaged throughout the Brillouin zone, as discussed above for Pb$_5$Ge$_3$O$_{11}$, such q-dependence would not be expected. A possible explanation is that just above the phase transition, the critical fluctuations may take the form of long-lived clusters. That is, there may exist a strong very low frequency component in the spectrum of the order parameter fluctuations. If the spatial extent of these clusters is $\geq q^{-1}$, then the scattering would be similar to that just below T_c. Thus, the results in this case are consistent with a scattering mechanism of a higher order, but involving only the zone-center phonons.

It should be pointed out that the *Raman* spectrum also gives evidence for the existence of strong critical fluctuations above T_c. Below the transition, several Raman lines become active, as expected. We have measured[17] the intensity of the strongest of these lines, near 140 cm^{-1}, quantitatively as a function of temperature near T_c. This was done *simultaneously* with the central peak observations discussed above, utilizing the scattering out the other side of the sample. This arrangement leaves no question as to the comparison of temperatures between the two sets of data. The results showed that the Raman intensity of the new line extrapolated to zero well *above* T_c. In fact, above T_c the central peak and Raman intensities behave very similarly. The same fluctuations which could be responsible for the central peak intensity above T_c would also cause the observed Raman intensity. In the latter case, there is no need to assume that these fluctuations are

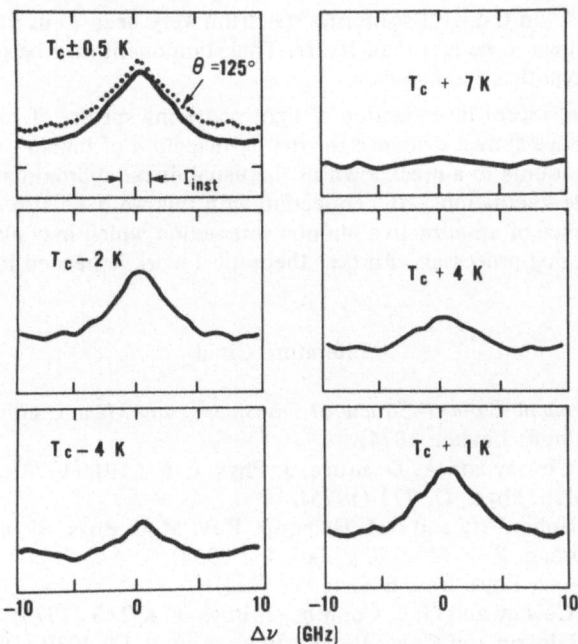

Fig. 3: Spectra obtained from $BaMnF_4$ near its transition to an incommensurate phase. The scattering geometry is right angle except where indicated. The intensity is found to persist well above the transition temperature T_c=247 K. The peak possesses observable intensity only for $\vec{q}$ near the bc plane.

especially long-lived or spatially extended, but the Raman observations are consistent with the central peak mechanism given in the preceding paragraph.

It is important to note that a similar result would occur if the *intensity* of the two-phonon process (at $\vec{k}+\vec{q}$ and $\vec{k}$) were strongly k-dependent. Anything which restricts the process to the zone center (in k-space) would have the same effect. The observed spectrum will be a complicated convolution of the density of states, the spectrum of the order parameter fluctuations, and the scattering efficiency, all of which may be functions of crystal wavevector $\vec{k}$. It is for this reason that a complete theory of the higher order scattering processes is so vital for understanding the $BaMnF_4$ results.

There is an additional question raised by the results in $Pb_5Ge_3O_{11}$. In this case, the lack of q-dependence makes a two-phonon process (difference scattering) a likely mechanism. However, the characteristic frequency observed is only 29 GHz~1 cm^{-1}. This is small for a zone-averaged phonon width of a typical branch, if one assumes that the observed spectrum will simply be a convolution of the one-phonon spectrum with itself. This may not be the case. The scattering process in this case involves pairs of phonons, annihilating one on the same branch at $\vec{k}+\vec{q}$ while creating one at $\vec{k}$. An attractive interaction between the phonons could create the narrower lineshape observed. It is also possible that an effect similar to that discussed in the preceding paragraph may restrict the values of $\vec{k}$ which participate in the scattering process. In fact, a very slow component has also been observed[11] in the light scattering spectrum very near T_c in $Pb_5Ge_3O_{11}$, whose width has been shown to be less than 10 Hz. This component may be related to the slowly relaxing clusters hypothesized above.

In conclusion, recent investigation of light scattering spectra of solids near structural phase transitions have shown evidence for the participation of higher order scattering processes. This amounts to a breakdown of the usual linear approximation for the scattering from soft mode fluctuations. In connection with this we have also conjectured about the possible existence of an attractive phonon interaction which may play a role in such higher order scattering processes. Further theoretical work is needed to fully understand both questions.

Literature Cited

1.   *Anharmonic Lattices, Structural Transitions, and Melting*, edited by T. Riste (Noordhoff, Leiden, 1974).
2.   R. A. Cowley and A. D. Bruce, J. Phys. C **6**, L191 (1973); K. G. Wilson, Rev. Mod. Phys. **47**, 773 (1975).
3.   P. C. Hohenberg and B. I. Halperin, Rev. Mod. Phys. **49**, 435-475 (1977).
4.   G. Shirane, Rev. Mod. Phys. **46**, 437 1974).
5.   C. P. Enz, Phys. Rev. B **6**, 4695 (1972).
6.   R. A. Cowley and G. J. Coombs, J. Phys. C **6**, 143 (1973).
7.   B. I. Halperin and C. M. Varma, Phys. Rev. B **14**, 4030 (1976).
8.   H. Schmidt and F. Schwabl, Phys. Lett. **61**, 476 (1977).
9.   K. B. Lyons and P. A. Fleury, J. Appl. Phys. **47**, 4898 (1976).
10.   P. A. Fleury, Comm. Sol. St. Phys. **IV**, 167 (1972).
11.   K. B. Lyons and P. A. Fleury, Phys. Rev. B **17**, 2403 (1978).
12.   J. F. Ryan and K. Hisano, J. Phys. C **6**, 566 (1973); K. Hisano and J. F. Ryan, Sol. St. Comm. **11**, 119 (1972); see also W. Muller-Lierheim, T. Suski, and H. Otto, Phys. Stat. Sol. B **80**, 31 (1977).
13.   R. A. Cowley, J. D. Axe, and M. Iizumi, Phys. Rev. Lett. **36**, 806 (1976).
14.   K. B. Lyons and H. J.Guggenheim, Sol. St. Comm., to be published.
15.   T. J. Negran and K. B. Lyons, to be published.
16.   R. N. Bhatt and W. L. McMillan, Phys. Rev. B **12**, 2042 (1975).
17.   K. B. Lyons and H. J. Guggenheim, to be published.

HIGH RESOLUTION X-RAY AND LIGHT SCATTERING SPECTROSCOPY

OF LIQUID CRYSTALS

P.S. Pershan, G. Aeppli, R.J. Birgeneau and J.D. Litster

Department of Physics and Center for Materials Science
and Engineering, Massachusetts Institute of Technology
Cambridge, Ma. 02139, U.S.A.

INTRODUCTION

In liquid crystals there exists a rich variety of phases with
varying orientational and translational order intermediate between
crystalline solid and isotropic liquid phases. These phases are
suited to test the role of symmetry, spatial dimensionality, and
thermal fluctuations on the properties of condensed phases that can
exist in nature. Uniaxial liquid crystals are particularly suitable
for this since the dominant collective modes cause orientational
fluctuations in the local dielectric tensor that scatter light
rather strongly. In this conference we shall summarize some of the
x-ray and light scattering experiments that have been carried out
at MIT and Harvard over the past few years.

To begin, we remind you of the two types of liquid crystal
systems. Thermotropic liquid crystals consist of what are roughly
cigar shaped organic molecules with quite anisotropic properties
such as dielectric polarizability and diamagnetic susceptibility.
A typical liquid crystal molecule is octyloxycyanobiphenyl (8OCB)
sketched below.

Phase transitions for these materials are principly studied as a function of temperature. Lyotropic liquid crystals are usually formed from mixtures of water and amphiphillic molecules such as soaps or lipids. A typical lipid, di-palmitoyl phosphatidyl choline (DPPC) is shown below.

$$C_{15}H_{31}-COO-CH_2$$
$$|$$
$$C_{15}H_{31}-COO-CH$$
$$|$$
$$C-PO_4^- -(CH_2)_2 \ NH_3^+$$

Phase transitions for these occur as a function of water content as well as temperature.

There are three principle uniaxial liquid crystalline phases. The nematic phase (N), which only exists in thermotropic systems, has orientational long range order of the molecules with no translational order beyond that of normal liquids. The smectic A (SmA) has the orientational order of the nematic and in addition has translational order in one dimension, i.e., a one dimensional density wave. The wave vector of the wave, q_0, is parallel to the molecular orientation (SmA). One commonly speaks of a layered structure, although for the thermotropic smectic-A the density wave is a more accurate description. Lyotropic smectic-A systems consist of alternate layers of water and bilayers of amphiphillic molecules in which the layers are well defined. The smectic A phase does not have long range translational order within the layer and is often discussed as though it were a stack of two dimensional liquids. The third uniaxial phase that we will discuss is the smectic B. At the present time the precise form of long range order in this phase is not determined. However, one might loosely describe it as a stack of two dimensional crystals. We will discuss both a model for this phase and ongoing x-ray experiments regarding this phase.

THE NEMATIC PHASE

General reviews on liquid crystals have been written by de Gennes and Chandrasekhar.[1,2] We restrict ourselves here to a few qualitative remarks on this phase. A rigorous hydrodynamic analysis[3] of the nematic order predicts two overdamped fluctuations $\delta \vec{n}$ of the director with relaxation rates $(1/\tau(q)) \simeq Kq^2/\eta$ where K is the phenomenological elastic tensor first proposed by Frank and Oseen[4,5] and η is a tensor dissipative parameter related to viscosity. Light scattering spectra can be calculated from the spectral densities of dielectric constant fluctuations.

$$<\delta\epsilon(q,0)\ \delta\epsilon(q,z)> \sim k_BT[Kq^2]^{-1}\ exp\text{-}\tau/\tau(q)$$

which become very large for small q and give nematics a very turbid appearance.

THE SMECTIC A PHASE

If one neglects compressability the lamellar character of the smectic-A phase together with the absence of long range positional order within the layers leads to an elastic energy density[3]

$$F = 1/2\ B(\partial u/\partial z)^2 + 1/2\ K\left[\frac{\partial^2 u}{\partial x^2} + \frac{\partial^2 u}{\partial y^2}\right]^2 \tag{1}$$

where u is the displacement of a layer along the uniaxial symmetry direction. The Frank-Oseen constant K is the same as for the nematic phase (i.e., $\delta n_x \sim -\partial u/\partial x$) and B is a constant that describes the elastic resistance to changing the mean thickness of a smectic layer.

The ratio $\sqrt{K/B}$ has the dimensions of length and away from critical regions it is typically of the same magnitude as the smectic A layer spacing. As a result it is very much easier to curve smectic layers than it is to change their thickness and this gives rise to some very interesting optical effects.

For example, a plane parallel laser beam at oblique angle to the layers in a typical smectic sample will be scattered into narrow crescents determined by the condition (k_z) incident = (k_z) scattered. In general there will be two such crescents since there are two polarizations for both incident and scattered light. The effect is due to the fact that for an elastic energy of the type shown the strain due to an defect spreads out in space as $x^2 + y^2 \sim z\sqrt{K/B}$ So long as $\sqrt{x^2+y^2} \lesssim$ the wavelength, the light cannot resolve the strain and scattering is as though the inhomogeneity induced by the defect extended over a column $\sim\lambda^2\sqrt{B/K}$ high. Since $\sqrt{B/K}$ is typically of the order of 10^6 to 10^7 cm^{-1} the $\Delta k_z \simeq 0$ selection rule is reasonably well obeyed. This type of scattering has been observed in both thermotropic[6,7], and lyotropic smectics and is[8] one of the simplest observations one can make to confirm the layered character of the phase. An even simpler observation, dating back to Bragg[9] and Friedel,[10] is the texture of samples observed under the optical microscope. Mathematical theorems exist to demonstrate that if the layers of a lammellar system are of fixed thickness, but if the curvature can be as high as desired one can fill all of space with smectics distorted into what is known as "focal conic structures". That is the layers curve slowly except along lines

that are either parabolae, elipses or hyperbolae and which are the
loci of cusps in the layer curvatures. In general these are complicated, unfathomable structures which are recognizable but impossible
to understand in detail. Under controlled conditions however, it
is possible to form extended two dimensional arrays[11] of confocal
parabolae. Characterization of these textures[12] in both thermotropic and lyotropic smectics is one of the few studies that have
been done in both types of liquid.

We mentioned earlier that many of the properties of liquid
crystals are intermediate between those of liquids and those of
crystals; sound propagation which can be studied by Brillouin scattering is an excellent example. Crystals have one longitudinal
and two shear-waves, all of which are underdamped with linear dispersion relations. At low frequencies liquids have only one underdamped sound wave (longitudinal), however the smectic-A has one
longitudinal and one shear-like sound propagation mode. The second
of these is particularly interesting since its velocity is an extremely anisotropic function of propagation direction[1], *viz.*

$$N_s^2(\theta) = (B_s/\rho) \cos^2\theta \sin^2\theta \tag{2}$$

where θ is the angle between the direction of propagation and the
uniaxial axis. This mode has been observed in both lyotropic and
thermotropic liquid crystals using the Brillouin technique.[13,14]

X-ray diffraction from the smectic-A phases reveals sharp peaks
indicative of the lamellar structure, however because the periodicity
is in only one dimension, in contrast to conventional crystals,
these are not true Bragg peaks. The mean squared fluctuations in
a layer position can be calculated from the elastic energy given above.

$$\langle u^2(\vec{r}) \rangle = \frac{kT}{(2\pi)^2} \frac{d^3q}{Bq_\parallel^2 + K_1 q_\perp^4} \tag{3}$$

The limits of integration can be chosen as $(2\pi/L) \le |\vec{q}| \le q_0$ to
obtain an approximate answer. (L is the size of the sample). One
readily obtains

$$\langle u^2(\vec{r}) \rangle \simeq \frac{kT}{4\pi(BK_1)^{1/2}} \ln(q_0 L) \tag{4}$$

Thus the SmA phase in three dimensions shows the same logarithmic
singularity from long wavelength Goldstone modes as one finds for
solids in two dimensions; one expects the SmA phase does not exhibit
true long range order, but rather the algebraic decay of correlations
predicted by Wegner and Jancovici and later developed in more detail
by Kosterlitz and Thouless.[15]

To study this by x-ray scattering, we recall that the x-rays measure the Fourier transform of the pair correlation function

$$G(\vec{r}) = \langle e^{iq_0[u(\vec{r})-U(0)]}\rangle \qquad (5)$$

This can be calculated from Eq.(3) in the harmonic approximation[16] to be proportional to

$$G(\vec{r}) \sim \frac{1}{(x^2+y^2)^\eta} e^{-\eta E_1\left(\frac{x^2+y^2}{4\lambda z}\right)}$$

where $\eta = kTq_0^2/8\pi\lambda B)$, $\lambda = (K_1/B)^{1/2}$ is the analogue of the penetration depth in a type I superconductor, and E_1 is the exponential integral. From the properties of E_1 it is readily seen that $G(\vec{r})$ does not extend to infinity (as it would for long range order) but has an anisotropic power law (algebraic) decay

$$G(\vec{r}) \sim \frac{1}{r_\perp^{2\eta}} \qquad r_\perp \gg r_{||} \qquad (7a)$$

$$\sim \frac{1}{r_{||}^\eta} \qquad r_{||} \gg r_\perp \qquad (7b)$$

This means the scattering from the SmA density wave in the Sm phase is not a Bragg peak, but a power law singularity, which has been verified experimentally[17] and is shown in Fig. 1.

THE NEMATIC-SMECTIC A PHASE TRANSITION

A hydrodynamic theory of the smectic-A does not distinguish between fluctuations in the average molecular orientations and the layer orientations since these are strongly coupled variables. However McMillan[18] and de Gennes[1] have demonstrated a profitable analogy between smectic ordering and superconductivity in which the molecular orientation is analogous to the vector potential. Others have since developed this farther.[19] In this analog the smectic layers are described as a static density wave $\delta\rho \sim |\psi|e^{i\phi}$, where $\vec{\nabla}\phi = q_0\hat{z} + q_0\vec{\nabla}u$, and $d = 2\pi/q_0$ defines the layer spacing. The amplitude $|\psi|$, analogous to the amplitude of the superconducting electron wave function, is the smectic order parameter.

Although there have been theoretical arguments[19] to the effect that the nematic to smectic-A transition should always be weakly first order, the existing experimental evidence is consistent with a second order transition in some materials and we will continue the discussion on that basis.

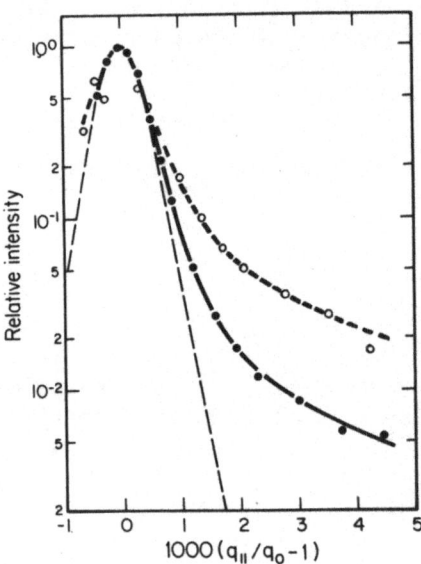

Fig. 1. X-ray structure factor for the SmA density wave in the
         SmA phase of 8OCB. Solid circles $\eta = 0.17$ ($t=9\times10^{-4}$),
         open circles $\eta = 0.38$ ($t=5\times10^{-6}$), dashed line is reso-
         lution.

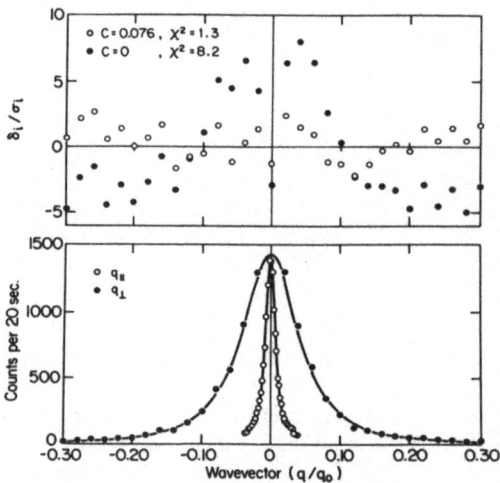

Fig. 2. Lower panel: typical x-ray scattering scan through smectic
         A peak ($0\ 0\ q_0$) in the nematic phase of 8CB.
         Upper panel: errors with and without the $q_\perp^4$ term in the
         cross section.

When the phase transition is approached from the nematic side pretransition effects, analogous to the fluctuation diamagnetism observed in superconductivity cause some of the Oseen-Frank constants K to have a critical behavior $K - K° = K'\xi$ where the critical length ξ diverges as some power $(T - T_c)^{-\nu}$. Since K is a tensor, and since the material is anisotropic there are two different lengths $\xi_{||}$ and $\xi_{\perp}$ and one of the constants is actually expected to diverge as $\xi_{||}^2/\xi_{\perp}$. If we ignore this difference for the moment the renormalization group calculation is isomorphic with that for superfluid He and $\nu \simeq 0.67$. These effects are readily observable in a light scattering experiment, and the diverging correlation lengths can be directly measured by x-ray scattering; we will present results below.

On the smectic side of the transition there are two different critical effects. Firstly the smectic constant B, that describes the elastic resistance to layer compression will go to zero, but in addition the non-hydrodynamic parameter D that couples the layer tilts to the molecular orientation axis will also go to zero. This last parameter is responsible for the fact that away from the critical region the smectic-A has one less hydrodynamic mode than the nematic. These effects can be studied by visible light.

The x-ray scattering intensity can be calculated in terms of the fluctuations in the density wave that constitute the smectic order parameter.

$$\delta\psi^2(\vec{q})> \frac{kT\ \chi_s}{1 + \xi_{||}^2(q_{||}\text{-}q_o)^2 + \xi_{\perp}^2 q_{\perp}^2} \qquad (8)$$

Here χ_s is a generalized susceptibility $(-1/a)$ for the smectic order parameter. Thus by measuring the intensity and width of the x-ray scattering peak one may determine χ_s, $\xi_{||}$ and $\xi_{\perp}$. A typical scan through the peak is shown in Fig. 2. An unexpected result is that the term $\xi_{\perp}^2 q_{\perp}^2$ in (8) must be replaced by $\xi_{\perp}^2 q_{\perp}^2 (1 + c\xi_{\perp}^2 q_{\perp}^2)$ in order to fit the profile of the $q_{\perp}$ scan - that is, the $q_{\perp}$ scans fall off more rapidly than a Lorentzian. The χ^2 for the fit is shown with and without the extra term. The relative importance of this term increases as $t = (T/T_c - 1) \rightarrow 0$ and ξ diverges. Our belief is that this term is a manifestation of divergent fluctuations in the phase of c which prevent the SmA phase from having true long range order; more about that presently. The experimental results of a series of M.I.T. experiments on pre-transitional SmA behavior in the nematic phase[20,21] are summarized below.

Three materials, cyanobenzylidene-octylozyaniline (CBOOA), 80CB, and octylcyanobiphenyl (8CB) were studied. These are all so-called "bilayer smectics", the value of q_o (0.179, 0.197, 0.198 Å^{-1}, respectively) corresponding to slightly less than twice the molecular length, indicating there is probably some antiferroelectric short range order of the molecules. When the x-ray scattering results are analyzed to determine power law singularities for the critical divergences one finds the critical exponents given in the table below ($5 \times 10^{-5} < t < 2 \times 10^{-2}$)

Exponent	CBOOA	8OCB	8CB
γ	1.30 ± 0.06	1.32 ± 0.06	1.26 ± 0.06
$\nu_{\parallel}$	0.70 ± 0.04	0.71 ± 0.04	0.67 ± 0.02
$\nu_{\perp}$	0.62 ± 0.05	0.58 ± 0.04	0.51 ± 0.04

These are the effective exponents obtained if one assumes a single power law divergence. The exponents obtained for γ and $\nu_{\parallel}$ are in satisfactory agreement with theoretical values (d=3, n=2, $\gamma=1.316$, $\nu=0.669$) and support de Gennes' helium analogy. However, the situation is less clear with the results for $\xi_{\perp}$. The greatest difficulty appears to be with 8CB. It should be emphasized that the different exponents correspond only to a rather small evolution in the ratio $\xi_{\parallel}/_{\perp}$ over three decades of t. The results are summarized in the table below.

Material	$\xi_{\parallel}/\xi_{\perp}$ at $t=10^{-2}$	$\xi_{\parallel}/\xi_{\perp}$ at $t=10^{-4}$
CBOOA	5.5 ± 0.5	8 ± 1
8OCB	5 ± 1	9 ± 1
8CB	4 ± 1	9 ± 1

In the worst case, 8CB, the ratio $\xi_{\parallel}/\xi_{\perp}$ changes by a factor 2.6 over $t = 2 \times 10^{-2}$ to $t = 5 \times 10^{-5}$; if the exponents were truly different this ratio must go to ∞ in the remaining 15 mK to T_C. Our experiments are not able to tell us if this is indeed the case.

Light scattering experiments have also been carried out for these three samples at M.I.T. and give results for $\xi_{\parallel}$ which are in good agreement with the x-ray data. Unfortunately, it was not possible to obtain reliable measurements of $\xi_{\perp}$ by light scattering, because of the large non-diverging background.

We summarize our present understanding of the Nematic to Smectic A phase transition. The N phase experiments are consistent with de Gennes' He analogue with an upper marginal dimensionality d* = 4, but we have to understand the evolution in $\xi_{\parallel}/\xi_{\perp}$. It is possible that there are indeed two lengths in the problem, but it may simply be that this evolution represents the influence of anisotropic critical behavior (because of the anisotropy K_i) predicted by Lubensky and Chen[22]. The term $q_{\perp}^4\xi_{\perp}^4$ observed in the x-ray cross section predicted by the SmA phase by Eq. (7a). If so, this represents the only indication in the N phase of the lack of true long range order in the SmA phase.

The SmA x-ray scattering represents the first direct experimental evidence of the algebraic decay of correlation functions predicted at lower marginal dimensionality. The elastic constants B and D are probably sensitive to the lack of true long range order and their anomalous critical behavior is probably associated with the divergence of long wavelength fluctuations in u. Since B is associated with layer compression ($\partial u/\partial z$) while D measures the force keeping the molecules normal to the layers (and is associated with their orientational order, which is truly long range, as well) it is perhaps not surprising the two constants have different critical behavior. These ideas have yet to be tested by quantitative calculations, however.

STRUCTURE OF THE SMECTIC B PHASE

The better ordered smectics (B,D,E,H etc.) have some type of translational order within the layers. Birgeneau and Litster[23] have proposed a model for these phases which is based on calculations for two-dimensional ordering carried out by Halperin and Nelson.[24] When the anisotropy of the two dimensional crystal lattice was explicitly considered, Halperin and Nelson found three phases could exist. To discuss these quantitatively one must define both translational and orientational order parameters, the orientational one refers to the orientation of nearest neighbor bonds between molecules (specified by the angle $\theta(\vec{r})$ with respect to some fixed axis). If $\vec{G}$ is a reciprocal lattice vector and $\vec{u}(\vec{r})$ the displacement of an atom from its lattice site then the positional order parameter is defined in the usual way

$$P(\vec{G},\vec{r}) = <e^{i\vec{G}[\vec{u}(\vec{r})-\vec{u}(0)]}>$$

For a hexagonal lattice one defines an orientational order parameter

$$O(\vec{r}) = <e^{i6[\theta(\vec{r})-\theta(0)]}>$$

Halperin and Nelson found a low temperature phase (i) in which $O(\vec{r})$ has true long range order while $P \sim r^{-\eta(G)}$ has the algebraic decay of the Kosterlitz-Thouless[15] topological order. Then, on warming there is an intermediate phase (ii) with short range positional order $P \sim e^{-r/\xi}$ and algebraic decay of $O \sim r^{-\eta(6)}$. Finally the system becomes a 2D liquid with short range order of both O and P. Birgeneau and Litster proposed to explain the various smectic phases by stacking up layers of these two dimensional phases. To understand the result, we need to recall that whenever a correlation function decays algebraically, the associated susceptibility is infinite. Stacking phase (i) with even an infinitesimal interaction between layers would therefore result in a 3D solid with true long range positional order. However stacking phase (ii) with a sufficiently weak interaction between the layers would result in a system with true long range order for O and short range order for P. This phase, depending upon the basic 2D crystal structure, tilt of the molecules with respect to the layers, and so on, is proposed to account for all of the well ordered smectic phases (SmB, SmH, etc.). Within this picture a quite natural transition to SmA or SmC phases occurs when O also becomes short range; the SmA and SmC phases are thus regarded as stacked two dimensional liquids. This is an attractive model which was consistent with the known experimental data when proposed. It explained the absence of higher order Bragg peaks and the rather diffuse nature of those peaks observed in SmB phases, while at the same time the bond orientational long range order explained how the rotational symmetry of the hexagonal lattice could be seen.

To test these ideas, we are now carrying out high resolution x-ray scattering studies of butoxybenzylidene-octylanaline (BBOA). The data are in process of being analyzed, and we cannot make quantitative statements until the data have been deconvoluted and the molecular form factor corrected for. Qualitatively we find resolution limited Bragg peaks for the SmA density wave (001) and (002), the latter ~10^{-3} times weaker. In the layer plane we also find resolution limited peaks, but the Debye-Waller factor allows us to observe only (100) and a barely detectable peak at (110). These peaks appear as pips on top of a relatively intense and fairly narrow diffuse background. We also observe cross peaks (10 1/2), (101), (10 3/2), (102), and (10 5/2) consistent with a three dimensional hexagonal close packed structure.

On the other hand we have not yet analyzed the data sufficiently to either confirm or deny that there are true Bragg peaks. For example, if the x-ray structure consisted of some type of algebraic singularity convolution with a resolution function of finite width would also yield what would appear to be a resolution limited peak sitting on top of a broad background. In either case, however, the order must extend over many thousands of Angstroms in all three dimensions. A final answer to the nature of the smectic B phase must wait further analysis.

This work was supported in part by the National Science Foundation under grants DMR-7680895 and DMR-7823555 and in part by the Joint Services Electronics Program under contract No. DAAG-29-78-C-0020.

REFERENCES

1. P. G. de Gennes, The Physics of Liquid Crystals, Oxford University Press, 1974.
2. S. Chandrasekhar, Liquid Crystals, Cambridge University Press, 1977.
3. P. C. Martin, O. Parodi and P. S. Pershan, Phys. Rev. A6, 2401 (1972)
4. C. W. Oseen, Trans. Faraday Soc. 29, 883 (1933).
5. F. C. Frank, Disc. Faraday Soc. 25, 19 (1958).
6. N. A. Clark and P. S. Pershan, Phys. Rev. 30, 3 (1973).
7. R. Ribotta, G. Durand, and J. D. Litster, Sol. State Comm. 12, 27 (1973).
8. L. Powers and N. A. Clark, Proc. Nat. Acad. Sci. (USA) 72, 840 (1975).
9. W. H. Bragg, Nature 133, 445 (1934).
10. G. Friedel, Ann. Phys. (Paris) 19, 273 (1922).
11. C. S. Rosenblatt, R. Pindak, N. A. Clark and R. B. Meyer, J. de Physique 38, 1105 (1977).
12. S. A. Asher and P. S. Pershan, J. de Physique 40, 11 (1979).
13. York Liao, N. A. Clark and P. S. Pershan, Phys. Rev. Letters 30, 639 (1973).
14. J. P. LePesant, L. Powers, and P. S. Pershan, Proc. Nat. Acad. Sci. (USA) 75, 1792 (1978).
15. J. M. Kosterlitz and D. J. Thouless, J. Phys. C6, 118 (1973); F. J. Wegner, Z. Phyzik 206, 465 (1967); B. Jancovici, Phys. Rev. Letters 19, 20 (1967).
16. A. Caillé, Comples rendus Ac. Sc. Paris 274B, 891 (1972).
17. J. Als-Nielsen, R. J. Birgeneau, M. Kaplan, J. D. Litster, and C. R. Safinya, Phys. Rev. Letters 39, 1668 (1977).
18. W. McMillan, Phys. Rev. A4, 1238 (1971).
19. B. I. Halperin and T. C. Lubensky, Sol. St. Comm. 10, 753 (1972).
20. J. D. Litster, J. Als-Nielsen, R. J. Birgeneau, S. S. Dana, D. Davidov, F. Garcia-Golding, M. Kaplan, C. R. Safinya, and R. Schaetzing, J. de Physique 40, C3-339 (1979).
21. D. Davidov, C. R. Safinya, M. Kaplan, S. S. Dana, R. Schaetzing, R. J. Birgeneau, and J. D. Litster, Phys. Rev. B19, 1657 (1979)
22. T. C. Lubensky and Jing-Huei Chen, Phys. Rev. B17, 336 (1978); Jing-Huei Chen, T. C. Lubensky, and D. R. Nelson, Phys. Rev. B, in press.
23. R. J. Birgeneau and J. D. Litster, J. de Physique Lettres 39, L-399, (1978).
24. B. I. Halperin and D. R. Nelson, Phys. Rev. Letters 41, 121, 519 (1978).

OPTICS AND ELECTRO-OPTICS OF CHIRAL SMECTICS

V.A. Belyakov and V.E. Dmitrienko

All Union Research Institute for Physical-Technical
and Radiotechnical Measurements
Moscow, USSR

Optical methods are traditionally and fruitfully used in the investigation of liquid crystals[1-3]. They permit one to elaborate the structure of liquid crystals and their changes under external agents. Naturally these methods ought to be used also for the investigation of new kinds of liquid crystals with poorly known properties such as chiral smectics. These latter may exhibit ferroelectric properties[1,2,4] and appear very promising in applications. There are some papers on the optics of chiral smectics[3,5-11], but nevertheless further experimental and theoretical investigation is required.

In the present paper the optics and electro-optics of chiral smectics are developed in the framework of kinematical diffraction theory. In this approach-as is well known- the optical properties of chiral smectics may be readily connected with their dielectric tensor ε. The corresponding connections are for smectics undistorted as well as distorted by the applied electric field. For the latter case the variations of ε caused by the electric field are also found. The differences in the optic and electro-optic properties of chiral smectics and cholesterics are discussed.

I. Variation of the chiral smectic dielectric tensor with the applied field

The helical pitch of chiral smectics is comparable to the

wavelength of visible light, and light diffraction by the helix is possible. The unusual optical properties of chiral smectics are the consequence of this diffraction as in the case of cholesterics[3].

The dielectric tensor of undistorted chiral smectics has the form

$$
\hat{\varepsilon} = \begin{pmatrix} \varepsilon_{11}+\varepsilon_a\cos2\phi & \varepsilon_a\sin2\phi & \varepsilon_a'\cos\phi \\ \varepsilon_a\sin2\phi & \varepsilon_{11}-\varepsilon_a\cos2\phi & \varepsilon_a'\sin\phi \\ \varepsilon_a'\cos\phi & \varepsilon_a'\sin\phi & \varepsilon_{33} \end{pmatrix} \tag{1}
$$

where

$$
\varepsilon_{11} = (\varepsilon_1+\varepsilon_2\cos^2\theta+\varepsilon_3\sin^2\theta)/2; \quad \varepsilon_{33} = \varepsilon_2\sin^2\theta+\varepsilon_3\cos^2\theta;
$$

$$
\varepsilon_a = (\varepsilon_1-\varepsilon_2\cos^2\theta-\varepsilon_3\sin^2\theta)/2; \quad \varepsilon_a' = (\varepsilon_3-\varepsilon_2)\sin\theta\cos\theta \tag{2}
$$

θ and ϕ are the local polar and azimuthal angles of one of the principal axes (axis 3) of the dielectric tensor relative to a fixed coordinate system. The x and y axes of this system lie in the plane of the smectic layer and the z axis is perpendicular to it. The orientation of the principal axes of the dielectric tensor is determined by the orientation of the long molecular axis in the layer. The axis I lies in the layer plane, ε_1, ε_2 and ε_3 are the principal values of ε. The angle θ is constant throughout the volume of smectics, the azimuthal angle ϕ changes from layer to layer. Its dependence on z is

$$
\phi = 2\pi z/p_0 \tag{3}
$$

where p_0 is the helix pitch. The tensor $\hat{\varepsilon}$ is taken here as a periodic function of z, and may be represented by the following Fourier expansion:

$$
\hat{\varepsilon}(z) = \sum_{n=-\infty}^{\infty} \hat{\varepsilon}_n \exp[in\tau z] \tag{4a}
$$

where $\tau = 2\pi/p_0$. From eqs. (1) and (3) it follows that for undistorted chiral smectics only $\varepsilon_0, \varepsilon_{\pm1}$ and $\varepsilon_{\pm2}$ differ from zero[3]

Let us find the variations of ε caused by an electric field applied to the chiral smectic crystal. For the electric field orthogonal to the helix axis the corresponding variations may be found analogously to the case of cholesterics[1,2]. It will be shown that these variations are very different for smectics with

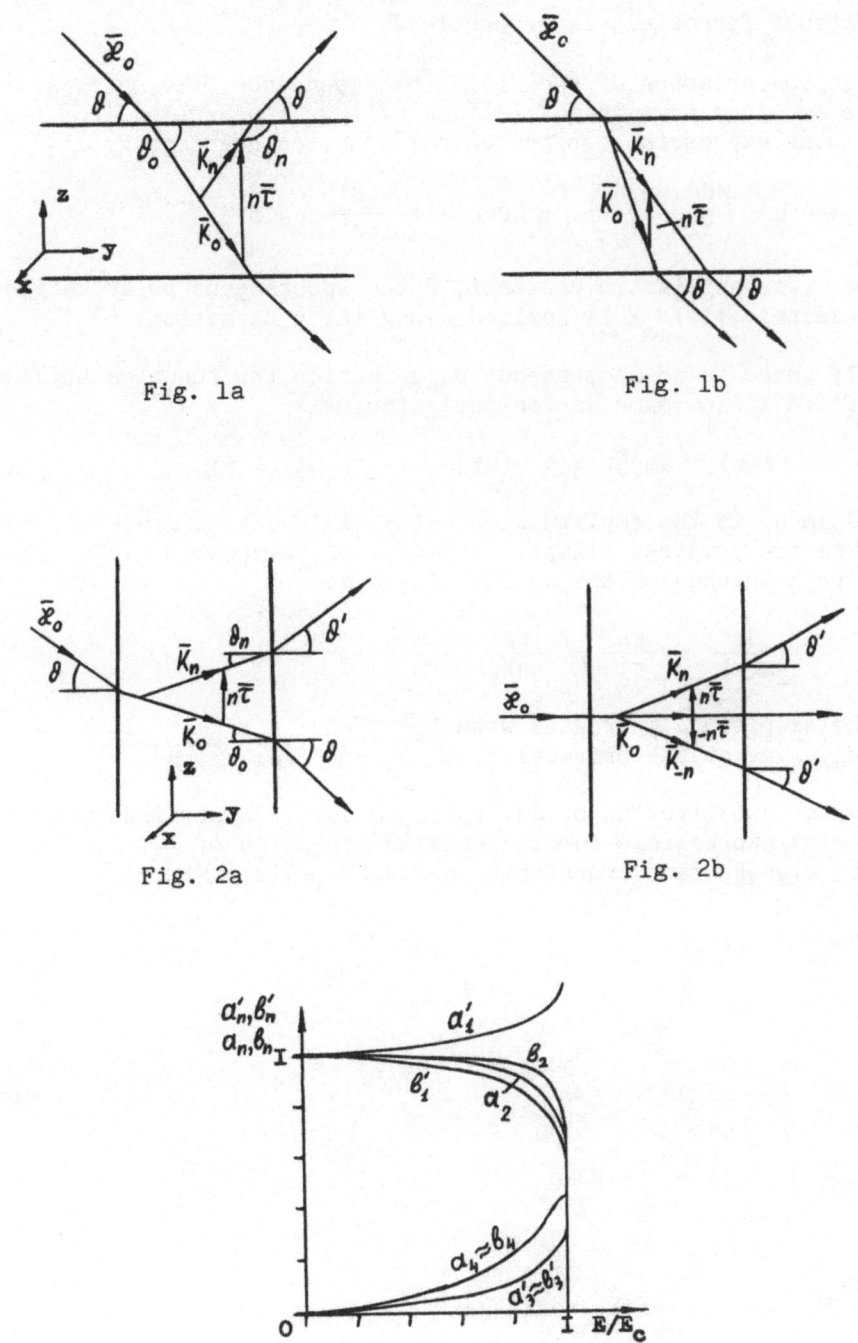

Fig. 1a

Fig. 1b

Fig. 2a

Fig. 2b

Fig. 3. The field dependence of $\hat{\varepsilon}_n$ (see Refs 7 and 8 for nonferroelectric smectic).

or without ferroelectric properties.

In the presence of the field the dependence of ϕ on z is different from that in eq. (3) and will be found below from the following expression for the chiral smectics free energy[1,2,4]:

$$F = \frac{B_3}{2} \left(\frac{d\phi}{dz} - \frac{2\pi}{p_o} \right)^2 + EP\cos\phi + \frac{\varepsilon_a E^2}{16\pi} \cos 2\phi \qquad (4b)$$

where B_3 is an elastic constant, P the spontaneous polarization; the electric field E is applied along the y direction.

If there is no spontaneous polarization the function $\phi(z)$ in the field is the same as for cholesterics[1,2]

$$\phi(z) = am \left[\pi^2 z / p_o E(k) \right] \qquad (5)$$

where am(u) is the amplitude of u $(1_2, 13)$, $u = \pi^2 z/p_o E(k)$, and E(k) is the complete elliptic integral of the second kind. The modulus k depends on the applied field as:

$$\frac{k}{E(k)} = \frac{Ep_o}{\pi^2} \sqrt{\frac{|\varepsilon_a|}{4\pi B_3}} = \frac{E}{E_c} \qquad (6)$$

The helix pitch p increases with the field: $p = 4p_o E(k)K(k)/\pi^2$ and diverges at the critical field $E_c = (\pi^2/p_o)\sqrt{4\pi B_3/|\varepsilon_a|}$

After substitution of eq. (5) into eq. (1) one finds the following expressions for the Fourier expansion of ε:
   a) even Fourier transforms, n=2,4,6,... (see Fig. 3)

$$\hat{\varepsilon}_o = \begin{pmatrix} \varepsilon_{11} + \varepsilon_a \beta_o & 0 & 0 \\ 0 & \varepsilon_{11} - \varepsilon_a \beta_o & 0 \\ 0 & 0 & \varepsilon_{33} \end{pmatrix} ;$$

$$\hat{\varepsilon}_n = \frac{\varepsilon_a}{2} \begin{pmatrix} \beta_n & -ia_n & 0 \\ -ia_n & -\beta_n & 0 \\ 0 & 0 & 0 \end{pmatrix} ; \hat{\varepsilon}_{-n} = \hat{\varepsilon}_n^* \qquad (7)$$

where $\beta_o = 1 - 2 \left[K(k) - E(k) \right]/k^2 K(k)$;

$$\beta_n = \frac{2\pi^2 n \sqrt{q^n}}{k^2 K^2(k)(1-q^n)} ; a_n = \beta_n \frac{1-q^n}{1+q^n}$$

b) odd Fourier transforms, n=1,3,5,... (see Fig. 3)

$$\hat{\varepsilon}_n = \frac{\varepsilon_a'}{2} \begin{pmatrix} 0 & 0 & \beta_n' \\ 0 & 0 & -ia_n' \\ \beta_n' & -ia_n' & 0 \end{pmatrix} \tag{8}$$

where

$$\beta_n' = \frac{2\pi\sqrt{q^n}}{kK(k)(1+q^n)}; \qquad a_n' = \beta_n' \frac{1+q^n}{1-q^n}$$

In eqs. (7), (8) $K(k)$ is the complete elliptic integral of the first kind, $q = \exp[-\pi K \sqrt{1-k^2}/K(k)]$. It follows from eqs. (7), (8) that in the weak field $\hat{\varepsilon}_n$ rapidly decreases with increasing n: $\hat{\varepsilon}_n \sim \varepsilon_a(E/E_c)^{n-2}$ for n=2,4,...and $\hat{\varepsilon}_n \sim \varepsilon_a(E/E_c)^{n-1}$ for n=1,3,... ($E \ll E_c$).

Let us examine now smectics in the presence of spontaneous polarization. Symmetry considerations show that the spontaneous polarization vector $\vec{P}$ lies in the smectic layer and is orthogonal to the long molecular axis[4]. Typically the mechanism of molecular orientation in the field due to the spontaneous polarization is dominant[1,2,4] and it is possible to neglect the orientation mechanism previously discussed, which is due to the dielectric anisotropy ε_a. If the dielectric anisotropy is neglected it follows from eq. (4) that the distortion of the helical structure may be described by eqs. (5), (6), with the following modifications. The quantity $\varepsilon_a E^2/16\pi$ must be replaced by PE, ϕ by $\phi/2$ and z by z/2. The resulting expression for $\phi(z)$ is

$$\phi(z) = 2\text{am}[\pi^2 z/2p_o E(k)] \tag{9}$$

where the modulus k is determined from the following equation

$$\frac{k}{E(k)} = \frac{2p_o}{\pi^2} \sqrt{\frac{PE}{\beta_3}} \sqrt{\frac{E}{E_c^*}} \tag{10}$$

with a new critical field $E_c^* = \pi^4\beta_3/4p_o P$

Substitution of eq. (9) into eq. (1) leads to the following expressions for the trigonometrical functions in (1)

$$\sin \phi = 2\text{sn}(\tfrac{u}{2})\text{cn}(\tfrac{u}{2}); \quad \sin 2\phi = 4\text{sn}(\tfrac{u}{2})\text{cn}(\tfrac{u}{2})[\text{cn}^2(\tfrac{u}{2})-\text{sn}^2(\tfrac{u}{2})];$$

$$\cos \phi = \text{cn}^2(\tfrac{u}{2})-\text{sn}^2(\tfrac{u}{2}); \quad \cos 2\phi = 1-8\text{sn}^2(\tfrac{u}{2})\text{cn}^2(\tfrac{u}{2}),$$

where sn(u) and cn(u) are the Jacobian elliptic functions[12] The Fourier transforms of $\hat{\varepsilon}$ now have the following form (see Fig. 4):

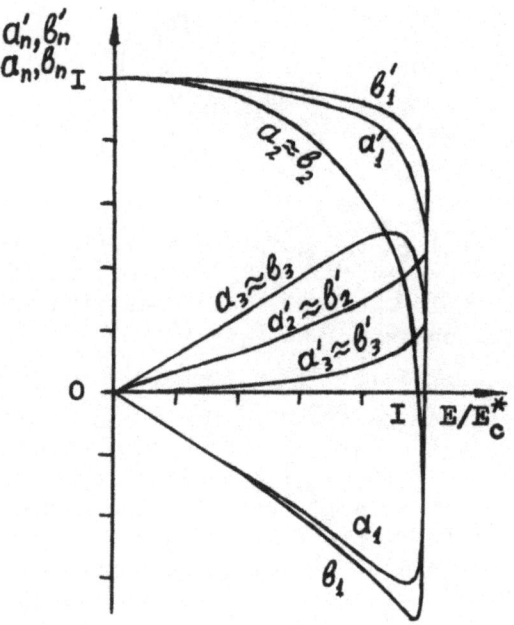

Fig. 4. Same as Fig. 3 for ferroelectric smectics (see Refs 11 and 12).

$$\hat{\varepsilon}_o = \begin{pmatrix} \varepsilon_{11} + \varepsilon_a\beta_o & 0 & \varepsilon_a'\beta_o' \\ 0 & \varepsilon_{11} - \varepsilon_a\beta_o & 0 \\ \varepsilon_a'\beta_o' & 0 & \varepsilon_{33} \end{pmatrix};$$

(11)

$$\hat{\varepsilon}_n = \tfrac{1}{2} \begin{pmatrix} \varepsilon_a\beta_n & -i\varepsilon_a a_n & \varepsilon_a'\beta_n' \\ -i\varepsilon_a a_n & -\varepsilon_a b_n & -i\varepsilon_a'a_n' \\ \varepsilon_a'\beta_n' & -i\varepsilon_a'a_n' & 0 \end{pmatrix}; \hat{\varepsilon}_{-n} = \hat{\varepsilon}_n^*$$

where n=1,2,3,4,... and

$$\beta_{\circ} = -\frac{1}{3} [\, 1 + 4\beta_{\circ}'\, (2 - k^2)/k^2\,]$$

$$\beta_n = \frac{2\beta_n'}{3k^2} [\, \frac{n\pi}{K(k)}\,]^2 - 2(2-k^2)\; ; \quad a_n = \beta_n\, \frac{1-q^{2n}}{1+q^{2n}}$$

$$(12)$$

$$\beta_{\circ}' = 1 - 2\, [\, K(k) - E(k)\,]\, /k^2 K(k);$$

$$\beta_n' = \frac{4\pi^2 n q^n}{k^2 K^2(k)\, (1-q^{2n})}\; ; \quad a_n' = \beta_n'\, \frac{1-q^{2n}}{1+q^{2n}}$$

Note that in weak fields $\varepsilon_n \sim \varepsilon_A (E/E_C)^{n-2}$ (for $n > 2$). For the purpose of application it may be essential that the distortion of the helix by an electric field is greater for smectics with the spontaneous polarization because $E_C^* \ll E_C$.

In conclusion of this section note that eqs. (7), (8), (11) are obtained in assumption that the tilt angle θ does not change in the applied field, the smectic layers do not bend and the only possible deformation is a change of the aximuthal angle ϕ.

2. Light diffraction by chiral smectics

The qualitative structural differences of chiral smectics in comparison with cholesterics reveal themselves in the form of the dielectric tensor $\hat{\varepsilon}(z)$ (the dielectric tensor $\hat{\varepsilon}(z)$ for cholestetics can be obtained from eq. (1) by setting $\theta = \pi/2$, $\varepsilon_1 = \varepsilon_2$). In particular, the period of $\hat{\varepsilon}(z)$ for chiral smectics is equal to the pitch p rather than p/2 for cholesterics. The latter leads to additional diffraction reflections of light in chiral which is not present in cholesterics[3,5].

The theory of the optical properties of chiral smectics has been discussed in several papers[3,5,6,9,11]. Numerical calculations of the light diffraction by the smectic helix was carried out in[5]. The papers[3,11] were devoted mainly to the first order reflections.

Here the main attention is paid to the comparison of the optical properties of distorted and undistorted chiral smectics. The kinematical diffraction theory gives the simplest way to make

such a comparison and a clear picture of light diffraction in
chiral smectics results.

Let us examine the light diffraction by a chiral smectic
sample. The scattering cross section in the kinematical
approximation has the well known form[1,3]:

$$\frac{d\sigma}{d\Omega_{\vec{K}_N}} = \left(\frac{\omega^2}{4\pi c^2}\right)^2 |\vec{e}_n^* \hat{\varepsilon}_n \vec{e}_o|^2 |\int \exp[i(\vec{K}_o + n\vec{\tau} - \vec{K}_n)\vec{r}]d\vec{r}|^2 \qquad (13)$$

where $\vec{K}_o$, $\vec{e}_o$; $\vec{K}_n$, $\vec{e}_n$ are the wave and polarization vectors of
the incident and scattered waves respectively, ω is the frequency
of light and the integration runs over the sample.

The cross section (13) as a function of the incidence angle θ
(or frequency ω) achieves a sharp maximum if the Bragg condition

$$\vec{K}_n = \vec{K}_o + n\vec{\tau} \qquad |\vec{K}_n| \simeq |\vec{K}_o| \qquad (14)$$

is obeyed. Due to strong birefringence of smectics the values and
directions of $\vec{K}_o$, $\vec{K}_n$ in eq. (14) are different for different
polarizations (eigen polarizations) of the light. Moreover, due
to the complex form of $\hat{\varepsilon}_n$ polarization mixing takes place in the
diffraction scattering[3,5,11].

If the usual boundary conditions are taken into account and
the Fourier expansion for $\hat{\varepsilon}$ used, the following expressions for
the relation of the intensities of the diffracted and incident
waves I_d and I_o, respectively, are found from eq. (13)

$$\frac{I_d}{I_o} = |\vec{e}_n^* \hat{\varepsilon}_n \vec{e}_o|^2 \frac{\omega^4}{c^4} \frac{\sin^2[(K_{oz} + n\tau - K_{nz})L/2]}{K_{nz}^2 (K_{oz} + n\tau - K_{nz})^2} \qquad (14a)$$

$$\frac{I_d}{I_o} = |\vec{e}_n^* \hat{\varepsilon}_n \vec{e}_o|^2 \frac{\omega^4}{c^4} \frac{\sin^2[(K_{oy} - K_{ny})L/2]}{K_{ny}^2 (K_{oy} - K_{ny})^2} \qquad (14b)$$

$$K_{ny} = \sqrt{K_n^2 - K_{ox}^2 - (K_{oz} + n\tau)^2}$$

where eqs. (14a) and (14b) relate to the cases of the helix axis
normal (fig. 1) and parallel to the sample surface, respectively;
the polarization vectors $\vec{e}_o$ and $\vec{e}_n$ correspond to the eigen
polarizations of the waves with wave vectors $\vec{K}_o$ and $\vec{K}_n$, L is the
sample thickness.

The connection between the angles ϕ and ϕ' (fig. 2) is given by
the known relation (n=0, $\pm$ 1, $\pm$2,...)

$$\sin \theta' = \sin \theta + n\tau c/\omega \tag{15}$$

In conclusion of this section we note that the validity of these formulae is restricted by the usual kinematical condition

$$(I_d/I_o) \ll 1 \quad \text{i.e.} \quad |\vec{e}_n^* \hat{\varepsilon}_n \vec{e}_o|^2 \; \omega^2 L^2/c^2 \ll 1 \tag{16}$$

The condition (16) is fulfilled for many situations.

3. Optics of undistorted chiral smectics

In an undistorted chiral smectic the refractive indexes for ordinary ($\vec{\sigma}$ -polarization) and extraordinary ($\vec{\pi}$ -polarization) beams may be found from well known equations. To examine the polarization dependences of the scattered beam one needs to assume e_o and e_n to be the eigen polarizations ($\vec{\sigma}$ or $\vec{\pi}$) in the factor $|\vec{e}_n^* \hat{\varepsilon}_n \vec{e}_o|$. The corresponding expressions for this factor are then

$$(\vec{\sigma}\hat{\varepsilon}_n \vec{\sigma}) = (\hat{\varepsilon}_n)_{xx} = \varepsilon_a \beta_n/2$$

$$(\vec{\pi}_n \hat{\varepsilon}_n \vec{\sigma}) = \sin\theta_n (\hat{\varepsilon}_n)_{xy} - \cos\theta_n (\hat{\varepsilon}_n)_{zx} = (-i\varepsilon_a a_n \sin\theta_n - \varepsilon'_a \beta'_n \cos\theta_n)/2$$

$$(\vec{\sigma}\hat{\varepsilon}_n \vec{\pi}_o) = \sin\theta_o (\hat{\varepsilon}_n)_{xy} - \cos\theta_o (\hat{\varepsilon}_n)_{xz} = (-i\varepsilon_a a_n \sin\theta_o - \varepsilon_a \beta_n \cos\theta_o)/2 \tag{17}$$

$$(\vec{\pi}_n \hat{\varepsilon}_n \vec{\pi}_o) = \sin\theta_o \sin\theta_n (\hat{\varepsilon}_n)_{yy} + \cos\theta_o \cos\theta_n (\hat{\varepsilon}_n)_{zz} - \sin(\theta_o + \theta_n)(\hat{\varepsilon}_n)_{yz} =$$

$$[-\varepsilon_a \beta_n \sin\theta_o \sin\theta_n + i\varepsilon'_a a'_n \sin(\theta_o + \theta_n)]/2$$

(for θ_0 and θ_n see figs. 1 and 2). From eq. (17) it follows that in undistorted chiral smectics only the first and the second order reflections exist ($\hat{\varepsilon}_n = 0$ if $n > 2$). The first order $\vec{\sigma}$ -polarized light is scattered to $\vec{\pi}$ -polarized and vice versa because the only nonzero factors are $(\vec{\pi}_1 \hat{\varepsilon}_1 \sigma) = \cos\theta_1 \, \varepsilon'_a/2$ and $(\vec{\sigma}\hat{\varepsilon}_1 \vec{\pi}_o) = -\cos\theta_o \, \varepsilon'_a/2$. For light propagating along the helix the first order diffraction is absent ($\cos \theta_o = \cos \theta_1 = 0$).

Polarization dependences for the second order diffraction scattering are described by the following expressions

$$(\vec{\sigma}\hat{\varepsilon}_2 \sigma) = (\hat{\varepsilon}_2)_{xx}; \quad (\vec{\pi}_2 \hat{\varepsilon}_2 \vec{\sigma}) = \sin\theta_2 (\hat{\varepsilon}_2)_{xy}; \quad (\vec{\sigma}\hat{\varepsilon}_2 \vec{\pi}_o) = -\sin\theta_o (\hat{\varepsilon}_2)_{xy}$$

$$(\vec{\pi}_2 \hat{\varepsilon}_2 \vec{\pi}_o) = \sin\theta_o \sin\theta_2 (\hat{\varepsilon}_2)_{yy}; \quad (\hat{\varepsilon}_2)_{xx} = i(\hat{\varepsilon}_2)_{xy} = (\hat{\varepsilon}_2)_{yy} = \varepsilon_a/2 \tag{18}$$

The second order reflection is analogous to the diffraction in

cholesterics. In particular, for light propagation along the
helix only one circular polarization undergoes diffraction, right
for a right helix and left for a left one.

As follows from eqs. (17), (18) there are two refection orders
with different polarization properties. For small tilt angle $\theta \, \hat{\varepsilon}_1 \sim \theta$
and $\hat{\varepsilon}_2 \sim \theta^2$ i.e. the first order reflection is stronger than the
second one. Note that the first order reflection in chiral
smectics has no analog in cholesterics because it arises from
periodicity equal to p/2, which is absent in cholesterics.

4. Chiral smectics in an external field

As one can see from eqs. (7), (11) for $\hat{\varepsilon}_0$ in an external field
chiral smectics became biaxial. However, the higher order
reflections which arise due to the higher order terms in the
Fourier expansion of $\hat{\varepsilon}$ (see (7), (8), (11)) are more essential for
light diffraction.

Examine first the smectics without spontaneous polarization.
If the direction of light propagation lies in the plane made by
the helix axis and field direction, the $\vec{\sigma}$ and $\vec{\pi}$ polarizations are
eigenpolarizations as in the absence of the field. The
polarization properties of all odd reflections are then determined
by the following factors (see eq. (14)

$$(\vec{\pi}_n \hat{\varepsilon}_n \vec{\sigma}) = -\cos\theta_n \ (\hat{\varepsilon}_n)_{xz} = -\cos\theta_n \varepsilon_a' \beta_n'/2 ; (\vec{\pi}_n \hat{\varepsilon}_n \vec{\pi}_o) = (\vec{\sigma} \hat{\varepsilon}_n \vec{\sigma}) = 0 \qquad (19)$$

$$(\vec{\sigma} \hat{\varepsilon}_n \vec{\pi}_o) = -\cos\theta_o \ (\hat{\varepsilon}_n)_{xz} = -\cos\theta_o \varepsilon_a' \beta_n'/2 ,$$

where n=$\pm$1, $\pm$3,... and $\hat{\varepsilon}_n$ is given by eq. (8). The polarization
properties of all odd reflections are analogous to the first order
reflection in undistorted smectics. In particular all odd
reflections vanish for the light propagation along the helix axis.

For the even order reflections eqs. (18) are valid if one
replaces $\hat{\varepsilon}_2$ by $\hat{\varepsilon}_n$ from eq. (7). These reflections are analogous
to higher order reflections in cholesterics distorted by the
external field [3]   and do not vanish for light propagation along
the helix axis. If the field E is small enough, $\hat{\varepsilon}_{2m} \sim \varepsilon_a (E/E_C)^{2m-2}$
and $\hat{\varepsilon}_{2m+1} \sim \varepsilon_a (E/E_C)^{2m}$ i.e. the reflection intensity falls with
increasing m.

The specific feature of ferroelectric smectics is the
existence of both even and odd order reflections for the light
propagation along the helix distorted by the field. The
polarization dependences are described in this case by the
following expressions:

$$(\vec{\sigma}\hat{\varepsilon}_n\vec{\sigma}) = \varepsilon_a\beta_n/2; \quad (\vec{\pi}_n\hat{\varepsilon}_n\vec{\pi}_o) = -\varepsilon_a\beta_n/2$$

$$(\vec{\sigma}\hat{\varepsilon}_n\vec{\pi}_o) = (\vec{\pi}_n\hat{\varepsilon}_n\vec{\sigma}) = -i\varepsilon_a a_n/2$$

(20)

For a small tilt angle θ the expressions for $\hat{\varepsilon}_n$ become simpler if one neglects the terms proportional to θ^2:

$$\hat{\varepsilon}_n = \frac{(\varepsilon_3-\varepsilon_2)\,\theta}{2}\begin{pmatrix} 0 & 0 & \beta_n' \\ 0 & 0 & -ia_n' \\ \beta_n' & -ia_n' & 0 \end{pmatrix}$$

(21)

In this case the polarization properties of all orders of reflection are the same as in the first order without field.

5. Conclusion

The present investigation shows that optical and electro-optical properties of chiral smectics are rather informative and complex. With properties similar to those of cholesterics[1,3] chiral smectics exhibit some qualitative differences. The change of the optical properties in an electric field is different for ferroelectric and nonferroelectric smectics and is connected with the different structural changes in these cases.

The higher order reflections are the most specific manifestation of the field influence, especially for light propagation along the helix axis. In the last case all orders of reflections exist for ferroelectric smectics but only the even ones for nonferroelectric smectics. The even and odd reflections have, in general, different polarization properties that enable their experimental differentiation.

The optical measurements of chiral smectics may be used for investigations of their structural and ferroelectric properties. It is worthwhile to note that the above examined dielectric and ferroelectric regimes of helix deformation may be achieved in the sample if one used a.c. fields of different frequences. For low frequency the spontaneous polarization plays the main role, for high enough frequency the distortion of the helix is determined by the dielectric anisotropy only[1,2].

Note that results obtained for the distortion of chiral smectics by the electric field may be also used for the distortion caused by other factors, for example by shear (8) or by boundary influences. If structural changes are known the light diffraction in smectics may be calculated in the kinematical approximation with the help of the formulae given.

References

1. de Gennes, P.G. The Physics of Liquid Crystals, Clarendon Press, Oxford, 1974.
2. Blinov, L.M. Electro-and Magnetooptics of Liquid Crystals, Nauka, Moscow, 1978.
3. Belyakov, V.A., Dmitrienko, V.E., Orlov V.P., Usp. Phys. Nauk, $\underline{127}$, 221, 1979.
4. Pikin, S.A., Indenbom, V.L.,Usp. Fiz. Nauk, $\underline{125}$, 251, 1978.
5. Berreman, D.W., Mol. Cryst. and Liquid Cryst., $\underline{22}$, 175, 1973.
6. Parodi, O., J. de Phys. (Colloq), $\underline{36}$, CI-325, 1975.
7. Brunet, M., J. de Phys. (Colloq), $\underline{36}$, CI-321, 1975.
8. Pieranski, P., Guyon, E., Keller, P., Liebert, L., Kuczynski, W., Mol. Cryst. and Liquid Cryst., $\underline{38}$, 275, 1977.
9. Taupin, D., Guyon, E., Pieranski, P. J. de Phys., $\underline{39}$, 406, 1978.
10. Suresh, K.A., Chandrasekhar, S., Mol. Cryst. and Liquid Cryst., $\underline{40}$, 133, 1977.
11. Garoff, S., Meyer, R.B., Barakat, R., J. Opt. Soc. Am., $\underline{68}$, 1217, 1978.
12. Whittaker, E.T., Watson, G.N., A Course of Modern Analysis, University Press, Cambridge, 1927.

THE SIZE, SHAPE AND POLYDISPERSITY OF MICELLES OF AMPHIPHILLIC MOLECULES

G. Benedek, N. Mazer, P. Missel, C. Young and M.C. Carey

Department of Physics and Center for Materials Science
and Engineering, Massachusetts Institute of Technology
Cambridge, Massachusetts 02139

Using the method of Light Scattering Spectroscopy, we have measured the size, shape and polydispersity of sodium dodecyl sulfate micelles in the regime of high salt and high detergent concentrations. From our measurements, we have deduced the chemical potentials associated with the interaction between detergent monomers in cylindrical and spherical portions of the micelle, and also the dependence of these chemical potentials on salt concentration.

RESONANCE RAMAN STUDIES OF VISUAL PIGMENTS

Robert Callender

Physics Department
City College of City University of New York
New York, N.Y. 10031

INTRODUCTION

Visual pigments are composed of a small chromophore (absorption center) called retinal (the aldehyde of vitamin A) covalently linked to surrounding protein, called opsin. The pigments are situated in specialized membranes. The best studied, rhodopsin, is found in velebrate rods, the cells of the retina responsible for low light level vision as opposed to color vision. The purple membrane protein of a baterial cell called halobacterium halobium also contains retinal as its chromophore surrounded by a colorless protein. The absorption of light by visual pigments causes, eventually, a neural response giving rise to vision. Light absorption by the purple membrane results in protons being pumped across the cell wall of the bacterium; the energy of this electral gradient is then used to produce available chemical energy for the cell in terms of high energy chemical bonds (formation of ATP).

After light absorption by the chromophores of the two pigments, a successive set of spectrally distinct, temperature-dependent states are produced. The only action of light absorption by rhodopsin is to form a spectrally red shifted pigment, called bathorhodopsin, in less than 6 picoseconds. Rhodopsin has a lower free energy than bathorhodoosin. Photon energy is converted to chemical energy in the rhodopsin to bathorhodopsin transition; this chemical energy is then available to drive subsequent chemical reactions finally producing neural excitation. Another pigment, isorhodopsin, is not found in nature but also forms bathorhodopsin upon light absorption. Its chromophore is composed of a different geometric form of retinal, i.e. 9-cis retinal, than that of rhodopsin, i.e. 11-cis retinal (see Figure 1). The purple membrane

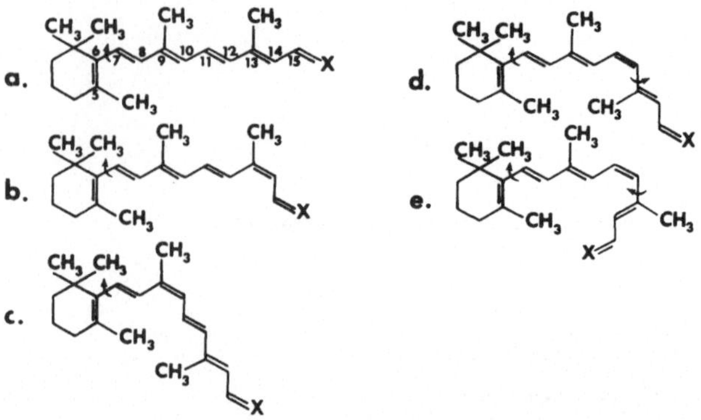

Figure 1. Conformations of various isomers of retinal (X=O), its Schiff base (X=N), and its protonated Schiff base (X=NH$^+$). (a) all-trans; (b) 13-cis; (c) 9-cis; (d) 11-cis, 12 s-trans; (e) 11-cis; 12-s-cis. Arrows indicate flexible bonds whose equilibrium configuration is not planar.

pigment of halobacterium halobium is often called bacteriorhodopsin since its chromophore is the same as rhodopsin and since its chromophore is covalently joined to surrounding protein by a Schiff base linkage, i.e. a -C=N- bonding (see Figure 1). In some sense, the function of both chromophores can be said to be the same, namely to convert a significant fraction of light energy to chemical energy. Several excellent review articles have written on these systems (Ebrey and Honig, 1975; Honig, 1978).

Resonance Raman spectroscopy has provided a great deal detailed information concerning the structure and structural changes of the chromophore in these two pigment systems (for reviews see Callender and Honig, 1977; Mathies, 1979). The Raman cross-section of the chromophore vibrational modes are greatly enhanced when the incident light frequency lies in the visible, since the light is in resonance with the absorption structure of the chromophore, relative to protein modes, as the surrounding protein is colorless. Five to seven orders of magnitude in enhanced Raman cross-section can be realized. In addition, water (the ubiquitous biological medium) has a very low Raman cross section and thus does not give a troublesome background spectrum. Thus the observed Raman signal is due to modes of the chromophore only and can be analyzed in terms of structure. As will be apparent below, the Raman structure is quite sensitive to the geometric form of retinal. In addition, our results below are key, we believe, in pointing towards the molecular mechanism by which light energy is initially converted to chemical energy. Resonance Raman spectroscopy has found wide applicability in the study of other biologically interesting chromophores (for reviews see Spiro, 1974; Warshel, 1977; and Johnson and Petricolas, 1976) in addition to visual pigments and bacteriorhodopsin.

We present here the Raman spectrum of bathorhodopsin. When this is compared to the spectrum of rhodopsin that has previously been measured (Callender et al., 1976; Mathies et al., 1976), a great deal can be said concerning the molecular mechanism involved in what is known as the primary event in vision, the rhodopsin to bathorhodopsin transformation.

In the next section we consider briefly the central experimental problem associated with Raman measurements of visual pigments and that is the extreme photosensitivity of the sample. Special techniques have been developed to, in fact, overcome this problem. In the section on Results and Discussion we present and discuss Raman data of bathorhodopsin. Using this data, a model for the rhodopsin to bathorhodopsin transition and the corresponding transition in bacteriorhodopsin is proposed. The central feature of this model is a mechanism of energy storage based on charge separation in the interior of the protein caused by a geometrical change of the chromophore.

SAMPLE PHOTOLABILITY

The Raman effect, even for resonance enhanced cross-sections, is an extremely weak phenomenon. The chromophore absorption cross-sections of visual pigments is about 10^{-16} cm^2/mol., about eight orders of magnitude larger than the resonance Raman cross-section of the most intense Raman active mode. The purpose of visual pigments is the absorpe light and modify its chemical structure; and it does this with a quantum yield approaching one (0.67 to be exact). It is clear that it is much more probable for a rhodopsin molecule to absorb light, thus be effectively "destroyed" relative to the original sample, than to Raman scatter a photon. Two techniques have been developed to allow well defined Raman results from extremely photosensitive samples.

The "pump-probe" technique can be used for two or more sample species are interconvertable by light. Oseroff and Callender (1974) used this method to study the Raman spectral features of rhodopsin, bathorhodopsin, and isorhodopsin. By lowering sample temperature to liquid nitrogen, thermal transitions in the rhodopsin scheme are prevented, and these three pigments are rapidly interconverted under laser irradiation. The composition is determined at equilibrium by the laser frequency (through the absorption cross-sections and quantum yields which can be separately measured; see Oseroff and Callender, 1974). Thus, a weak "probe" laser beam is used to produce the Raman signal, and a relatively stronger "pump" laser beam simultaneously irradiating the sample controls sample

composition. Using digital methods and computer subtraction
techniques, the Raman spectrum of one species can be isolated from a
series of multi-component spectra with varying compositions. This
is the technique used here to obtain the bathorhodopsin spectrum.
(See Oseroff and Callender, 1974; Callender, 1978; Erying and
Mathies, 1979; and Aton et al., 1979, for more details and
applications).

The second method developed for these systems is also applicable
to other samples in solution is the flow technique (Callender et
al., 1976; Mathies et al., 1976). This technique involves flowing
solution samples through the irradiating focussed laser beam with a
velocity sufficient to insure that any given molecule has a low
probability of absorbing a photon. Thus sample in the laser-sample
interaction area is, of course, always absorbing light and being
degraded; but new fresh material is replenishing degraded sample
sufficiently fast to maintain nearly pure starting material. This
technique is very similar to flowing systems in dye lasers and
rotating cells of Raman spectrometer where heated sample is removed
from the laser beam being replaced by cooler material.

RESULTS AND DISCUSSION

Figure 2A-2C shows Raman spectra produced by probe laser
irradiation at 476.2 nm (of the same power for each spectrum) with
and without additional coincident pump irradiation at different
frequencies. The photostationary state composition under these
conditions are also given in Figure 2. The spectrum of
bathorhodopsin from these three spectra can be obtained in the
following way. The subtraction of an appropriate fraction (1.08) of
the spectrum of Figure 2B from that of Figure 2A leaves a spectrum
composed of a positive contribution from bathorhodopsin and a
negative contribution from isorhodopsin with the contribution from
rhodopsin cancelling out. This results in the spectrum of Figure
2D. By adding an appropriate amount (0.36 of Figure 2C) of the
isorhodopsin spectrum of figure 2C, we isolate the bathorhodopsin
spectrum of Figure 3. It should be pointed out that the greatest
uncertainty in this procedure is the composition of the
photostationary mixtures, which have been determined here (see
above; Aton et al., 1979) to no better than ± 5 percent. Thus,
significant remnants of the other pigments can contribute to the
bathorhodopsin spectrum. In testing various worst case
possibilities by using pigment concentrations at the limits of
experimental error, we found that the bathorhodopsin spectrums was
generally insensitive to the uncertainties in the pigment
concentration although there was a small uniform increase or
decrease in intensity of the other Raman bands relative to the main
band at 1538 cm^{-1}. A more detailed analysis will be published
elsewhere (Aton et al., 1979). Eyring and Mathies (1979) have also

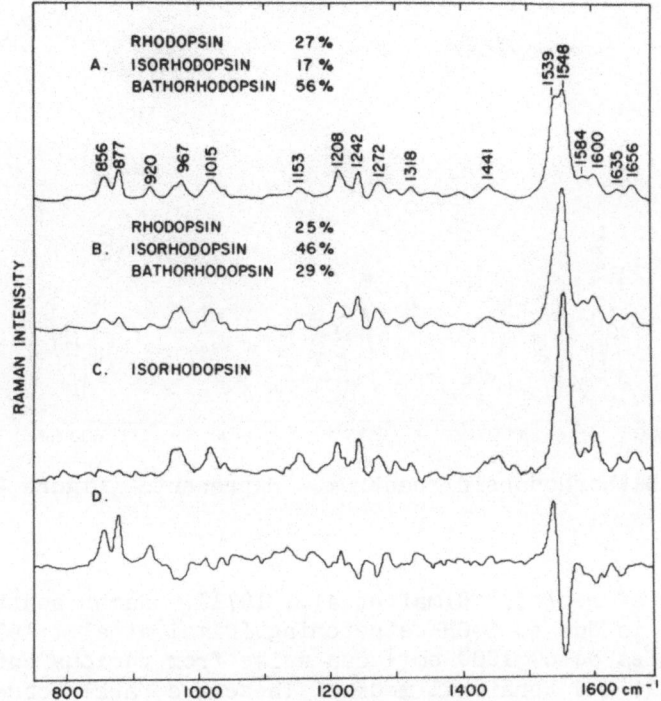

Figure 2. The Raman data and one subtraction step leading to the bathorhodopsin spectrum. All spectra were taken with a 476.2 nm probe, at 7.3 cm^{-1} resolution, and at a sample temperature of 80K. The sample composition are given in the figure. (A) 476.2 nm probe beam alone, (B) simultaneous 580 nm pump beam applied at seven times pump/probe ratio, (C) simultaneous 568.2 nm pump beam at 25 times pump/probe ratio, and (D) A-1.08B. A, B, C are scaled for the same input power and D is scaled by a factor of three larger.

recently reported a bathorhodopsin spectrum using techniques similar to those used here. The spectrum obtained here in Figure 3 and theirs agrees quite well.

Qualitatively the Raman structure of these pigments is not difficult to understand. The absorption structure of retinal responsible for the colors of visual pigments is due to ground state pi to excited state pi electronic transitions. Thus, the observed normal modes which have Raman resonantly enhanced cross-sections are those that couple effectively with the pi electron system. The strongest band (Figure 3) at 1538 cm^{-1} is due to a C=C stretch of the polyene backbone chain of retinal (see Figure 1). The bands between 1100 - 1400 cm^{-1} are due to C-C stretches and C-C-H and C-C-C bends. This region has been called the fingerprint region because the Raman pattern in this region is very sensitive to the

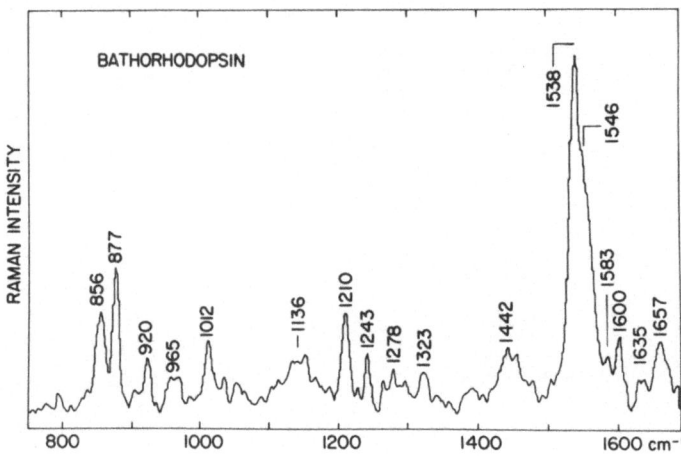

Figure 3. The Bathorhodopsin spectrum. Represents (figure 2) +
0.28 (figure 2).

geometric form of retinal (Rimai et al., 1971). Raman scattering
near 1010 cm^{-1} is due to C-CH$_3$ stretching (Rimai et al., 1971).
Lower frequencies below 1000 cm^{-1} can arise from various out of
plane, skeletal, and torsional modes. These are rarely observed in
pigments, presumably due to broadening mechanisms, but can be
observed in crystals (unpublished data). The details of the normal
mode structure are, at present, unknown although theoretical work is
rapidly progressing (see Warshel, 1977). Most of our
interpretations of the Raman spectra arise from comparisons of
solution model compound spectra to pigment spectra.

One outstanding question (Honig et al., 1976) concerning these
pigments is whether the retinal-protein linkage, -C=N-, contains a
proton covalently bonded to the nitrogen, i.e. -C=NH$^+$-. The
presence of the proton could cause a significant pi electron
delocatization and help explain why pigments absorb in the visible
as opposed to the ultravioldt since solution absorption spectra of
retinals containing this end group move from the ultraviolet to the
blue with addition of this proton. Model compound spectra of
retinals show a band due to C=N stretching at Ca. 1625 cm^{-1} and
one due to C=NH$^+$ stretching at Ca. 1655 cm^{-1} (Heyde et al.,
1971). A band at 1655 cm^{-1} is observed in rhodopsin (Lewis et
al., 1973; Oseroff and Callender, 1974; Callender et al., 1976;
Mathies et al., 1976; see Figure 2). In addition Oseroff and
Callender (1974) showed that in deuterated model chromophore and
rhodopsin, i.e. a C=ND$^+$ linkage, the band moves downward from Ca.
1655 cm^{-1} to 1630 cm^{-1}. This 25 cm^{-1} downward shift in
frequency upon deuteration has been taken as proof that the
retinal-protein linkage is protonated, carrying one extra positive
charge.

A problem, however, with the data of Oseroff and Callender (1974) is that the actual measurement was done on phototstationary state mixtures containing, as here, rhodopsin, isorhodopsin, and bathorhodopsin. The data was not analyzed in terms of the individual pigments as is presented here. It is clear from Figure 3 that bathorhodopsin contains a band at 1654 cm^{-1}; and, in addition, we find using the same analysis applied to deuterated samples that this peak moves downward to 1628 cm^{-1} (with no other significant changes in the bathorhodopsin spectrum; Aton et al., 1979). We then conclude that the retinal of bathorhodopsin is linked to its protein by a $-C=NH^+-$ entity. Furthermore, applications of the same technique to isolate the rhodopsin Raman spectrum shows that the effect of deuteration on the $-C=NH^+-$ mode is the same (Narva and Callender, to be published).

This approximately 25 cm^{-1} downward shift in frequency of the 1654 cm^{-1} band upon deuteration is quite large. Oseroff and Callender (1974) showed that a shift of just this size could be understood in terms of a simple reduced mass calculation applied to a $C=NH$ (with the mass of the hydrogen added to the nitrogen). However, the assumption that the $C=N$ band is coupled to various $C=C$ chain vibrations is almost certainly true, and this leads to the expectation that the observed deuterium effect will be considerably smaller than that calculated for the hypothetical diatomic oscillator. In fact, ^{15}N substituted retinals are shifted by only 15 cm^{-1} relative to ^{14}N compounds (Lewis et al., 1978), and this effect is only a simple change in mass.

We have recently (Aton et al., 1979) carried out a normal mode calculation applied to retinals whose end group is $C=NH^+$ using force constants typical to polyene systems (Gravin and Rice, 1971). While the results are only semi-quantitative, they indicate that the origin of the 25 cm^{-1} shift can be understood only if a significant coupling between the $C=N$ stretching mode and $C=N-H$ bending mode exists. In fact, this coupling can also be used to understand another puzzling feature of this system. The question is why do retinals containing as an end group $C=N$ show a lower frequency for this mode than the $C=NH^+$ mode (Heyde et al., 1971) since the added proton would be expected to pull electrons towards it and so decrease the $C=NH^+$ force constant.

Both effects are qualitatively easy to understand. The $C=N-H$ bend is at near 1250 cm^{-1}, and this increases the frequency of the more energetic vibration (mode-mode repelling) so that the frequency of $C=NH^+$ stretching lies above $C=N$ stretching. The effect of $C=N-D$ bending motion is much smaller because the bending frequency here is only 947 cm^{-1} (with our force constants) and the coupling weakens with increasing frequency difference of the modes. Thus, we obtain a 25 cm^{-1} downward shift upon deuteration. Our coupling parameter used to give these effects was one typical to covalent bonds (Aton et al., 1979).

We thus obtain the important conclusion that the retinal of
bathorhodopsin is linked to surrounding protein by a -C=NH$^+$-
linkage and that the proton is covalently bonded to the nitrogen.
Since rhodopsin shows the same Raman results with respect to the
C=NH$^+$ band for both protonated and deuterated samples (Narva and
Callender, to be published), we conclude that the bonding
characteristics of this end group, particularly the state of
protonation, is unchanged in the rhodopsin to bathorhodopsin
transformation.

Furthermore, the data for bathorhodopsin (Figure 3) indicates
quite strongly that there has been a geometrical change of the
retinal in the rhodopsin to bathorhodopsin transition. This
conclusion is based on the fact that the Raman bands in the
"fingerprint region", 1100-1400 cm^{-1}, is quite different from that
of rhodopsin (Callender et al., 1976; Mathies et al., 1976; Narva
and Callender, to be published); and as discussed above, this region
has been shown to be sensitive to the geometry of retinal. While
Raman bands of rhodopsin are quite close to those of bands of 11-cis
retinal (with the end group being C=NH$^+$; see Figure 1) in solution
(Mathies et al., 1977), the bathorhodopsin Raman bands are not in
good correspondence with any Raman spectrum of the isomers that can
be dissolved in solution (see Figure 1). Thus, while a major
geometric change has likely taken place in the rhodopsin to
bathorhodopsin transformation, the isomeric form of the chromophore
of bathorhodopsin is one not found in solution but some intermediate
structure.

MOLECULAR MODEL OF RHODOPSIN TO BATHORHODOPSIN TRANSFORMATION

Figure 4 summarizes a molecular model for the rhodopsin to
bathorhodopsin transformation. The starting point for the model is
that the chromophore of rhodopsin is linked to protein through a
-C=NH$^+$- bond and, of course, has an 11-cis geometry (far left
panel of Figure 4) as shown by the Raman data. Since buried charges
in proteins generally appear as members of an ion pair in a salt
bridge, there is presumably a negative counter-ion, anchored to the
protein, balancing the chromophores extra positive charge (Honig and
Ebrey, 1976; Honig et al., 1976). From the Raman data above and
from a number of other arguments (Hubbard and Kropf, 1958; Yoshizawa
and Wald, 1963; Rosenfeld et al., 1977; Green et al., 1977), the
primary photochemical event (the formation of bathorhodopsin) is
assumed to be an isomerization, a major geometrical rearrangement,
of the chromophore which breaks the salt bridge (middle panel of
Figure 4). This isomerization process is viewed here as taking
place in the excited state of the chromophore after light
absorption. Since the Raman data really require that the proton
associated with the C=N end group accompany the isomerization (see
above), there exists at this point a large unbalanced charge

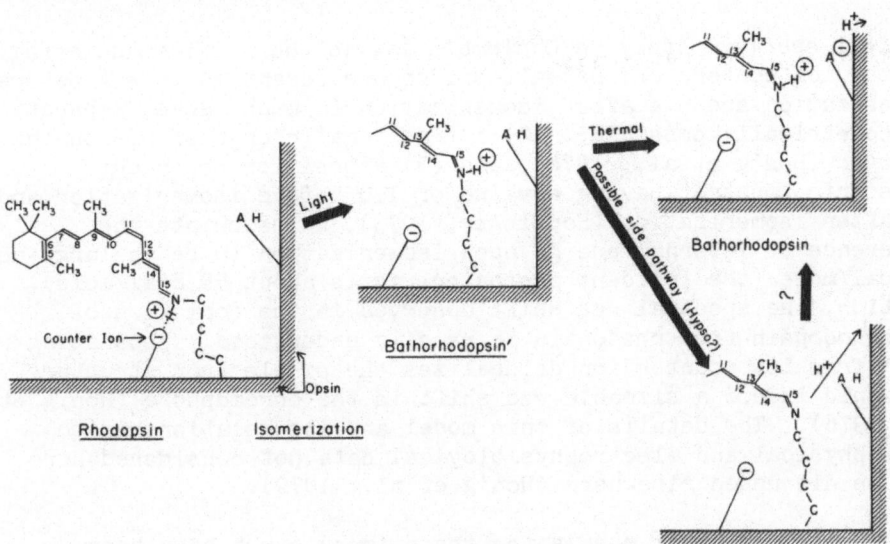

Figure 4. A model for the early events in visual excitation. The
11-cis chromophore of rhodopsin is depicted with its Schiff base
forming a salt bridge with a negative counter-ion. The
photochemical event is an isomerization about the 11-12 double bond
in rhodopsin (probably about the 13-14 bond in bacteriorhodopsin)
but any isomerization in any direction will produce charge
separation as shown in the first step in the figure. The pk's of
the Schiff base as well as those of other groups on the protein such
as AH are strongly affected by photoisomerization since a salt
bridge is broken, causing proton movement. Hypso is a pigment
sometimes observed (not discussed here) and could be explained as a
possible side reaction (see Honig et al., 1979). For
bacteriorhodopsin, the isomerization is trans-cis rather than
cis-trans but all other events are assumed to be equivalent.

separation within the protein's interior. As a final step in this
model, we assume some rearrangement of protons of the protein, but
in the vicinity of the chromophore as a ground state process, in an
incomplete attempt by the protein to form ion pairs (far right panel
of Figure 4). This last step forms bathorhodopsin. That proton
movements should occur resulting from changes in pk's of protein
groups near the disrupted salt bridge is reasonable, and picosecond
data (Peters et al., 1977) have strongly indicated that proton
translocations in the rhodopsin to bathorhodopsin transformation do
occur. The one indicated in Figure 4 is an example; many
possibilities suggest themselves.

Apart from satisfying the experimental data on this system, this
model offers a natural explanation of the requirement that a
substantial fraction of the incident photon energy be converted to

chemical energy. Applying Coulomb's law to the problem and assuming
(1) the ion centers of the salt-bridge are separated by 3 Å before
isomerization and 5 Å after isomerization (a much larger separation
is geometrically possible), (2) a fractional change of 0.5 on the
nitrogen (Honig et al., 1976), and (3) a position dependent
dielectric constant having a value of 1.0 before isomerization and
2.5 after isomerization (Hopfinger, 1973), we calculate the
difference of internal energy upon isomerization to be as large as
40 Kcal/mole (the incident photon energy is about 55 Kcal/mole). In
addition, the spectral red shift observed in the formation of
bathorhodopsin from rhodopsin is readily understood. Separating
$C=NH^+$ from its counter ion delocalizes the pi electron structure
and would induce a sizeable red shift in the chromophore (Honig et
al., 1976). The details of this model and its relationship to
psychophysical and electrophysiological data not considered here
will be discussed elsewhere (Honig et al., 1979).

A number of other models for the primary event have been
recently proposed. Some of these models (Peters et al., 1977; Van
der Meer et al., 1976; and Favrot et al., 1979), involve a change in
the bonding properties of the hydrogen on $C=NH^+$ and appear to be
inconsistent with the Raman results above (see also Erying and
Mathies, 1979). Two other models (Lewis, 1978; Warshel, 1978) are
consistent with the Raman data, but assume that proton movement is
the primary step and isomerization is a later process. This appears
to be inconsistent with picosecond data and other evidence (Honig et
al., 1979).

In summary the model presented here involves a photochemically
induced change separation covered by a geometric change in the
chromophore. It provides a general mechanism for the conversion of
light into chemical energy. It is easily generalized to
bacteriorhodopsin, although here the geometrical change of the
retinal chromophore would be trans to cis rather than cis to some
distorted trans as depicted in Figure 5 since the primary pigment of
bacteriorhodopsin contains a trans chromophore (Pettei et al.,
1977). In addition, it would provide a simple explanation of proton
pumping in bacteriorhodopsin since proton translocation is an
essential ingredient of the model. Delbruck (1976) has recently
pointed out that chlorophyll makes use of its rigidity to achieve
rapid transfer of an electron and preclude a back reaction while
retinal and other flexible chromophores such as phytochrome function
by imposing a conformational change on the protein. We believe our
model now extends these concepts by emphasizing the common goal of
both types of systems which is to generate the separation of charge.

We would like to thank Prof. B. Honig and T. Ebrey for many
stimulating conversations and joint work. This work was supported
by grants from the National Science Foundation
(PCM77-06728) and the City University Faculty award program.

REFERENCES

Aton, B., Doukas, A.G., Narva, D., Callender, R.H., Dinur, U., and Honig, B. (1979). Resonance Raman Studies of the Primary Photochemical Event in Visual Pigments. Biophysical J., in the press.

Callender, R.H., Doukas, A., Crouch, R., and Nakinishi, K. (1976). Molecular Flow Resonance Raman Effect from Retinal and Rhodopsin. Biochem. 15: 1621.

Callender, R.H., and Honig, B. (1977). Resonance Raman Studies of Visual Pigment. Ann. Rev. Biophys. and Bioeng. 6: 33.

Callender, R.H. (1978). Techniques of Resonance Raman Spectroscopy of Photoreactive Systems, in: Resonance Raman Spectroscopy as an Analytical Tool, A.J. Melveger, ed., Franklin Institute Press, Philadelphia.

Delbruck, M. (1976). Light and Life III. Carlsberg Research Commun. 41: 299.

Doukas, A.G., Aton, B., Callender, R.H., and Ebrey, T. (1978). Resonance Raman Studies of Bovine Metarhodopsin I and II. Biochem. 17: 2430.

Ebrey, T.G., and Honig, B. (1975). Molecular Aspects of Photo-receptor Function. Q. Rev. Biophys. 8: 124.

Eyring, G., and Mathies, R. (1979). Resonance Raman Studies of Bathorhodopsin: Evidence for a Protonated Schiff Base Linkage. Proc. Natl. Acad. Sci (USA) 76: 33.

Favrot, J., Leclercq, J.M., Roberge, R., Sandorfy, C. and Vocelle, D. (1978). Intermolecular Interactions in Visual Pigments.The Hydrogen Bond in Vision.Photochem. Photobiol. 29:99.

Gavin, R.M., and Rice, S.A. (1971). Correlation of pi-electron Density and Vibrational Frequencies of Linear Polyenes. J. Chem. Phys. 55: 2675.

Green, B., Monger, T., Alfano, R., Aton, B., and Callender, R.H. (1977). Cis-Trans Isomerization of Rhodopsin Occurs in Picoseconds. Nature 269: 179.

Heyde, M.E., Gill, D., Kilponen, R.G., and Rinai, L. (1971). Raman Spectra of Schiff Bases of Retinal (Models of Visual Photoreceptors). J. Am. Chem. Soc. 93: 6776.

Honig, B., Greenberg, A.D., Dinur, U., and Ebrey, T.G. (1976). Visual Pigment Spectra: Implications of the Retinal Schiff Base. Biochem. 15: 4593.

Honig, B. (1978). Light Energy Transduction in Visual Pigments and Bacteriorhodopsin. Ann. Rev. Phys. Chem. 29: 31.

Honig, B., Ebrey, T., Callender, R.H., Dinur, U., and Ottolenghi, M. (1979). Photoisomerization, Energy Storage, and Charge Separation: A Model for Light Energy Transduction in Visual Pigments and Bacteriorhodopsin. Proc. Natl. Acad. Sci. (USA), in the press.

Hopfinger, A. (1973). Conformational Properties of Macro-molecules. Academic Press, New York, see pp. 59-63.

Hubbard, R., and Kropf, A. (1958). The Action of Light on

Rhodopsin. Proc. Natl. Acad. Sci. (USA) 44: 130.

Johnson, B.B., and Peticolas, W.L. (1976). The Resonance Raman
     Effect. Ann. Rev. Phys. Chem. 27: 465.

Lewis, A., Fager, R.S., and Abrahamson, E.W. (1973). Tunable
     Laser Resonance Raman Spectroscopy of the Visual Process: I. The
     Spectrum of Rhodopsin. J. Raman Spectr. 1: 145.

Lewis, A., Marcus, M.A., Ehrenberg, B., and Crespi, H. (1978).
     Experimental Evidence for Secondary Protein-Chromophore
     Interaction at the Schiff base linkage in Bacteriorhodopsin.
     Molecular Mechanism for Proton Dumping. Proc. Natl. Acad. Sci.
     (USA) 75: 4642.

Lewis, A. (1978). The Molecular Mechanism of Excitation in
     Visual Transduction and Bacteriorhodopsin. Proc. Natl. Acad.
     Sci. (USA) 75: 549.

Mathies, R., Oseroff, A.R., and Stryer, L. (1976). Rapid-Flow
     Resonance Raman Spectroscopy of Photolabile Molecules: Rhodopsin
     and Isorhodopsin. Proc. Natl. Acad. Sci. (USA) 73: 1.

Mathies, R., Freedman, T.B., and Stryer, L. (1977). Resonance
     Raman Studies of the Conformation of Retinal in Rhodopsin and
     Isorhodopsin. J. Mol. Biol. 109: 367.

Mathies, R. (1979). Biological Applications of Resonance Raman
     Spectroscopy in the Visible and Ultraviolet: Visual Pigments,
     Purple Membrane, and Nucleic Acis, in: Chemical and Biochemical
     Applications of Lasers,C.B. Moore, ed., Academic Press, New York.

Oseroff, A.R., and Callender, R.H. (1974). Resonance Raman
     Spectroscopy of Rhodopsin in Retinal Disk Membranes. Biochem.
     13: 4243.

Peters, K., Applebury, M.L., Rentzepis, P.M. (1977). Primary
     Photochemical Event in Vision: Proton Translocation. Proc.
     Natl. Acad. Sci. (USA) 74: 3119.

Pettei, M.J., Yudd, A.P., Nakanishi, K., Henselman, R., and
     Stoeckenius, W. (1977). Identification of Retinal Isomers
     Isolated from Bacteriorhodopsin. Biochem. 16: 1955.

Rinai, L., Gill, D., and Parsons, J.1. (1971). Raman Spectra
     of Dilute Solutions of Some Stereoisomers of Vitamin A Type
     Molecule. J. Am. Chem. Soc. 93: 1353.

Rosenfeld, T., Honig, B., Ottolenghi, M., and Ebrey, T.G.
     (1977). Cis-Trans Isomerization in th Photochemistry of
     Vision. Pure Appl. Chem. 49: 341.

Spiro, T. (1974). Biological Applications of Resonance Raman
     Spectroscopy: Haem Protein. Acc. Chemical Res. 7: 339.

Van der Meer, K. Mulder, J.J.C., and Lugtenberg, J. (1976).
     A New Facet in Rhodopsin Photochemistry. Photochem. Photo Biol.
     24: 363.

Warshel, A. (1977). Interpretation of Resonance Raman Spectra
     of Biological Molecules. Ann. Rev. Biophys. and Bioeng. 6: 273.

Warshel, A. (1978). Charge Stabilization Mechanism in Visual
     and Purple Membrane Pigments.Proc. Natl. Acad. Sci.(USA) 75:2558.

Yoshizawa, T., and Wald, G. (1963). Prelumirhodopsin and the
     Bleaching of Visual Pigments. Nature (London) 197: 1279.

A NANOSECOND PROBE OF HEMOGLOBIN DYNAMICS USING TIME RESOLVED RESONANCE RAMAN SCATTERING

J. M. Friedman and K. B. Lyons

Bell Laboratories

Murray Hill, New Jersey 07974

INTRODUCTION

Hemoglobin (Hb) is one of the most extensively studied biological molecules;[1,2] nevertheless, the structural basis for its activity is still uncertain. Each of the four protein chains that comprise the Hb superstructure contains an iron prophyrin which can bind ligands such as O_2 and CO. The intriguing property of Hb is that the ligation process can induce an overall change in the quaternary structure (orientation of the four chains with respect to each other) which results in a dramatic alteration in the affinity of the remaining binding sites. One approach to the study of this effect is to remove or to add ligands to Hb on a time scale that is fast with respect to conformational changes in order to follow the dynamics of the subsequent structural or electronic changes within the protein. Numerous kinetic studies[3-18] of this kind using transient absorption as a probe have been carried out using photolysis to rapidly remove ligands such as O_2 or CO from Hb. The majority of these studies probe processes such as ligand recombination which occur on microsecond or longer time scales. Recently there have been transient absorption studies[8,9,15,16,17,18] which reveal the presence of nanosecond and picosecond transient species. On these time scales the photolysis process and the structural trigger for the conformational changes can be studied. All but one of these high speed studies[18] are single wavelength measurements which greatly limits interpretation. Furthermore, the structural and electronic basis for a given change in the absorption spectrum is difficult to determine because of the diffuseness of the porphyrin spectra. Raman spectra on the other hand are spectrally sharp and in the case of hemoglobin many of the Raman spectral lines

have been well characterized.[19,20] We have undertaken a study of
Hb in which we generate resonance enhanced Raman spectra of transient species that appear subsequent to photolysis. Using a mode
locked Nd:YAG laser in conjunction with nitrogen laser pumped dye
laser we can probe transient species on a time scale of nanoseconds
or longer. In this paper in addition to reviewing and discussing
previously reported transient Raman spectra[21] that we obtained with
a single pulse technique, we describe the double pulse techniques
which we have recently developed.

Single Pulse Experiments

     Excitation of Hb using visible light generates a Raman spectrum
that is resonantly enhanced due to the intense porphyrin absorption
bands. The enhancement affects specifically the Raman lines closely
associated with porphyrin group. Excitations resonant with the α
and β transitions in the yellow and green result in anomalously
polarized and depolarized Raman spectral lines that are attributable
primarily to transitions involving non-totally symmetric vibrational
modes of porphyrin ring, whereas blue excitations into the Soret
band generate strongly enhanced polarized spectra indicative of
transitions involving totally symmetric modes. Several of the
higher frequency (1300-1650 cm^{-1}) Raman lines have been shown to be
sensitive to the spin state of the iron,[20,22] the center to pyrrole
nitrogen distance[19] and the amount of backbonding into the porphyrin
π* orbitals.[20] In addition, Raman lines attributable to oxyhemo-
globin, carbonmonoxyhemoglobin (HbCO) and deoxyhemoglobin are
readily distinguishable. With this information it is possible to
plan a systematic study of the structural and electronic dynamics
associated with iron porphyrin subsequent to photodissociation. We
can anticipate that both the photodissociation process and the
quaternary structure related processes are amenable to study via
this approach insofar as these processes have an effect upon the
resonantly enhanced modes.

     The first attempts to probe Hb using transient Raman spectro-
scopy involved single pulse experiments.[21,23] We reported[21] results
obtained using as an excitation and probe source frequency-doubled
(5320Å) 20mJ pulses of 10 nsec duration from a Nd:YAG laser as an
excitation source. The first 5% of the pulse was sufficient to
photodissociate the HbCO within the irradiated volume of sample
assuming that the photoexcited HbCO evolves into either the photolyzed products or the relaxed re-excitable unphotolyzed HbCO on a
subnanosecond time scale. The first assumption has been experimentally demonstrated[15,16,18] whereas the latter is as yet uncertain.
The spectra were recorded on an optical multichannel analyzer which
will be described in the next section.

     The resulting spectrum resembles the Raman spectrum of deoxyHb
when resonantly enhanced with 5320Å excitation. This 10 nsec

spectrum does not contain any spectral features assignable to thermalized HbCO which indicates that either the entire sample of HbCO has been photodissociated on this time scale or there is a population of perturbed HbCO that does not manifest itself in this 10 nsec spectrum. Double pulse experiments are currently underway to distinguish between the two alternatives.

The deoxyHb-like spectrum obtained at 10 nsec indicates that the structural rearrangements of the porphyrin ring associated with the switch from a liganded to an unliganded iron porphyrin occur on a nanosecond or faster time scale. In particular, the anomalously polarized Raman peak which occurs at 1556 cm^{-1} in deoxyHb and at 1585 cm^{-1} in HbCO has been shown[19] to be correlated linearly with the porphyrin core size (center to pyrrole nitrogen distance). A plot of core size versus frequency for a wide range of porphyrins yields a slope of $-.002A/cm^{-1}$ for this Raman line. The occurrence of a deoxyHb like frequency in the 10 nsec spectrum indicates that the $\sim.05\AA$ increase in core size in going from HbCO to deoxy-Hb is essentially complete within 10 nsec. Similarly the appearance in the 10 nsec spectrum of a deoxy-Hb-like frequency at 1601 cm^{-1} as opposed to 1630 cm^{-1} for HbCO indicates that major electronic and spin rearrangements have also occurred on this time scale. The latter mode is known to be sensitive to both the spin state of the iron and backbonding to π^* orbital of the porphyrin ring.[20,22]

Although the resonance Raman spectrum of the 10 nsec transient closely resembles that of deoxy-Hb, a comparison of the two Raman spectra taken with the Nd:YAG-OMA system under identical conditions, reveals that there are frequency differences. The ap core size marker band which is also an iron spin state marker band is either unchanged or slightly shifted (1-2 cm^{-1}) to a lower frequency in the transient; however, the 1606 cm^{-1} (deoxy-Hb) band which is sensitive both to spin and π^* backbonding is red shifted in the transient by approximately 5 cm^{-1}. These frequency shifts may originate from any of several interesting processes.

In this experiment we are generating unliganded iron porphyrin on a time scale that is short compared to the microsecond or longer rearrangement time for protein quaternary structure. Consequently, at 10 nsec the iron porphyrin although unliganded is interacting with quaternary structure (r) associated with the fully liganded Hb (HbCO), whereas in deoxy-Hb the iron porphyrin is in the milieu of the T quaternary structure. This difference in heme environments might account for the observed differences in the Raman spectrum. Alternatively or in addition, the unliganded iron porphyrin in the transient might be electronically perturbed by CO which although photolyzed off the heme could remain with in the heme pocket on this time scale. There is also the possibility that there is a long (>nsec) ground state recovery time for the iron

porphyrin subsequent to photoexcitation of HbCO due to bottlenecks
arising from spin state changes. The transient spectrum would then
reflect the vibrational modes associated with an electronic state
potential other than that of the ground state.

One way to deconvolve the effects of quaternary structure upon
the unliganded porphyrin would be to compare the Raman spectra of
deoxy-Hb (T structure) and a chemically modified deoxy-Hb that is
stabilized in the R quaternary structure. Shelnutt et al.[24] have
recently report d the results of such an experiment. Using CW
Raman Difference Spectroscopy (RDS), they found that those Raman
lines that are sensitive to π^* backbonding are red shifted by
1-2 cm^{-1} when deoxy-Hb is stabilized in the R structure. A comparison of these RDS shifts to those obtained from the single pulse
transient experiments as well as preliminary double pulse results[25]
utilizing a Soret band resonance reveals a similarity in the pattern
of shifts. Although the transient species appears to have larger
shifts than those reported in the RDS study, the patterns of the
spectral shifts are qualitatively very similar which indicates that
at least some of the effects observed in the transient spectrum are
due to porphyrin-protein interactions. Double pulse experiments,
of the experimental design described below may be able to determine
if these shifts evolve on the time scale of the R-T configurational
changes.

Experimental Apparatus

The detection system is the same for both the single and double
pulse experiments. It consists of a Spex 1401 double grating monochromator interfaced to an optical multichannel analyzer (OMA). The
exit slit of the monochromator is removed and the exit plane then
focused onto the fiber optic input plane of the first stage image
intensifier of the OMA. After two stages of image-intensification,
the spectrum is recorded by an SEC vidicon tube, interfaced to an
HP2100A minicomputer. This tube has significant image storage
capability, which allows multiple pulses (up to 100) to be integrated internally, thus producing a shot-noise limited signal to
noise ratio. The resolution of the Spex-OMA system depends upon
the magnification of the imaging optics. At minimum magnification,
in first order, the system can cover a range up to 700 cm^{-1} with a
resolution of $\sim$8 cm^{-1}. In this study a range of 350 cm^{-1} was
employed. With 200μ slits, the resulting resolution was again
$\sim$8 cm^{-1}.

For the single pulse experiments, a YAG laser, the Holobeam
500QG, was employed alone. The 1.064μ output of nominal 100mJ
energy was doubled to produce 10 nsec pulses at 5320A with a pulse
energy of 20mJ and a repetition rate of 10pps. Focused to a 0.7 mm
spot, this pulse produced sufficient intensity to bleach the
scattering volume in less than 1 nsec. In fact, the first 5% of

the pulse was sufficient to bleach all the molecules present. The remaining 95% of the pulse then served as a probe to excite the resonant Raman scattering (RRS) spectrum, collected in right angle scattering from a rotating sample cell. By the use of a Kerr shutter with a 3 nsec switching time, this scattered light could be gated so as to observe the first or second half of the pulse, thereby increasing the time resolution to about 5 nsec.

For the double pulse experiment, the YAG laser described above is used in conjunction with a nitrogen-pumped dye laser, the Molectron UV14. The wide tuning range of this system enables the probing of a number of absorption bands, thus increasing the information available from the RRS spectra. The two lasers can be timed with respect to each other with an accuracy of ±3 nsec. The delay can be electronically controlled out to 1 msec. Hence it is possible to study the evolution of the sample over 5 decades in time after photolysis. It is necessary, of course, to reduce the dye laser intensity to a low value (e.g. 50μ per pulse) in order to avoid rephotolyzation of the sample. The RRS spectrum excited by the dye laser is collected in back-scattering geometry from a flowing sample cell. This improved cell design allows (1) the use of strongly absorbed excitation frequencies, (2) precise temperature control of the sample, and (3) control of the atmosphere inside the flow system.

In the double pulse experiment, then, we prepare the system in a few nsec by bleaching the entire scattering volume, focusing the YAG pulse to 0.7 mm. We can then vary both the probe wavelength and the probe delay as experimental parameters to map out the behavior after photolysis. The spectra associated with various peaks in the transient absorption spectrum may thus be observed and the species responsible for the absorption peaks be more fully characterized. This very powerful technique, not previously used due to its complexity, should yield a wealth of information about the structural changes accompanying photolysis of HbCO.

REFERENCES

1. R. G. Shulman, J. J. Hopfield and S. Ogawa, Quart. Rev. Biophys. 8, 3 (1975).
2. J. M. Baldwin, Prog. Biophys. Mol. Biol. 29, 225 (1975).
3. Q. H. Gibson, Biochem. J. 71, 293 (1959).
4. Q. H. Gibson and E. Antonioni, J. Biol. Chem. 242, 4678 (1967).
5. F. A. Ferrone and J. J. Hopfield, Proc. Nat. Acad. Sci. USA 73, 4497 (1976).
6. T. Reed, J. Bunkenberg and B. Chance In "Probe of Structure and Function of Macromolecules and Membranes", Vol. II, p.335, Acad. Press, New York (1971).
7. J. A. McCray, Biochem. Biophys. Res. Commun. 47, 187 (1972).

8.   B. Albert, R. Banerjee and L. Lindquist, Biochem. Biophys.
     Rev. Commun. 46, 913 (1972).

9.   B. Albert, R. Banerjee and L. Lindquist, Proc. Nat. Acad. Sci.
     USA 71, 558 (1974).

10. E. Antonione, N. M. Anderson and M. Brunori, J. Biol. Chem.
     247, 319 (1972).

11. R. H. Austin, K. W. Beeson, L. Eisenstein, H. Fraunfelder and
     I. C. Gunsalns, Biochemistry 14, 5355 (1975).

12. C. A. Sawicki and Q. H. Gibson, J. Biol. Chem. 251, 1533 (1976).

13. C. A. Sawicki and Q. H. Gibson, J. Biol. Chem. 252, 7538 (1977).

14. W. A. Saffran and Q. H. Gibson, J. Biol. Chem. 252, 7955 (1977).

15. C. V. Shank, E. R. Ippen and R. Bersohn, Science 193, 50 (1976).

16. L. J. Noe, W. G. Eisert and P. M. Rentzepis, Proc. Nat. Acad.
     Sci. USA 75, 573 (1978).

17. B. I. Greene, R. M. Hochstrasser and R. B. Weisman, In "Proc.
     of the Topical Meeting on Picosecond Phenomena", C. V. Shank
     and E. Ippen, ede, Springer-Verlag (1978).

18. B. I. Greene, R. M. Hochstrasser, R. B. Weisman and W. A. Eaton,
     Proc. Nat. Acad. Sci. USA 75, 5255 (1978).

19. L. D. Spaulding, C. C. Chang, N.-T. Yu, and R. H. Felton,
     J. Am. Chem. Soc. 97, 2517 (1975).

20. T. G. Spiro and J. M. Burke, J. Am. Chem. Soc. 98, 5482 (1976).

21. K. B. Lyons, J. M. Friedman and P. A. Fleury, Nature 275,
     565 (1978).

22. P. Stein, J. M. Burke and T. G. Spiro, J. Am. Chem. Soc. 97,
     2304 (1975).

23. R. B. Srivastava, M. W. Schuyler, L. R. Dosser, F. J. Purcell
     and G. Atkinson, Chem. Phys. Lett. 56, 595 (1978).

24. J. A. Shelnutt, D. L. Rousseau, J. M. Friedman and S. Simon,
     Proc. Nat. Acad. Sci. USA in press.

25. J. M. Friedman and K. B. Lyons, to be published.

RECENT RESULTS IN FOUR-PHOTON SPECTROSCOPY OF CONDENSED MEDIA

S. A. Akhmanov, L. S. Aslanyan, A. F. Bunkin
F. N. Gadzhiev, N. I. Koroteev and I. L . Shumai

Chair of Optics
Moscow State University
Moscow, USSR

§1. INTRODUCTION

Great progress has been achieved in the development of nonlinear spectroscopy methods over the last few years. Of prime importance is the fact that nonlinear spectroscopy enables one not only to obtain some new information on the substance under investigation but also to obtain much more precisely with far better resolution and sensitivity spectroscopic data on cross-sections, line shapes, line positions, etc. inherent in traditional laser spectroscopy. The achievements of the nonlinear spectroscopy of atoms and molecules are well known and it is a vital task to develop the nonlinear spectroscopy methods of condensed medium.

Over the last years various researches achieved great successes in this field. A number of coherent nonlinear spectroscopy methods has been used to solve fairly complicated problems in spectroscopy of solids. All these methods make use of the basic principles developed originally for Raman spectroscopy and applied later for the investigation of other types of resonances. All methods are based on the third order nonlinear susceptibility $\chi^{(3)}(\omega)$ dispersion measurements. The nonlinear cubic polarization

$$\vec{P}^{NL} = \hat{\chi}^{(3)}\vec{E}\,\vec{E}\,\vec{E} \tag{1}$$

gives rise to a vast number of four-photon interaction and self-action prosesses. All these processes can be effectively used for spectroscopic applcations.

Research in this field started as early as 1965 - 1967 [1 - 4] when four-photon processes were used for measurements of resonances in $\chi^{(3)}$ due to Raman active molecular vibrations and phonon modes in crystals. Already then the data enabled one to verify a number of parameters of Raman active modes.

The next stage in this field was closely connected with the wide application of tunable lasers. Several reports [5 - 7] presented at the Montreal conference on quantum electronics in 1972 demonstrated the advantages of coherent Raman spectroscopy. The measurement of the third order susceptibility $\chi^{(3)}$ dispersion was used to obtain information about the Raman active modes through the four-photon process

$$\omega_{s,a} = \omega_{probe} \mp (\omega_1 - \omega_2) \quad . \tag{2}$$

Here ω_{probe}, ω_1, ω_2 are the frequencies of laser, probe and pump waves. The intensity dispersion has been registered when $\omega_1 - \omega_2 \simeq \Omega$, Ω being the Raman resonance frequency. In [6] this coherent nonlinear spectroscopy method was called the method of Active Raman Spectroscopy. Byer [8] pointed out the advantage of anti-Stokes scattering and introduced the now widely used term CARS.

Coherent Raman spectroscopy achieved great progress within the next years. On the one hand, known methods were applied to study Raman resonances in solids, liquids and gases; on the other hand, new methods were developed intensively.

Among the new methods suggested were Raman Induced Kerr Effect [9], Polarization Raman Spectroscopy and Coherent Raman Ellipsometry [10], the CW Coherent Raman Gain technique [11], OHD RIKES [12], and nonlinear Raman interferometry [13]. In addition nonlinear spectroscopy methods developed originally for studying Raman resonances were successfully used for studying one and two-photon resonances of other types.

All of these methods make use of only two nonlinear optical effects of the cubic nonlinear polarization (1). To the first type of these effects we attribute processes of new wave generation and generation of radiation at new frequencies.

When laser radiation of the form:

$$\vec{E} = \sum_{\ell=1}^{3} \vec{E}_\ell = \sum_{\ell=1}^{3} \vec{e}_\ell A_\ell \exp\{i[\omega_\ell t - \kappa_\ell r]\} \tag{3}$$

(frequencies of two or three waves may coincide) is sent into the medium, the cubic polarization (1) cause generation of new waves, different from the original either in frequency (when frequencies of the pump waves do not coincide) or in wave vector (when all the waves have equal frequencies —— see, for example, [14]).

Unique spectroscopic information can be obtained by measuring
the dispersion of amplitude, phase or the state of polarization
of this new wave when the pump wave frequency or its combination are
scanned in the vicinity of the resonance under investigation.
We'll call these methods the methods of Active Spectroscopy. As was
pointed out above there are three variants of Active spectroscopy.
They are: 1. Amplitude Active Spectroscopy.
            2. Polarization Active Spectroscopy.
            3. Phase Active Spectroscopy.
The second type of effect is connected with interaction of only
two laser waves propagating in the nonlinear medium.

The influence of the wave with a frequency ω_2 on the wave
with a frequency ω_1 is described according to (1) by the
polarization component

$$\vec{P}^{NL}(\omega_1) = \hat{\chi}^{(3)}(\omega_1, \omega_2, \omega_1, -\omega_2) \, \vec{E}_1 \, \vec{E}_2 \, \vec{E}_2^* \quad . \tag{4}$$

Therefore, the dielectric constant variance induced by the wave with
a frequency ω_2 is given by

$$\Delta\epsilon^{NL}(\omega_1) = 4\pi\chi^{(3)}(\omega_1, \omega_2, \omega_1, -\omega_2) \, \vec{E}_2 \, \vec{E}_2^* \quad . \tag{5}$$

Thus, the variance in amplitude, state of polarization and phase
of the wave with a frequency ω_1, when the frequency ω_2 is tuned so
that some combination of the frequencies ω_1 and ω_2 is scanned
through the resonance ($\omega_1 - \omega_2$ for Raman resonance, $\omega_1 + \omega_2$ for
two-photon resonance) can be used to get spectroscopic information.
This kind of spectroscopy can naturally be called a form of modulation spectroscopy, the modulation being carried out through optical
nonlinearity.

The number of papers published in this field has increased
rapidly. Reviews of theoretical and experimental results can be
found in [15, 16].

In fact now we have the situation in which a researcher
working in this field has an opportunity to choose any method of
nonlinear spectroscopy most appropriate to the problem under
study. We believe this is illustrated by the results obtained in
the Nonlinear Optics Laboratory of the University of Moscow.

We present here examples of several complicated problems in
the field of Raman spectroscopy and the allowed electron
transition spectroscopy in liquids and solids which were solved
by four-photon nonlinear spectroscopy methods. They are:
  1. The investigation of Raman resonance broadening in cryogenic
     liquids (this problem was solved by the amplitude CARS
     method).

2. The inhomogeneously broadened Raman bands resolution in
   liquids, including the resolution of Raman band structure
   modified in the presence of solvent electrons (this problem
   was solved by the method of Coherent Raman Ellipsometry).
3. The resolution of the allowed electron resonance structure
   in liquids.
4. The registration of weak Raman lines in liquids (polarization spectroscopy methods can be used to solve this problem
   most effectively).

§2. CW HIGH RESOLUTION CARS SPECTROSCOPY OF CRYOGENIC MIXTURES

The lineshape of the polarized isotropic components of Raman
spectrum in liquids contains important information on vibrational
and rotational relaxation. Nevertheless theoretical and experimental study of vibrational relaxation in liquids was started only
in the last few years [17 - 23].

In the first part of this paper we report the results of the
experimental study of bandwidth, lineshape and position of super-
narrow Raman line of liquid nitrogen ($\Delta\nu_R$ = 0.058 $\pm$ 0.002 cm^{-1})
dissolved in liquid CO, CH_4, O_2, Ar and Kr. A stable single mode
Ar-ion laser (Spectra-Physics model 165-08) and Ar-ion laser pumped
single frequency tunable CW dye laser (Spectra-Physics model 580 A)
have been used in a traditional coherent Anti-Stokes Raman spectroscopy (CARS) arrangement to achieve high spectral resolution (about
70 MHz) and accuracy in measurements of the Raman band-width and line
center position. The spectral resolution determined by a convolution of two laser lineshapes was limited only by the laser
linewidths (less then 40 MHz).

The picosecond pulsed CARS technique has also been used
recently to investigate vibrational relaxation in cryogenic liquids
[23], but this method provides less spectroscopic information when
compared with CW CARS since it does not permit accurate measurement
of Raman lineshape and solvent-induced lineshift.

All the experiments in liquid N_2 and solutions have been
carried out at constant temperature T = 78K. The details of the
experimental arrangement are given elsewhere [21, 22].

Figure 1 shows the experimental dependence of the Raman
Q-branch bandwidth of liquid N_2 dissolved in various cryogenic
solvents on the molar concentration of solutions.

Figure 2 shows the experimental value of the solvent-induced
Raman lineshift of N_2 in various solvents as a function of molar
concentration of N_2 in solutions. The N_2 linewidth behaviour is
quite different in various solvents. While in CO the N_2 linewidth

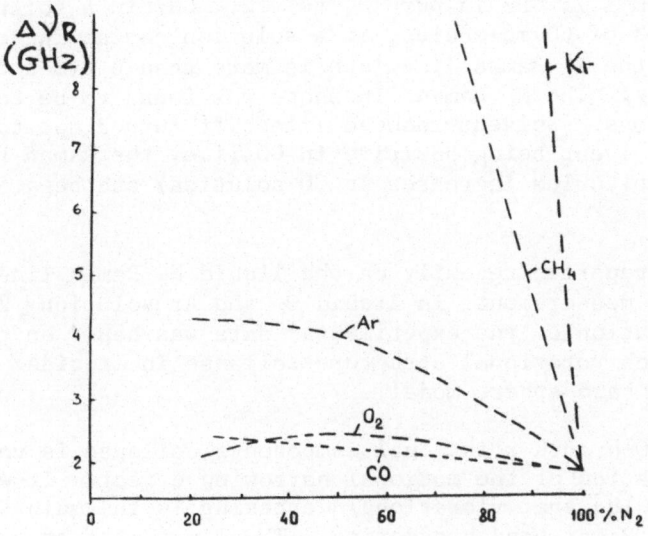

Fig. 1. The behaviour of the Raman linewidth of liquid N_2 dissolved in liquid CO, O_2, CH_4, Ar, and Kr as a function of molar concentration of solutions.

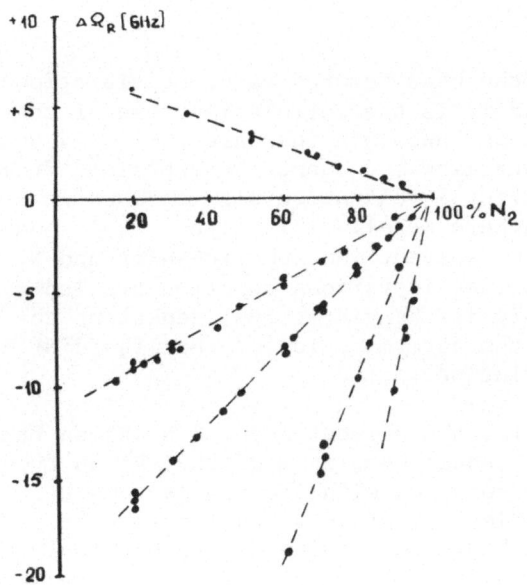

Fig. 2. Solvent-induced Raman frequency shift of N_2 in solutions.

increases from 1.75 GHz in pure N_2 to   2.4 GHz in a solution
containing 70% of CO molecules, in a solution containing only 10%
Kr molecules the N_2 Raman linewidth is more than 6 times that of
pure liquid N_2. The N_2 Raman lineshape was found to be Lorentzian
in all solutions. Solvent-induced lineshift turned out to depend
strongly on solvent being positive in CO (i.e. the Raman frequency
shift of N_2 molecules increases in CO solution) and negative in
other solvents.

We have reported recently on the liquid N_2 Raman linewidth
and lineshift measurements in liquid O_2 and Ar solution [21, 22].
The interpretation of the experimental data was based on the theory
of the Q-branch rotational structure collapse in liquids [24, 25]
and the rough hard sphere model.

But the Q-branch rotational components collapse is not the
only manifestation of the motional narrowing effect. It was shown
recently [17, 18] that vibrational dephasing is the main source of
the liquid N_2 Raman band broadening. The limit of fast modulation
[17] of the vibrational frequency due to the intermolecular
interaction and the molecular motion is valid for liquid nitrogen,
thus leading to the narrow Lorentzian Raman bandshape. This effect
is analogous to Dicke's narrowing [26] and the Q-branch rotational
components collapse and leads to the expression for the bandwidth
characteristic of motional narrowing

$$\Gamma = 2\tau_c < \Delta \omega^2 > .$$ (6)

Here $< \Delta\omega^2 >$ is the mean squared value of vibrational frequency
fluctuations, and τ_c is the correlation time of these fluctuations.
The liquid N_2 Raman bandwidth increase in liquid CH_4 and Kr solutions is quite striking and cannot be explained from the point of
view of the rotational relaxation slowing down in accordance with
the rough hard sphere model. This circumstance and an apparent
correlation in the solvent induced lineshift and N_2 Raman bandwidth increase values in various solution has led us to the
conclusion that it is the vibrational dephasing that plays the
leading role in the determination of the liquid N_2 Raman band
parameters in solutions.

To account for the solvent-induced N_2 Raman frequency shift
in solutions one should keep in mind that Raman frequency shift in
solution must be compared with that in gaseous N_2. The difference
in Raman frequencies for gaseous and liquid nitrogen is equal to
4.5 cm^{-1} [27] with 0.5 cm^{-1} being due to motional narrowing
[24, 25].

Both the solvent-induced Raman frequency shift and line
broadening are determined by the intermolecular interaction

influence on vibrational frequency. We believe it to explain the correlation in N_2 Raman line broadening and shift in different solution. Moreover, the N_2 line broadening and shift in solutions correlate well with the Lennard-Jones potential parameters [28] for the solvent molecules. Unfortunately, we were unable to make precise quantitative calculations of the N_2 Raman band parameters in solutions because the radial distribution functions are unknown for solutions studied.

The energy relaxation and resonant vibrational energy exchange [17, 18] are known to contribute to the vibrational relaxation in liquids too. The energy relaxation in pure liquid N_2 [19] and solutions [20] was found to be a very slow process and thus to give insignificant contribution to the polarized vibrational Raman bandwidth.

Oxtoby et al. [18] have shown that resonant vibrational energy exchange can cause N_2 line broadening in solution. This effect can account for approximately 15% of the observed N_2 Raman line broadening in liquid CO solution. Its contribution is small in liquid CH_4 and Kr. We believe the experimental results of measurement with a very high spectral resolution of the liquid N_2 Raman line position and bandwidth in solutions to be consistent with a motional narrowing effect which has its origin both in rotational j-diffusion and fast modulation of vibrational frequency due to intermolecular interaction and relative molecular motion in liquid. It should be pointed out that both rotational and vibrational relaxation processes need to be taken into account in creating a theory of the isotropic Raman lineshape in liquids.

§3. COHERENT ELLIPSOMENTRY OF RAMAN SCATTERING

The coherent ellipsometry of Raman scattering is a polarizational version of CARS. This technique enables one to measure the dispersion of the polarization parameters of the coherently scattered light (i.e. the ratio of the polarization ellipse axes b/a and the angle of orientation of the major axis of this ellipse ψ') instead of measuring the dispersion of the amplitude characteristic of CARS. Coherent ellipsometry has a number of advantages over conventional CARS. First of all it permits one to determine independently the dispersion of real and imaginary parts of the third order susceptibility (b/a $\sim$ Im$\chi^{(3)}$, ψ' $\sim$ Re$\chi^{(3)}$) with a relative error $10^{-3} \sim 10^{-4}$ inaccessible in amplitude CARS. At the same time, the coherent ellipsometry provides an opportunity to resolve inhomogeneously broadened Raman bands, namely, it enables one to resolve Raman lines with center displacement less than the homogeneous bandwidth. Also important is the circumstance that polarization of scattered light is independent of its amplitude thus being independent of amplitude fluctuations of laser pulses. The

ellipticity parameters can be measured with   high accuracy.
Hence, the full width at half maximum Γ, depolarization ratio ρ,
$\bar{\chi}^R_{1111}/\chi^{(3)NR}_{1111}$ can be determined accurately.   Here $\chi^{(3)NR}_{1111}$ is an
electronic contribution to the third order susceptibility $\chi^{(3)}_{ijk\ell}$,
and $\bar{\chi}^R_{1111}$ is the amplitude value of the resonant contribution
to $\chi^{(3)}$ proportional to the Raman cross-section [10].

The ellipticity b/a maximum position is shifted from the
center of the spontaneous Raman line by value dependent on
$\chi^{(3)R}_{1111}/\chi^{(3)NR}_{1111}$.   This peculiarity results in different frequency
shifts of particular lines with different spectroscopic parameters
and enables to resolve the inhomogeneous Raman band [29].   Several
factors can improve the resolution: a) the maximum value of
ellipticity is proportional to $(\bar{\chi}^R_{1111}/\chi^{(3)NR}_{1111})\sin\theta$ *) and weak
lines with a small $\bar{\chi}^R_{1111}$ but large θ can easily be discriminated
from the background of strong line with large $\bar{\chi}^R_{1111}$ but small θ;
b) an important factor for discrimination of overlapped lines is
the direction of the field vector $E_a(\omega_a = 2\omega_1 - \omega_2)$ rotation
(when $\rho < 1/3$ the vector rotates clockwise).   Computer calculations demonstrate the possibility to resolve the inhomogeneously
broadened band even in the case when the centers of particular
lines coincide.

The experimental arrangement for spectra resolution by coherent
ellipsomentry method is given elsewhere [10].   We have studied
hydrogen bonded liquids (HNO_3, H_3PO_4), chlorobenzene and
cyclohexane.   Coherent ellipsometry was used also to detect deformation of the Raman spectrum of solvent in the presence of solvent
electron (e_s).

Figure 3 shows the dispersion of the elliptic polarization
parameters of the anti-Stokes wave when the difference frequency
$\omega_1 - \omega_2$ is scanned in the vicinity of the Raman resonance 1305 cm^{-1}
of concentrated nitric acid.   The spontaneous Raman spectrum
obtained with a resolution of 1 cm^{-1} is shown above for comparison.
In coherent ellipsometry the band is resolved into two components.
The complex structure of the Raman bands in acids confirms a concept
of aqueous hydrogen bonded complexes in these solutions.

Coherent ellipsometry can be successfully used to resolve the
inhomogeneously broadened bands in hydrocarbons.   For example, the
coherent ellipsometry spectrum clearly shows three components near
the Raman active vibration 1445 cm^{-1} of cyclohexane the central one
being depolarized ($\rho > 1/3$) and two others polarized ($\rho < 1/3$).
These components are difficult to resolve in spontaneous Raman

---

*)Here θ is an angle between the polarization vectors of resonant
   P_R and nonresonant P_{NR} components of scattered light and is
   determined by the symmetry of Raman tensor of chosen vibration.

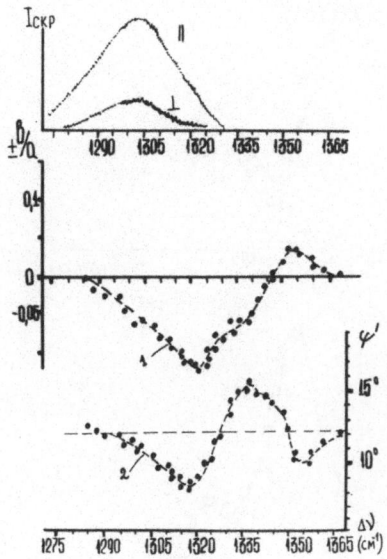

Fig. 3.   The Coherent Ellipsometry spectra in concentrated HNO_3
of the Raman band 1305 cm^{-1}. Points show experimental
values of b/a (1) and ψ'(2). Shown on top are polarized
and depolarized spontaneous Raman spectra.

spectra [30, 31]. The ellipsometry spectrum of the 1577 cm^{-1}
line of monochlorobenzene shows two components [29]. Spectroscopic
parameters of particular components of the Raman band can be
obtained by computer simulation.

One of the most important problems of physical chemistry is
the investigation of the nature of the solvent electron (e_S).
(An extra electron localized within the medium is called a solvent
electron.) This electron moves in a potential hole deep enough to
provide the existence of discrete energy levels [32].

The solvent electron was studied intensively since 1962 but
still little is known on its microstructure. The peculiarity of the
solvent electron consists in the fact that solvent molecules
participate directly in its creation. It means that important data
can be obtained from molecular characteristics of solvent in the
presence of e_S. We have studied e_S in hexamethilphosforthreeamid
(HMPA). The solvent electron was created electrochemically. The
solvent electron is stable enough in HMPA and e_S concentration
could be varied extensively. LiCl and $NaClO_4$ have been used as
conductive salts. Measurements have been carried out with LiCl
at $-7°C$ and with $NaClO_4$ at $-4°C$. The variation of e_S concentration
during the spectrum recording did not exceed 30% at these
temperatures. The e_S concentration used was $5 \cdot 10^{-5}$ M/L.

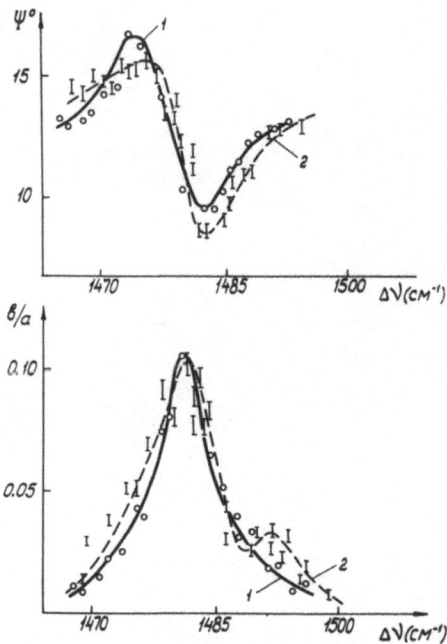

Fig. 4.   The dispersion of the major ellipse direction ψ' and the
          ellipticity b/a in the vicinity of the Raman line
          1486 cm^{-1} (deformation vibration of C–H bonds) in pure
          HMPA (1) and in the presence of solvent electron (2) as a
          function of difference frequency $\omega_1 - \omega_2$.

Measurements were carried out at the Raman lines 1486 cm^{-1}
(C–H vibration) 1207 cm^{-1} (P=O) and 1067 cm^{-1} (C–N) in the presence
of P_s and without it. Figure 4 shows the ellipsometry spectra of
the Raman line 1486 cm^{-1}. In the presence of e_s there appears an
additional maximum in the ellipticity spectrum and the spectrum of
ψ' becomes asymmetrical. The corresponding measurements in the
vicinity of 1207 cm^{-1} and 1067 cm^{-1} Raman lines show no deformation
of the lineshape in the presence of e_s. Hence one can suppose that
CH$_3$ groups of HMPA participate in e_s cluster formation. This
assumption is consistent with presentation of HMPA molecule as a
dipole with positive charge carried by hydrogen atoms of CH$_3$ groups.

§4. THE COHERENT ELLIPSOMETRY OF ELECTRON RESONANCES

Coherent ellipsometry can be successfully applied to study
resonances of any nature including one and two–photon electron
resonances [33]. The expressions for the ellipticity parameters
of the coherently scattered light in the case of single one–photon
electron resonance are the following:

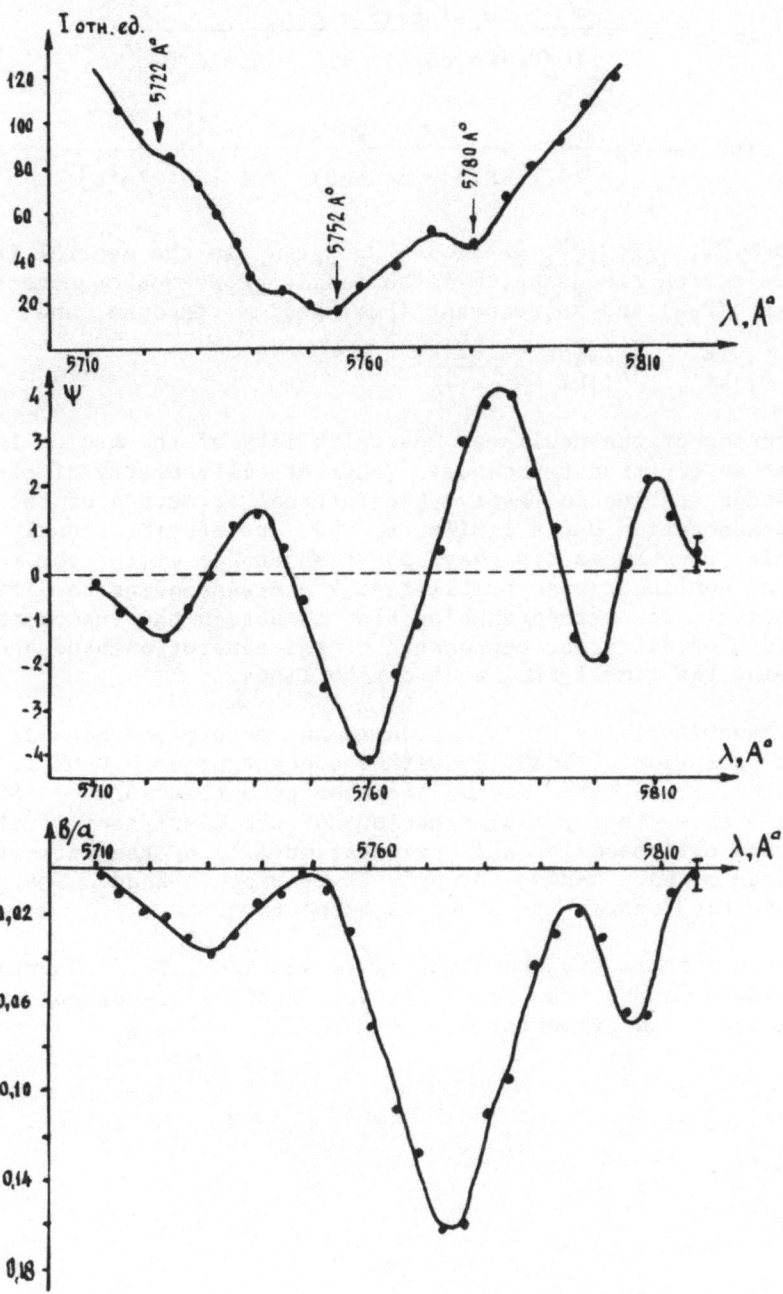

Fig. 5. The Coherent Ellipsometry spectrum of the absorption band 5700 – 5800 Å of aqueous solution of Nd(NO₃)₃.

$$\tan 2\psi' = -\frac{2\alpha \sin\theta \ (\Delta + \alpha \cos\theta)}{[\ (\Delta + \alpha \cos\theta)^2 + 1 - \alpha^2\sin^2\theta\ ]}\ ,$$

$$\sin 2 \left(\text{arctg } \frac{b}{a}\right) = \frac{2\alpha \sin\theta}{[(\Delta + \alpha \cos\theta)^2 + 1 + \alpha^2\sin^2\theta]}\ . \tag{7}$$

Here $\alpha \sim \overline{\chi}^E_{\overline{1}111}/\chi^{3}_{111}\text{NR}$; $\Delta = (\omega_2 - \Omega_E)T_2$, Ω_E is the central frequency of an electron resonance, Θ is the angle between the vectors of resonant (P_{ER}) and nonresonant (P_{NR}) medium response, and

$$\chi^{(3)E}_{ijk\ell} = \chi^{(3)NR}_{ijk\ell} + \frac{\overline{\chi}^E_{ijk\ell}}{-i+\Delta} \tag{8}$$

is a tensor of the nonlinear susceptibility of the medium for the case of an electron resonance. Coherent ellipsometry of electron resonances enables to resolve the internal structure of the one-photon absorption bands similar to the case of vibrational molecular resonances discussed above and to determine the values of resonant nonlinear susceptibilities χ^E corresponding to different components. The method enables also to obtain the "homogeneous" bandwidth of different components of the absorption band and to determine the time T_2 for each of the lines.

Experimentally we have studied the absorption band of the aqueous solution of $Nd(NO_3)_3$ with a concentration 0.1 M/l. The frequency ω_2 was tuned within the absorption band 5700 – 5800 Å. Figure 5 shows the typical behaviour of the dispersion of ellipticity (b/a) and ellipse major axis orientation (ψ') of the anti-Stokes wave when ω_2 was scanned through the absorption band. The upper curve on Fig.3 shows the absorption spectrum.

The authors wish to thank A. V. Vannikov, T. S. Zhuravleva, S. G. Ivanov and R. Yu. Orlov for very helpful discussion and assistance in experiments.

REFERENCES

1. P. Maker and R. Terhune, Phys. Rev. A137, 801 (1965).
2. J. A. Giordmaine and W. Kaiser, Phys. Rev. 144, 676 (1966).
3. N. Bloembergen et al., IEEE J. of Quantum Electronics QE-3, 197 (1967).
4. J. Coffinet and F. De Martini, Phys. Rev. Letters 22, 60 (1969).
5. N. Bloembergen et al., Rept. at 7th Intern. Quantum Electronics Conference, Montreal, Canada, 1972.
6. S. A. Akhmanov et al., ibid.
7. J. J. Wynne, ibid.
8. R. L. Byer, R. F. Begley, and A. B. Harvey, Appl. Phys. Letters 25, 387 (1974).
9. D. Heiman, R. Hellwarth, M. Levenson, and G. Martin, Phys. Rev. Letters 36, 189 (1976).
10. S. A. Akhmanov, A. F. Bunkin, S. G. Ivanov, and N. I. Koroteev, Zh. Eksperim. i Teor. Fiz. 74, 1272 (1978).
11. A. Owyoung, Opt. Commun. 22, 323 (1977).
12. J. J. Song, J. H. Lee, and M. Levenson, "Low Frequency Dispersion of the Third Order Nonlinear Susceptibility" Digest of Techn. Papers, X Intrn. Quantum Electr. Conf. May 29-June 1, 1978, Atlanta, U. S. A.
13. A. Owyoung and P. S. Peercy, J. Appl. Phys. 48, 674 (1977).
14. A. Maruani, J. L. Oudar, E. Batifol, and D. Chemla, Phys. Rev. Letters 41, 1372 (1978).
15. S. A. Akhmanov and N. I. Koroteev, Soviet Phys. - Usp. 123, 423 (1977).
16. M. D. Levenson, Phys. Today 30, No. 5 (1977).
17. W. G. Rotschild, J. Chem. Phys. 65, 455, 2958 (1976).
18. D. W. Oxtoby, D. Levesque, and J. J. Weis, J. Chem. Phys. 68, 5528 (1978).
19. S. R. J. Brueck and R. M. Osgood, Chem. Phys. Letters 39, 568 (1976).
20. W. F. Calaway and G. E. Ewing, J. Chem. Phys. 63, 2842 (1975).
21. S. A. Akhmanov, F. N. Gadjiev, N. I. Koroteev, R. Yu. Orlov, and I. L. Shumai, JETP Letters 27, No. 5, 243 (1978).
22. S. A. Akhmanov et al., Vestnik Moskovskogo Universiteta Seria Fyzyka, Vol. 19, No. 4, 25 (1978).
23. H. M. Hesp, J. Langelaar, D. Belelaar, and J. van Voorst, Phys. Rev. Letters 39, 1376 (1977).
24. S. I. Temkin and A. I. Burshtein, Pis'ma Zh. Eksperim. i Teor. Fiz. 24, 99 (1976).
25. S. R. J. Brueck, Chem. Phys. Letters 50, 516 (1977).
26. J. L. Gersten and H. M. Foley, J. Opt. Soc. Am. 58, 933 (1968).
27. G. B. Grun, A. K. McQuillan, and B. P. Stoicheff, Phys. Rev. 180, 61 (1969).
28. J. O. Hirshfelder, C. F. Curtiss, and R. B. Bird, "Molecular Theory of Gases and Liquids" (J. Wiley, New York, London, 1954).

29. L. S. Aslanyan, A. F. Bunkin, and N. I. Koroteev, Opt. i
     Spektroskopiya 46, 165 (1979).
30. L. M. Sverdlov, M. A. Kovner, and E. P. Krainov, "Vibrational
     Spectra of Complex Molecules" (in Russian) (Nauka, M., 1970).
31. D. A. Ramsey and G. B. B. M. Sutherland, Proc. Roy. Soc.
     (London) A190, 245 (1947).
32. A. V. Vannikov, Soviet Chem. Usp. 44, 1931 (1975).
33. L. S. Aslanyan, A. F. Bunkin, and N. I. Koroteev, Pis'ma
     v Zh. Techn. Fiz. 4, 1177 (1978).

RECENT PROGRESS IN FOUR-WAVE MIXING SPECTROSCOPY IN CRYSTALS

N. Bloembergen

Division of Applied Sciences
Harvard University
Cambridge, Massachusetts 02138

1. INTRODUCTION

The general framework for describing the large variety of non-linear optical phenomena caused by an electric polarization cubic in the electric field amplitudes

$$P_i(\omega_4,\underline{r}) = \frac{1}{2} \chi^{(3)}_{ijk\ell}(-\omega_4,\omega_1,-\omega_2,\omega_3) E_j(\omega_1) E_k^*(\omega_2) E_\ell(\omega_3)$$
$$\exp[i(\underline{k}_1 - \underline{k}_2 + \underline{k}_3)\cdot\underline{r} - i\omega_4 t] + \text{c.c.} \tag{1}$$

was introduced in 1962. In media with inversion symmetry this is the lowest order nonvanishing electromagnetic response. This non-linearity describes a coupling between four electromagnetic waves.[1] In general, each of the four waves has its own frequency, wave vector and polarization direction. The polarization vectors are denoted by $\hat{e}_1$, $\hat{e}_2$, $\hat{e}_3$ and $\hat{e}_4$, respectively, and the nonlinear scalar coupling coefficients

$$\chi^{NL}_{1234} = \hat{e}_1 \hat{e}_2^* : \chi^{(3)}(-\omega_4,\omega_1,-\omega_2,\omega_3) : \hat{e}_3 \hat{e}_4^*$$

$$\chi^{NL}_{ij} = \hat{e}_i \hat{e}_i^* : \chi(-\omega_i,\omega_i,\omega_j,-\omega_j) : \hat{e}_j \hat{e}_j^*$$

are introduced. With the wave vector mismatch $\Delta kz = \underline{k}_1 - \underline{k}_2 + \underline{k}_3 - \underline{k}_4$, the four coupled complex amplitude equations take the form

$$\frac{\partial E_1}{\partial z} + \frac{1}{v_{g1}} \frac{\partial E_1}{\partial t} = 2\pi i(\omega_1/n_1 c)\left[\chi^{NL}_{1234} E_2 E_3^* E_4 \exp(-i\Delta kz)\right.$$
$$\left. + \sum_{j=1}^{4} \chi^{NL}_{1j} E_1 E_j E_j^*\right]$$

$$\frac{\partial E_2}{\partial z} + \frac{1}{v_{g2}} \frac{\partial E_2}{\partial t} = 2\pi i (\omega_2/n_2 c) \left[\chi_{1234}^{NL} E_1 E_3 E_4^* \exp(i\Delta kz) \right.$$

$$\left. + \sum_{j=1}^{4} \chi_{2j}^{NL} E_2 E_j E_j^* \right]$$

$$\frac{\partial E_3}{\partial z} + \frac{1}{v_{g3}} \frac{\partial E_3}{\partial t} = 2\pi i (\omega_3/n_3 c) \left[\chi_{1234}^{NL} E_1^* E_2 E_4 \exp(-i\Delta kz) \right. \qquad (2)$$

$$\left. + \sum_{j=1}^{4} \chi_{3j}^{NL} E_3 E_j E_j^* \right]$$

$$\frac{\partial E_4}{\partial z} + \frac{1}{v_{g4}} \frac{\partial E_4}{\partial t} = 2\pi i (\omega_4/n_4 c) \left[\chi_{1234}^{NL} E_1 E_2^* E_3 \exp(+i\Delta kz) \right.$$

$$\left. + \sum_{j=1}^{4} \chi_{4j}^{NL} E_4 E_j E_j^* \right]$$

In many important cases two (or more) of the waves may be degenerate in frequency, and/or wave vector, and/or polarization. The coupling is especially strong if the conditions of energy and momentum conservation are satisfied,

$$\omega_4 = \omega_1 - \omega_2 + \omega_3 \quad \text{and} \quad \underset{\sim}{k}_4 = \underset{\sim}{k}_1 - \underset{\sim}{k}_2 + \underset{\sim}{k}_3 \qquad (3)$$

In general, the nonlinear susceptibility $\chi^{(3)}$ is a fourth-rank tensor, and has 81 tensor elements $\chi_{ijk\ell}^{(3)}$. This number is, of course, drastically reduced by symmetry. For example, in the cubic symmetry $\overline{4}3m$, there are only four independent elements. In an isotropic fluid there are only three. These numbers may be further reduced by frequency degeneracies. Each element of $\chi^{(3)}$ consists of a sum of 48 terms. Explicit expressions for these have been published[2], and each term has a typical form, with three resonant factors in the denominator

$$\chi^{(3)} = \frac{1}{6} NL\hbar^{-3} \sum_{gk,n,j}$$

$$\frac{\mu_{gk} \mu_{kn} \mu_{nj} \mu_{jg} \rho_{gg}^{(o)}}{(\omega_{kg}-\omega_1-i\Gamma_{kg})\{\omega_{ng}-i\Gamma_{ng}-(\omega_1-\omega_2)\}\{\omega_{jg}-i\Gamma_{jg}-(\omega_1-\omega_2+\omega_3)\}}$$

$$+ \text{ 47 other terms} \qquad (4)$$

where N is the number of particles per unit volume, μ_{gk} is the electric dipole matrix between states g and k, etc., $\hbar\omega_{gk}$ is the energy difference between this pair of states, and Γ_{kg} is the damping of the off-diagonal element of the density matrix, corresponding to the homogeneous width of the one-photon transition.

The different terms are distinguished by the time ordering of the photon creation and annihilation processes and the damping mechanism. The evolution of the density matrix operator can be obtained from standard higher order time-dependent perturbation theory. T. K. Lee and coworkers [3,4] have applied a diagrammatic approach which facilitates a systematic accounting of the various terms. It is important to consider separately the evolution of <bra| and |ket> state functions, since the material system is also subjected to random interactions which lead to damping. Therefore each term of $\chi^{(3)}$ becomes a complex quantity. It is often possible to single out one or several resonant terms in $\chi^{(3)}$ and lump the remaining terms in a nonresonant contribution.

Various examples in Fig. 1 include the nonresonant process, two-photon resonant processes, and combinations of one-photon and two-photon resonant processes. Terms resonant at the combination frequencies, $\omega_1 - \omega_2$, $\omega_3 - \omega_2$ and $\omega_1 + \omega_3$, correspond to Raman type processes and two-photon absorption, respectively. If there are no one-photon resonant terms, then $\chi^{(3)}$ may be written as

$$\chi^{(3)} = \chi^{NR}\left[1 + \frac{\alpha_R}{\omega_R - (\omega_1 - \omega_2) - i\Gamma_R} + \frac{\alpha_{R'}}{\omega_{R'} - (\omega_3 - \omega_2) - i\Gamma_{R'}} + \frac{\alpha_E}{\omega_E - (\omega_1 + \omega_3) - i\Gamma_E} \right] \tag{5}$$

The observed generated intensity at ω_4 is proportional to $\left| \chi^{(3)} \right|^2$. Here ω_R and $\omega_{R'}$ are resonant frequencies for a Raman transition and $\hbar\omega_E$ is the energy of an excitation reached by a two-photon absorption

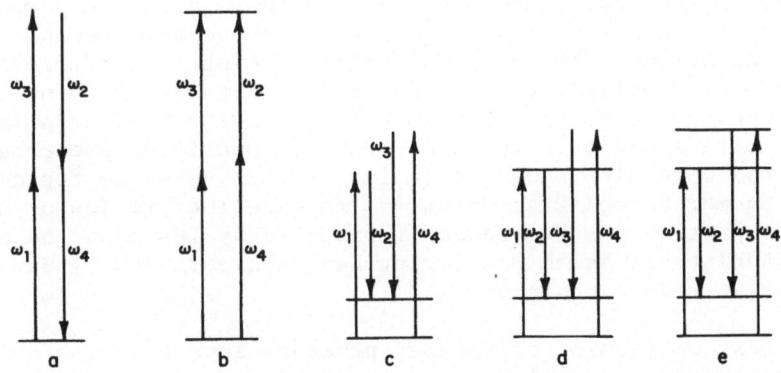

Fig. 1. Various resonant situations in light generation at the combination frequency $\omega_4 = \omega_1 - \omega_2 + \omega_3$ by a material system in the ground state |g>. (a) nonresonant parametric mixing; (b) two-photon absorption resonant mixing; (c) Coherent Antistokes Resonant Scattering (CARS); (d) one-photon resonant CARS; (e) all four one-photon transitions are resonant.

process. Far away from any resonance it is correct to describe this as a parametric process in which one quantum each at ω_1 and ω_2 is destroyed, and one quantum each is added to the beams at ω_3 and ω_4. Exactly at a Raman resonance $\omega_1 - \omega_2 = \omega_R$, a different language is more appropriate. The imaginary part of $\chi^{(3)}(-\omega_4, \omega_1, \omega_2, -\omega_3)$ should be considered as an interference term in the Raman transition probability between the two states with energy difference $\hbar\omega_R$. This transition can be accomplished by the absorption of $\hbar\omega_1$ and the emission of $\hbar\omega_2$ but also by the absorption of $\hbar\omega_4$ and the emission of $\hbar\omega_3$. This point has also been noted by Taran,[5,6] but was not taken into account in the analysis by Anderson.[7]

Recently, experimental attention has also been devoted to one-photon resonant terms in $\chi^{(3)}$. In that case the medium becomes absorbing at one or more of the frequencies ω_1, ω_2, ω_3 and ω_4. One may distinguish situations in which two or even three factors in the denominator are simultaneously resonant. The simultaneous occurrence of a one-photon resonance and a Raman resonance leads to coherent resonant Raman scattering and resonant CARS.

Furthermore, a variety of distinct polarization geometries must be considered. The various nonresonant and resonant terms in $\chi^{(3)}$ will exhibit different tensorial properties. Consequently the polarization properties of the light generated at ω_4 will depend not only on the polarization directions of the incident beams at ω_1, ω_2 and ω_3, but also on the frequencies. This is exploited, for example, in nonlinear ellipsometry by Akhmanov and coworkers[8] and also in the Raman Induced Kerr Effect Scattering[9] and polarization spectroscopy.[10]

While the general framework sketched above is quite compact, it describes a rather bewildering array of phenomena, because of the many variations offered by different combinations of frequencies, wave vectors and polarization directions. Fortunately, a number of excellent reviews have been published. A concise general survey of nonlinear optics has been given by Shen.[11] Nonlinear spectroscopy, in atoms and molecules and especially crystals, was the topic of an E. Fermi Summer School Proceeding.[12] An excellent review of coherent Raman spectroscopy was recently prepared by Levenson and Song.[10] Coherent Antistokes Raman Scattering has been reviewed by Taran, and others at recent conferences.[5,6]

The next section of this paper presents some examples of nonlinear spectroscopy of solid state excitations, although the formalism is also applicable to atomic gases and molecular fluids. The choices are somewhat arbitrary and limitations of space and time prohibit a comprehensive survey. A final section is devoted to the topic of conjugate wave front generation and (nearly) degenerate four-wave mixing, because during the past two years this topic has received much attention, both in condensed matter and in vapors.

2. EXAMPLES OF NONLINEAR WAVE MIXING SPECTROSCOPY IN CRYSTALS

A. Diamond

Many investigations have been made with "three-wave mixing", for which $\omega_1 = \omega_3$ and $\underset{\sim}{k}_1 = \underset{\sim}{k}_3$ in Eqs. (1-3). Light generated at the frequency $\omega_4 = 2\omega_1 - \omega_2$ in the direction of the wave vector $\underset{\sim}{k}_4 = 2\underset{\sim}{k}_1 - \underset{\sim}{k}_2$ is detected. When $\omega_1 - \omega_2$ is resonant with a Raman active mode of the material, ω_4 may be called the antistokes frequency. Figure 2 shows the intensity of the generated light as $\omega_1 - \omega_2$ is tuned through the Raman active optical phonon mode in diamond.[13] Note the destructive interference between the resonant and nonresonant term in Eq. (5) in the high frequency side of the Raman resonance. The experimental curves can be fitted by a Lorenzian shape

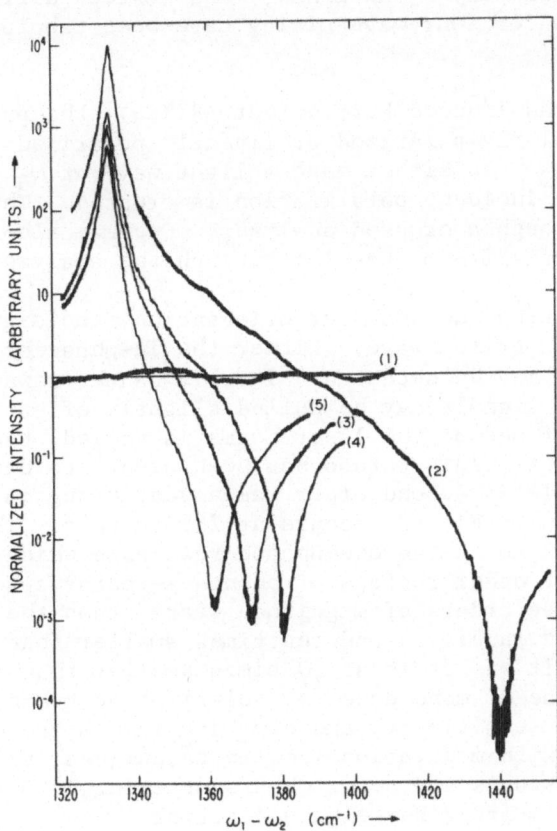

Fig. 2. Intensity of light at $2\omega_1 - \omega_2$ generated by two incident beams ω_1 and ω_2 in diamond, as a function of $\omega_1 - \omega_2$. The five curves are for different orientations of the polarization direction (after Levenson and Bloembergen[13]).

out to 120 cm^{-1} from the Raman-active optical phonon resonance, i.e. more than one hundred times its width. Furthermore, the data provide an accurate calibration of the nonresonant and resonant part. When $2\omega_1$ exceeds the band gap, the term for two-photon absorption processes in Eq. (5) makes a contribution. This effect is most noticeable through the filling-in of the minimum.[13]

The five different curves are for different polarizations of the three waves. If all wave vectors are parallel to a cubic axis, no Raman resonance occurs. It is seen that the Raman resonance displays a different anisotropy than the nonresonant part. If the waves at ω_1 and ω_3 have polarizations of 45° with respect to each other, the wave at ω_4 becomes elliptically polarized and its major axis rotates through 180° as $\omega_1 - \omega_3$ is varied through the Raman resonance.[13] Such polarization effects have been exploited in the development of nonlinear ellipsometry by Akhmanov,[8] and sixteen different polarization situations for four-wave mixing have been tabulated by Levenson and Song.[10]

In the Raman induced Kerr effect (RIKES), the pump wave at ω_1 is either circularly polarized or linearly polarized at 45° to the probe wave at ω_3. In either case a light wave at ω_3 polarized orthogonal to the incident polarization is created. This light may be detected through a crossed analyzer. One may also detect an antistokes signal $2\omega_1 - \omega_2$ passing through the analyzer.

By off-setting the analyzer orientation, the signal may be mixed with a comparison wave. Either the in-phase or 90° out-of-phase component may be detected. In this manner unwanted nonresonant background signals may be nulled slightly off-resonance. As the frequency of one of the laser beams is varied, the desired signal stands out. This method has been used[14] to demonstrate convincingly stimulated second order Raman scattering in diamond. This resonance, shown in Fig. 3, occurs at 2668.6 cm^{-1}. There is a distinct shift from twice the one-phonon resonance, which may be ascribed to the fact that other regions of phonon k-vector space are involved. The peak is three orders of magnitude weaker than the first order Raman peak shown in Fig. 4 and ten times smaller than the nonresonant contribution. It was at least 20 times smaller than the fluorescent background in the diamond used.[14] Polarization techniques can thus give remarkable sensitivity, and c.w. dye lasers are now being used, in combination with modulation lock-in techniques, to detect both coherent Raman stokes and antistokes scattering,[15] Brillouin scattering[16] and stimulated Rayleigh scattering.

B. Cuprous Chloride

This crystal has $\overline{4}3m$ symmetry and lacks a center of inversion. Thus, the polariton mode is both Raman and infrared active, and has a characteristic dispersion curve. Resonance may be achieved by

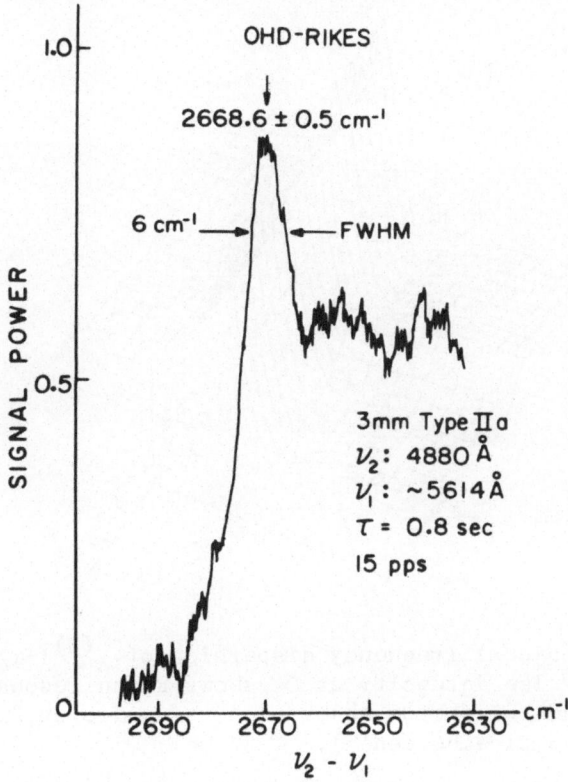

Fig. 3.   Stimulated coherent second order antistokes scattering in
diamond (after Eesley and Levenson[14]).

either varying $\omega_1 - \omega_2$, or by varying the angle, so that $\underset{\sim}{k}_1 - \underset{\sim}{k}_2$
changes at fixed ω_1 and ω_2. The latter method was first used by
De Martini[17] to measure the polariton dispersion in GaP.

Since $\chi^{(2)}$ does not vanish, second harmonic generation and
other lower order processes occur. The $\chi^{(3)}$ processes should be
carefully distinguished from two-step processes. Kramer and Bloembergen[18] have observed the generation of intensity at $2\omega_1 - \omega_2$, where
both ω_1 and ω_2 could be independently varied. For $2\omega_1$ in the vicinity of the sharp Z_3 exciton resonance at 386.4 nm wavelength, and
simultaneously $\omega_1 - \omega_2$ in the vicinity of the infrared polariton
resonance at 210 cm^{-1}, a characteristic two-dimensional frequency
dispersion of $\chi^{(3)}$ is observed, as shown in Fig. 4. The nonresonant
nonlinearity may thus be compared directly with that of the exciton-
polariton and phonon-polariton excitations. Dispersive features
in strongly absorbing ultraviolet and infrared regions are observable
by utilizing light beams only in the visible transparent region.
Thus the temperature dependence of the exciton damping could be observed, while the excitons were created far away from the surface.

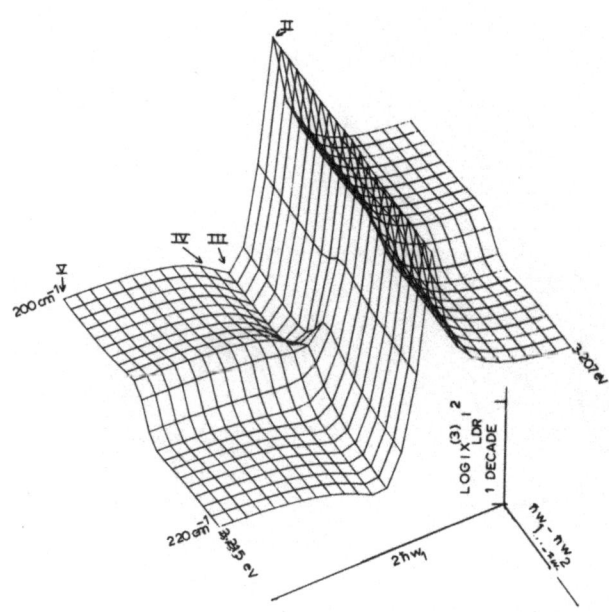

Fig. 4.   Two-dimensional frequency dispersion of $\chi^{(3)}(-\omega_4,\omega_1,-\omega_2,\omega_1)$
          in CuCl.   The intensity at ω_4 shows Raman resonance at
          $\omega_1- \omega_2$ and two-photon absorption resonance at $2\omega_1$ (after
          Kramer and Bloembergen[18]).

          Frohlich[19] measured the upper branch exciton dispersion curve
by two-photon absorption about a decade ago.   More recently two-
photon absorption has been used to create biexcitons.[20]   This process
is remarkably effective because the required one-photon energy is
only slightly below the single exciton resonance.   There is consequently a giant enhancement of $\chi^{(3)}$.   The lineshape of the biexciton
excitation has features that are being attributed to a Fano-type
interference of the biexciton level with the conduction band continuum.   The biexciton can be created with momenta between zero and
$2k_1$, depending on the angle between the two incident photons at ω_1.
The fluorescent decay channels include decays into one photon and
a transverse or longitudinal single exciton.   The fluorescent spectra
have features that vary with the intensity of the incident light at
ω_1.   This has recently been interpreted as a Bose condensation of
the biexcitons at high pump intensities.[20]

C.   Two-Photon Absorption Coefficients in Ionic Crystals

          Several of the same techniques that have contributed to progress
in Raman spectroscopy may also be used to improve two-photon spectroscopy.   In particular, the two-photon cross section may be com-

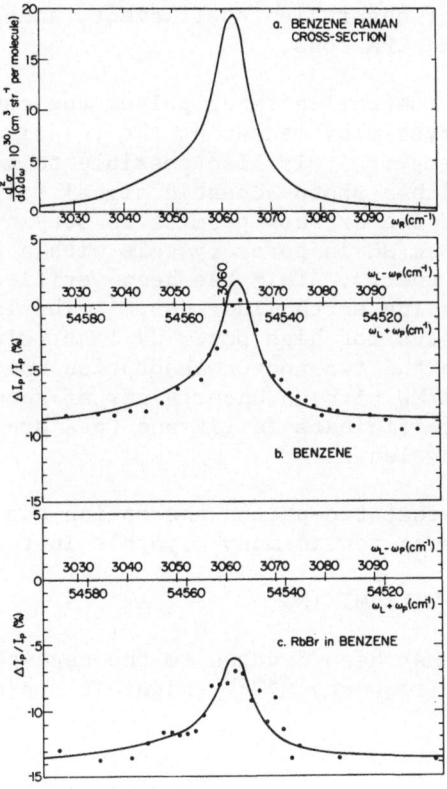

Fig. 5. Two-photon absorption loss in RbBr compared with the
Raman gain in benzene (after Prior and Vogt[21]).

pared directly with a known Raman cross section. The use of composite samples is advantageous to calibrate a two-photon cross
section in terms of the known Raman cross section of a molecular
fluid, such as benzene. Quite recently Prior and Vogt[21] have so
measured the two-photon cross section in some alkali halides, as
well as in benzene itself. The experimental results are shown in
Fig. 5. The strong pump pulse was at the second harmonic of a ruby
laser. Part of this beam was used to pump a UV dye laser to obtain
a signal beam at ω_2. The Raman gain and the two-photon absorption
loss depend in precisely the same manner on the intensities of the
two beams. Thus the influence of temporal and spatial fluctuations
in the laser pulses is largely eliminated in this comparison method.

The intensity in the samples varies as

$$\frac{dI_2}{dz} = -\alpha \, I_2 - \beta \, I_1 \, I_2 \tag{6}$$

where α is the linear absorption coefficient, and β is the algebraic sum of Raman gain and TPA loss.

With carefully controlled laser pulses the coefficient β may also be measured directly by measuring the relative attenuation as a function of intensity. It is also possible to measure the non-linear absorption with a photo-acoustic signal following the laser pulse. The state of the art now permits to set an upper limit on β of about 3×10^{-6} cm/MW in pure crystals with a band gap larger than the two-photon energy. This has been verified for pure alkali halides and for alkaline earth fluorides.[22] The latter materials are suitable candidates for high power UV laser windows. For example, at 266 nm wavelength the two-photon absorption coefficient in NaCl is $\beta = 3.5 \times 10^{-3}$ cm/MW with an uncertainty of about 25 percent, but the corresponding coefficients in LiF and CaF_2 are less than 10^{-5} cm/MW at the same wavelength.[23]

It is expected that two-photon absorption edges will be measured with considerable precision in many crystals in the near future.

3. DEGENERATE FOUR-WAVE MIXING

Much attention has been devoted to the case that all four light waves have the same frequency.[24-30] Figure 6 depicts the choice of

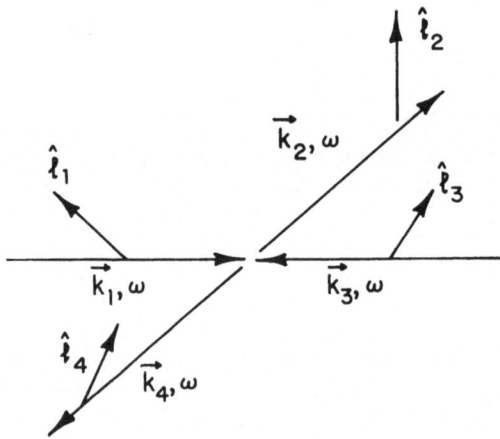

Fig. 6. Degenerate four-wave mixing, used in phase conjugate, frequency selective reflection.

wave vectors, $k_1 = -k_3$ and $k_4 = -k_2$. In the strong standing wave pump field of beams 1 and 3, a signal beam incident at k_2 produces an output wave in the reflected direction with an amplitude proportional to

$$E_4 \propto \chi^{(3)}(-\omega,\omega,-\omega,\omega)\ E_1\ E_2^*\ E_3 \tag{7}$$

Thus the backward wave has the conjugate phase of the input wave E_2. The backward wave k_4 will retrace the input beam k_2 exactly, regardless of the phase distortions (time reversal). This permits real time phase corrections, and the effect may be regarded as real time holography.[26] First the waves k_1 and k_2 create a diffraction pattern with corresponding changes in the complex index of refraction. The wave E_3 is diffracted by this grating to yield the reconstructed wave E_4. This same wave is also obtained as the diffraction of beam 1 from the grating created by the interference of beams 2 and 3. The two diffraction gratings have different periodicities, determined by $|k_1 - k_2|^{-1}$ and $|k_3 - k_2|^{-1} = |-k_1 - k_2|^{-1}$, respectively. The effect of the two gratings may be separated by choosing the polarization directions[30] as shown in Fig. 6. The effect has been observed in atomic vapors, in molecular fluids and in solids. Off-resonance a phase grating may be produced. In liquid CS_2, for example, this is caused by partial orientation of the molecules which have a strong anisotropic polarizability. The effect is especially strong near strong absorption lines or at laser emission lines. It has been demonstrated in CO_2 lasers[28,29] at 10 μm wavelength, as well as in solid state laser materials such as ruby[25] and neodymium.[27] In these cases the grating is formed by the spatial modulation of the population in the excited state due to saturation. If the decay of the gratings is due to spatial diffusion, the grating with the small period will decay more rapidly. Thus transient measurements would also permit the separation of the two gratings and a determination of the diffusion constants.[30]

The method can also be applied to the case that $\omega_2 = \omega_1 + \Delta\omega = \omega_3 + \Delta\omega$. In that case $\omega_4 = \omega_1 - \Delta\omega$. Momentum matching is only approximately conserved. But if the length of the interaction region ℓ satisfies the condition $2n\Delta\omega c^{-1}\ell \ll 1$, this imposes no limitation. The sharp resonance of $\chi^{(3)}(-\omega,\omega,-\omega,\omega)$ in atomic vapors leads to frequency-selective reflection filters.

Jacobson and Shen[16] have shown that if $\Delta\omega$ corresponds to the Brillouin shift of an acoustic wave, considerable enhancement of scattering into ω_4 (and ω_2) may occur. Suitable choice of polarization directions again gives excellent discrimination against background signals even for $\Delta\omega$ as small as 0.1 cm^{-1}. Inelastic Rayleigh scattering may also be probed by these new techniques.

In conclusion, four-wave mixing spectroscopy is in a period of active growth. New contributions to spectroscopy of condensed matter, as well as to atomic spectroscopy, are being obtained. The wide variety of manifestations all fit into a general framework outlined in this brief review.

REFERENCES

1. J. Armstrong, N. Bloembergen, J. Ducuing and P. S. Pershan, Phys. Rev. 127:1918 (1962).

2. N. Bloembergen, H. Lotem and R. T. Lynch, Jr., Indian J. Pure and Appl. Phys. 16:151 (1978).
    The following misprint should be corrected in the expression for $K_1(\Omega_u, \Omega_v)$ on page 157: "ω_{kg}" on the right hand side should read "ω_{kj}".

3. S. Y. Yee, T. K. Gustafson, S. A. J. Druet and J. P. Taran, Opt. Commun. 23:1 (1977).

4. S. Y. Yee and T. K. Gustafson, Phys. Rev. B 18:1597 (1978).

5. J. P. Taran, in: "Laser Spectroscopy III," J. L. Hall and J. L. Carlsten, eds., Springer, Berlin (1977), p. 315, and references quoted therein.

6. J. P. Taran, in: "Tunable Lasers and Applications," A. Mooradian, T. Jaeger and P. Stokseth, eds., Springer, Berlin (1976), p. 378.

7. H. C. Anderson, private communication, also quoted in L. A. Carreira, L. P. Goss and T. B. Malloy, J. Chem. Phys. 69: 855 (1978). R. S. Hudson and H. C. Anderson, "Molecular Spectroscopy," a specialist periodic report, Burlington House, London (1978), Vol. 5, p. 142.

8. S. A. Akhmanov, A. F. Bunkin, G. G. Ivanov and N. I. Koroteev, J.E.T.P. 74:1272 (1978).

9. D. Heiman, R. W. Hellwarth, M. D. Levenson and G. Martin, Phys. Rev. Lett. 36:189 (1976).

10. M. D. Levenson and J. J. Song, "Coherent Raman Spectroscopy", Springer, Berlin, to be published.

11. Y. R. Shen, Rev. Mod. Phys. 48:1 (1976).

12. N. Bloembergen, ed., "Nonlinear Spectroscopy, Course 64 of the E. Fermi International School of Physics," North-Holland Publishing Co., Amsterdam (1977).

13. M. D. Levenson and N. Bloembergen, Phys. Rev. B 10:4447 (1974).

14. G. L. Eesley and M. D. Levenson, Opt. Lett. 3:178 (1978).

15. A. Owyoung, Opt. Commun. 16:266 (1976).

16. A. G. Jacobson and Y. R. Shen, Appl. Phys. Lett. 34:464 (1979).

17. See F. De Martini, ref. 11, p. 319, and references quoted therein.

18. S. D. Kramer and N. Bloembergen, Phys. Rev. B 14:4654 (1976).

19. D. Frohlich, B. Staginnus and E. Schonherr, Phys. Rev. Lett. 19:1032 (1967); D. Frohlich, E. Mohler and P. Wiesner, Phys. Rev. Lett. 26:554 (1971).

20. L. L. Chase, N. Peyghambariam, G. Grynberg and A. Mysyrowicz, Phys. Rev. Lett. 42:1231 (1979).

21. Y. Prior and H. Vogt, Phys. Rev. B June 15 (1979).
22. P. Liu, W. L. Smith, H. Lotem, J. H. Bechtel, N. Bloembergen
     and R. S. Adhav, Phys. Rev. B 17:4620 (1978).
23. P. Liu, R. Yen and N. Bloembergen, Appl. Opt. 18:1015 (1979).
24. R. W. Hellwarth, J. Opt. Soc. Am. 67:1 (1977).
25. P. F. Liao and D. M. Bloom, Opt. Lett. 3:4 (1978).
26. A. Yariv, Opt. Commun. 25:23 (1978).
27. A. Tomita, Appl. Phys. Lett. 34:463 (1979).
28. R. C. Lind, D. G. Steel, M. B. Klein, R. L. Abrams, C. R.
     Giuliano and R. K. Jain, Appl. Phys. Lett. 34:457 (1979).
29. R. A. Fisher and B. J. Feldman, Opt. Lett. 4:140 (1979).
30. D. S. Hamilton, D. Heiman, J. Feinberg and R. W. Hellwarth,
     Opt. Lett. 4:124 (1979).

COHERENT RAMAN ELLIPSOMETRY OF LIQUID WATER: NEW

NEW DATA ON THE VIBRATIONAL STRETCHING REGION

N. I. Koroteev*, M. Endemann and R. L. Byer

Applied Physics Department
Stanford University
Stanford, California 94305

I. INTRODUCTION

The unambiguous resolution of close and overlapping lines is one of the most important and difficult problems in spectroscopy. In spontaneous Raman spectroscopy of condensed substances the solution to this problem is a curve fitting procedure involving a formal resolution of observed bands into a number of symmetrical lines of chosen shape and intensity. This technique, however, suffers from the lack of uniqueness as a consequence of the lack of resolution.

A well known and important example of the lack of uniqueness is the variety of models proposed to describe the spontaneous Raman lineshape of liquid water in the 3200-3600 cm^{-1} stretching vibrational region.[1-5] Each of the models gives nearly identical shape to the polarized and depolarized Raman bands, but yet are clearly different from each other in number, position, half-widths, lineshapes and intensities of the individual components.

We show that active control of the polarization in Coherent Anti-Stokes (Active) Raman Spectroscopy (CARS), which is in fact a variant of Coherent Raman Ellipsometry (CREM), offers the spectroscopist a new tool with which to probe the inner structure of broad Raman bands. We apply the technique to the 3400 cm^{-1} band of liquid water and resolve spectral details that can be used to verify existing models of this band and thus the structure of water in the liquid state.

---

* On sabbatical leave from the Department of Physics
    Moscow State University, Moscow 117234, U.S.S.R.

II.   RESOLUTION OF OVERLAPPING BAND COMPONENTS

In the coherent Raman ellipsometry method polarization conditions are chosen to cause destructive interference between closely adjacent or overlapping band components. The ability of CREM to resolve components was discussed and experimentally used to resolve the doublet structure of the 1305 cm^{-1} Raman line of an aqueous solution of HNO_3.[6-7] Here we develop this technique in more detail with emphasis on broad featureless bands.

Consider a simple example of a pair of closely spaced Raman lines with Lorentzian lineshapes and identical linewidths and intensities but with slightly different depolarization ratios (i.e. $\rho_1 = 0.32$ and $\rho_2 = 0.35$). Figure 1a shows the resulting spontaneous Raman bandshape when the two components are closer than their FWHM linewidths. The bandshape becomes indistinguishable from the individual components and hence the band is unresolved. A measurement of the depolarization ratio through the band does not improve the situation due to the close depolarization ratios ρ_1 and ρ_2. A similar unresolved bandshape occurs for ordinary CARS where the frequence dispersion of $|\chi_{1111}^{(3)}(\omega_a; \omega_1, \omega_1, -\omega_2)|^2$ is measured as shown in Fig. 1b.

However, polarization CARS does allow the bandshape to be resolved as shown in Fig. 1c. Here the polarization vectors $\bar{e}_1$ and $\bar{e}_2$ of the linearly polarized waves at frequencies ω_1 and ω_2 make an angle $\phi = 70°$ with respect to each other. A polarization analyzer is used in the anti-Stokes beam and is set near the position to suppress the non-resonant background.[7] The three contributions to the nonlinear polarization at $\omega_a = 2\omega_1 - \omega_2$ are

$$\bar{p}(3) (\omega_a) = \chi_{1111}^{(3)NR} \bar{p}_{NR} + \chi_{1111}^{(3)R1} \bar{p}_{R1} + \chi_{1111}^{(3)R2} \bar{p}_{R2} \qquad (1)$$

where

$$\bar{p}_{NR} = 2\bar{e}_1 (\bar{e}_1, \bar{e}_2^*) + \bar{e}_2^*(\bar{e}_1, \bar{e}_1) \qquad (2)$$

gives the polarization of the non-resonant background and

$$\bar{p}_{R1,R2} = 3(1 - \rho_{R1,R2}) \bar{e}_1 (\bar{e}_1, \bar{e}_2^*) +$$

$$3\rho_{R1, R2} \bar{e}_2^* (\bar{e}_1, \bar{e}_1) \qquad (3)$$

determines the polarization of each of the Raman resonances. If the normal of the transmission plane of the analyzer is set between $\bar{p}_{R1}$ and $\bar{p}_{R2}$, their projections on this plane are of opposite sign thus leading to destructive interference. It is not important for

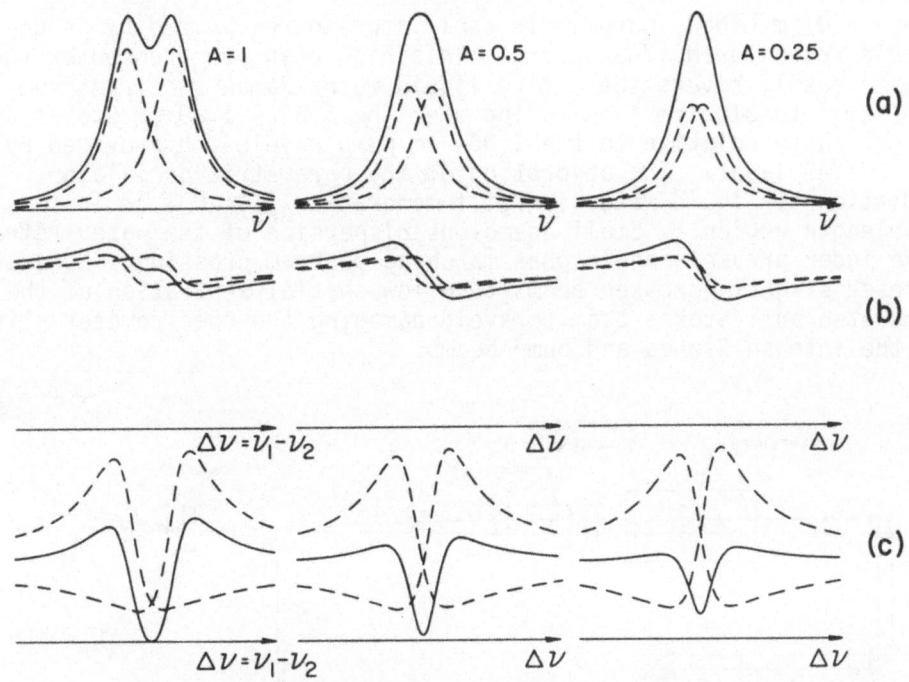

Fig. 1.   Pair of overlapping lines in a) spontaneous Raman spectroscopy; b) ordinary CARS; c) polarization CARS. The
parameter of curves is the relative distance between the
line centers, $A = (\Omega_1 - \Omega_2)/(FWHM)$. Resulting curves are
shown by solid lines, individual components are indicated
with dashed lines. In Fig. 1c, the angle ε between $\bar{p}_{NR}$
and the normal of the analyzer transmission plane is
chosen to be $\varepsilon = 0.14°$; $\bar{\chi}^{(3)R1}_{1111} / \chi^{(3)NR}_{1111} = 0.1$.

the resolution of closely spaced components whether $\bar{p}_{NR}$ is between
$\bar{p}_{R1}$ and $\bar{p}_{R2}$ or outside of this range. However, better contrast in
the interference is achieved if the analyzer is set such that peak
values of the projections of $\chi^{(3)R1}_{1111} \bar{p}_{R1}$ and $\chi^{(3)R2}_{1111} \bar{p}_{R2}$ vectors onto
the analyzer transmittance plane are of the same order as that of
the $\chi^{(3)NR}_{1111} \bar{p}_{NR}$ vector. The interference condition can be
"actively" controlled by rotation of the analyzer. Thus the closely spaced lines that compose the band can be resolved in polarization CARS even though they remain unresolved in spontaneous Raman
and ordinary CARS spectroscopy.

III. MEASUREMENTS

The experimental setup for applying the polarization CARS
technique to liquid water is shown in Fig. 2. The broad Raman
band of liquid water led to the selection of the widely tunable

1.4 - 4.0 µm LiNbO$_3$ parametric oscillator source pumped by an unstable resonator ND:YAG laser.[8] This high peak power computer tuned
source easily covers the entire liquid water Raman spectrum from
2900 cm^{-1} to 3100 cm^{-1} by tuning over the 1.53 - 1.89 µm Stokes wavelength range relative to the 1.064 µm pump wavelength provided by
the Nd:YAG laser. The absorption of the parametric oscillator
radiation by liquid water is still moderate $\ell_{ABS} \simeq 0.1$ cm in this
wavelength region.[9] Small anamolous dispersion of the water refractive index prevents angle phasematching by beam crossing. However,
we used slightly crossed beams to allow spatial separation of the
generated anti-Stokes beam to avoid damaging the spectrometer slits
by the intense Stokes and pump beams.

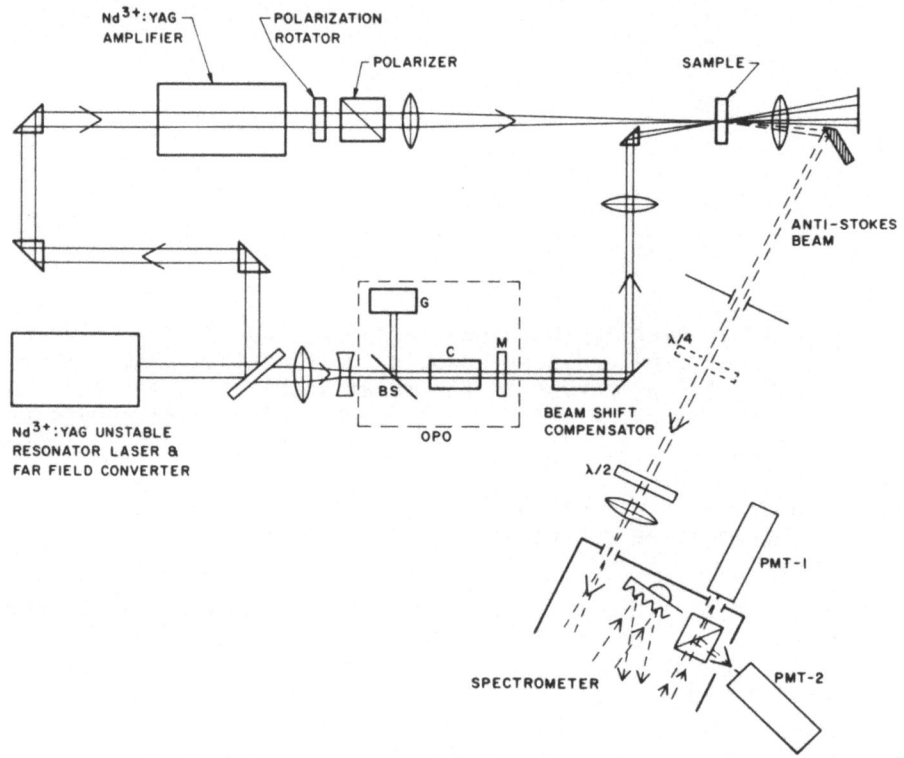

Fig. 2.   Schematic of the measurement apparatus showing the unstable
          resonator Nd:YAG laser pumped LiNbO$_3$ OPO tunable source,
          which consists of a 5 cm LiNbO$_3$ crystal (C), grating (G),
          output coupler (M) and input beamsplitter (BS). A quartz
          crystal rotator and Glan polarizer were used to set the
          polarization of the 1.064 µm pump beam at $\phi = 70°$ to the
          OPO radiation. $\lambda/4$ and $\lambda/2$ are Fresnel rhomb dispersionless
          quarter and half wave plates. PMT-1 and 2 are the signal
          and reference photomultiplier detectors.

Typical beam powers at the sample were 3.5 MW at 1.064 μm and
0.1 MW at the Stokes wave. Both beams were focussed to a diameter
of 0.5 mm inside the sample. The cell was constructed using thin
0.2 mm glass windows to decrease anti Stokes generation in the glass.
The 7 mm cell thickness was much greater than the water absorption
depth for Stokes radiation to avoid anti-Stokes generation in the
exit cell window. Measurements were made at room temperature on
singly distilled water. The temperature rise due to absorbed optical power was estimated to be less than 5°C.

A PDP-11 minicomputer was used to synchronously tune the
$LiNbO_3$ optical parametric oscillator (OPO)[10]   and the grating
spectrometer, which was set at the anti-Stokes wavelength.

The computer also collected data from signal and reference
channels on each laser pulse at a 10 Hz rate, calculated their ratio
and averaged over 50 pulses. Figure 3 shows a schematic of the
computer control and signal processing system. The statistical
deviation of the ratio was then calculated and all data were stored
on disk. After each spectral scan the computer normalized the data
to the spectral response of the RCA 7265 photomultiplier detectors,
spectral and polarization response of the 1 meter grating spectrometer and plotted the resulting points. Spectral scans were taken
in 15 cm^{-1} steps resulting in scan times of nearly half an hour.

Attention was paid to systematic changes in the anti-Stokes
signal due to dispersion of the coherence length and absorption of
the Stokes beam. In water, ℓ_{coh} = 0.6 mm at 2900 cm^{-1} and 0.2 mm
at 4100 cm^{-1} where ℓ_{coh} = $\pi/\Delta k$. The systematic variation of the
anti-Stokes signal as well as fluctuations in its intensity caused
by small intensity and direction instabilities of pumping beams
were normalized by ratioing the signal transmitted by the analyzer
to the orthogonally polarized component rejected by the analyzer
thus exploiting one of the ellipsometry techniques in CARS[7,11]
recently developed by Oudar et.al.[12]

IV.   RESULTS

For each angle position of the analyzer we took a polarization
CARS spectrum with and without a quarter wave plate in the anti-
Stokes beam. This provided complete information on the change in
elliptical polarization of the CARS signal[12] and gave spectra that
were equivalent to the two dispersion curves, ellipticity and major
axis inclination angle dispersion, usually studied by Coherent Raman
Ellipsometry (CREM).[7,11]

Figure 4 shows selected examples of the observed spectra at
various setting angles of the analyzer. Here the analyzer angle
ε is referred to the normal to the non-resonant background polar-
ization vector $\bar{p}_{NR}$. The measured angle between $\bar{p}_{NR}$ and $\bar{e}_1$ was

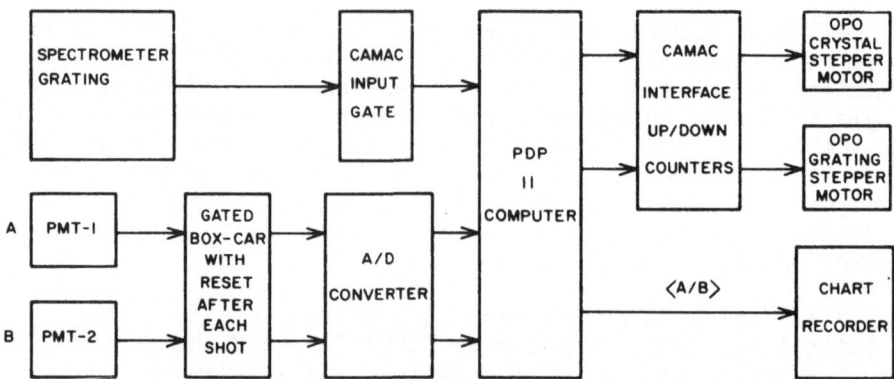

Fig. 3.   Schematic of the computer controlled tunable source
          and the data processing system.

ϕ_{exp} = 46 ± 0.5° compared to the calculated angle of ϕ_{Calc} = tan^{-1}
(1/3 tan ϕ) = 42.4° at ϕ = 70° assuming that Kleinman's symmetry
applies for '$\chi_{ijk\ell}^{(3)NR}$.[11]   No non-resonant background signal was observed for the analyzer angle set to ε = 0°.   In the present geometry, ε > 0 corresponds to the vector $\bar{p}_{NR}$ lying between $\bar{e}_1$ and the
normal to the analyzer transmission plane.   Finally, all spectra
were taken several times to verify their reproducibility.

     The spectra obtained and shown in Fig. 4 can be used to de-
termine the four dispersion curves of the real and imaginary parts
of the two independent components of the third order non-linear
susceptibility tensor '$\chi_{1111}^{(3)}$ (ω_a, ω_1, ω_1, $-\omega_2$) and '$\chi_{1221}^{(3)}$ (ω_a, ω_1,
ω_1, $-\omega_2$) since each of the spectra can be described by an algebraic
combination of the four susceptibility components.[6]   The results
of such a calculation will allow a direct comparison of the polarization CARS and spontaneous Raman spectra since the dispersion
curves of Im $\chi_{1111}^{(3)}(\Delta\omega)$ and Im $\chi_{1221}^{(3)}(\Delta\omega)$ coincide with the polarized
and depolarized spontaneous Raman band spectra.

     Moreover, the calculation of the susceptibility components
could give experimental evidence for the validity of the Kramers-
Kronigs relation between the real and imaginary parts of the nonlinear susceptibility in the vicinity of a Raman resonance of a
complicated lineshape which is of general interest.[7,15]

     Figure 5 shows the computer simulated polarization CARS spectra
of liquid water using the data of various proposed models.   For
these calculations we have taken the number of components, their
positions, FWHM, integrated intensities and depolarization ratios
from models proposed by Murphy and Bernstein[3] and Scherer and
others[5] as well as an early model due to Schultz and Hornig.[1]   In
addition, we have taken into account the non-resonant background

CARS signal generated by the front glass window of the CARS cell. This is important since the strong dispersion of the coherence length of water causes changes in the relative values of water and glass signals across the Raman band, thus distorting the spectra. For simplicity all lineshapes were assumed Lorentzian. The fitting parameter for the curves was the ratio of the peak value of the imaginary part of the resonant Raman susceptibility tensor $\chi^{(3)R}_{1111}$, to the non-resonant component $\chi^{(3)NR}_{1111}$, of water. The closest

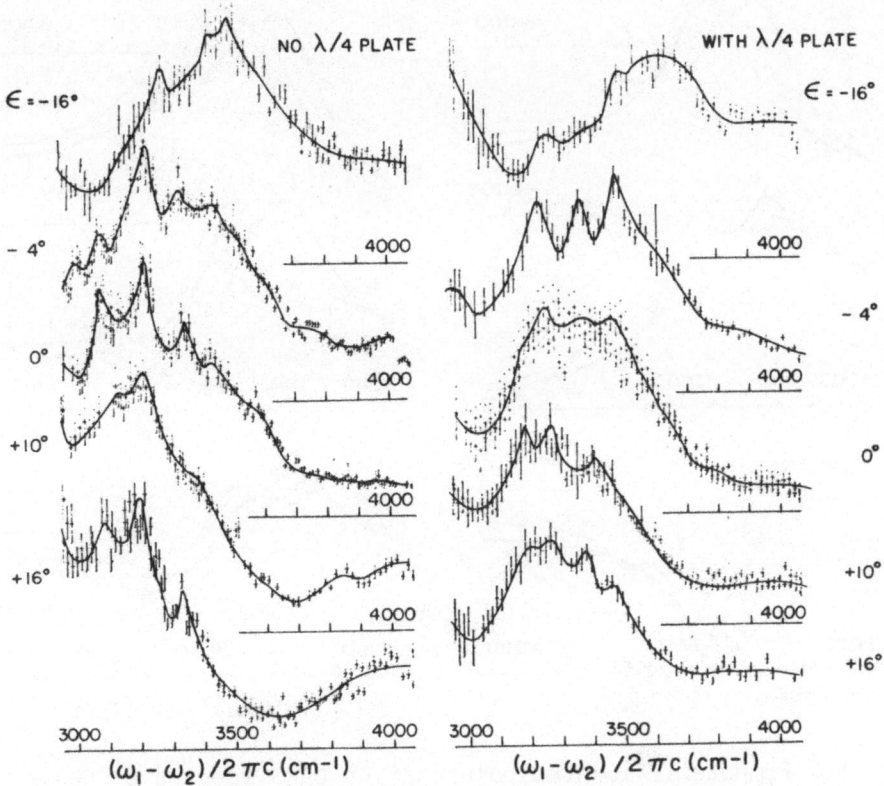

Fig. 4. Polarization CARS spectra of liquid water for various angles ε between $\bar{p}_{NR}$ and the normal of the analyzer transmission plane as a parameter. The spectra on the left show the dispersion of the ratio of the signal and reference photomultiplier signals, averaged over 50 pulses per point along the rms deviation. The spectra on the right show the same, but with a quarter wave plate placed in the anti-Stokes beam with its principle axis at an angle of -3° with $\bar{p}_{NR}$. Solid lines are drawn through experimental points to make it easier to follow.

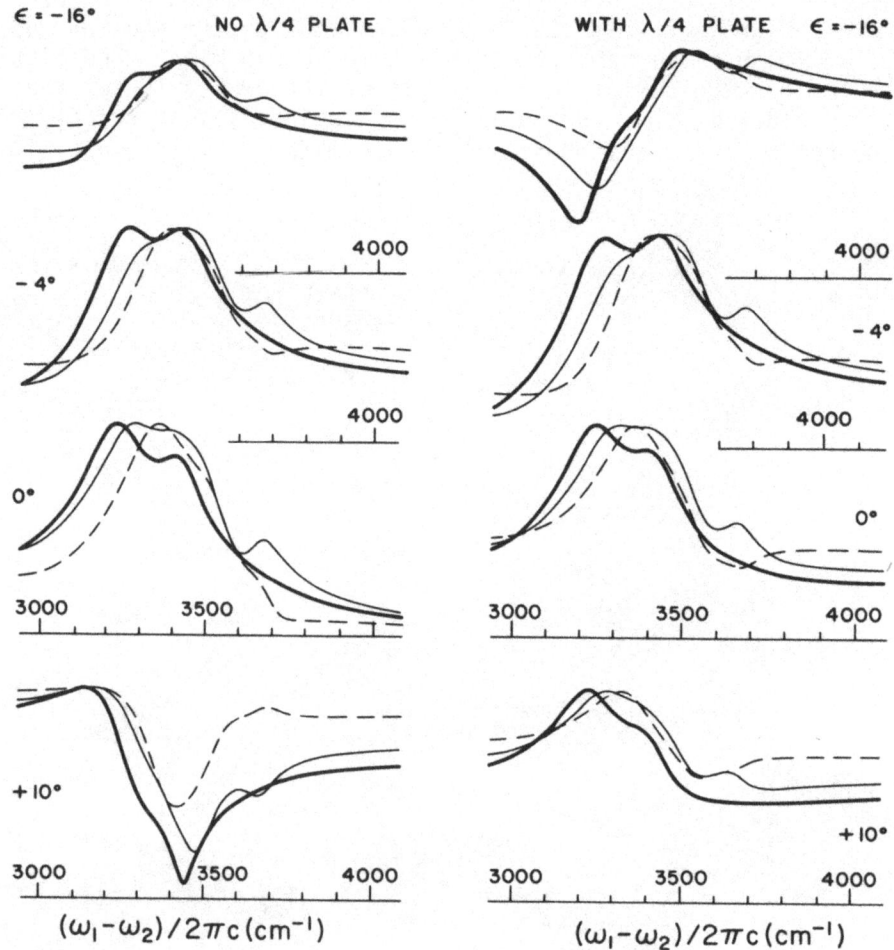

Fig. 5.   Computer simulated polarization CARS spectra of water.
          All conditions are equivalent to that of Fig. 4. Heavy
          solid lines: Murphy and Bernstein's model;[3] light
          solid lines: Scherer and others model;[5] dashed lines:
          Schultz and Hornig's model.[1] $\bar{\chi}^{(3)R}_{1111}$ / $\chi^{(3)}_{1111}$ NR = 0.65.
          Dispersion of ℓ_{coh} in water was calculated using data
          on water refractive index. The ratio of non-resonant
          cubic susceptibilities of water and glass $(\chi^{(3)NR}_{1111})$ glass/
          $(\chi^{(3)}_{1111}NR)$ water = 0.63 was taken from Levine and Bethea's
          papers,[14] (see also [15]), Table 14.1).

qualitative agreement of calculated and observed spectra was achieved with $\bar{\chi}_{1111}^{(3)R} / \chi_{1111}^{(3)NR} \approx 0.65$ which is three times less than reported by Itzkan and Leonard in their early CARS studies of H_2O.[13] Although the calculated spontaneous Raman spectra are nearly identical for the models this is clearly not the case for the calculated polarization CARS spectra. The Schultz and Hornig model results in curves qualitatively inconsistent with observed spectra at $\varepsilon > 0$. Murphy and Bernstein's model gives spectra which appear to more closely match the experimental ones. However, there is significant quantitative disagreement between theory and experiment for all of the models considered.

V. CONCLUSION

In conclusion we have shown that spectra obtained with polarization CARS are more informative and provide significantly improved resolution of broad complex Raman bands. In fact, the spectra obtained with polarization CARS are two dimensional images of a four dimensional curve which describes the spectroscopic 4-D space with coordinates Im $\chi_{1111}^{(3)}$ ($\Delta\omega$), Im ' $\chi_{1221}^{(3)}$ ($\Delta\omega$), Re $\chi_{1111}^{(3)}$ ($\Delta\omega$) and Re $\chi_{1221}^{(3)}$ ($\Delta\omega$) of the Raman resonance under investigation. The frequency detuning $\Delta\omega$ is the parameter of the 4-D curve. The changing polarization conditions can be interpreted as changing the experimenter's viewpoint of the Raman resonance. This allows a more detailed investigation of the Raman resonance without any distortion of the spectrum.

Using polarization CARS we have resolved, for the first time, the fine structure of the water vibrational Raman band by actively controlling the polarization condition. This experiment clearly demonstrates the advantages of Coherent Raman Ellipsometry,[7,11] which is sometimes, and properly, called Coherent Active Raman Spectroscopy.

ACKNOWLEDGEMENTS

We would like to thank M. A. Henesian, M. Duncan and J. Unternahrer for help in different stages of the experiment and J. K. Oudar and Y. R. Shen for helpful discussion and making available a copy of their paper[12] prior to publication. One of us, (Nikolai Koroteev), acknowledges fruitful discussions with S. A. Akhmanov, R. W. Hellwarth and M. D. Levenson. This work was partially supported by the National Science Foundation and partially by the International Research and Exchanges (IREX) program.

REFERENCES

1. J. W. Schultz, D. F. Hornig, J. Phys. Chem. 65, 2131 (1961).

2. G. E. Walrafen, in "Water, A Comprehensive Treatise", ed. by F. Franks, vol. 1, Plenum Press, New York, N. Y. Ch. 5, 1972.

3. W. F. Murphy and J. M. Bernstein, J. Phys. Chem. 76, 1147 (1972).

4. K. Cunningham and P. A. Lyons, J. Chem. Phys. 59, 2132 (1973).

5. J. R. Scherer, M. K. Go and S. Kint, J. Phys. Chem. 78, 1304 (1974).

6. N. I. Koroteev, paper presented at Fifth Vavilov Conference on Nonlinear Optics, Novosibirsk, June 1977 (unpublished). A. F. Bunkin, M. G. Karimov, N. I. Koroteev, Vestnik Moskovskogo Universiteta, Ser. Fizika, v. 19, p. 3 (1978), (in Russian).

7. S. A. Akhmanov, A. F. Bunkin, S. G. Ivanov and N. I. Koroteev, Zh. Eksp. i Teor, Fiz. 74, 1272 (1978), [Sov. Phys. - JETP, 47, 667 (1978)].

8. R. L. Herbst, H. Komine and R. L. Byer, Opt. Commun. 21, 5, (1977).

9. W. Luck, Fortschr. Chem. Forsch. Bd. 4, 653 (1964).

10. R. L. Byer and R. L. Herbst, "Parametric Oscillation and Mixing", in Topics in Applied Physics, vol. 16: Nonlinear Infrared Generation, ed. Y. R. Shen, Springer-Verlag, p. 87-137 (1977).

11. S. A. Akhmanov and N. I. Koroteev, Usp. Fiz. Nauk, 123, 405 (1977), [Sov. Phys. -Uspekhi, 20, 899 (1977].

12. J. L. Oudar, R. W. Smith and Y. R. Shen, Appl. Phys. Letts. (in press).

13. I. Itzkan, D. A. Leonard, Appl. Phys. Letts. 26, 106 (1975).

14. B. F. Levine, C. G. Bethea, J. Chem. Phys. 65, 2429 (1976). C. G. Bethea, Appl. Optics, 14, 2435 (1975).

15. R. W. Hellwarth, "Third Order Nonlinear Susceptibility of Liquids and Solids", Progr. in Quant. Electr. Pergamon, Oxford - New York, 1977, vol. 5, Part I.

TIME-RESOLVED COHERENT ANTI-STOKES RAMAN SCATTERING IN WEAKLY DISORDERED MOLECULAR CRYSTALS

R.M. Hochstrasser and I.I. Abram

Department of Chemistry and
Laboratory for Research on the Structure of Matter
University of Pennsylvania
Philadelphia, Pa. 19104

INTRODUCTION

The presence of structural or substitutional disorder in a crystal alters the structure of its states and modifies its transport properties. Traditionally both the experimental and the theoretical study of the effects of disorder have dealt with time-independent measurements in the frequency domain (such as spectral or specific heat determinations and density of states calculations) even when time-processes (such as relaxation or transport) are of interest[1].

We have undertaken a time-domain approach to the problem by studying the effects of disorder on the coherent vibrational states of molecular crystals. The experimental procedure involving coherent vibrational states has been extensively described in the literature[2]. It consists of preparing a coherent vibrational state in the system under study by use of stimulated Raman scattering and subsequently monitoring the time-evolution of its coherent amplitude by time-resolved coherent anti-Stokes Raman scattering. The molecular crystals used in our experiments were $\alpha-N_2$[3] and $p-H_2$[4]. The energy associated with the coherent excitations corresponds to the first quantum of the intra-molecular vibrational mode (2328 cm^{-1} for N_2 and 4150 cm^{-1} for H_2).

The dominant type of disorder in the nitrogen crystals studied was structural; it depended on the method of preparation of the crystal and could not be controlled in a systematic way. In the case of $p-H_2$ the dominant type of disorder was substitutional and consisted of $o-H_2$ impurities. Variation of the

447

impurity concentration between 0.2% and 2.4% provided a systematic
way to study the effects of disorder on the time-evolution of
the coherent amplitude. The presence of disorder was found to
cause a decay of coherent crystal states created and detected
through optical processes. The decay of the anti-Stokes intensity
(corresponding to the <u>square</u> of the coherent amplitude) was
monitored over 5 or 6 decades as a function of time. In every
case we studied, the decay was found to be significantly non-
exponential, the effective decay rate decreasing with time.

    Coherent states in matter are usually understood in terms
of the Bloch-vector formalism developed initially for magnetic
resonance situations. Loss of coherence is viewed as a statistical or dynamical phase randomization of the collection of
Bloch-vectors representing the coherent state. We have developed
a representation of extended coherent excitations in solids in
terms of local (molecular) Bloch-vectors. Delocalization of the
excitation is taken into account explicitly in the local time-
evolution. We have applied this representation to the problem
of coherence loss in disordered crystals. The details of the
observed decay in our experiments can be understood within this
formalism as arising from the quantum mechanical nature of the
propagation (delocalization) process of excitations through the
disordered crystal. In this presentation we discuss the qualitative concepts that enter in the consideration of coherence loss
in extended crystal states of weakly disordered molecular crystals.

COHERENT STATES IN PERFECT CRYSTALS

    In molecular crystals, intermolecular interactions cause
intramolecular vibrational states to delocalize and to form a
band of extended vibrational excitons (vibrons) each corresponding
to a definite wavevector k. In solid N_2 the vibron band is
1 cm^{-1} wide while in solid H_2 its width is 4 cm^{-1}. In both
cases the state accessible by stimulated Raman scattering (k=0)
lies at the low-energy end of the band.

    In a perfect crystal the coherent vibrational wavepacket
created by stimulated Raman scattering can be described in the
extended state representation as a minimum-uncertainty superposition of the number eigenstates of the k = 0 vibron, which
in the small amplitude approximation can be written as

$$|| \beta \gg_k = e^{-N|\beta|^2/2} \sum_{n=0}^{\infty} \frac{(\sqrt{N}\beta)^n}{\sqrt{n!}} \quad | n \gg_k \qquad (1)$$

where N is the number of molecules in the crystal and $| n \gg_k$

is the extended crystal state containing n excitations and having wavevector k. $| n \gg_k$ is a crystal eignstate to a good approximation for $n \ll N$. The mean number $\bar{n}$ of excitation quanta in the crystal in $\bar{n}=N|\beta|^2 \ll N$.

Alternatively, in the local-state representation the coherent wavepacket can be described as a product-state in which each molecule (j) in the crystal is found in the same superposition of its ground and excited vibrational states $| 0 >_j$ and $| 1 >_j$ respectively. In the small-amplitude approximation $\| \beta \gg_k$ can be written to second order in β for each molecule:

$$\| \beta \gg_k = \prod_{j=1}^{N} \{ (1- \frac{|\beta|^2}{2})|0>_j + \beta e^{ikr_j} |1>_j \} \qquad (2)$$

Because of the anharmonicity of the intramolecular vibration of N_2 and H_2, only the first excited vibrational state of each molecule need be considered, so that the two-level description of eq. (2) is adequate.

Using the Feynman-Vernon-Hellwarth representation[5], each molecule may be associated with a Bloch-vector representing its coherent excitation in a manner analogous to the formalism of coherent states in magnetic resonance situations. The vector is defined by its projections along three mutually perpendicular axes in Bloch-space, corresponding to the expectation values of the population-difference operator $\sigma_j^z = (a_j^+ a_j - a_j a_j^+)/2$ and of the operators corresponding to the real and imaginary parts of the polarization $\sigma_j^x = (a_j^+ + a_j)/2$ and $\sigma_j^y = (a_j^+ - a_j)/2i$. A macroscopic measurement of coherent amplitude of the extended coherent state corresponds to the determination of the geometric sum of x-y projections of all local Bloch-vectors in the crystal.

In the extended-state representation, the time-evolution of the coherent wavepacket is easily understood: the wavefunction of eq. (1) varies its phase cyclically at the frequency of the k-th exciton state.

In the local-state representation the time-evolution of the coherent wavepacket and of the coherent amplitude are viewed in terms of the individual time-evolutions of the local Bloch-vectors. In the absence of intermolecular interactions, molecular excitations cannot delocalize. Each local Bloch-vector in this case effects a precession around $\hat{z}$-axis of its site of residence at the precession frequency of the free molecule. The extended coherent state then time-evolves cyclically at the free-molecule frequency. In view of the delocalized nature of intramolecular excitations in molecular crystals the model of isolated local

Bloch-vectors must be modified in order to account for the delocalization process. Delocalization can be viewed within the context
of two different models.

(1) The simplest view of the delocalization process consists
of a particle-like, classical motion of the excitation quanta:
An excitation resides on a given site for a time τ and then hops
to an adjacent site where again it resides for a time τ and so on.
While the excitation resides on a site the Bloch-vector of that
site precesses at the free-molecule frequency. When the excitations
hops from one site to another, the Bloch-vector of the initial site
decays, while the Bloch-vector of the new site of residence "tips
up" and starts precessing, retaining the phase memory of the previous site of residence. The overall time-evolution of an extended
coherent state in this model is identical to that of isolated
molecules: the local Bloch-vectors precess at the frequency of the
free molecules.

(2) Delocalization consists of a wave-like quantum mechanical
motion of the individual excitation throughout the crystal. In
order to view the quantum mechanical features of the delocalization process we may examine the time-evolution of a crystal state
in which a single molecule is prepared in a small amplitude
superposition of its ground and excited states, while all other
molecules in the crystal are left in their ground state. The
Bloch-vector of the reference site effects a precession around
$\hat{z}$ - axis at the molecular frequency, while at the same time it
decays because of delocalization. In the rotating-frame of the
free molecule the motion of the local Bloch-vector is due ex-
clusively to delocalization. The effect of delocalization on the
time-evolution of the local Bloch-vector is easiest understood
for a one-dimensional crystal with nearest-neighbor interactions:
In the rotating frame of the free-molecule, the Bloch-vector
approaches the ground-state $|0>$ in an asymptotic oscillatory
fashion. The projection of the Bloch-vector and the x-y plane
corresponds to the probability-amplitude that an excitation quantum
is resident on the reference site and is given by a zeroth-order
Bessel function $g(t) = J_0 (2\gamma t)$ for a one-dimensional crystal,
where γ is the nearest-neighbor interaction energy. In the
formalism of magnetic resonance this motion of the Bloch-vector may
be thought as the result of a time-dependent effective 'delocalization field" along the $\hat{y}$ - axis applying a torque on the Bloch-vector
and driving it periodically from the $+\hat{x}$ to the $-\hat{x}$ direction and
vice-versa. As the residence amplitude of excitation decays at the
reference site the Bloch-vectors of all sites are "tipped up" in
a wave-like fashion, indicating that at any time there is a finite
probability amplitude of each site being visited by the delocalized
excitation quantum. When an extended coherent state is prepared
in a crystal, all molecules are found in the same superposition of
their ground and excited states. In that case, as the residence

amplitude of an excitation decays at a given site because of delocalization, the probability amplitude of visitation of the
reference site by other excitations increases with the proper
phase, so that the local Bloch-vector effects a precession at the
frequency of the extended state. In this model therefore the
extended coherent state time-evolves cyclically at the proper
frequency.

COHERENT STATES IN DISORDERED CRYSTALS

The simplest model of disorder consists of a random variation
of molecular excitation energies in an otherwise perfect crystal.
For the case of substitutional disorder molecular excitation
energies may assume one of two descrete values (the host or
impurity energies) while for structural disorder we may take site
energies as forming a narrow (Gaussian) distribution around the
mean molecular energy, with variance D^2. We examine the case in
which the spread of molecular energies is smaller than the width
of the exciton band.

A coherent wavepacket prepared through a laser process has a
definite wavevector, and can thus be represented by equations
analogous to (1) or (2). However, since disorder destroys the
translational symmetry of the lattice, states of definite wavevector are not crystal eigenstates in this case. The time-
evolution of the coherent wavepacket of eq. (1) in a disordered
crystal involves a loss of its minimum - uncertainty (coherent)
characteristics. The details of this time-evolution depend on the
particular superposition of crystal energy-eigenstates which
constitute a wavevector state at t = 0. They can be relatively
easily understood, however, in the local representation of eq. (2),
as this representation interfaces directly with the formalism of
coherent processes used in magnetic resonance through the Bloch-
vector language.

If there were intermolecular interactions the local Bloch-
vectors would precess at the individual mismatch frequencies. For
the case of structural disorder this would give rise to a
Gaussian decay of the coherent amplitude,

$$G(t) = \exp\left(-D^2 t^2/2\right) \tag{3}$$

like the phenomenon of free-induction decay in any inhomogeneous
system with a Gaussian distribution of transition frequencies.
For the case of substitutional disorder, in which the crystal is
composed of two molecular species with two distinct transition
frequencies, the coherent amplitude would display quantum beats
and its magnitude would vary cyclically at the difference

frequency. The inclusion of intermolecular interactions and the possibility of delocalization for intramolecular excitations, change the time-evolution of the observable coherent amplitude.

In the simplest model of delocalization, the particle-like classical motion of an excitation quantum causes an individual excitation to sample many precessional frequencies within the experimental time-scale, as it moves from site to site. Associating a Bloch-vector with each moving excitation (rather than with each site) we may describe its time-evolution at any time as a precession at the frequency of its site of residence. Visitation of sites with randomly distributed frequencies produces a random deviation (advancement or retardation) of the precession phase of each Bloch-vector from the mean. Visitation of many sites produces on the average a restoration of any initial phase advance or retardation, indicating that the overall coherent amplitude decays at a slower rate when delocalization is permitted. This is analogous to the phenomenon of "motional narrowing" observed in magnetic resonance. For the case in which the average phase advance during a residence time τ is small ($D\tau \ll 1$) the random phase-deviations give rise to an exponential decay of the overall coherent amplitude given by

$$G(t) = \exp\left(-D^2\tau t/2\right) \qquad\qquad\qquad (4)$$

Our experimental results, however, show a significant deviation from exponential behavior, indicating that the particle-like classical model for delocalization of the excitation is not adequate. Calculation of the quantum mechanical time-evolution of the local Bloch-vectors for the $k = 0$ state of the N_2 or H_2 vibron bands produces a decay of the overall coherent amplitude which is slower than exponential and closely resembles our experimental results. Unlike the case of the perfect crystal, however, the quantum mechanical motion of the local Bloch-vectors in a disordered crystal cannot be represented by a simple model. The reason is that the distribution of precessional Hamiltonians does not commute with the delocalization Hamiltonian (since delocalization couples the Bloch-vectors to each other) so that the precessional and the delocalization motions of the Bloch-vectors cannot be considered separately.

For the purpose of illustration, however, we may introduce a classical feature in the time-evolution of the local Bloch-vectors and consider it as being decomposable into two types of motion: a precession about the vertical axis at the frequency of the molecule on which the excitation resides, and a motion with components parallel to the vertical axis due to intermolecular interactions (delocalization). We consider the case of a one-dimensional crystal with nearest neighbor interactions in

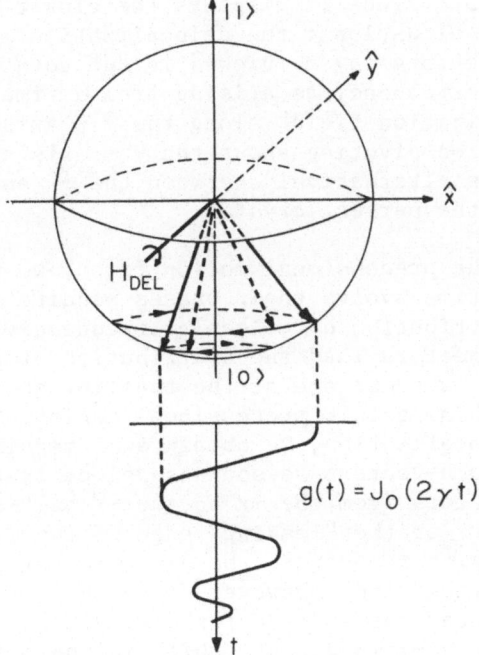

Figure 1.   Model for the quantum-mechanical time-evolution of
             local Bloch-vectors in a weakly disordered crystal:
             The distribution of local precessional frequencies
             given rise to a spread of the local Bloch-vectors.
             In the rotating frame of the mean molecular frequency,
             delocalization drives the local Bloch-vectors from
             the $+\hat{x}$ to the $-\hat{x}$ half-space. Delocalization inverts
             the order of precessing local Bloch-vectors so that
             in the subsequent time-evolution their spread
             decreases producing an echo-like effect. The net
             effect of the quantum mechanical aspect of delocal-
             ization is a slower rate of spreading of the local
             Bloch-vectors.

which we have excited coherently the k=π/2 exciton state whose
energy lies at the center of the exciton band. For this state
in the perfect crystal, the frequency of precession of the local
Bloch-vectors is identical to that of the free molecule. Although
this case is never encountered in optical experiments it is the
easiest to illustrate since it displays the closet analogies to
magnetic resonance situations: the delocalization motion of the
individual Bloch-vectors may be viewed in the rotating frame of
the mean molecular frequency as arising from a time-dependent
effective "delocalization field" along the $\hat{y}$ - axis and constituting
an oscillatory motion pivoting about the $\hat{y}$ - axis and driving the
local Bloch-vectors alternatively between the $+\hat{x}$ and $-\hat{x}$ half-spaces
as in the case of the perfect crystal.

Considering the precessional motion first, we note that as
the Bloch-vectors time-evolve their phases acquire a spread
because of the distribution of molecular frequencies. The
faster-precessing vectors lead the distribution of phases while
those which precess slowest are at the trailing end of the
distribution. Combining this precessional motion with the motion
resulting from delocalization, we obtain an inversion of the
ordering of the Bloch-vectors as soon as delocalization drives
the local Bloch-vectors from the $+\hat{x}$ to the $-\hat{x}$ half-space. That
is the Bloch-vectors of the "leading" edge of the distribution
are found behind (with respect to the direction of precession)
those of the "trailing" end. However since the former precess
faster than the latter the spread of phases diminishes with time
giving rise to an echo-like effect, which is the direct manifestation of the quantum mechnical (wave-like) nature of exciton motion.

The extent and time-dependence of this exciton-echo depends
on the spatial phase relationships of the initial coherent state
i.e. its wavevector and the lattice topology. Consideration of
the exciton-echo gives rise to a decay behavior which depends on
the phase structure of the coherent state under consideration.
For the k = 0 state in close-packed lattices (fcc like N_2 or hcp
like H_2) the exciton-echo gives rise to a component of the
coherent amplitude which decays at a rate slower than exponential.

REFERENCES

1. R.J. Elliot, J.A. Krumhansl and P.L. Leath, The theory and
        properties of randomly disordered crystals and related
        physical systems, Rev. Mod. Phys. 46:465 (]974).
2. A. Laubereau and W. Kaiser, Vibrational dynamics of liquids
        and solids investigated by picosecond laser pulses,
        Rev. Mod. Phys. 50:607 (1978).

3. I.I. Abram, R.M. Hochstrasser, J.E. Kohl, M.G. Semack and D.White, Coherence loss for vibrational and librational excitations in solid nitrogen J. Chem. Phys. in press.

4. I.I. Abram, R.M. Hochstrasser, J.E. Kohl and D. White, to be published.

5. R.P. Feynman, F.L. Vernon, Jr. and R.W. Hellwarth, Geometrical Representation of the Schrödinger equation for solving Maser problems, J. Appl. Phys. 28:49 (1957).

INTENSE LIGHT RESONANCE SCATTERING: SPECTRA AND PHOTON CORRELATIONS

P.A. Apanasevich

Institute of Physics of the Byelorussian Academy
of Sciences, Leninsky Pr. 70, Minsk, 220602, USSR

The advent of lasers opened new possibilities in
investigation of resonant light scattering and stimulated the
development of nonlinear theory of spectral and correlation
properties of this interesting phenomenon. Some predictions of
that theory for light scattering by atoms, molecules, and ions
perturbed by random influence of the environment are discussed
qualitatively in this paper, which is based on investigations
carried out in the Institute of Physics of the Byelorussian
Academy of Sciences[1-4].

We begin with consideration of resonant scattering of weak
light by a perturbed atom. In this case the processes of resonant
light scattering can be presented as a sum of elementary
transitions shown schematically by different arrows in Fig. 1.
Affected by the monochromatic light wave with amplitude $\vec{E}_o$ and
frequency ω_0 resonant with the unforbidden transition $A \leftrightarrow B$, the
atom is excited from the state B to the coherent state possessing
energy $W_C = W_B + \hbar\omega_0$ and wave function proportional to the
product $\psi_A E_o$ of the wave function of the excited resonance
state A and the incident light amplitude. The probability of that
excitation is given by

$$A = |V|^2 \Gamma_{AB}^\varepsilon / [\varepsilon^2 + (\Gamma_{AB}^\varepsilon/2)^2] , \qquad (1)$$

where $V = \vec{P}_{AB}\vec{E}_o/\hbar$ is the energy of interaction of the incident
wave and the atom dipole moment $\vec{P}_{AB}$ for the resonant transition.
$\varepsilon = \omega_{AB} - \omega_o$ is the frequency detuning of the resonance. The parameter

$$\Gamma_{AB}^{\epsilon} = \Delta_{AB}^{\epsilon} + \Gamma_{A}^{\epsilon} + \Gamma_{B}^{\epsilon}, \; \Gamma_{K}^{\epsilon} = \sum_{J} (A_{KJ} + d_{KJ}^{\epsilon}) \tag{2}$$

describes the spectral width in Eq. (1) and can be treated as a total probability of the coherent state decay. The decay is initiated by the atom-electromagnetic vacuum interaction causing spontaneous transitions followed by emission of light (wavy arrows in Fig. 1), and by the atom-environment interaction resulting in nonoptical transitions (curved arrows). The term Γ_{K}^{ϵ} gives the probability of the coherent state decay due to spontaneous (A_{KJ}) and radiationless (d_{KJ}^{ϵ}) transitions of atom from the level K to the levels J. The value Δ_{AB}^{ϵ} describes the atom's transition from the coherent state C to the nearest eigenstate A. The aforementioned transition is caused by an adiabatic (diagonal to the atomic states) part of the atomic perturbation by its environment and its probability Δ_{AB}^{ϵ} is proportional to the spectral density of the perturbation correlation function at the frequency ϵ [4]. In contrast to this, the probabilities d_{KJ}^{ϵ} contain information on high-frequency atomic perturbations by the environment, i.e., on perturbations possessing the components of the frequency $(\omega_{KJ} - \epsilon)$. The dependence of spontaneous transition probabilities A_{KJ} on ϵ can be neglected because $A_{KJ}^{\epsilon} = A_{KJ}^{o}$ x $(\omega_{KJ} - \epsilon)^{3}/\omega_{KJ}^{3}$, and in the resonance case $|\epsilon| \ll \omega_{KJ}$.

Spontaneous atomic transitions from the coherent state to the initial one result in photon emission, their frequency and phase being coincident with those of the incident light. These photons produce elastic or coherent (Rayleigh) scattering, often called, however, resonance fluorescence. The latter title was introduced by Weisskopf in his classic work on the quantum theory of this phenomenon[5]. Spontaneous atomic transitions from the coherent state to the states J, different from the state B, result in production of resonance (Raman) scattering at frequences $(\omega_{o} - \omega_{JB})$ with spectral widths Γ_{JB}^{o} equal to the widths of the transitions J ↔ B. Spectrally integrated intensities $I_{\omega_{o} - \omega_{JB}}$ of resonance scattering lines (including elastic scattering corresponding to J = B), are proportional to the probabilities A_{AJ} of spontaneous transitions at which they are emitted.

As was already noted, an atom with the probability Δ_{AB}^{ϵ} can make transitions from the coherent state to the eigenstate A. Such transitions fully destroy phase correlations of the atomic state with the light incident on it, i.e. lead to the incoherent population of the level A. Therefore, the subsequent spontaneous transitions produce lines of intrinsic atomic fluorescence, characterized by the frequencies, ω_{AJ}, and the spectral widths Γ_{AJ}^{o}. The center of these lines are shifted by a value of $(-\epsilon)$ from the lines of Raman and coherent (at J = B) scattering.

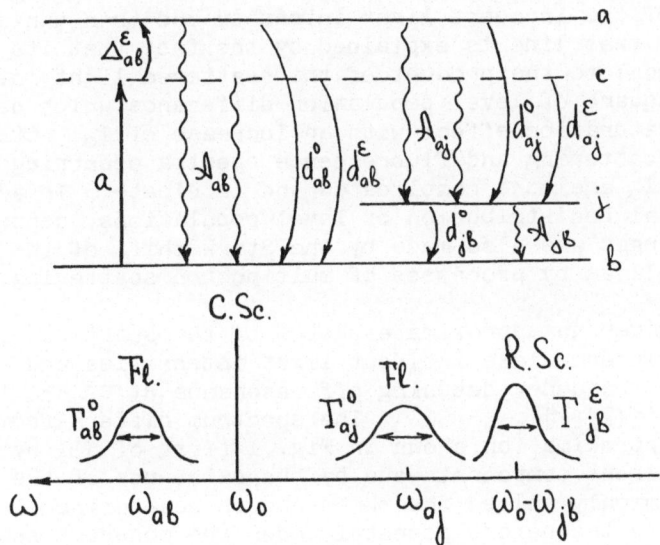

Fig. 1.   Transition scheme and spectrum of weak light scattering
          on atom perturbed by environment.

As can be readily stated, their intensities are proportional
to $\Delta^{\varepsilon}_{AB} A_{AJ}/\Gamma^{\circ}_{A}$; i.e. they are $\Delta^{\varepsilon}_{AB}/\Gamma^{\circ}_{A}$ times the
intensities of the correponding scattering lines. Here, Γ°_{A} is a
total probability of the level A decay.

It should be noted that the ratio $I_{\omega_{AJ}}/I_{\omega_{\circ}-\omega_{JB}} = \Delta^{\varepsilon}_{AB}/\Gamma^{\circ}_{A}$ is of
considerable interest for the experimental measurement of $\Delta^{\varepsilon}_{AB}$,
as a function of ε, i.e. for the investigation of low-frequency
atomic perturbations by environment. Recently such investigations
of strontium vapors in argon were carried out by Carlsten et al.[6].
They measured the ratio of the intensities of fluorescence and
Rayleigh scattering as a function of the frequency detuning of
resonance.

Let me now pass on to the problem of resonant scattering of intense monochromatic light.

As follows from theory[1-4] and experiments[6,7] an increase in the scattered light intensity causes considerable changes in the scattered light spectrum. The coherent scattering, for instance, flashes up at first, achieves its maximum brightness at some value of the incident light intensity and then vanishes. The behaviour of that line is explained by the fact that its intensity is proportional to the product of the scattered light intensity I_0 and the square of level population difference which decreases due to the saturation effect with an increase of I_0. Other changes in scattering and fluorescence spectra occurring with an increase of I_0 are more complicated and intricate. In addition to substantial redistribution of level populations, contributions to these changes are also made by the Stark shift of the resonance levels as well as by processes of multiphoton scattering.

Fig. 2 gives an approximate sketch of the scattered light spectrum at intermediate incident light intensities and considerable frequency detuning off resonance [at $\varepsilon^2 \gg |V|^2$ $(\Gamma^\circ_{AB})^2$; $|V|^2 \gtrsim (\Gamma^\circ_{AB})^2$]. The spectrum differs from that in the linear approximation shown in Fig. 1 first of all by a shift of the incoherent components and by the existence of the line at $2\omega_0 - \omega_{AB}$ commonly called the three-photon scattering line, as well as by the incoherent pedestal under the coherent scattering line. As a result of the Stark shift of resonance levels by a value of δ, equal here to $|V|^2/\varepsilon$, the atomic transition frequencies take, correspondingly, values of: $\omega'_{AB} = \omega_{AB} + 2\delta$; $\omega'_{AJ} = \omega_{AJ} + \delta$, and $\omega'_{JB} = \omega_{JB} + \delta$.

Both multiphoton light scattering and scattering processes by atoms in an excited state result in the formation of components at $2\omega_0 - \omega'_{AB}$, and the pedestal under the coherent scattering line. Some of these processes are shown in Fig. 3. The lowest multiphoton process contributing to the lines at $2\omega_0 - \omega'_{AB}$ is the three-photon process in which two photons disappear from the incident wave, an atom moves to a coherent excited state and a scattered photon is emitted with a frequency ω near $2\omega_0 - \omega'_{AB}$. This gives a contribution proportional to the square of the scattering light intensity ($\sim I_0^2$).

In case of intense light scattering by unperturbed atoms, the three-photon process is inevitably accomplished with second photon emission at frequency ω' satisfying the conditions $\omega' = 2\omega_0 - \omega \approx \omega'_{AB}$ or $\omega' \approx 2\omega_0 - \omega - \omega'_{JB} \approx \omega_{AJ}$. Due to this the whole process of scattering has the four-photon form. The process of two-photon scattering by an excited atom can take place between the processes of three-photon scattering and emission of the fourth photon. Thus, the

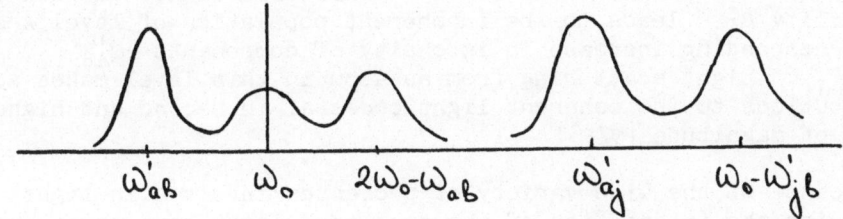

Fig. 2. Spectrum of moderate intensity light scattering, i.e. at $\varepsilon^2 \gg |V|^2$, $(\Gamma_{AB}^o)^2$ and $|V|^2 \gtrsim (\Gamma_{AB}^o)^2$.

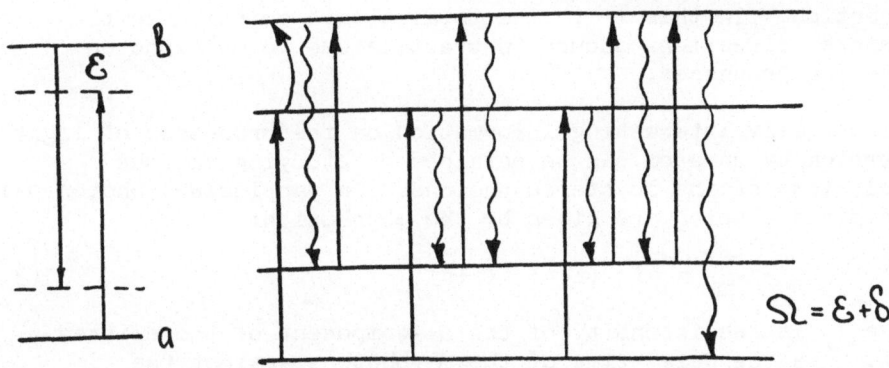

Fig. 3. Schemes of some multiphoton transitions and scattering processes on an excited atom.

four-photon scattering process is transformed into a six-photon
process, leading to the formation of components with frequencies
$\omega'_{AB}, 2\omega_o - \omega'_{AB}$ and the incoherent component near frequency ω_o
       The contribution of this process to the various scattered
light components is proportional to the cube of the incident light
intensity ($\sim I_o^3$).

     In case of light scattering from perturbed atoms multiphoton
scattering processes can be interrupted at any step of their
development. Their interruption by transitions connected with the
probability $\Delta_{AB}^{\varepsilon}$ leads to the incoherent population of level A and
the corresponding increase in intensity of components ω'_{AB}
and ω'_{AJ} . Light scattering from an atom in this level makes some
contributions to the coherent light pedestal in second and higher
orders of magnitude $|V/\varepsilon|^2 \sim I_o$.

     Because of the wide variety of processes involved in light
conversion the intensities of the scattered light components have
a complicated dependence on the parameters characterizing the
atom-light and atom-environment interactions. These dependences
are greatly simplified, however, in the limit of high incident
light power and low frequency detunings. In this case the
intensity ratios for different components take on a universal form
independent of the peculiarities of the atom-environment
interaction. In this limit the total atomic emission in the
transverse directions occurs in practice due to multiphoton
scattering processes.

     Especially interesting information on the processes of light
conversion by an atom can be obtained by studying various
correlations of the scattered photons. In particular, photon pair
coincidences, which are given by the expression:

$$G_{\alpha\beta}^{\tau} = <I_\alpha (t) \, I_\beta (t+\tau) > \qquad\qquad (3)$$

(where I_α is the intensity of the α-component of the emitted
light, τ is the delay time of the β-photons against the
α-photons.) Both the total light emitted by a separate atom
transition and the separate lines of this light may be regarded as
a component in (3). The pair correlations of the same component
are determined by the expression (3) at $\alpha = \beta$.

     Calculation of the function $G_{\alpha\beta}^{\tau}$ for radiation scattered by
a single atom shows that the τ dependence of the function is
determined by level population kinetics, i.e. by the change in
population of the state from which the β-photons are emitted when at
time zero the atom was in the final state for the α-photon
emitting process. Contrary to this the emission spectrum (the
Fourier transform of the correlation function of the field

amplitude) is determined by the kinetics of motion of the atomic
dipole moment.

From the above mentioned peculiarity of the function $G_{\alpha\beta}^{\tau}$ it
follows that at $\tau \to 0$ the pair coincidence number can decrease to
zero, i.e. one can observe the effect opposed to that of photon
bunching discovered by Hanbury-Brown and Twiss[8]. The only
exception to the rule occurs when the final state for the first
photon emission coincides with the initial state for the second
photon emission, as occurs, for example, in two-photon emission.
In this case at $\tau = 0$ the function $G_{\alpha\beta}^{\tau}$ has its peak value;
i.e. the bunching of the α and β photons takes place.

The effect of the photon antibunching in resonance scattering
was first predicted by Carmichael and Walls[9]. Mandel and
co-workers observed the effect experimentally in a weak sodium
atomic beam[10].

The typical dependences of $G_{\alpha\beta}^{\tau}$ on τ for the total light
emitted by separate transitions under investigation are shown in
Fig. 4. In this case, as a rule, the function tends to its
large τ limit which equals the product of the intensities of
fluxes in question, not monotonically but with oscillations at the
Rabi frequency $\Omega = \varepsilon(1 + 4|V|^2/\varepsilon^2)^{\frac{1}{4}}$. The dependence
of $G_{\alpha\beta}^{\tau}$ on τ for various combinations of three components of the
resonant transition is illustrated in Fig. 5. Here the indices +,
0, and − denote the components with frequencies $\omega_o + \Omega$, ω_o,
and $\omega_o - \Omega$, respectively. In this case the values of $G_{\alpha\beta}^{t}$
tend to their large τ limit monotonically according to the
exponential law, with the exponential index:

$$\Gamma = [\varepsilon^2(\Gamma_A^o + \Gamma_B^o) + \Gamma_{AB}^{\Omega} |V|^2]/\Omega^2. \qquad (4)$$

The absence of oscillations in pair coincidence counting of such
photons is due to the spectral separation of the lines. For
separation of the spectral lines possessing widths of the order of
Γ_{AB}^o and of line spacing Ω a spectral selector whose
transmission band $\Delta\omega$ satisfies the conditions $\Gamma_{AB}^o < \Delta\omega \ll \Omega$
is required. Using this selector leads to the averaging of the
selected fluxes with respect to times of the order of $\Delta\omega^{-1}$,
i.e. the selector cuts off entirely the Ω frequency oscillations.

As is seen from Fig. 5 antibunching must be observable for
photons of every line $\omega_o \pm \Omega$ and bunching must occur for photons
of different lines $\omega_o + \Omega$ and $\omega_o - \Omega$. This is in full
correspondence with the above formulated rule. The bunching
effect is provided by the multiphoton scattering processes
delivering simultaneously the photons into both lines to
within $\Delta\omega^{-1}$. There is no photon correlation for line
combinations different from those represented in Fig. 5.

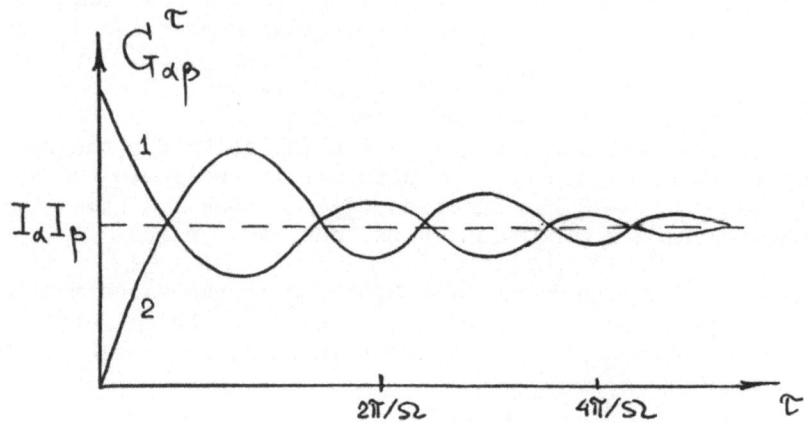

Fig. 4. Dependence of pair coincidence number on τ for total
          fluxes: the photon bunching (1) and antibunching (2)
          effects.

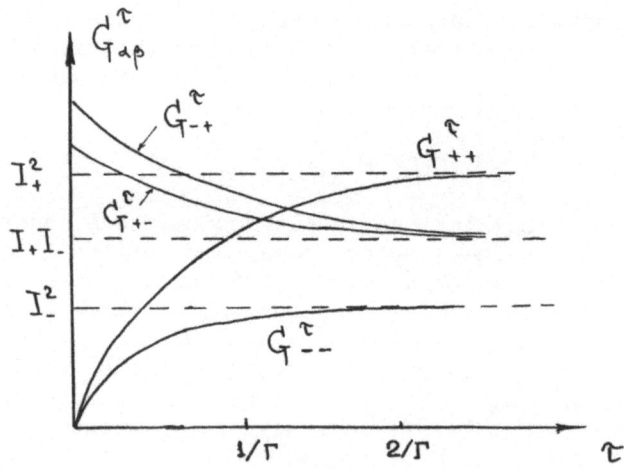

Fig. 5. Dependence of pair coincidence number on τ for photons
          from lines of resonance transition.

REFERENCES

1. P.A. Apanasevich, Opt. and Spectr., (USSR) $\underline{14}$; 612, 1963; $\underline{16}$, 709, 1964; Izv. Akad. Nauk SSSR, Ser, Fiz.-Mat. Nauk., $\underline{32}$, 1299, 1968.

2. P.A. Apanasevich, S. Ya. Kilin, Zh. Prikl. Spectrosk., $\underline{24}$, 738, 1976; $\underline{29}$, 252, 1978; Phys. Lett. $\underline{62A}$, 83, 1977, J. Phys. B: Atom Molec. Phys., $\underline{12}$, 83, 1979.

3. S. Ya. Kilin, Preprint No. 152, Institute of Physics, Byelorussian Academy of Sciences, USSR, 1978.

4. P.A. Apanasevich, Essentials of Theory of Light-Matter Interaction., Nauka i Tekhnika Press, Minsk, 1977.

5. V. Weisskopf, Ann. Phys. $\underline{9}$, 23, 1931.

6. J.L. Carlsten, A. Shoke, M.G. Raymer, Phys. Rev. A$\underline{15}$, 1029, 1977.

7. R.E. Grove, F.Y. Wu, S. Ezekiel, Phys. Rev. Lett. $\underline{35}$, 1426, 1975; Phys. Rev. A$\underline{15}$, 227, 1977.

8. R. Hanbury Brown, R.Q. Twiss. Nature, 177, 27, 1956; Proc. Roy. Soc. London A 242, 300 and A 243, 291, 1957.

9. H.J. Carmichael, D.F. Walls, J. Phys. B$\underline{9}$, 43, 1976; B$\underline{9}$, 1199, 1976.

10. H.J. Kimble, M. Dagenais, L. Mandel, Phys. Rev. Lett. $\underline{39}$, 691, 1977.

INTENSITY EFFECTS IN RESONANCE LIGHT SCATTERING*

B.R. Mollow

Department of Physics
University of Massachusetts
Boston, Massachusetts 02125

When low intensity monochromatic light is incident upon an isolated stationary atom initially in its ground state, the spectrum of the light scattered in a two step excitation-deexcitation process which returns the atom to its ground state must be a δ-function at the incident field frequency. Quantum mechanically, this result follows at once from energy conservation, since in the low intensity limit only one photon is absorbed from the incident field, and only one is emitted under the stated conditions. From a classical point of view, one may understand the δ-function spectrum for the scattered field by thinking of the atom as consisting of one or more harmonic oscillators (a picture which is useful under weak excitation), and recognizing that the induced atomic electric dipole moment must oscillate at the same frequency as the incident field which drives the atom.

Under near resonance conditions, important modifications in the scattered field spectrum appear when the intensity of the incident light is increased to the point where saturation effects begin to appear. These modifications are nonlinear with respect to the incident field intensity, and result from processes in which more than one photon is absorbed and (under near resonance coditions) an equal number emitted, so that though energy is still conserved in the process as a whole, the energy of any one scattered photon, while lying near that of an incident photon, need not equal it exactly. The intensity dependent spectrum contains, in addition to a δ-function or coherent component, broadened incoherent components separated by the Rabi frequency

$$\Omega' = (\Omega^2 + \Delta^2)^{1/2},$$

where Ω is the power broadening parameter $E\mu_{10}/\hbar$ and Δ is the detuning of the laser from resonance. (One way of understanding the incoherent spectral terms is to think of the field-induced atomic dipole moments are suffering amplitude modulations due to intensity dependent modulations of the atomic populations which in turn are readily understood in the light of the formal equivalence between any two level system and an effective spin system.)

For the case in which the laser field couples the atomic ground state to a single excited state and the damping is purely radiative, the solution has been found in full generality by Mollow[1]. The result is valid for arbitrary laser intensity and detuning, subject only to the innocuous restriction that the power-broadened linewidth and the detuning be small compared to the optical laser frequency. The spectrum under these conditions is fully symmetrical about the laser frequency, and is given quite generally and exactly by the formula[1]

$$\widetilde{g}(\nu) = \widetilde{g}^{\sim}(\nu-\omega),$$

$$\widetilde{g}^{\sim}(\nu) = |\bar{\rho}_{10}|^2 2\pi\delta(\nu) + \bar{\rho}_{11}\kappa\Omega^2(\nu^2+\Omega^2/2 + \kappa^2)/|f(i\nu)|^2,$$

in which

$$\bar{\rho}_{10} = i\Omega(\kappa/2 + i\Delta)/2(\Omega^2/2 + \Delta^2 + \kappa^2/4),$$

$$\bar{\rho}_{11} = \Omega^2/4(\Omega^2/2 + \Delta^2 + \kappa^2/4),$$

$$|f(i\nu)|^2 = \nu^2(\nu^2-\Omega^{\sim2}-5\kappa^2/4)^2 + \kappa^2(4\nu^2 - \Omega^2/2 - \Delta^2 - \kappa^2/4)^2,$$

ω is the laser frequency, and κ is the Einstein A-coefficient for the transition in question. (Theoretical analyses of the same problem have been carried out from a variety of different points of view[2-8], and all have led to exactly the same results as Mollow[1].)

Ample experimental confirmation of these theoretical predictions has been obtained by Ezekiel and coworkers[9] and Walther and coworkers[10], extending earlier work by Stroud and coworkers[11] on spectral measurements of the sodium D_2 line in an atomic beam.

When the atomic relaxation is due in part to collisional effects which can be treated in the impact approximation, the intensity dependent scattering spectrum can be evaluated[12-14] by means of simple generalizations of the methods used to treat the radiative case. (One of the earliest treatments of the collisional spectrum, due to Newstein[15], contains limiting forms (corresponding to high saturation) of the solutions for the model treated more fully in Ref. 12.) Reasonable agreement between theory[13,14] and experiment in the case of the collision-modified spectrum has been obtained by Carlsten et al.[16] , in measurements of light scattering from strontium vapor in a cell containing argon buffer gas.

*Supported by the National Science Foundation

References

1. B.R. Mollow, Phys. Rev. 188, 1969 (1969)
2. G. Oliver, E. Ressayre, and A. Tallet, Nuovo Cimento Lett. 2, 777 (1971).
3. B.R. Mollow, Phys. Rev. A12, 1919 (1975).
4. B.R. Mollow, J. Phys. A8, L130 (1975).
5. S. Swain, J. Phys. B8, L437 (1975).
6. H.J. Carmichael and D.F. Walls, J. Phys. B9, 1199 (1976).
7. H.J. Kimble and L. Mandel, Phys. Rev. A13, 2123 (1976).
8. C. Cohen-Tannoudji and S. Reynaud, J. Phys. B10, 345 (1977).
9. F.Y. Wu, R.E. Grove, and S. Ezekiel, Phys. Rev. Lett. 35, 1426 (1975); R.E. Grove, F.Y. Wu, and S. Ezekiel, Phys. Rev. A15, 227 (1977).
10. H. Walther, in Proceedings of the Second Laser Spectroscopy Conference, Megeve, France, 1975 (Springer, Berlin,1975); W. Hartig, W. Rasmussen, R. Schieder, and H. Walter, Z. Phys. A278, 205 (1976).
11. F. Schuda, C.R. Stroud, Jr., and M. Hercher, J. Phys. B7, L198 (1974).
12. B.R. Mollow, Phys. Rev. A2, 76 (1970).
13. B.R. Mollow, Phys. Rev. A5, 2217 (1972).
14. B.R. Mollow, Phys. Rev. A15, 1023 (1977).
15. M. Newstein, Phys. Rev. 167, 89 (1968).
16. J.L. Carlsten, A. Szoke, and M.G. Raymer, Phys. Rev. A15, 1029 (1977).

THE SPONTANEOUS DIFFRACTION OF LIGHT BY RESONANCE ATOMS

A.P. Kazantsev

Landau Institute For Theoretical Physics,
Academy of Sciences
USSR, Moscow

In the present paper we shall discuss the correlation properties of scattered emission. The space-time correlators of light are determined by the permittivity and the relevant correlators of the medium. A rarefied gas of resonance atoms is considered as the scattering medium. In this case all the correlators are easily calculated.

An external field sets up an effective diffraction grating in the medium. Therefore, a greatly anisotropic component also appears in scattered light superimposed upon the background of the isotropic component. In addition, an anomalous correlator of the amplitudes of the scattered field appears in the external field.

By varying the spatial structure of the external field, we can separate and measure the desired correlator of the medium. The medium is assumed to be optically transparent, that is

$$2\pi k\ell \ \ \mathrm{Im} \ \chi < 1$$
$$\varepsilon = 1 + 4\pi\chi \tag{1}$$

Here ε and χ are the permittivity and the susceptibility of the medium, ℓ is the dimension of the region of interaction with light, and k is the wave number. According to condition (1), the scattered light can be found on the basis of perturbation theory.

The external field $E_0(\bar{r},t)$ is considered to be monochromatic and consists of several plane waves with wave vectors k_i and amplitudes E_i:

$$\varepsilon_0(\bar{r},t) = E_0(\bar{r}) \exp(-i(\omega_0 + \Delta_0)t) + c.c$$
$$E_0(r) = \sum_i \exp(i\bar{k}_i\bar{r}), \ |k_i| \ c = \omega = \omega_0 + \Delta_0 \tag{2}$$

where Δ_0 is minor detuning relative to the transition frequency ω_0. (Note that $|\bar{k}_1|$ is independent of i).

We shall also divide the operator of the scattered field into positive and negative frequency parts $E_s(\bar{r},t) \exp(-i\omega_0 t) + E_s^+(\bar{r},t) \exp(i\omega_0 t)$. Let us introduce the correlation functions

$$u(\bar{r},t_1;\bar{r}_2 t_2) = \frac{c}{4\pi} < E_s^+(\bar{r},t_1)\ E_s(\bar{r},t_2) >$$

$$v(\bar{r}_1,t_1;\bar{r}_2,t_2) = \frac{c}{4\pi} < E_s(\bar{r}_1,t_1)\ E_s(\bar{r}_2,t_2) > \qquad (3)$$

$$v*(\bar{r}_1,t_1;\bar{r}_2,t_2) = \frac{c}{4\pi} < E_s^+(\bar{r}_2,t_2)\ E_s^+(\bar{r}_1,t_1) >$$

The angular brackets signify averaging over the photon vacuum field and over the states of the medium.

It is obvious that $u(\bar{r},t_1;\ \bar{r},t_2)$ determines the energy flux and the spectrum of scattered emission. It is exactly this quantity that is usually determined in practice.

The correlator $v(\bar{r}_1,t_1;\ \bar{r}_2,t_2)$ vanishes for thermal radiation sources. It appears only in an external coherent field. Below we consider the conditions for which it could be observed.

Scattered emission consists of two kinds of contributions:

$$u = u_1 + u_2;\ v = v_1 + v_2 \qquad (4)$$

The first addend here is due to scattering by density fluctuations. It is proportional to the particle concentration $n (u_1, v_1 \sim n)$. The second addend is due to collective diffraction effects, and $u_2, v_2 \sim n^2$.

In the Fraunhofer zone we have

$$u_1(r,t_1;r,t_2) = \frac{ck_0^4}{4\pi r^2}\ \sum_i < p_i^+(t_1 - \frac{r}{c})\ p_i(t_2 - \frac{r}{c}) > \qquad (5)$$

$$v_1(r_1,t_1;r_2,t_2) = \frac{ck^4}{4\pi r^2}\ \exp(ik_0(r_1 + r_2))\ \times$$

$$\times\ \sum_i \exp\left\{ik_0\left[\frac{\bar{r}_1\bar{r}_i}{r_1} + \frac{\bar{r}_2\bar{r}_i}{r_2}\right]\right\} < p_i(t_1 - \frac{r_1}{c})\ p_i(t_2 - \frac{r_2}{c}) > \qquad (6)$$

The summation is over the particles of the medium, $p_i(t)$ is the operator of the dipole moment of i-th atom. For the problem of the resonance fluorescence in an external field, the correlator

(5) has already been calculated (1-3). The spontaneous emission of atoms is due to quantum fluctuations of the electromagnetic field $E(\bar{r},t)$. Consequently, the total field acting on the atoms is $E_0(\bar{r},t) + E(\bar{r},t)$, where $\mathbf{E}(\bar{r},t) = \sum_k \exp((-i\,\omega_0 + \Delta_k)t + i\bar{k}\bar{r})E_k$ + Herm. Conj.

$$\hat{E}_k = \sqrt{\frac{2\pi\hbar\omega_0}{V}}\,\hat{a}_k \tag{7}$$

The symbols a_k^+ and a_k stand for the creation and annihilation operators of a photon with momentum $\hbar k$, and V for the volume of interaction of the atoms with the field.

Let us represent the operator of the dipole moment of a unit volume of medium in the form $(P_0(\bar{r},t) + P(\bar{r},t)) \exp(-i\omega_0 t) + h.c.$, $P_0(\bar{r},t)$ is the dipole moment induced by the external field, and $P(\bar{r},t)$ is the moment due to zero-point oscillations of the field $E(\bar{r},t)$. The spontaneous dipole moment can be written in the form

$$\hat{P}(\bar{r},t) = \sum_k \{\alpha_k(\bar{r}) \exp(-i\Delta_k t + i\bar{k}\bar{r})\,\hat{E}_k +$$
$$+ \beta_k(\bar{r}) \exp(i(\Delta_k - 2\Delta_0)t - i\bar{k}\bar{r})\,E_k^+ \tag{8}$$

The coefficients α and β are c-numbers. They describe the response of the medium in the external field $E_0(\bar{r})$ to the weak field $\exp(-i\Delta_k t + i\bar{k}\bar{r})^{E_k}$. In the absence of an external field β vanishes, while α is the conventional polarizability of the medium. In the presence of an external field, the coefficient β is proportional to E_0^2, and it describes combination scattering. The final result can be expressed through the Fourier components $\alpha_k(\bar{r})$ and $\beta_k(\bar{r})$, which we shall designate by $a_k(\bar{q})$ and $b_k(\bar{q})$:

$$u_2(\bar{r},t_1;\bar{r},t_2) = \frac{ck_0^4\hbar\omega_0}{2r^2} \int \frac{d^3k}{(2\pi)^3}\ \left|bk[(2\bar{k}-\bar{k}_0)\,\frac{\bar{r}}{r}\,]\,\right|^2\ x$$
$$= 2k - k_0$$
$$x \exp(i(2\omega - \omega_k)(t_1 - t_2))\ x \tag{9}$$
$$x \exp(-i\Delta_0(t_1 + t_2))$$

$$v_2(r_1,t_1;r_2,t_2) = \frac{ck_0^4\hbar\omega_0}{2r^2} \int \frac{d^3k}{(2\pi)^3} a_k(\frac{\overline{k}\overline{r}_1}{r})b_k\left[(2\overline{k}-\overline{k}_0)\frac{\overline{r}_2}{r_2}\right]$$

$$\exp(-i(\Delta_k - \Delta_0)(t_1 - t_2)$$

$$\exp(i\overline{k}_0(\overline{r}_1 + \overline{r}_2)) \tag{10}$$

We shall now give some estimates which were obtained based on the formulas for the correlators $u_{1,2}$ and $v_{1,2}$.

PLANE RUNNING WAVE

We shall first consider the very simple case of a plane running wave $E_0(\overline{r}) = E_0 \exp(i\overline{k}_0\overline{r})$. Such a wave sets up a diffraction grating with the period π/k_0 in the medium. Hence, the anisotropic component of the scattered emission is directed forward, along the external field that has passed the beam.

The field spontaneously emitted forward differs from the incident field in its spectral composition and in its polarization. We can therefore separate the scattered field from the external one. Let us compare the intensity of the spontaneously diffracted radiation with that of isotropic emission within the solid angle of the beam δO_b:

$$\frac{u_2 r^2 \delta O_b}{u_1 r^2 \delta O_b} \sim \frac{W \, \text{Im} \, \chi}{\delta O_b} \tag{11}$$

Here $W = \left|\dfrac{\overline{d}\overline{E}_0}{\hbar(\Delta_0 + i\gamma)}\right|^2$ is the saturation parameter, γ is the width of atomic resonance; $\delta O_b \sim (2\pi/k_0\ell)^2$ is the angular divergence of the beam.

It is obvious that the quantity (11) may be very large in fields that are not very weak. Thus, a spontaneously scattered field has a sharp peak in the direction of propagation of the external field. From a practical viewpoint, however, cases in which the diffracted emission does not coincide in direction with the incident radiation are more interesting.

STANDING WAVE

The case of a standing wave is unique, since collective spontaneous emission of the atoms is possible in any direction. This effect is similar to the recently discovered phenomenon of the reflection of a weak signal from a non-linear medium in the field of a standing light wave (4-8).

In this case we have

$$\frac{u_2}{u_1} \sim Wk\ell \, \mathrm{Im}\chi \qquad (12)$$

If the field is strong, $W \sim 1$, and the path of the photons is comparable with the size of the medium, then the velocity of collective emission is comparable with the velocity of photon scattering on density fluctuations. The anomalous correlator $v(\bar{r}_1, t_1; \bar{r}_2, t_2)$ differs from zero only for beams propagating in opposite directions. If the origin of coordinates is placed at the center of the scattering medium, then we should have $\bar{r}_2/r_2 = -r_1/r_1$. For order of magnitude, we have

$$\left| v_1(\bar{r}, t_1; -\bar{r}, t_2) \right| \sim \left| u_1(\bar{r}, t_1; \bar{r}, t_2) \right|$$

$$\frac{v_2}{v_1} \sim k\ell \, \mathrm{Im}\chi \qquad (13)$$

Thus, correlators that are non-linear with respect to density become significant for any direction of the scattered field.

CONICAL DIFFRACTION

Let us now assume that the external field consists of two plane waves with vectors k_1 and k_2 propagating at an angle to each other. Diffraction gratings with the periods $\pi / |\bar{k}_1 - \bar{k}_2|$ and $\pi / |\bar{k}_1 + \bar{k}_2|$ will be formed in the medium. Let a weak field with wave vector k_3 fall on such a grating. A diffracted field with the vector k_4 will appear if the Bragg condition $\bar{k}_4 = \bar{k}_3 \pm (\bar{k}_1 - \bar{k}_2)$ or $\bar{k}_4 = -\bar{k}_3 + \bar{k}_1 + \bar{k}_2$ is satisfied. This condition signifies that the incident and the scattered beam must be on the surface of one of the cones whose axis is determined by the vector $\bar{k}_1 + \bar{k}_2$, $\bar{k}_1 - \bar{k}_2$, and $\bar{k}_2 - \bar{k}_1$. It is natural to call such a case of diffraction,

forced diffraction of light. It has already been studied in
(7-10). In the case being considered, it is assumed that a test
signal is absent, and the spontaneous emission is amplified on the
surface of the cone with axis ($\bar{k}_1 + \bar{k}_2$). The beams on the
surface of the cone that are symmetrical to its axis contribute to
the correlator v ($\bar{r}_1, t_1$; $\bar{r}_2, t_2$). The energy of such
diffracted radiation forms a fraction of the total scattered
energy of the order of W $\chi''/\cos(\theta/2)$, where θ is the angle
between the vectors $\bar{k}_1$ and $\bar{k}_2$.

EMISSION IN A LINE

Amplification of the spontaneous emission only in a certain
direction is possible with certain special geometry of the
external field. Assume that the external field consists of a
standing wave with wave vector $\bar{k}_1$ and a running wave with wave
vector $\bar{k}_2$. We have the following condition of phase matching in
the fifth order of perturbation theory with respect to the
external field: $2\bar{k}_1 + 2\bar{k}_2 - \bar{k}_3 = \bar{k}_4$. If the vectors $\bar{k}_1$
and $\bar{k}_2$ make an angle of 120°, then the vector $\bar{k}_3 = \bar{k}_4$, and
$\bar{k}_3$ is directed along the bisector of this angle.

Let us estimate the intensity of such emission numerically.
The saturation parameter near resonance for a Doppler broadened
line shape is about unity in a field with an intensity of 10^4
W/cm^2. On condition that $k\ell\chi'' \sim 1$ and $\ell \sim 0.1$ cm, we find that
the intensity of emission diffracted in a definite direction is
about 1 W/cm^2.

We thus see that by changing the spatial structure of the
external field, we can observe different correlation functions of
scattered emission. In addition to u_1, other correlators allow
us to obtain additional information about the properties of a
scattering medium.

REFERENCES

1. S.G. Rautian, I.I. Sobelman, Zh. eksp. i teor. fiz., $\underline{41}$, 456
   (1961).
2. B.R. Mollow, Phys. Rev. $\underline{A2}$, 76, (1970); $\underline{A12}$, 1919.
3. A.P. Kazantsev, Zh. eksp. i teor. fiz., $\underline{66}$, 1229 (1974).
4. A. Yariv, Opt. Commun., $\underline{21}$, 49 (1977).
5. R.W. Hellwarth, J. Opt. Soc. Am., $\underline{67}$, 1 (1977).
6. D. Grichkovsky, N.S. Shierin, R.J. Bennet, Appl. Phys. Lett.,
   $\underline{33}$, 805 (1978).
7. D.W. Pohl, S.E. Schwarz, V. Irniger, Phys. Rev. Lett., $\underline{31}$, 32
   (1973).

8. D.W. Pohl, V. Irniger, Phys. Rev. Lett., $\underline{36}$, 480 (1976).

9. S.A. Akhmanov, N.I. Koroteev, Zh. eksp. i teor. fiz., $\underline{67}$, 1306 (1974).

10. M.D. Levenson, N. Bloembergen, Phys. Rev., $\underline{B10}$, 4447 (1974).

GIANT RAMAN SCATTERING BY MOLECULES ADSORBED ON METALS: AN OVERVIEW

E. Burstein, C. Y. Chen and S. Lundquist[*]

Physics Department and Laboratory for Research on the
Structure of Matter, University of Pennsylvania[+]
Philadelphia, PA

INTRODUCTION[‡]

The observation by Fleischmann, et al in 1974[1] of a strong
Raman scattering (RS) in the visible by pyridine molecules adsorbed
on a Ag electrode that had been roughened electrochemically by multiple oxidation-reduction cycles to increase the surface area, and
the subsequent demonstration in 1977 by Jeanmaire and Van Duyne[2]
and by Albrecht and Creighton[3] that the RS by pyridine molecules
adsorbed on a Ag electrode following a single electrochemical
oxidation-reduction cycle is greater by a factor of 10^5 to 10^6 than
that of pyridine molecules in neat pyridine or in aqueous solution,
has attracted widespread attention of theorists and experimentalists.
Strongly enhanced RS has since then been observed for CN^- and a
number of other molecules adsorbed on Ag and an appreciably weaker
RS has also been observed for molecules adsorbed on Cu, Pt and,
more recently, on Au[4-19].

Although a complete understanding of the "giant" RS by
molecules adsorbed on metals has not yet been achieved, there has
been meaningful progress, both experimentally and theoretically,
in elucidating various aspects of this unusual phenomenon. There
is now considerable evidence that surface roughness on a submicroscopic scale plays a crucial role in the enormous enhancement of
the RS by molecules adsorbed on a Ag electrode and in the greatly
enhanced inelastic light scattering by charge carrier excitations
in the metal. In fact to our knowledge, no one has as yet observed
RS in the visible by a monolayer of pyridine or of any other
"transparent" substance on a smooth Ag surface. Furthermore an
enhanced RS which is comparable in intensity to that observed for
pyridine on electrochemically processed Ag has been observed for

iso-nicotinic acid adsorbed at the surface of thin ($\sim$ 100 Å) evaporated Ag island (e.g. aggregated) films. There is also evidence
that giant RS occurs only when the molecules are chemisorbed on the
metal, i.e. close proximity of the molecules to the metal per se is
not sufficient. A number of theoretical models have been proposed
for the giant RS by adsorbed pyridine and CN$^-$ on Ag, which are stated
by their authors to be capable of yielding enhancements of 10^5 to
10^6 on a smooth surface. However, it is evident, from the key role
played by surface roughness, and from other indications, that the
overall enhancement involves a number of contributions, some of
which may be quite specific to the particular metal adsorbed molecule system. We intend in this paper to review the key experimental
data that have been obtained by various groups (including our own),
to discuss key theoretical models that have been proposed, and
finally to provide an assessment of the physics underlying the phenomenon of giant RS by adsorbed molecules on metals.

EXPERIMENT[‡]

RS by a monolayer of "transparent" molecules is generally much
too weak to be observed by ordinary techniques. Accordingly the observation of even a very weak signal due to molecules adsorbed on a
metal is an indication of an appreciable metal-mediated enhancement
of the RS by the molecules. In the case of giant RS by pyridine
adsorbed on a Ag electrode after an optimum electrochemical oxida-
tion reduction cycle, one observes a RS signal of 10^5 counts per sec
when using 50 mW of excitation at 5145 Å.

An enhanced RS by adsorbed molecules on metals has been
reported for a variety of molecular species (e.g. pyridine, cyano-
pyridine and other nitrogen heterocyclics, crystal violet, CN$^-$,
SCN$^-$, CO$_3^-$, Cl$^-$, CO, p-pyridine carboxaldehyde, benzoic acid, etc.)
that are adsorbed on various metals (e.g. Ag which causes by far the
largest enhancement, Cu, Au and Pt) by various procedures (e.g.
electrochemical and chemical deposition form solution, vapor deposition in high vacuum and the use of thin metal island films as
substrates and as overlayers). In the case of molecules such as
crystal violet and methyl orange which are absorbing in the visible,
the "resonance" enhanced RS cross-sections are further enhanced
when the molecules are adsorbed on Ag.

The in-situ RS by pyridine and by CN$^-$ adsorbed on Ag electrodes
have been studied more extensively than that of other systems. The
very strong enhancement of the RS by the adsorbed molecules is not
very sensitive to the polarization of the incident and scattered
radiation, nor to the angles of incidence and scattering, and the
radiation scattered by the totally symmetric vibration modes, unlike that observed for the molecules in solution, is appreciably
depolarized[2]. The RS by pyridine and CN increases to a maximum and
then decreases as the charge transferred in the anodization part of

the electrochemical cycle increases[8,13,14]. Furthermore, the enhanced RS by the adsorbed molecules is accompanied by a strong scattering continuum which extends well beyond 4000 cm^{-1} (Fig. 1)[14]. The scattering continuum also increases to a maximum at essentially the same charge-transfer at which the RS by the molecules is a maximum (Fig. 2). The scattering continuum, like the enhanced RS by the molecules, is relatively insensitive to the polarization of the incident and scattered radiation, and to the angles of incidence and scattering. It is similar in character to the inelastic light scattering continuum that is exhibited by a rough Ag surface in the absence of adsorbed molecules. It is accordingly attributed to inelastic light scattering by particle-hole pair excitations in metal via $(p \cdot A)^2$ processes which are made possible by the breakdown in momentum conservation caused by surface roughness.

Scanning electron microscope and Auger electron spectroscopy studies[10,15] have shown that the electrochemical processing of the Ag electrode acts to "clean" the surface, and thereby to enable the molecules to adsorb onto the electrode and, for weak to moderate electrochemical processing, to produce surface roughness on a submicroscopic (≈ 100 Å) scale. That the surface roughness introduced by the electrochemical processing is on a submicroscopic scale accounts for the fact that Pettinger, et al[15] find no indication in their electro-reflectance studies of the Ag electrode surface of any surface roughness-induced excitation of surface plasmons by p-polarized EM radiation, even after moderately strong electrochemical processing.

Although there were initially some indications that an enhanced RS also occurs when pyridine is adsorbed on a smooth Ag surface without electrochemical (or chemical) processing[12], there is now evidence for both pyridine and CN$^-$ that the RS by adsorbed molecules on a smooth metal substrate is, in fact, not observable, and that it is the submicroscopic roughness that is introduced by the electrochemical processing which is responsible for the enhancement of the RS by the molecules[14]. The data indicate that the RS by adsorbed molecules on a smooth Ag surface is weaker, by a factor of at least 10^3 to 10^4, than the RS by the molecules adsorbed on an optimally electrochemically processed Ag electrode.

As noted by Van Duyne[10], the most prominent peaks in the Raman spectra for pyridine adsorbed on Ag are those due to the totally symmetric A_1 (u ∥ z) modes which are infrared as well as Raman active. Well defined Raman peaks due to B_1 (u ∥ x) and B_2 (u ∥ y) modes also appear in the spectra. The various peaks only exhibit small shifts in frequency from the corresponding peaks in the spectra of neat pyridine or of pyridine in aqueous solution. However, there does seem to be some preferential enhancement of the A_1 modes relative to the B_1 and B_2 modes. The metal enhanced RS by the totally symmetric modes is strongly depolarized. This is also

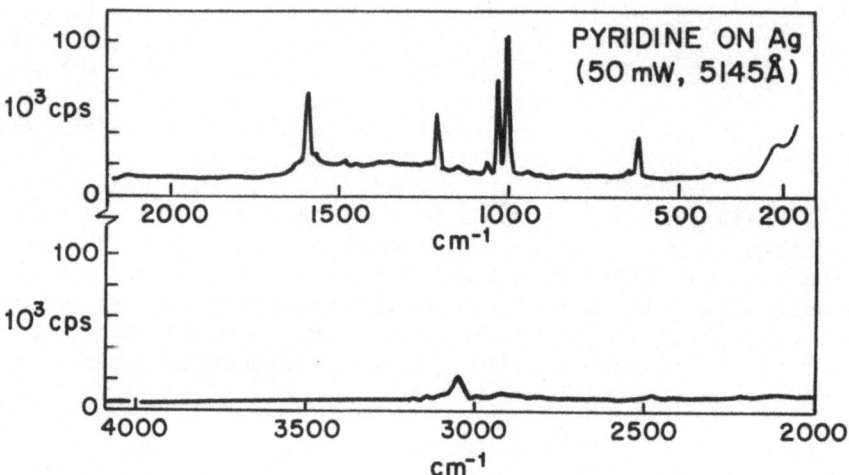

Fig. 1. Raman spectrum of pyridine adsorbed on an evaporated Ag film after an anodization charge-transfer equivalent to ~40 mono-layers of Ag. The spectrum was obtained using 50 mW of 5145 Å excitation. (Chen, et al, Ref. 14).

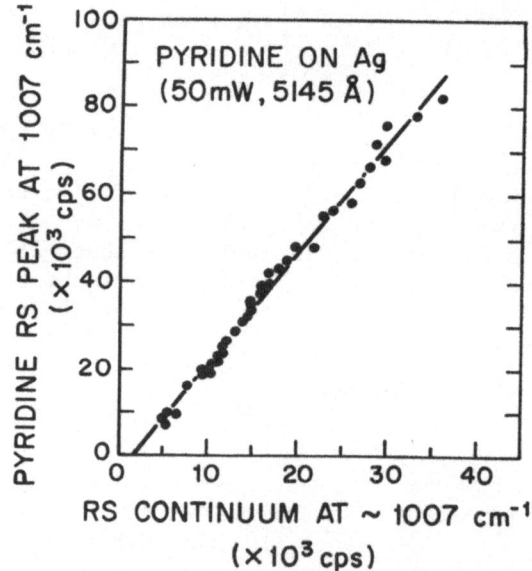

Fig. 2. Plot of the intensity of the pyridine peak at 1007 cm^{-1} versus the intensity of the light scattering continuum at ~1007 cm^{-1} with increasing anodization charge-transfer up to an equivalent of ~50 monolayers of Ag. (Chen, et al, Ref. 14)

true for the RS by the C-N vibration mode of adsorbed CN⁻ on Ag,
which is strongly infrared as well as Raman active, and for the
scattering continuum which accompanies the RS by the adsorbed molecules.

The moderately strong peak which appears at 216 cm⁻¹ in the
spectra for pyridine adsorbed on Ag is attributed to a Ag-N vibration,
in accord with the proposal put forth by Van Duyne[10] that the pyridine is adsorbed on Ag via the N "lone pair". The moderately
strong Raman peak which appears at 226 cm⁻¹ in the spectrum for CN⁻
adsorbed on Ag is correspondingly attributed to a Ag-C vibration, on
the reasonable assumption that the CN⁻ group is bonded to Ag via a
C bond[8].

The wavelength dependence of the Raman spectrum of pyridine
adsorbed on a Ag electrode has been measured by various investigators
with somewhat different results. Thus, the early data obtained by
Jeanmaire and Van Duyne[2] indicated that there was no appreciable
variation from an ω^4 dependence. Later data obtained by Pettinger,
et al[15] and by Creighton, et al[16] indicated that the intensity of
the peaks due to the A₁ modes at 1008 cm⁻¹, 1026 cm⁻¹ and 1036 cm⁻¹,
when normalized to an ω^4 dependence, increased by an order of magnitude when the wavelength of the incident radiation increased from
4500 Å to 6500 Å. Pettinger, et al[15] have also reported that after
the electrochemical oxidation-reduction cycle, the differential reflectance spectrum ($\Delta R/R$ versus λ) of the electrochemically processed
Ag electrode exhibits a broad minimum at ~7500 Å, which they suggest
may be due to optical absorption by a Ag-pyridine-Cl⁻ complex. More
recently Van Duyne[10] also reported a similar trend for the 1008 cm⁻¹
and 1037 cm⁻¹, but noted that the deviations from the ω^4 law were
smaller than those reported by Pettinger, et al and by Creighton,
et al. On the other hand, he noted that the peak at 1215 cm⁻¹ followed a rigorous ω^4 dependence and that some of the observed deviations from an ω^4 law were artifacts of laser damage.

An important contribution toward the elucidation of the nature
of the enhanced RS by molecules adsorbed on Ag electrodes has been
made by Moscovits[17] who suggests that the enhanced RS by the molecules in the visible is a direct result of the excitation of the
transverse collective electron resonances (also termed conduction
electron resonances) of sub-microscopic bumps on the surface of the
electrochemically processed Ag. By modeling the sub-microscopic
bumps as a layer of metal spheres on a flat metal substrate, he
showed that the transverse collective electron resonance could account for the minimum in the $\Delta R/R$ versus λ spectrum and for the
wavelength dependence of the RS which were obtained by Pettinger,
et al[15]. Chen et al[14] have also emphasized the role played by the
sub-microscopic surface roughness introduced by the electrochemical
processing of the Ag electrode. They point out that the excitation
of the transverse collective electron resonances leads to an

enhancement of the incident electric field at the adsorbed molecule.
They note also that the fact that the transverse collective electron
resonances can be excited by both s- and p- polarized EM radiation
accounts in part for the insensitivity of the enhanced RS to the
polarization and to the angles of incidence and scattering. They
note also that the sub-microscopic surface roughness causes a breakdown in the conservation of momentum and, thereby, an enhanced radiative excitation and recombination of particle-hole pairs in the
surface region of the metal.

Although there is still considerable interest in the RS by
molecules adsorbed on metal electrodes, for example, with regard to
the dependence of the scattering intensity of the different A_1 modes
of pyridine on electrode potential, and the role played by the
composition of the electrolyte, there is increasing interest in
other procedures for adsorbing molecules at metal surfaces. The
vapor deposition of molecules onto clean surfaces in ultra-high
vacuum is obviously the most interesting one from the point of view
of obtaining adsorbed molecules that can be adequately charaterized.
However, the observation by Wood and Klein[11] of the RS by CO molecules adsorbed, at pressures greater than 10 Langmuirs, on evaporated
Ag films at liquid N_2 temperature is the only result for vacuum deposited molecules reported thus far.

Two other procedures have been used recently with considerable
success, namely, the use of a thin evaporated metal film, and in
particular a thin evaporated metal island (aggregated) film, as the
metal overlayer on molecules initially adsorbed on a dielectric substrate, and the use of a thin evaporated metal island film as the
substrate for the adsorbed molecules:

Tsang and Kirtley[18] have observed the RS by 4-pyridine carboxaldehyde and by para-substituted benzoic acid chemisorbed, via the
carboxylate group, on the thin oxide layer of an Al film and over-
layed with a thin evaporated Ag film which also served as the top
electrode of an $Al/Al_2O_3/$ metal electron tunneling structure. They
showed moreover, on the basis of inelastic tunneling data for this
structure, that there was no more than a monolayer of molecules at
the Ag overlayer film. Although they were able to observe the inelastic electron tunneling spectrum of benzoic acid, they were unable
to observe any RS by the benzoic acid. They found that the RS by
the molecules was further enhanced when the structure was formed on
a substrate coated with an evaporated film of CaF_2 to produce surface roughness. They were also able to observe RS by the molecules
when Cu was used as the overlayer metal, but were unable to observe
RS when Pb, Al, Sn and Au were used.

In experiments carried out at the University of Pennsylvania,
which were aimed at elucidating the role played by sub-microscopic
surface roughness, Chen, et al[19] have used thin (25 to 100 Å)

evaporated Ag island films as the evaporated metal film overlayer on iso-nicotinic acid (4-pyridine carboxylic acid) and on benzoic acid which were initially deposited on glass from aqueous solution (via chemisorption of the carboxylate group), and also as the metal substrate for the adsorption of the molecules from aqueous solution (via the chemisorption of the carboxylate group at the oxide, or sulfide, monolayer that forms when the Ag island film is exposed to air). The RS data obtained for these configurations are quite striking. When the Ag island film was used as the overlayer, a strongly enhanced RS was observed for iso-nicotinic acid, which is comparable in intensity to that observed for pyridine on an Ag electrode (Fig. 3) but no RS was observed for benzoic acid, in accord with the results obtained by Tsang and Kirtley. On the other hand, when the Ag island film was used as the substrate a strong RS was observed for benzoic, as well as for iso-nicotinic acid (Fig. 4). The fact that benzoic acid exhibits a strongly enhanced RS when it is chemisorbed via the carboxylate group to the Ag island film substrate, and none when the benzene ring is in close proximity but not chemisorbed to the Ag island film overlayer, suggests that chemisorption plays an important role in the enhancement process. The Raman spectra for iso-nicotinic acid and benzoic acid adsorbed on Ag island films are similar to the spectra of pyridine adsorbed on an Ag electrode. Thus the spectra for the Ag island films exhibit a strong scattering continuum which extends beyond 4000 cm^{-1}. In addition the RS by the molecules and the scattering continuum are insensitive to the polarization of the incident and scattered radiation and to the angles of incidence and scattering, and are appreciably depolarized. The oblique incidence transmission spectrum of a 50 Å Ag island film with adsorbed iso-nicotinic, which is within the experimental uncertainty the same as that for the film without the adsorbed molecules, exhibits a characteristic broad minimum at ~7500 Å due to excitation of the "transverse" (e. g. parallel to the film) collective electron resonance of the film, and a narrow minimum at ~3300 Å due to excitation of the "perpendicular" collective electron resonance of the film. Preliminary data in the visible on the wavelength dependence of the Raman peaks of iso-nicotinic acid indicate that they are consistent with the broad absorption peak of the Ag island film at 7500 Å.

Chen, et al have also observed RS for iso-nicotinic acid and benzoic acid on a 50 Å Au island film substrate and for iso-nicotinic acid with a 50 Å Au island film overlayer. The spectra for the Au island film were weaker by a factor of about 100 than the corresponding spectra for the Ag island films. The RS by the molecules adsorbed on the Au island films was not observable below 5200 Å, and increased appreciably in strength on increasing the excitation wavelength from 5308 Å to 6471 Å. These results are consistent with the fact that the transmission spectrum of the evaporated 50 Å Au island film on glass exhibits a broad minimum at ~7000 Å, which is due to the transverse collective electron resonance, and a decrease in

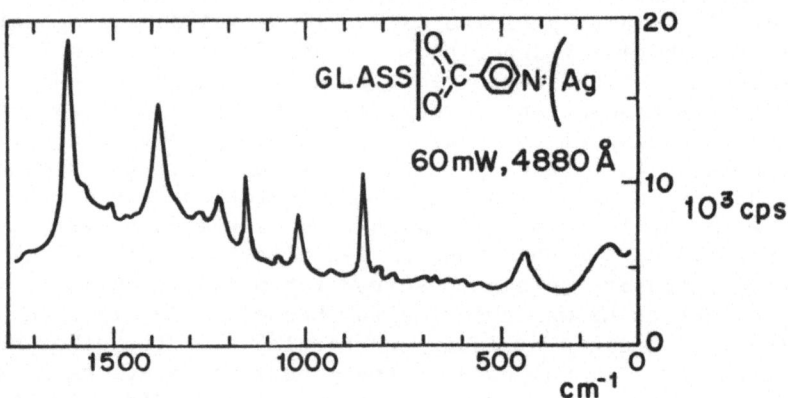

Fig. 3. Raman spectrum of iso-niotinic acid supported on glass with a 50 Å Ag island film as overlayer. The spectrum was obtained using 60 mW of 4880 Å excitations. (Chen et al, Ref. 19)

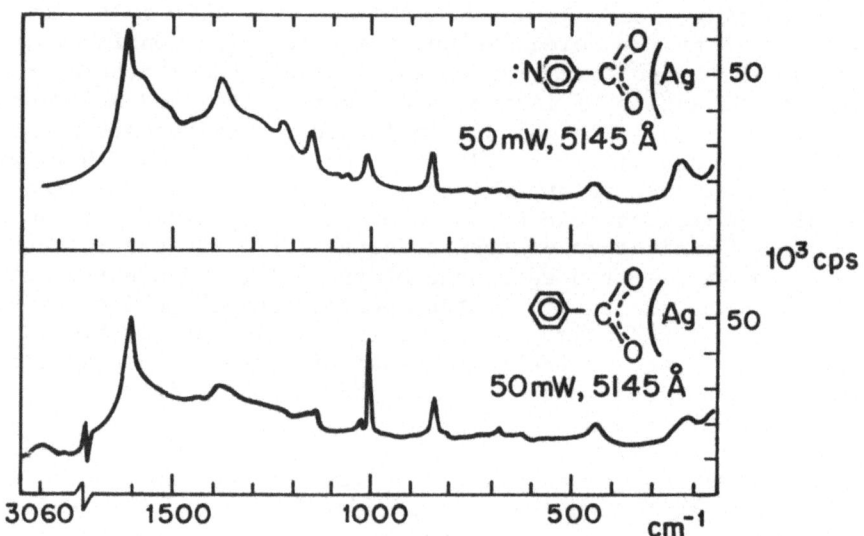

Fig. 4. Raman spectra of iso-nicotinic acid and benzoic acid adsorbed from aqueous solution on a 50 Å Ag island film substrate. The spectra were obtained using 50 mW of 5145 Å excitation. (Chen et al, Ref. 19).

transmission below 5200 $\overset{\circ}{A}$, which corresponds to the onset of "interband transitions".

THEORY

   The RS tensor of the adsorbed molecules on metals can be viewed
macroscopically as involving contributions of the form $(\partial X_A/\partial Q_j)Q_j$,
$(\partial X_S/\partial Q_j)Q_j$ and $(\partial X_{A-S}/\partial Q_j)Q_j$, where X_A, X_S and X_{A-S} are the electric susceptibilities of the adsorbed molecules (adsorbate), the
metal (substrate) and the adsorbate-substrate complex respectively,
and Q_j is the normal coordinate of the vibration mode involved in
the RS, representing either the atomic displacement co-ordinate u_j
and the electric field E_j set up by the atomic displacements. The
objective for theory is to ascertain the microscopic processes that
are responsible for the giant RS cross-section of adsorbed molecules
on metals and for the dependence of the RS cross-section on the nature of the metal, the type of molecule, and the nature of the bonding of the molecule to the metal.

   The microscopic models that have been proposed, to account for
the enhancement of the RS by adsorbed molecules on metals, fall essentially into four categories: a) modulation of the reflectivity
(e.g. electric susceptibility) of the metal; b) effects of the
image dipole field; c) excitation of electron-hole pairs in the surface region of the metal; and d) excitation of collective electron
resonances of the rough metal surface.

a) Otto[20] has suggested that the electronic susceptibility of the
metal is modulated by the local (coulomb) field E_j set up by the
vibration modes of the adsorbed molecules, assumed to be in a regu-
lar array, by mechanisms that play a role in electro-reflectance,
i.e. the RS is due to $(\partial X_S/\partial E_j)E_j$. He suggests that surface roughness
should enhance the contributions from the large wave vector Fourier
components of the local field of the vibrating charges of the molecules.

   McCall and Platzman[21] have more recently proposed that the
dominant contribution to the moduclation of the electric susceptibility comes from the displacements u_b of the atom that is chemically
bonded to the metal substrate, which they suggest modulates the
charge (q_t) that is transferred in the formation of the metal-atom
bond, i.e. the RS is due to $(\partial X_S/\partial q_t)(\partial q_t/\partial u_b)u_b$. They point out
that in the case of CN^- adsorbed on Ag, essentially all of the scat-
tering by the Ag-C and the C-N vibration modes comes from the
modulation of the Ag dielectric constant by the displacements of the
C atom. Using this model, they obtain an estimated value of 0.15
for the ratio of the total integrated line intensity of the Ag-C
vibration peak to that of the C-N vibration peak, which is close to
the observed ratio of 0.1 derived from the data obtained by Furtak[8]
for CN^- adsorbed on a Ag electrode.

b) King, et al[22] propose that the electric field at the adsorbed molecule is greatly enhanced by the contribution from the image dipole that is induced in the metal by the electronic excitation of the molecule, and that the electronic polarizability of the adsorbed molecule is thereby greatly enhanced. The electric field at the molecule due to the image dipole is given by

$$\mu \quad = \quad \alpha_{loc} E_{loc} \quad = \quad \alpha_{loc}(\bar{E} + E_{image}(r))$$

$$E_{image}(r) \quad = \quad \frac{(\tilde{\varepsilon}_m - \varepsilon_a)\mu}{(\tilde{\varepsilon}_m + \varepsilon_a)4r^3} \quad = \quad (\tilde{\beta}/4r^3)\mu$$

where r is effective distance from the electronic excitation dipole to the metal surface; $\tilde{\varepsilon}_m$ and ε_a are the dielectric constants of the metal and the adjacent medium respectively, μ is the dipole moment of the electronic excitation; α_{loc} is the local electronic polarizability; E_{loc} and $\bar{E}$ are the local and macroscopic electric field at the molecule, respectively. The macroscopic electronic polarizability of the adsorbed molecule defined by $\mu = \alpha\bar{E}$ is accordingly given by

$$\tilde{\alpha} \quad = \quad \frac{\alpha_{loc}}{1 - (\tilde{\beta}/4r^3)\alpha_{loc}}$$

Thus the electronic polarizability of the molecule is enhanced by a factor of $(1 - \tilde{\beta}\alpha_{loc}/4r^3)^{-1}$. King, et al. conclude that, by using what they consider to be reasonable values for the various parameters, a Raman intensity enhancement of several order of magnitude is possible at physically obtainable separations between the metal surface and the adsorbed molecule.

Efrima and Metiu[23] suggest that the enhancement of the RS by adsorbed pyridine on Ag is in fact due to the displacement of the electronic excitation of pyridine to lower frequencies by the electric field at the adsorbed molecule, which is induced by the presence of the metal. Thus they attribute the enhancement to a "metal surface induced resonant RS" by the adsorbed molecules. To obtain a more appropriate expression for metal-induced field at the molecules, they calculate the effects of the van der Waals (e.g. dispersive) interaction of the electronic excitations of the molecule with electronic excitations of the metal. Within the framework of a local dielectric constant, the local field at the molecule due to its interaction with the metal is that set up by its image dipole, and the expression for the induced dipole moment takes the form

$$\mu(\omega) \quad = \quad \tilde{\alpha}(\omega)\bar{E}(\omega) \quad = \quad \frac{f_o e^2 \bar{E}(\omega)}{m(\omega_o^2 - \omega^2 - 2\omega\Delta_s(\omega) - 2i\omega\gamma_o)}$$

where ω_o, γ_o and f_o are the resonant frequency, damping constant and oscillator strength of the electronic excitation in the absence of the image dipole field, and $\Delta_s(\omega) = \Delta_s(\omega) + i\Gamma_s(\omega)$ is a self-energy term due to E_{image}, whose real and imaginary parts shift and broaden the electronic resonance. There is of course, also a size-able contribution to the local field (and self-energy) from the induced dipoles at neighboring molecules[24]. $\Delta_s(\omega)$ and $\Gamma_s(\omega)$ are given by

$$\tilde{\Delta}_{s/\!/}(\omega) \quad = \quad \frac{\tilde{\Delta}_{s\perp}(\omega)}{2} \quad = \quad \frac{4\pi\varepsilon_1 f_o e^2 \tilde{\beta}(\omega)}{m\omega_o^2 (2r)^3}$$

where the subscripts $/\!/$ and $\perp$ indicate the orientation of μ rela-tive to the surface. Thus the magnitudes of Δ_s and Γ_s are deter-mined by the magnitude of r^{-3} and by the magnitudes of the real and imaginary parts respectively of $\tilde{\beta} = (\tilde{\varepsilon}_m - \varepsilon_a)/(\tilde{\varepsilon}_m + \varepsilon_a)$. As pointed out by Efrima and Metiu the enhanced electronic polarizability model of King, et al is the low frequency counterpart of their more general model. One the other hand their model is actually an extension of the theoretical model of Philpott[25] in which the frequency and broadening of the electronic excitation of a molecule at the surface of a metal is obtained from the explicit interaction of the elec-tronic excitation of the molecules with the surface plasmons of the metal.

Efrima and Metiu note that the magnitude of the enhancement at the shifted resonance, i.e. at $\omega^2 = \omega_o^2 - 2\omega\Delta_s(\omega)$ is determined by $\Gamma_s(\omega)^{-4}$ and therefore by $Im\tilde{\varepsilon}(\omega)$. They account for the fact, that the wavelength dependence of the RS by pyridine does not exhibit a sharp resonance, by suggesting that the adsorbed molecules are not all at the same distance r from the surface and, therefore, exhibit a dis-tribution of resonance frequencies. Finally they point out that, since $Im\tilde{\varepsilon}_m(\omega)$ is much smaller for Ag than for Cu and Au throughout the visible[26], Ag should yield a much greater enhancement of the RS by adsorbed molecules than Cu and Au.

c)   Burstein, et al[27] have suggested that the relevant RS processes

involve the radiative excitation and recombination of electron-hole
pair excitations of the metal/adsorbed molecule system. In the case
of smooth metal surfaces, the radiative excitation and recombination
of electrons and holes having a wide range of energies and momenta
in the surface region of the metal is due to the presence of the
surface and to local fields that are set up by the periodicity
of the lattice. The breakdown of momentum conservation by surface
roughness and the presence of adsorbed molecules, provide further
mechanisms for the excitation of electron-hole (e-h) pairs by incident EM ratiation. They proposed the following mechanisms as the
potentially important ones:

● The excitation of an e-h pair by the incident radiation is
transferred via coulomb interaction to the electronic excitations
of the adsorbed molecules and then back to an e-h pair which radiatively recombines. The RS process can be viewed as a modification
of the direct RS by the molecule in which the incident and scattered
photons are replaced by renormalized photons (e.g. polaritons)
which are admixed with e-h pair excitations. It can be formulated
in terms of the RS of polaritons by the adsorbed molecules via the
coulomb interaction of the electronic excitation of the molecules
with the e-h pair excitation content of the polaritons. It can
also be expressed in terms of the contribution from the e-h pair
excitations to the local field at the adsorbed molecule.

● The excited e-h pairs, whose wave functions extend beyond the
surface of the metal, are inelastically scattered via the Frohlich
(coulomb) interactions with the vibrating charges of the adsorbed
molecules and then radiatively recombine. This is one of the
microscopic processes that contribute to $(\partial \chi_S / \partial E_j) E_j$. The inelastic
scattering of the e-h pairs is similar in its description to the
scattering process that is involved in inelastic electron tunneling[28] except that the carriers are scattered back into the same
metal.

● The RS process involves a charge-transfer excitation of the
metal-adsorbed molecule complex. In the limit of weak coupling
between the metal and the adsorbed molecule, the charge-transfer
excitation can be viewed as taking place via the excitation of an
e-h pair in the metal, followed by the hopping (e.g. tunneling) of
the excited electron (hole) into and back from a virtual bound
state of the molecule. The temporary trapping of the electron
(hole) in the virtual bound state induces a nuclear relaxation of
the molecule, and when the electron (hole) leaves the virtual
bound state the molecule is left in its ground electronic state,
but a different vibrational state. The electron charge-transfer
process, in effect, corresponds to an inelastic scattering of the
electron via a negative ion resonance of the adsorbed molecule[29].
In the case of strong coupling, the relevant "one-electron" states

are an admixture of the electronic states of the metal and molecule,
with the electron located predominantly in the metal (molecule) in
the ground state and predominantly in the molecule (metal) in the
excited state. This is one of the microscopic processes that contribute to $(\partial \chi_{AS}/\partial Q_j)Q_j$.

• The EM radiation is inelastically scattered via A^2 and $(p \cdot A)^2$
processes by e-h pair excitations [30] which are coupled via coulomb
interaction with the vibrational excitations of the molecules. The
process, which corresponds to scattering by e-h pair polarons, can be
viewed as an extension of the RS by e-h pair excitations involving
an additional step in which the excitation of the e-h pair is transferred to a vibrational excitation of the adsorbed molecules. The
cross-section for this mechanism, like that for the Frohlich scattering mechanism, depends on the magnitude and orientation of the
vibrating charge displacements, but does not depend directly on the
electronic states of the adsorbed molecule (or metal-adsorbed molecule complex).

    The mechanisms differ in their dependence on the electronic
states of the metal/molecule system and therefore differ in their
dependence on the wavelength of the incident ratiation. The
"Frohlich scattering" and "e-h pair polaron" scattering processes
will not exhibit any appreciable dependence on wavelength. The
"excitation-transfer" mechanism will yield a peak in the scattering
intensity at the electronic excitation resonances of the adsorbed
molecule. The "charge-transfer" mechanism, on the other hand, will
exhibit a "resonance onset" when the energy of the incident photons
approaches the energy separation between the virtual band state $\mathcal{E}_{VBS}$
and the Fermi level, i.e. when $\hbar\omega$ approaches $|\mathcal{E}_{VBS} - \mathcal{E}_F|$, followed
by a weak variation with frequency when $\hbar\omega$ exceeds $|\mathcal{E}_{VBS} - \mathcal{E}_F|$. It
is similar to the onset that occurs in inelastic electron tunneling[28].

d)   As noted earlier, Moscovits[17] has suggested that the giant RS
of molecules on a Ag electrode is due to the excitations of the
transverse collective electron resonance of sub-microscopic bumps
on the metal surface. However he did not address the question as
to how the excitation of the resonance is transferred to (and back
from) the electronic excitation of the molecules. Chen, et al[14]
have pointed out that the excitation of the transverse collective
electron resonance of the sub-microscopic rough Ag surface leads to
an increase of the electric field at the adsorbed molecules and
thereby to an enhancement of the direct RS by the adsorbed molecules.

    We will consider here the more tractable case of the enhanced
RS by adsorbed molecules on a thin evaporated metal island film[19],
since metal island (e.g. aggregated) films have been studied quite
extensively, both experimentally and theoretically[31]. For simplicity
we will assume that the metal islands a) are ellipsoids of revolution
with their symmetry axis oriented perpendicular to the surface,

b) are uniform in size, and c) are uniformly distributed. For an isolated island the frequencies of the collective oscillations parallel and perpendicular to the surface of the film are, in the absence of damping, given by

$$\omega_{\parallel,\perp}^2 = \frac{4\pi L_{\parallel,\perp} n e^2}{m\varepsilon_{\parallel,\perp}^0(\omega)}$$

where $L_\parallel$ and $L_\perp$ are the $\parallel$ and $\perp$ geometry depolarization factors for the ellipsoids; n and m are the electron density and effective mass; and $\varepsilon_{\parallel,\perp}^0(\omega)$ is the frequency dependent contribution to the dielectric constant from other (e.g. interband) electronic excitations.

As shown by Yamaguchi, et al[32] the collective electron resonances of the array of islands can be derived by treating the electric moments of the islands, whose dimensions are very small compared to the wavelength, as point dipoles, and taking into account the contributions to the local field at the point dipole of an island from neighboring dipoles and from its image dipole within the dielectric substrate. The net effect of these contributions to the local field is to shift and broaden the transverse collective electron resonance of the island, and also to increase the electric field at the adsorbed molecules.

The electronic polarizability of an isolated island is, on the basis of the Drude "free electron gas" model, given by

$$\tilde{\alpha}_{\parallel,\perp}^{loc} = \frac{Vne^2}{m(\omega_{\parallel,\perp}^2 - \omega^2 - i\omega\gamma_{\parallel,\perp})}$$

where V is the volume of the island, and $\gamma_{\parallel,\perp}(\omega)$, is the damping constant of the collective electron oscillations. $\gamma_{\parallel,\perp}(\omega)$ is appreciably larger than the damping constant of the bulk metal due to the relative increase in scattering of the electrons by the surface and to increased Landau damping of the collective excitation by particle-hole pair excitations resulting from the quantization of the electronic states[33].

The dipole moment induced in an island by EM radiation is given by

$$\mu_{\parallel,\perp}(\omega) = \tilde{\alpha}_{\parallel,\perp}^{loc}(\omega) E_{\parallel,\perp}^{loc}(\omega) = \tilde{\alpha}_{\parallel,\perp}^{loc}(\omega)\left\{\bar{E} + (E_{id} + E_{nd})_{\parallel,\perp}\right\}$$

$$\mu_{\parallel,\perp}(\omega) = \frac{\tilde{\alpha}^{loc}_{\parallel,\perp}(\omega)\ \bar{E}_{\parallel,\perp}(\omega)}{1 - (\delta_{id}(\omega) + \delta_{nd}(\omega))_{\parallel,\perp}\tilde{\alpha}^{loc}(\omega)_{\parallel,\perp}}$$

where $\bar{E}$ is the macroscopic (e.g. average) electric field; and $E_{id} = \delta_{id}\mu$ and $E_{nd} = \delta_{nd}\mu$ are the contributions to the local electric field at the island from the image dipole (id) and from the neighboring dipoles (nd), respectively. δ_{id} depends on the distance of μ from the substrate and on the dielectric constants of the substrate and the medium adjacent to the substrate. δ_{nd} depends on the distances from μ to the surrounding dipoles and on the dielectric constant of the medium adjacent to the substrate.

The macroscopic polarizability α, defined by $\mu = \alpha\bar{E}$ is, therefore, given by

$$\tilde{\alpha}_{\parallel,\perp}(\omega) = \frac{\tilde{\alpha}^{loc}_{\parallel,\perp}(\omega)}{1 - \delta_{\parallel,\perp}(\omega)\tilde{\alpha}^{loc}_{\parallel,\perp}(\omega)} \quad ; \quad \delta_{\parallel,\perp} = (\delta_{id} + \delta_{nd})_{\parallel,\perp}$$

Correspondingly, the local electric field at an island is given by

$$E^{loc}_{\parallel,\perp}(\omega) = \frac{\bar{E}_{\parallel,\perp}(\omega)}{1 - \delta_{\parallel,\perp}(\omega)\tilde{\alpha}^{loc}_{\parallel,\perp}(\omega)}$$

Yamaguchi, et al show that $\delta_{\parallel}\alpha^{loc}_{\parallel}$ is positive, which is in accord with the observed shift of the transverse collective resonance to lower frequencies, and that $\delta_{\perp}\alpha^{loc}_{\perp}$ is negative and small, in accord with the observed smaller shift of the perpendicular collective resonance to higher frequencies. They derive the following expression for the dielectric response of the metal island film:

$$\epsilon_{\parallel}(\omega) - 1 = \frac{\Phi}{F_{\parallel}(\omega) + g(\omega) + \Delta g(\omega)}$$

$$1 - \frac{1}{\epsilon_{\perp}(\omega)} = \frac{\Phi}{F_{\perp}(\omega) + g(\omega) + \Delta g(\omega)}$$

$$g(\omega) = Re\left(\frac{1}{\tilde{\epsilon}_m(\omega - 1)}\right) \quad ; \quad \Delta g(\omega) = \Sigma_i I\left(\frac{1}{\tilde{\epsilon}_m - 1}\right)_i$$

where ϕ is a filling factor that is determined by the size of the islands and by the distance between islands; $F_{//,\perp} = L_{//,\perp} - \delta_{//,\perp}/4\pi$ is an effective depolarization factor that takes into account the contribution to the local field from E_{id} and E_{nd}, as well as the depolarization field which depends on the shape of the island; and $\varepsilon_m(\omega)$ is the frequency dependent dielectric constant of the isolated metal island whose imaginary par involves the contributions, designated by the subscript i, to the damping from bulk, surface and e-h pair scattering processes, i.e., $\Delta g = \Delta g_{bulk} + \Delta g_{surf} + \Delta g_{e-h}$. The frequency of the transverse resonance $(\Omega_{//})$ corresponds to the pole of $\text{Im } \varepsilon_{//}(\omega)$ and the corresponding frequency of the perpendicular resonance $(\Omega_\perp)$ corresponds to the pole of $\text{Im}(1/\varepsilon_\perp(\omega))$. In real situations the islands are random in size and in spatial distribution, and furthermore are not necessarily ellipsoidal in shape. This causes an appreciable broadening of the transverse collective resonance of the film, but does not have as large an effect on the perpendicular resonance.

The sizeable absorption of EM radiation by the metal island film, when the transverse collective electron resonance is excited, is a direct consequence of the increased magnitude of $E_{//loc}$. Since $E_{//loc}$ is also the electric field that acts on the adsorbed molecules, the increase in $E_{//loc}$ leads to an increase in the excitation of the molecules by the incident radiation and, thereby, to an enhancement of the direct RS bv the molecules. It should be emphasized that in a real metal island film $E_{//loc}$ will vary considerably from one island to another because of random size and spatial distribution of islands.

The RS "enhancement factor" $\eta = (E_{//loc}/\overline{E})^2$ at the transverse collective resonance $\Omega_{//}$, which corresponds to $g(\omega) = L_{//} - \delta_{//}(\omega)/4\pi$, is given by

$$\eta = \frac{\Gamma_{//}^2 + (4\pi\Delta g)^2}{(4\pi\Delta g)^2}$$

Using values for $\Delta g(\Omega_{//})$ and $\delta_{//}(\Omega_{//})$ which can be obtained from the frequency and depth of the transmission minimum of the 50 Å Ag island film at $\Omega_{//}$, we estimate η to be ~ 120.

We also expect that the coupling of the scattered electric field from the molecule will also enhance the intensity of the scattered radiation, i.e., the metal island film serves as an efficient antenna for the scattered radiation as well as for the incident radiation. In this respect, the role played by transverse collective electron resonances in the enhancement of the RS by the adsorbed molecules is analogous to the role played by surface EM modes in the enhancement of the RS by adsorbed molecules at the metal-air

interface of an ATR prism configuration[34]. The overall enhancement
of the RS by the adsorbed molecules on a 50 Å Ag island film, that
results from the coupling of the incident and scattered radiation
to the transverse collective resonance of the film, is thus likely
to be 10^3 or greater.

The sub-microscopic roughness of either the Ag electrode or the
Ag island film causes a breakdown in wave vector conservation and,
thereby, leads to an increase in the radiative excitation and recombination of e-h pairs. The strong scattering continuum which is exhibited by electrochemically processed Ag surface and by Ag island
films is accordingly attributed to inelastic light scattering by
e-h pair excitations in the metal via surface roughness-induced
$(p \cdot A)^2$ processes for which $k_{e-h} \neq k_i \pm k_s$[14]. Since the incident radiation creates real e-h pairs, an appreciable part of the scattered
radiation may actually correspond to luminescence, i.e. to e-h pair
recombination radiation. The absence of any detectable Fano-type
interference between the RS by the adsorbed molecules and the scattering continuum, is an indication that this may in fact be the case.
As a further consequence of the increased radiative excitation and
recombination of e-h pairs, there will also be an increase in the
e-h excitation content of the renormalized incident and scattered
photons and, thereby, an increase in the effectiveness of the metal
substrate to transfer excitation to (and back) from the adsorbed
molecules (or metal-adsorbed molecule complex). Finally we note
that in the case of large surface roughness, which would for example occur as a result of strong electrochemical processing, the
incident and scattered radiation would also be coupled to surface-
EM waves and, under these circumstances, there could be a further
enhancement of the RS by the adsorbed molecules.

The various theoretical models that have been proposed, and for
that matter are continuing to be proposed, invariably attribute the
giant enhancement of the RS by adsorbed molecules on metals to a
single mechanism that is, moreover, applicable to molecules adsorbed
on a smooth surface. It is now clear, in view of the role played
by sub-microscopic roughness of the Ag electrode that the overall
enhancement by a factor of 10^5 to 10^6 is a result of a combination
of mechanisms. The increased coupling of the incident and scattered radiation with the adsorbed molecule via the collective electron resonances of the sub-microscopicable roughened Ag surface will
obviously further increase the enhancement to be expected from a
given microscopic model.

DISCUSSION

The absence of specific information about the electronic states
of the metal-adsorbed molecule system is at present the major deficiency of efforts to elucidate the enhanced RS by adsorbed molecules
on metals. There has been considerable progress in understanding

the nature of the energy levels of molecules adsorbed on metals[35].
Thus even when the molecules in contact with the metal are only
physisorbed", there is a very appreciable shifting and broadening of
the electronic levels and thereby the electronic excitations of the
adsorbed molecule due to interference (e.g. "resonance") between the
one-electron states of the molecule with the broad continuum of one-
electron states of the metal. At the distances required for sizeable
enhancement by the image dipole mechanism[22-24], the shifting and
broadening of the electronic excitation of the molecule by "interference" is very likely considerably larger than that due to the
dispersive interaction of the electronic excitations with surface
plasmons or even with e-h pair excitation. There is moreover some
indication, from the RS data which Chen et al[19] have obtained for
benzoic acid and iso-nicotinic acid using Ag island films (both as
overlayer and a substrate), that chemisorption is an important ingredient in the enhanced RS by molecules adsorbed on metals.

Chemisorption causes further, even more drastic, modifications
of the energy levels of the metal-adsorbed molecule complex. In a
typical situation the strong mixing of the molecular levels with the
continuum of states of the metal leads to the formation of bonding
and anti-bonding states with splittings of up to several eV. When
the bonding levels are occupied, i.e. below ε_F, and the anti-bonding
levels are empty, electronic transitions become possible, which may
lie in the visible. The shifts in the unoccupied energy levels of
the adsorbed molecules may in some cases be large enough to pull
down a molecular unoccupied (e.g. electron affinity) level to the
region of the Fermi level or below. The broadened level will then
be partly or completely filled by the transfer of charge from the
metal (i.e., the bonding is partly ionic). This provides additional
mechanisms for optical transitions. Such transitions and those between the bonding and anti-bonding levels will however be very broad,
and, as a consequence, will not introduce any readily observable resonance structure in the wavelength dependence of the RS by the adsorbed molecules. This is consistent with the fact that the observed
wavelength dependence of molecules adsorbed on an electrochemically
processed Ag electrode, or on a 50 Å Ag island film, is essentially
that due to the dielectric response of the sub-microscopically rough
Ag surface. An appreciable increase in the oscillation strength of
the electronic excitations of the adsorbed molecules is another possible consequence of the appreciable admixture of the wave function
of the energy levels of the molecules with those of the metal.

Finally, with regard to the observability of RS by molecules on
metals, it should be emphasized that the inability to observe a RS
by a particular molecule may simply be due to the fact that the
molecule may not be adsorbed (chemisorbed) on the metal. Since sub-
microscopic surface roughness of the substrate (e.g. electrode or
metal island film) increases the chemical "reactivity" of the surface, it increases the probability that a given molecule will be

adsorbed (chemisorbed) on the metal surface.

ACKNOWLEDGEMENTS

We wish to acknowledge valuable discussions with B. I. Lundquist and W. Plummer about the chemisorption of molecules on metals.

REFERENCES

*Department of Mathematical Physics, Chalmers Institute of Technology, Goteborg, Sweden.

+Research supported by ARO-Durham and by NSF through the University of Pennsylvania Materials Research Laboratory.

‡For a review of the Raman scattering by molecules adsorbed on metal electrodes, see R. P. Van Duyne[10].

1. M. Fleischmann, P.J. Hendra and A.J. McQuillan, Chem. Phys. Lett. 26, 123 (1974).

2. D. L. Jeanmaire and R. P. Van Duyne, J. Electroanal. Chem. 84, (1977).

3. M. G. Albrecht and J. A. Creighton, J. Am. Chem. Soc. 99, 5215 (1977).

4. A. J. McQuillan, P. J. Hendra and M. Fleischmann, J. Electroanal. Chem. 54, 253 (1975).

5. R. P. Cooney, E. S. Reid, M. Fleischmann and P. J. Hendra, J. Chem. Soc. Faraday Trans.I, 73, 169 (1977).

6. R. P. Cooney, E. S. Reaid, P. J. Hendra and M. Fleischmann, J. Am. Chem. Soc. 99, 2002 (1977).

7. A. Otto, Surf. Sci. 75, 392 (1978).

8. T. E. Furtak, Solid State Commun. 28, 903 (1978).

9. G. Hagen, B. Simic Glavaski and E. Yeager, J. Electroanal. Chem. 88, 269 (1978).

10. R. P. Van Duyne, Chemical and Biological Applications of Lasers, C. B. Moore, Vol. 4, Ch. 5 (1978).

11. T. H. Wood and M. V. Klein, J. Vac. Sci and Tech. 16, 459 (1979)

12. R. M. Hexter, Proc. US-Japan Seminar on Inelastic Light Scattering, Solid State Commun. 32 (1979).

13. B. Pettinger and V. Wenning, Chem. Phys. Lett. 56, 253 (1978).

14. C. Y. Chen, E. Burstein and S. Lundquist, Proc. US-Japan Seminar, Solid State Commun. 32, (1979).

15. B. Pettinger, V. Wenning and C. M. Kolb, Ber. Bunsenges. Phys. Chem. 82, 1326 (1978).

16. J. Creighton, M. Albrecht, R. Hester and J. Matthew, J. Chem. Phys. Lett. 5, 55 (1978).

17. M. Moscovits, J. Chem. Phys. 69, 4159 (1978) and Proc. US-Japan Seminar, Solid State Commun. 32 (1979).

18. J. C. Tsang and J. Kirtley, Solid State Commun. 30, 617 (1979).

19. C. Y. Chen, I. Davoli and E. Burstein (to be published).

20. A. Otto, Proc. Conf. on Vibrations in Adsorbed Layers, Julich, Germany (1978).

21. S. L. McCall and P. M. Platzman (to be published).

22.   F. W. King, R. P. Van Duyne and G. C. Schatz, J. Chem. Phys. 69,
          4472 (1978).
23.   S. Efrima and H. Metiu, Chem. Phys. Lett. 60, 59 (1978); J. Chem.
          Phys. 70, 1602, 2297, 1939 (1979).
24.   G. L. Eesley and J. R. Smith, Solid State Commun. (in press).
25.   M. R. Philpott, J. Chem. Phys. 62, 1812 (1975).
26.   P. P. Johnson and R. W. Christy, Phys. Rev. B6, 4370 (1972).
27.   E. Burstein, Y. J. Chen, C. Y. Chen, S. Lundquist and
          E. Tosatti Solid State Commun. 29, 565 (1979).
28.   J. Kirtley, D. J. Scalapino and P. K. Hansma, Phys. Rev. B14,
          3177 (1976).
29.   S. F. Wong and G. J. Schultz, Phys. Rev. Lett. 35, 1429 (1975).
30.   E. Burstein, A. Pinczuk and S. Buchner, Proc. 14th Int. Conf.
          on Physics of Semiconductors, ed. by B. L. H. Wilson
          (Institute of Physics, Bristol, 1979) p. 1231.
31.   See for example, P. Rouard and A. Meessen "Optical Properties
          of Thin Films", Progress in Optics XV, ed. E. Wolf (North
          Holland, 1977), p. 79-137, and references therein.
32.   T. S. Yamaguchi, S. Yoshida and A. Kinbara, Thin Solid Films
          21, 173 (1974).
33.   A. Kawabata and R. Kubo, J. Phys. Soc. Jap. 21, 1765 (1966).
34.   Y. J. Chen, W. P. Chen and E. Burstein, Phys. Rev. Lett. 36,
          1207 (1976).
35.   B. I. Lundquist, Hjelmberg and O. Gunnarson, "Photoemission and
          the Electronic Properties of Surfaces", ed. by B. Feuerbacher,
          B. Fitton and R. F. Willis (John Wiley & Sons, Chichester,
          1978) Chapter 5.

RAMAN SPECTROSCOPY OF MOLECULAR MONOLAYERS IN

INELASTIC ELECTRON TUNNELING SPECTROSCOPY JUNCTIONS

J. C. Tsang and J. R. Kirtley

IBM Thomas J. Watson Research Center
Yorktown Heights, New York 10598

The intensity of Raman scattering from molecules in solution suggests that the observation of Raman scattering from molecular monolayers should be very difficult given the small number of molecules involved. However, recent experimental work on Raman scattering from molecules adsorbed on the surface of metal electrodes in electrochemical cells has shown that strong Raman scattering can be obtained from a monolayer of adsorbed molecules.[1,2,3] It has been estimated that the Raman cross section of a pyridine molecule adsorbed on an electrochemically prepared Ag surface in a KCl solution is six orders of magnitude larger than for the same molecule in solution.[2] Although this large enhancement has stirred considerable experimental and theoretical interest, its microscopic origins remain unknown.[4,5,6]

We have used metal-insulator-molecule-metal structures suitable for Inelastic Electron Tunneling Spectroscopy (IETS) studies to investigate surface enhanced Raman scattering.[7] All previous work on surface Raman scattering has involved the surfaces of metals in solution. Many techniques normally used to probe a molecule metal interface cannot be applied to this system in situ. Our choice of the solid IETS system was dictated by its stability in both air and vacuum, which allows electron-microscopy and radioactive tracer studies to be used on the actual Raman samples. In addition, these structures allow comparison of the surface Raman spectra with the IETS spectra of the same samples. IETS is a spectroscopic tool of proven sensitivity to monolayer and submonolayer molecular coverages.[8,9] As a result, we present here new information about the physical mechanisms responsible for anomalous Raman scattering.

We have shown that strong Raman scattering can be obtained from a molecular monolayer adsorbed on the oxide in a metal-oxide-metal tunnel

junction structure.[7] We compare the Raman spectrum obtained from a roughened Al-AlO$_x$-4 pyridinecarboxaldehyde(4 py-COH)-Ag junction with the IETS
spectrum obtained from such a junction and with the Raman spectrum obtained
from pyridine adsorbed from the solution onto Ag in an electrochemical cell.
By evaporating different thicknesses of CaF$_2$ onto the glass substrate before
laying down our junctions, we can control the roughness of our structures.[10,11]
The intensity of the molecular Raman scattering increases monotonically with
increasing surface roughness. We compare these results with roughness-induced
changes in the intensity of light emission from the decay of surface plasmons
excited by the inelastic tunneling of electrons.[12]

Raman scattering studies of our tunnel junctions show a weak excitation
wavelength dependence for laser energies between 1.8 and 2.5 eV. However,
the Raman cross section of our doped silver junctions depends strongly on
excitation energy between 2.8 and 3.5 eV, decreasing rapidly with increasing
energy for energies near 3.5 eV. We then show that these results suggest the
surface plasmon of the metal is the electronic intermediate state in the Raman
scattering process. Such an interpretation is consistent with both the surface
roughness and excitation wavelength dependence of the surface Raman scattering and the fact that strong molecular scattering has only been observed for
selected molecules adsorbed on Ag.

While we can show that the coupling of light to the surface plasmons of
the metal is necessary for the observation of strong Raman scattering from
molecules, we can also show that it is not sufficient. The failure to observed
surface enhanced Raman scattering from roughened Ag tunnel junctions doped
with molecules such as benzoic acid shows that any Raman scattering in these
systems is at least three orders of magnitude weaker than the scattering we
observe in 4 py-COH doped junctions. We have previously attributed this
qualitative difference in behavior to quantitative differences in the interaction
between the molecule and the metal surface.[7]

All of our experiments were performed on IETS junctions evaporated on
glass substrates. An opaque film (40 nm) of Al was evaporated in 10^{-6} torr
vacuum and the top surface oxidized in the presence of water vapor. The
junctions were doped from the solution with both aldehydes and acids such as
benzoic acid and 4 py-COH.[9] It has been shown that both the acid and the
aldehyde chemisorb on the oxide. The COH or COOH groups dissociate to
form OCO$^-$ radicals bonded to the oxide surface. The excess dopant evaporates
off the oxide surface. After the junction is doped, a metallic overlayer is
evaporated onto the junction. This layer is about 20 nm thick. It is electrically
continuous so that IETS measurements can be performed and is also semi-
transparent so that light can penetrate the junction region. Metals used as the
counterelectrode include Ag, Au, Cu, Pb, In and Al.

The inelastic electron tunneling spectra were obtained using a conventional tunneling bridge and a voltage modulation of between 1 and 2 meV. The IETS measurements were made at 2. and 4.2 K. The Raman scattering measurements were made using discrete lines from both Ar^+ and Kr^+ lasers. The spectra were obtained in the backscattering geometry with the laser light incident on the sample at a grazing angle on the counterelectrode (20 nm thick) side of the junction. The scattered light was collected and analyzed by a conventional Raman monochromator. The Raman measurements were made both in air and under vacuum and between 300 K and 1.8 K.

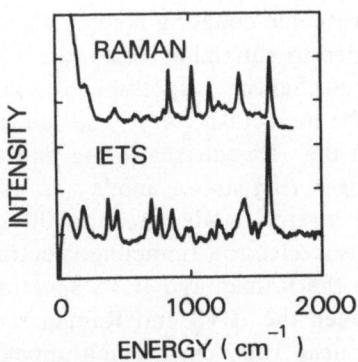

Fig. 1. Raman and Inelastic Electron Tunneling Spectra of an Al-AlO$_x$-4 py-COH-Ag junction.

In Fig. 1a, we show the Raman spectrum of an Al-AlO$_x$-4 py-COH -Ag tunnel junction. In Fig. 1b, we show the 4.2 K inelastic tunneling spectrum obtained from such a junction. Both the Raman and IETS results show directly that we are observing scattering from a molecular monolayer. 4 py-COH has a C=O bond associated with the aldehyde group attached to the carbon nitrogen ring. The chemisorption of the molecule onto the oxide surface breaks this bond and creates an O-C-O complex involving the oxide surface. Spectroscopically, this results in the disappearance of a strong line at about 1720 cm^{-1} which is due to the stretching of the C=O bond and the appearance of a pair of new lines at 1380 and 1550 cm^{-1} which arise from the symmetric and antisymmetric vibrations of the O-C-O surface complex. These last two features will be broadened by the presence of the oxide surface. All of these changes are in fact observed and we see no sign of any C=O vibrations.

Our oxidized aluminum films have simple, well defined surfaces. This is in contrast to studies of the surfaces of the electrodes used in the electrochemical cell surface scattering which show relatively large surface areas and have been characterized as sponge-like. As a result, our spectra shown unambiguously that conventional Raman spectroscopy can detect the presence of a molecular monolayer on a simple surface.

A comparison of the Raman and IETS spectra in Fig. 1 shows a number of interesting features. All of the structures observed in Raman scattering are also observed by the IETS. However, there are lines which are seen by IETS but are not seen in Raman scattering. This can be due to selection rules involving the symmetry of the molecule or to mode dependent differences in the microscopic coupling responsible for the anomalous scattering. Since 4 py-COH bonded to aluminum oxide is a relatively low symmetry system, this suggests that the mechanism responsible for molecular Raman scattering does not couple to all the molecular vibrational modes in the same way. The strongest structure in both the Raman scattering and the IETS spectrum is the 1600 cm^{-1} carbon-nitrogen ring stretch mode. In addition, we find (not shown in Fig. 1) that the C-H stretch modes near 3000 cm^{-1}, which can be readily observed in the inelastic electron tunneling spectra of many organic molecules, are very weak in both the Raman and IETS spectra of 4 py-COH.[7] The only major disagreement between the IETS and Raman scattering results is in the position of the structure near 1000 cm^{-1} which appears at 1016 cm^{-1} in the Raman scattering and at 970 cm^{-1} in the IETS scattering. If we consider only the lines which we observe in the Raman scattering, we find a strong similarity between the IETS spectra and the Raman spectra with respect to both the positions of the molecular vibrations and their relative intensities.

The Raman spectrum shown in Fig. 1 can also be compared to the results published by Jeanmaire and van Duyne.[2] While we see many of the lines they observe, the dominant structure in their spectrum is the line near 1005 cm^{-1}. They observe strong scattering from the C-H vibration. Experimental measurements of the Raman spectrum of a solution of 4 py-COH using our experimental system suggest that the surface enhanced Raman signals we observe are three to four orders of magnitude larger than would be expected from our solution data. While this is nominally smaller than the enhancement reported by Jeanmaire and van Duyne, an accurate comparison of these results should take into account the facts that our results are obtained in a geometry where the laser light must pass through 20 nm of silver before seeing the molecules and that our actual surface areas are probably smaller than those of Jeanmaire and van Duyne.[11] If we account for these factors, then our observed enhancements are close to those reported by Jeanmaire and van Duyne.[2]

Previous experimental work on Raman scattering from Ag in CN solutions and in electrochemical cells suggests that the strength of the scattering is strongly dependent on the surface roughness of the silver.[2] At least part of the six order of magnitude enhancement in the Raman cross section of pyridine on Ag has been attributed to the increase in the area of the silver surface due to the electrochemical cycling of the electrode.[11] We have studied the Raman scattering from IETS junctions laid down on films of CaF$_2$ which were evaporated onto our glass substrates. Endriz and Spicer have shown that 100 nm thick films of Al evaporated on even the smoothest available glass substrates have a surface roughness d with an rms value of 1 nm.[10] The evaporation of CaF$_2$ onto the glass substrate produces additional surface roughness which can be replicated by a metal film evaporated on the CaF$_2$. It has been shown that a 100 nm thick film of Al evaporated on a 90 nm film of CaF_2 shows d of 3 to 4 nm.[10] The surface roughness increases monotonically with increasing CaF_2 thickness for thicknesses above 10 nm. In Fig. 2, we show the Raman spectra of Al-AlO$_x$-4 py-COH -Ag junctions evaporated on varying thicknesses of CaF_2. We find that the appearance of the Raman spectra does not depend on the CaF_2 thickness but that the absolute intensities of spectra are strongly dependent on the CaF_2 thickness. This is shown in Fig. 3 where we plot the dependence on CaF_2 film thickness of the intensities of strong Raman lines at 1600 and 1016 cm^{-1}. We also plot in Fig 3 the dependence on CaF_2 film thickness of the intensity of light emitted by the decay of surface plasmons excited by tunneling electrons across an insulating barrier.[12] Also shown in Fig. 3 is the number of molecules adsorbed on the rough oxide surfaces as obtained from radioactive tracer measurements made with junctions fabricated on CaF_2 films and doped with special radioactively labeled benzoic acid. The observed increase in the intensity of the Raman scattering cannot be attributed to the increase in the number of adsorbed ions on our roughened junctions. Curve 3b was obtained from the results of McCarthy and Lambe.[12] In their experiments, inelastic electron tunneling across a metal-insulator-metal structure was used to excite the surface plasmons of the structure. By deliberately roughening the surfaces of these junctions, they were able to radiatively couple the normally non-radiative surface plasmons to light. The strength of this coupling[13] is proportional to d^2 and McCarthy and Lambe showed that the intensity of the

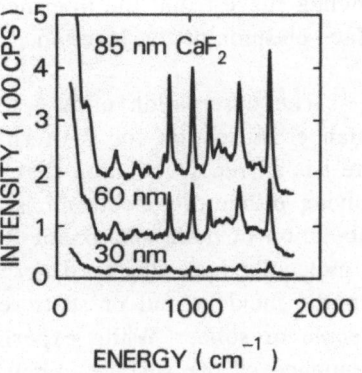

Fig. 2. Raman spectra of Al-AlO$_x$-Ag junctions doped with 4 py-COH and laid down on three different thickness of CaF_2.

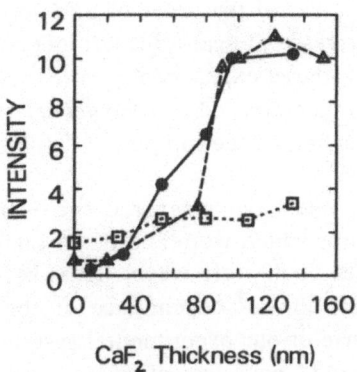

Fig. 3. Dependence of a)the intensity of the surface Raman scattering (dots and solid line), b)the intensity of the light emission by inelastic electron tunneling (open triangles and broken line) and c) the number of adsorbed molecules (squares and dotted line) on CaF_2 film thickness.

plasmon luminescence increases with increasing junction roughness.   The similarities in both the shapes and the absolute magnitudes of the surface roughness enhancement curves in Fig. 3 for the intensity of the Raman scattering and the light emission from surface plasmons excited by inelastic electron tunneling suggest that the intermediate state involved in Raman scattering is the surface plasmon of the junction.

The observation of a 4 to 6 order of magnitude enhancement in the Raman cross section for scattering from molecules adsorbed on a metal substrate has stirred speculation that this enhancement is due to a resonant process involving either the electronic states of the molecule, of the metal, or of a combination of both.[6] Resonant Raman scattering experiments from both solids and molecules have observed abrupt enhancements of the Raman cross section when the incident and or scattered light is close to a sharp resonance of the molecule or solid.   While experimental results on the excitation wavelength dependence of the surface enhanced Raman scattering agree that there is no sharp excitation wavelength dependence in the Raman cross section, they disagree on the exact excitation wavelength dependence.   Jeanmaire and van Duyne have reported that the intensity of the molecular Raman scattering shows no excitation energy dependence beyond the normal ω^4 dependence for excitation energies between 1.8 and 2.8 eV.[2] Creighton et al.[3] and Pettinger et al.[14] found that the surface scattering, after correcting for the ω^4 dependence of the Raman cross section, decreased monotonically with increasing excitation energy over the same range of energies.   Their observed decrease in the scattering cross section was of the order of a factor of 10 and therefore largely compensated for the increase in scattering efficiency due to the ω^4 term in the Raman intensity. All of these experiments were performed using pyridine adsorbed on Ag in electrochemical cells.   However, there were differences in the details of the preparation of the samples.

We have measured the Raman cross section of 4 py-COH doped Ag junctions which were laid down on 70 nm and 90 nm films of CaF_2. We used a number of discrete wavelengths between 1.8 and 3.5 eV. Our results, normalized for the ω^4 dependence of the Raman cross section and the instrumental response of our experimental system by a separate measurement of the intensity of the 470 cm^{-1} line of quartz, are shown in Fig. 4. We find that the Raman cross section for our samples increases slightly with increasing excitation energy near 2.5 eV and then decreases rapidly with increasing excitation energy near 3.5 eV. The intensity of the sharp Raman lines due to the molecular vibrations of 4 py-COH decreases rapidly with increasing excitation energy for energies above 2.5 eV (wavelengths less than 500 nm).The broader structure which underlies the 4 py-COH structure has been attributed by Otto[15] to Raman scattering from a surface carbonate adsorbed on the silver from the atmosphere increases with increasing energy below 2.8 eV (wavelenths greater than 460 nm) and decreases for shorter wavelengths.

The surface plasmon energy of bulk Ag is near 3.6 eV. However, Davis has shown that the surface plasmon energy for a thin film of silver evaporated on a 3 nm thick oxide of a thick aluminum film is close to 3.4 eV.[16] The excitation wavelength dependence of the surface Raman scattering which we observe for energies between 2.0 and 3.5 eV is similar to the wavelength dependence of the light emission due to surface plasmons excited by tunneling electrons for bias voltages above the surface plasmon energy.[17] Lambe and McCarthy and Hansma and Broida[17] have found that the light emission increases slightly with increasing energy for energies well below the surface plasmon energy and decreases rapidly for energies just below the surface plasmon energy. In the absence of surface roughness, light cannot couple to the surface plasmons. The optical emission from surface plasmons from a rough surface is bounded at higher energies by the surface plasmon energy ω_{sp} and is zero for energies greater than ω_{sp}. It decreases at lower energies since the

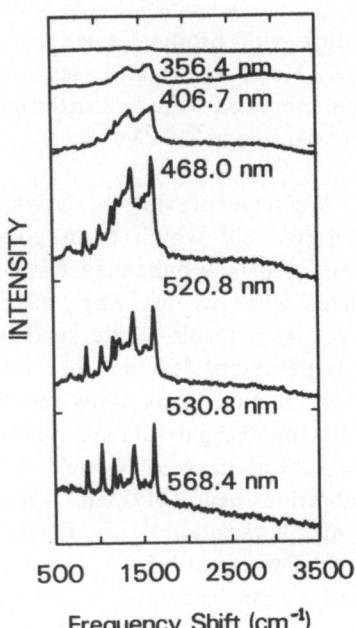

Fig. 4. Dependence on excitation wavelength of the Raman spectrum of an Al-AlO_x-Ag junction doped with 4 py-COH and evaporated on 80 nm of CaF_2.

surface plasmon density of states decreases with decreasing energy. The actual shape of the emission will depend on the details of the surface roughness since this provides the additional momentum needed to couple the light to the slower surface plasmon. Surface roughness characterized by a relatively long correlation length will produce surface plasmon emission which is weaker, and peaks at a lower energy than surface roughness of the same amplitude which is characterized by a relatively short correlation length.[13] Rendell et al. have calculated the functional dependence for the dipole moment in the resonance region of a metal sphere-aluminum film system and have derived the above behavior.[18] The sensitivity of the plasmon emission to both the magnitude of the surface roughness and its Fourier components means that the quantitative frequency dependence of the Raman scattering will depend sensitively on the details of the surface preparation.   This can explain the different results for the excitation wavelength dependence of the scattering in electrochemical cells obtained by Jeanmaire et al.[2], Creighton et al.[3] and Pettinger et al.[15] since changes in the electrode source material, the solution chemistry and the electrode anodization procedure will produce changes in the physical properties of the electrode surface.   However, in all cases, we would expect that the cross section for the surface enhanced Raman scattering on Ag will decrease dramatically for excitation energies near 3.5 eV.

We have previously shown that many molecules including benzoic acid and formic acid which form good IETS junctions with silver do not show observable aurface enhanced Raman scattering.   We have found that this is the case even when we use CaF_2 roughened films and excitation wavelengths near 2.7 eV.   As a result, while surface roughness and the use of a metal such as silver is necessary for the observation of enhanced Raman scattering, it is not sufficient. In Fig. 5, we show the IETS spectra of an Ag junction doped with 4 py-COH for energies between 100 cm^{-1} and 3500 cm^{-1}.   The intensity of the C-H vibrations near 3000 cm^{-1} in this junction is much weaker than the C-N ring vibrations near 1600 cm^{-1}.   Korman et al.[19] have explained this behavior in terms of the variation of the barrier height between the molecule and the metal surface.   The metal-insulator-molecule-metal structure is electrically modeled in terms of a three-interface, 5-parameter structure.   The parameters in this model are the widths of the oxide and molecule layers and the heights of the potential barriers at the metal-oxide, oxide-molecule and molecule-metal interfaces. Korman et al.[19] have found that the weak C-H vibrations seen in Fig. 5 can be attributed the presence of a metal molecule interface barrier height which is comparable to the bias and/or phonon energy.   In most commonly studied IETS systems, the metal molecule interface barrier height is of the order of 2-4 volts and the IETS intensities of the C-H vibrations are comparable to the intensities of the lower energy modes. This difference between 4 py-COH and other molecules such as formic acid presumably has its origin in the presence of the nitrogen lone pair of pyridine and suggest that these electrons play a role in the Raman scattering.

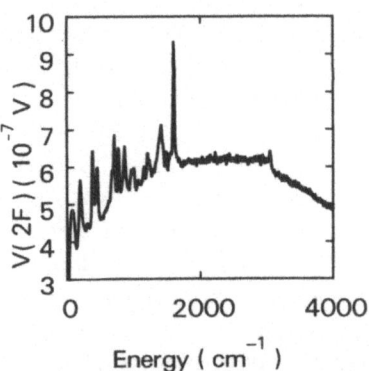

Fig. 5. The IETS spectra of an Al-AlO$_x$-Ag junction doped with 4 py-COH for frequency shifts up to 4000 cm^{-1}.

Just as we can show that the excitation of surface plasmons is necessary for the observation of strong molecular Raman scattering but not sufficient, we can also show that a low barrier height between the metal and the molecule is also necessary but not sufficient for the observation of strong scattering. We have attempted to study the Raman spectrum of roughened 4-acetyl benzoic acid doped Pb junctions. The IETS spectrum suggests that the Pb-4-acetyl benzoic acid potential barrier is of the order of 1 eV and therefore comparable to the Ag-4 py COH barrier. However, we were unable to observe any scattering from this system. This is consistent with our previous result since the surface plasmon in Pb is heavily damped by the presence of strong interband transitions in the visible.

The standard, perturbation theory treatment of Raman scattering involves two different matrix elements, one coupling the light to the electronic states and one scattering an electronic state from the elementary excitation in question. Our results suggest that the relevant matrix elements for surface Raman scattering are the matrix element for the coupling of light to the surface plasmons of the metal via surface roughness and the matrix element for the scattering of the surface plasmon via a modulation of the metal molecule interface by the molecular vibrations. The first of these has been treated in detail by Elson and Ritchie in their work on the optical absorption at a rough surface by surface plasmons.[13] The second has been treated in the approximation where the metal vacuum interface is taken to be abrupt and the molecule interacts with the metal via the long range electromagnetic field of the surface plasmon.[20] It has been shown that this coupling cannot explain our experimental results.[4] We suggest that the the modulation of the electron density at the metal surface by the molecular vibrations can produce a more substantial effect.

Theoretical treatments of the problem of surface enhanced Raman scattering have in general been based on classical models.[4] The interaction between the molecule and the metal surface has been treated in terms of the classic image charge and it has been found that a substantial enhancement in the Raman cross section is possible if the molecule is close enough to the metal surface. However, the distances required are comparable to or less than the normal interatomic distances and raise questions concerning the validity of the classical approximations. While models based on the classical image charge can explain why certain metals produce strong surface Raman scattering, it is not clear that they can explain why only certain molecules produce strong Raman scattering. Our observations on the frequency, surface roughness, metal and molecule dependence of the anomalous scattering plus its strong sensitivity to the immediate proximity of the molecule and the metal surface suggests that the surface plasmons of the metal and the chemical interaction between the metal and the molecule must be considered explicitly in any theory of this scattering.

ACKNOWLEDGEMENT. We thank J. A. Bradley and A. M. Torressen for invaluable experimental assistance.

REFERENCES

1. M. Fleishmann, P. J. Hendra and J. McQuillen, J. Chem. Soc. Chem. Commun. 3, 80 (1973).
2. D. L. Jeanmaire and R. P. van Duyne, J. Electroanal. Chem. 84, 1 (1977).
3. J. A. Creighton, M. G. Albrecht, R. E. Hester and J. A. D. Matthew, Chem. Phys. Lett. 56, 253 (1977).
4. F. W. King, R. P. van Duyne and G. C. Schatz, J. Chem. Phys. 69, 4472 (1978), S. Efrima and H. Metiu, Chem. Phys. Lett. 60, 59 (1978).
5. M. Moskovitz, J. Chem. Phys. 69, 4159 (1978).
6. E. Burstein, C. Y. Chen, S. Lundquist and E. Tossatti, Solid State Commun. 29, 567 (1979).
7. J. C. Tsang and J. R. Kirtley, Solid State Commun. 30,   (1979).
8. J. Lambe and R. C. Jaklevic, Phys. Rev. 165, 821 (1968).
9. P. K. Hansma and J. R. Kirtley Acc. Chem. Res. 11, 440 (1978).
10. J. G. Endriz and W. E. Spicer, Phys. Rev. B4, 4144 (1971).
11. J. F. Evans, M. G. Albrecht, D. M. Ulling and R. M. Hester, Surf. Sci. 75, L777 (1978)
12. S. L. McCarthy and J. Lambe, Applied Physics Lett., 30, 427 (1977).
13. J. M. Elson and R. H. Ritchie, Phys. Rev. B4, 4129 (1971).
14. A. Otto, Surf. Sci. 72, L392 (1978).
15. B. Pettinger, U. Wenning and D. M. Kolb, Ber Bunsenges Phys. Chem. 82, 329 (1978).
16. L. C. Davis, Phys. Rev. B-16, 2482 (1978).
17. J. Lambe and S. L. McCarthy, Phys. Rev. Lett. 37, 923 (1976) and P. K. Hansma and H. P. Broida, Applied Physics Lett. 32, 545 (1978).
18. R. W. Rendell, D. J. Scalapino and B. Muhlschlegel, Phys. Rev. Lett. 41, 1746 (1978).
19. C. S. Korman, J. C. Lau, A. M. Johnson and R. V. Coleman, Phys. Rev. B19, 994 (1979).
20. M. R. Philpott, J. Chem. Phys. 62, 1812 (1975).

A THEORY OF "GIANT RAMAN SCATTERING" BY ADSORBED MOLECULES ON METAL SURFACES

Shlomo Efrima, Tsofar Maniv and Horia Metiu[a]

Department of Chemistry
University of California, Santa Barbara
Santa Barbara, California 93106

A large number of recent experiments[1] on Raman scattering by adsorbed molecules have shown several unusual features. (1) The scattering intensity is, in some cases, four to six orders of magnitude larger than that expected on the basis of the known scattering cross section and the amount of material at the surface. (2) The ability of the metal substrate to induce this enhancement, varies greatly from one metal to the other. (3) The excitation spectrum (the scattered Raman intensity, I, as a function of the scattered frequency $\omega_i - \omega_v$) does not follow a well defined law; Jeanmaire and van Duyne[1a] find $I \propto (\omega_i - \omega_v)^4$, Albrecht and Creighton[1b] obtain that I is frequency independent, Creighton, Albrecht, Hester and Matthew[1c] report $I \propto (\omega_i - \omega_v)^2$ while Tsang and Kirtley[1d] observe a more complicated function. (4) Surface roughness seems to play an important role and induces a change of intensity higher than that expected on the basis of metal surface increase. (5) Finally, a "modest" enhancement (a few orders of magnitude) of the resonant Raman intensity has also been reported.[1a] Other interesting observations concerning the effect of the electrode potential and anodization have been reviewed recently by Efrima and Metiu.[2]

The purpose of the present article is to discuss these observations in terms of our recent theoretical work.[2-9] We take the point of view that in spite of chemisorbtion, one can usefully speak of the molecule and the metal as two separate optical entities; the spectroscopic properties of the molecule can be described by a local polarizability (which may be different from that of the free molecule) and that of the metal by its dielectric responce. It is further assumed that the induced molecular dipole, $\vec{\mu}(t)$, satisfies

Drude equation:

$$\ddot{\vec{\mu}} + \omega_0^2 \, \vec{\mu}(t) + 2\Gamma_0 \, \dot{\vec{\mu}}(t) = (e^2/m) \overleftrightarrow{f} \cdot \vec{E}_\ell (\vec{R}_m, t). \tag{1}$$

Here ω_0 is the frequency of the lowest lying electronic level, Γ_0 is a friction coefficient simulating all loss mechanisms not included explicitly in Eq. (1), $\overleftrightarrow{f}$ is the oscillator strength and $\vec{E}_\ell (\vec{R}_m, t)$ is the local electric field acting on the molecule (located at $\vec{R}_m$). Since the molecule is very close to the metal, a point dipole description of the molecular charge density is not accurate.[10] One should use a model in which the bond or the atoms have independent polarizabilities, thus taking into account all the multipoles. This can be easily done, but we retain Eq. (1) here, for the sake of simplicity.

The surface acts on the induced dipole through the local field $\vec{E}_\ell$, which consists of two parts:

$$\vec{E}_\ell = \vec{E}_p + \vec{E}_s. \tag{2}$$

The primary field, $\vec{E}_p$, is the total field incident upon the molecule, in the case when the molecule is not polarized. If macroscopic electrodynamics is used, $\vec{E}_p$ is computed with Fresnel equations. The secondary field is caused by the charge and current densities which the molecular dipole $\vec{\mu}$ induces in the metal. If macroscopic Maxwell equations are used, the field $\vec{E}_s$ can be computed from

$$E_s(R_m, t) = \int_{-\infty}^{+\infty} dt' \, \overleftrightarrow{G}_s (\vec{R}_m; t-t') \, d\vec{\mu}(t')/dt'. \tag{3}$$

Equations for $\overleftrightarrow{G}$ can be obtained by using a methodology presented by Tai.[11] The detailed results are presented in Refs. 4-5.

A very simple computational scheme yields now the scattered Raman intensity.[4-6] (a) We use (2) and (3) in (1) and solve for the induced dipole moment $\vec{\mu}(t)$. (b) Then we expand[12] the induced dipole moment in the amplitude of the normal modes. (c) Then we solve Maxwell equations to obtain the intensity of the radiation emitted by the dipole, at the frequency $\omega_i - \omega_v$ (ω_i = the frequency of the incident light, ω_v = the vibrational frequency).

The results of this calculation are discussed in detail by Efrima and Metiu.[4-5] We concentrate here on some of the physical aspects. Assuming, for simplicity, that the molecule is isotopic we obtain for the induced dipole[3-7]

$$\vec{\mu}(\omega) = (e^2/m) f \, \vec{E}_p \{\omega_0^2 - \omega^2 - 2\omega\Delta(\omega) - 2i\omega[\Gamma_0 + \Gamma_s(\omega)]\}^{-1} \tag{4}$$

with

$$\left.\begin{array}{c}\Delta(\omega)\\\Gamma(\omega)\end{array}\right\} = (e^2/m) f(4\Pi\varepsilon_1\varepsilon_0 z_m^3)^{-1}\omega^{-1}\left\{\begin{array}{c}\text{Re}\\\text{Im}\end{array}\right\}(\varepsilon_2(\omega)-\varepsilon_1)(\varepsilon_2(\omega)+\varepsilon_1)^{-1}. \quad (5)$$

Within this model the only possibility for a very large enhancement of the Raman scattering is the occurence of resonant scattering. The resonant frequency ω_r, is given by (see Eq. (4))

$$\omega_r \simeq \omega_0 - \Delta(\omega_r). \qquad\qquad\qquad (6)$$

If this is lower than the resonant frequency ω_0, of the free molecule (for a free molecule $\Delta(\omega) = \Gamma(\omega) = 0$), then by choosing an incident frequency ω_i satisfying

$$\omega_r < \omega_i < \omega_0 \qquad\qquad\qquad (7)$$

we can perform resonant Raman scattering on the surface molecules and non-resonant Raman on the free molecules present in the electrolyte. In our opinion this is the origin of the observed enhancement and we use for this the name of Surface Induced Resonant Raman Scattering (SIRRS). Numerical calculations[4-7] show that for all incident frequencies, except those very close to the surface plasmon frequency, ω_{sp}, $\Delta(\omega) > 0$ and the requirement formulated in Eq. (7) is satisfied. When ω_i is very close to ω_{sp}, $\Delta(\omega)$ becomes negative and $\omega_r > \omega_0$. We predict therefore that, in the case when the adsorbed molecules are all at the same distance from the surface, the giant SIRRS enhancement will disappear when ω_i sweeps through ω_{sp}. Since not all molecules are at the same distance, the effect will not be as dramatic, but we do expect a substantial intensity decrease. The physical reason for this is intuitively clear: the induced dipole oscillates at the frequency of the incident light (or at ω_i, plus or minus the vibrational frequencies) and when ω_i is very close to ω_{sp} the dipole loses energy to the surface plasmon, rather than emit it to the detector.

The SIRRS hypothesis seems to explain the great variation in the enhancing ability of different metals. Since SIRRS intensity is proportional to $(\Gamma_0 + \Gamma(\omega))^{-4}$ (that is, to the square of the derivative of $\vec{\mu}(\omega)$ with the normal mode coordinate, at the resonance frequency), it is controlled by the magnitude of $\Gamma(\omega)$. From Eq. (5) we see that the metal enters in the expression for $\Gamma(\omega)$ through Im $(\varepsilon_2(\omega)-\varepsilon_1)(\varepsilon_2(\omega)+\varepsilon_1)^{-1}$. Computation of this quantity for Ag, Hg, Cu, Au at various visible frequencies[7] show that this quantity is smallest for Ag and Hg and much larger for Cu and Au. Detailed calculations[7] yield for Ag and Hg an enhancement of 10^5-10^6 and Cu and Au of $10^3 \sim 10^2$ (depending on the incident frequency and the distance to the surface). This is in the range of the observed intensities, except for Hg for which we do not have data. We predict that if the Hg surface can hold molecules in a position such that their induced dipole is perpendicular to the surface, a large enhancement (smaller, but close to that of Ag) will be observed.

The SIRRS hypotheses can also explain qualitatively the fact that the excitation spectrum seems to lack any regularity. If all scattering molecules were located precisely at the same distance from the surface, the excitation spectrum will be a curve sharply peaked around the resonance frequency $\omega_0-\Delta$. However, the peak position, height and width are very sensitive to the distance of the molecule to the surface. The observed spectrum is then the average over all these sharp spectra, corresponding to all the possible distance. The result of such averaging is a much smoother spectrum, which depends strongly on the distribution function used for the distance to the surface. This function is not known and varies from system to system according to the nature of the binding, etc. Perhaps the distance is best defined in the inelastic tunneling junction configuration employed by Tsang and Kirtley[1d] and such systems should be used to test our results.

The fact that roughness influences strongly the dielectric response of the metal is rather well understood.[13] One could include such effects in the present theory but this has not yet been done. The role of roughness has been discussed by Moskovits.[14]

We point out that the SIRRS theory can be easily worked out for the case when the molecule is adsorbed on a metal film. Presumably the local environment is not sensitive to film thickness, while the solution of the Maxwell equations and, therefore, the resulting expressions for $\Gamma(\omega)$, $\Delta(\omega)$ and intensity, is. If one fits the experimental data on the semi-infinite solid, the theory can use the parameters thus obtained (distances, polarizability, etc.) to predict the dependence of the spectrum on the film thickness. This will be a rather detailed test of the SIRRS hypotheses. Furthermore, since we assume that the giant scattering is a resonance effect, this implies that it is closely related to the electronic absorbtion or fluorescence spectrum of the molecules located in the first monolayer. Ideally Raman experiments should be done in conjunction with fluorescence and electronic absorbtion studies.

Finally, we must emphasize that surface induced resonant scattering can also take place if the chemisorbed molecule forms a new electronically excited state with the metal, which has lower energy than the first excited state of the free molecule. At this point we cannot rule out this possibility.

We conclude this article with a survey of the errors involved in using the macroscopic Maxwell equations and discuss the possibility of replacing them by a more detailed, microscopic theory.[8-9]
(a) The dielectric constant of the metal near the surface is different from that of the bulk metal, and the additional "surface contribution" is very important in computing the interaction between the metal and a charge located close to the surface.
(b) The field exerted by the molecular charge upon the metal is

inhomogeneous. The degree of inhomogeneity is increased as the distance to surface gets smaller. Therefore, the high wave vector contributions to the dielectric constant of the metal should be taken into account. (c) The presence of the surface introduces an anisotropy in the dielectric constant. Due to this, the longitudinal and transverse components of the induced fields are coupled. As a result, transverse fields will induce longitudinal polarizations which will be sources of longitudinal fields.

The macroscopic theory uses an optically measured dielectric constant, which is isotropic, lacks the surface contribution and corresponds to $\vec{k} \approx 0$. All the effects discussed at (a)-(c) are, therefore, missing in the macroscopic calculations. Furthermore, the use of the boundary conditions compounds the errors, since these conditions are the result of a spatial coarse graining of the "microscopic" Maxwell equations[15] and are valid only when the field is detected by a macroscopic object or one that is far enough from the surface. This is not the case here. Another way of putting it is to observe that the macroscopic boundary conditions eliminate the effects of the inhomogeneous surface charge density profile.

It is therefore of interest to develop a microscopic theory of the electromagnetic fields near a metallic surface in the presence of external light sources and moving molecular charges.[8,9] The basic theory uses the formal apparatus of plasma theory[16] or many body theory of an electron gas in jellium.[17] We must take into account the presence of the surface and we use procedures developed by Beck and Celly,[18a] Newns[18b] and Feibelman.[18c] We obtain a generalization of their results in a form that is adapted to the problem at hand.[8,9]

We start with quantum Maxwell equations and compute the expectation values of the vector and scallar potentials:

$$\{-\frac{1}{c^2}\frac{\partial^2}{\partial t^2} + \nabla^2\}<F_\mu(\vec{r},t)> = -4\pi S_\mu^{ex}(\vec{r},t) - 4\pi<S_\mu(\vec{r},t)> \qquad (8)$$

Here the fields F_μ are defined by $F_0 = \Phi$, $F_1 = A_1$, ..., $F_3 = A_3$ where Φ is the scalar potential and $\vec{A}$ is the vector one. We also define the sources $S_0 = \rho(\vec{r},t)$ and $S_1 = j_1/c$, ..., $S_3 = j_3/c$, where ρ is the charge density and $\vec{j}$ the current density. We use Lorentz gauge. S_μ are the charge density and the current induced in the metal, and S_μ^{ex} are those of the external sources (molecules, lasers, etc.). According to the RPA approximation[18] the expectation values of the sources can be given by the linear response of the metal under the influence of the local fields:

$$<S_\mu(\vec{r},\omega)> = \sum_{\nu=0}^{3}\int d\vec{r}' \ Q_{\mu\nu}(\vec{r},\vec{r}';\omega)<F_\nu(\vec{r}',\omega)> \qquad (9)$$

The response tensor $Q_{\mu\nu}$ can be computed once the electron Green function G_0 is known. This is given by

$$G_0(\vec{r},\vec{r}',\vec{K},\omega) = \sum_{\vec{K},\kappa} e^{i\vec{K}\cdot(\vec{R}-\vec{R}')} \phi_\kappa(z)\phi_{\kappa'}(z') (L^{-1})$$

$$\left\{ \frac{f(\omega_{\kappa\vec{K}})}{\omega-\omega_{\kappa\vec{k}} + i\eta} + \frac{1-f(\omega_{\kappa\vec{K}})}{\omega-\omega_{\kappa\vec{k}} - i\eta} \right\} \tag{10}$$

Here L^2 is the area of the slab, $\vec{R}$ is the parallel component of $\vec{r}$, $\vec{K}$ is the parallel momentum and f is the Fermi function. The one electron basis set is obtained numerically, by solving the Lang-Kohn[20] self-consistent equation

$$\{-\frac{\hbar^2}{2m}\ \frac{d^2}{dz^2}\ V[n(z)]\}\phi_\kappa(z) = \epsilon_\kappa\phi_\kappa(z) \tag{11}$$

for the jellium surface, in the presence of the external molecular charges. The energy appearing in Eq. (10) is $\hbar\omega_{\kappa\vec{K}} = \epsilon_\kappa + \hbar^2 K^2/2m$.

The equations (8)-(11) form a closed set which can be solved on the computer.

The secondary field $\vec{E}_s$ computed by this procedure (by taking an oscillating dipole outside the metal and computing $\vec{A}$ and Φ (hence $\vec{E}$) at the position of the dipole) differs significantly from the macroscopic one, especially in regards to its dependence on the distance to the surface. This affects the distance dependence of Γ and Δ. At larger distances (3-4 Å), $\vec{E}_s$ is larger than predicted by a macroscopic calculation with a bulk dielectric constant computed with the same model. This is due to the interaction of the dipole with the electron "spill-out." At shorter distances, screening becomes important and $\vec{E}_s$ is much smaller than the "macroscopic" value. In particular, the divergence given by the macroscopic formula (Z_m^{-3} in Eq. (5)) is removed. Note that this behavior of the dynamic response parallels, not unexpectedly, the static behavior predicted by Lang and Kohn[20] and Applebaum and Hamann.[20]

We are now in the process[21] of studying numerically the microscopic corrections to the Fresnel formulae. The macroscopic theory uses an isotropic dielectric constant and as a result the electric field vector of a beam of light near the surface is transverse. The microscopic surface contribution to the dielectric response is however anisotropic. Or in different but equivalent terms, the current-charge density correlation function at the surface is not zero. Due to this, the transverse field of light induces a charge density at the surface, which radiates a longitudinal field. The total field incident on the molecule, $\vec{E}_p$, is the sum of the field coming from the source plus that radiated from the metal. The latter contains a longitudinal component, which is absent in the Fresnel formulae. Near the surface this additional field could be quite large (longitudinal fields are usually larger than the transverse ones) and as a result $\vec{E}_p$ may be larger than expected on the basis of the Fresnel equations. Since the scattered intensity is proportional to $\vec{E}_p \cdot \vec{E}_p^*$ this may be the cause for the enhancement of the resonant Raman scattering. We are testing now, numerically, this conjecture.

REFERENCES

a. Alfred P. Sloan Fellow
1. We quote here only a few of the most recent works:
    (a) D.J. Jeanmaire and R.P. Van Duyne, J. Electroanal. Chem. 66 235 (1975); 84 1 (1977); R.P. Van Duyne, J. Chim. Phys. (Paris) Supp. C5, 239 (1977)· (b) M.G. Albrecht and J.A. Creighton, J. Amer. Chem. Soc. 99 5215 (1977). (c) J.A. Creighton, M.G. Albrecht, R.E. Hester and J.A.D. Matthew, Chem. Phys. Lett. 55 55 (1978). (d) J.C. Tsang and J. Kirtley Anomalous Surface Enhanced Molecular Raman Scattering from Inelastic Tunneling Spectroscopy Junctions (preprint).
    (e) T.H. Wood and M.V. Klein, J. Vac. Sci. Technol. (to be published); D. DiLella, R. Lipson, P. McBreen and M. Moskovits (to be published); B. Pettinger and V. Wenning, Chem. Phys. Lett. 56 253 (1978); A. Otto, Surface Sci. 75 1392 (1978); R.P. Cooney, E.S. Reid, M. Fleischmann, P.J. Hendra, J. Chem. Soc. Faraday I, 73 1691 (1977).
2. S. Efrima and H. Metiu, Israel J. Chem. ·(to be published).
3. S. Efrima and H. Metiu, Chem. Phys. Lett. 60 59 (1978).
4. S. Efrima and H. Metiu, J. Chem. Phys. 70 1602 (1979).
5. S. Efrima and H. Metiu, J. Chem. Phys. 70 2297 (1979).
6. S. Efrima and H. Metiu, J. Chem. Phys. 70 1939 (1979).
7. S. Efrima and H. Metiu, Surface Sci. (submitted).
8. T. Maniv and H. Metiu, J. Chem. Phys. (to be published).
9. T. Maniv, G. Korzeniewski and H. Metiu (in preparation).
10. J. Kirtley, D.J. Scalapino and P.K. Hansma, Phys. Rev. B14 3177 (1976); S. Efrima and H. Metiu, Surface Sci. (submitted).
11. Chen-To Tai, Dyadic Green's Functions in Electromagnetic Theory, Intex., San Francisco, 1971.
12. Certain errors are involved in this simple procedure, They can be avoided as discussed in Ref. 6.
13. See a review in H. Raether, Phys. Thin Films, 9 145 (1977).
14. M. Moskovits, J. Chem. Phys. 69 4159 (1978) and preprint.
15. J.D. Jackson, Classical Electrodynamics, Wiley, New York, 1975; S.A. de Groot, The Maxwell Equations, North-Holland, Amsterdam, 1969.
16. D.F. DuBois, in Kinetic Theory, ed. W.E. Brittin, pp. 469, Gordon and Breach, New York, 1967.

17. (a) J.R. Schrieffer, Theory of Superconductivity, W.A. Benjamin, Reading, 1964. (b) A.A. Abrikosov, L.P. Gor'kov and I.Ye. Dzyaloshinskii, Quantum Field Theoretical Methods in Statistical Physics, Pergammon, N.Y., 1965. (c) A.L. Fetter and J.D. Walecka, Quantum Theory of Many-Particle Systems, McGraw Hill, N.Y., 1971.

18. D.E. Beck and V. Celli, Phys. Rev. B2 2955 (1970); D.M. Newns, Phys. Rev. B1 3304 (1970); P.J. Feibelman, Phys. Rev. B12, 1319 (1975).

19. D. Pines, Elementary Excitations in Solids, Benjamin, N.Y., 1964; P.M. Plazman and P.A. Wolff, Waves and Interactions in Solid State Plasmas, Academic Press, N.Y., 1973.

20. N.D. Lang and W. Kohn, Phys. Rev. B7 3541 (1972); J.A. Appelbaum and D.R. Hamann, Phys. Rev. B6 1122 (1972); E. Zaremba and W. Kohn, Phys. Rev. B13 2271 (1976); N.D. Lang, Solid State Phys. 28 2255 (1973).

21. T. Maniv and H. Metiu, in preparation.

CONCLUDING REMARKS

SERGEY A. AKHMANOV
Moscow State University

Let me say first of all that in spite of its long history,
light scattering is still a growing field. So —
          new physical ideas;
          new experimental methods; and
          new people, are still becoming involved.
Of course, this has been a good basis for the success of our
Seminar-Symposium.

I think that our Seminar-Symposium was very fruitful and
stimulating. Among others, I would like especially to mention new
ideas and new results, which were presented here in such fields as:
          Resonant processes in solids;
          Surface scattering phenomena
          Two-dimensional systems;
          Coherent spectroscopy.

I think that in the last field the gap between "linear" and
"non-linear" spectroscopists now is smaller than before — so at
the next Seminar-Symposium we should have a common language (I
believe it will be a language of non-linear spectroscopy!)

Let me say also that extremely important for the success of
our Seminar-Symposium was the warm and cordial hospitality of our
American friends, who were so cordial, warm and friendly. We could
not imagine this Seminar-Symposium without the permanent friendli-
ness, energy and spirit of cooperation of Professor Joseph Birman,
Professor Herman Cummins, and Professor Melvin Lax.

There is no doubt, that our Seminar-Symposium will make a
good contribution to scientific exchange and cooperation between
scientists of our two countries. I think that our Seminar-Symposia
are a good tradition. Now there are definite plans to organize
the next Soviet-American Light Scattering Seminar-Symposium in
1981 in our country in Tallin, or in Leningrad: both are beautiful
places!

So, we hope to see American participants at the Conferences
and Seminar-Symposia in our country soon. Thank you.

HERMAN Z. CUMMINS
City College of the City University of New York

On behalf of the Organizing and Program Committee, I want to
thank all of the speakers and participants for contributing to a
lively and exciting Seminar-Symposium.

As you all know, our committee organized this Seminar-
Symposium twice —— once for its originally scheduled occurrence in
May 1977, and, following an unanticipated two year postponement,
again in 1979. Although the postponement caused some inconvenience
and unfortunate duplication of effort, there have been several
exciting new areas included in the Seminar-Symposium program which
would not have been possible two years ago. Both the final session
and lively round-table discussion of surface enhanced Raman scattering and the recurring theme of two-dimensional phase transitions
represent such new areas.

The high level of interest in this Seminar-Symposium has amply
demonstrated two important facts: first, that light scattering
spectroscopy is very definitely alive and growing in new and unanticipated directions; second, that there is a strong community
of interest between Soviet and American scientists working in this
area who can all benefit greatly from the opportunity to exchange
ideas and results in the setting of such a binational meeting.
The historic strength of condensed matter physics in the USA and
the USSR combined with the new technologies of tunable and pulsed
high-power lasers have seen a rapid parallel development of research
in this area during the past ten years which provides an outstanding
opportunity for scientific cooperation. I firmly believe that continuation of these light scattering Seminar-Symposia together with
the establishment of cooperative research projects can provide many
benefits to science and scientists in both countries.

As we all recognize, however, our attempts to advance scientific cooperation must proceed within the context of problems and
conflicts present in our world. Issues of world peace and human
rights concern governments and citizens of both our countries and,
of course, also concern us as scientists. Therefore, inevitably,
the path of cooperation along which this Seminar-Symposium constitutes a milestone must be tortuous, marred by unexpected
obstacles. But if we are willing to pursue our goals of cooperation
and friendship with diligence despite the inevitable obstacles,
then we can reasonably hope to see the long-term rewards of our
efforts in an ongoing series of exchanges and mutually rewarding
scientific interactions.

As the formal sessions of the Seminar-Symposium end, our
Soviet colleagues are preparing to depart on a week of Post-Symposium
visits during which they will be guests at over thirty academic,
industrial and government laboratories throughout the United States.
During these visits, we hope that many of you will find time to
pursue the common scientific interests which the crowded schedule
of the Seminar-Symposium has allowed, unfortunately, all too little
time to discuss.

We thank the American scientists, many of whom are participating
in this Seminar-Symposium, for having agreed to serve as hosts
during these Post-Symposium visits. We wish our Soviet colleagues
an enjoyable and rewarding Post-Symposium week and a safe trip
home, and look forward with pleasure to future opportunities to
renew our scientific and personal contacts.

25 May 1979

USSR PARTICIPANTS

Vladimir M. Agranovich   Inst. of Spectroscopy, Moscow
Sergey A. Akhmanov   Moscow State University
Pavel A. Apanasevich   Inst. of Physics, Minsk
Viktor S. Bagayev   Inst. of Physics, Moscow
Vladimir A. Belyakov   Physico-Technical Inst., Moscow
Viktor A.S. Borovik-Romanov   Inst. of Phys. Problems, Moscow
Lyudmila A. Bureyeva   Spectroscopy Council, Acad. Sci., Moscow

Vladimir V. Hizhnyakov   Inst. of Physics, Tartu, Estonia
Iya P. Ipatova   Physico-Technical Inst., Leningrad
Aleksandr A. Kaplyanskii   Physico-Technical Inst., Leningrad
Aleksandr P. Kazantsev   Inst. of Theoretical Physics, Moscow
Nicolai I. Koroteev   Moscow State University (Visiting Stanford Univ. 1978-79)
David N. Klyshko   Moscow State University
Arkadiy P. Levanyuk   Inst. of Crystallography, Moscow
Sergey A. Permogorov   Physico-Technical Inst., Leningrad
Lev P. Pitaevskii   Inst. of Physical Problems, Moscow
Yuriy A. Popkov   Physico-Technical Inst. of Low Temperatures, Kharkov
Karl K. Rebane   Inst. of Physics, Tartu, Estonia
Lyubov A. Rebane   Inst. of Physics, Tartu, Estonia
Peter M. Saari   Inst. of Physics, Tartu, Estonia

USA PARTICIPANTS

Izo I. Abram   University of Pennsylvania
Govind Agrawal   City College of City Univ. of New York
Andreas C. Albrecht   Cornell University
Nabil Amer   University of California, Berkeley
Milivoj Belic   City College of City Univ. of New York
Michael I. Bell   National Bureau of Standards
Bernard Bendow   RADC/EISS., L.G. Hanscom Field
George Benedek   Massachusetts Institute of Technology
A. Nihat Berker   Harvard University

Joseph L. Birman	City College of City Univ. of New York
N. Bloembergen	Harvard University
Ralph Bray	Purdue University
Richard G. Brewer	IBM Research Laboratory, San Jose
Elias Burstein	University of Pennsylvania
R. L. Byer	Stanford University
Robert Callender	City College of City Univ. of New York
Manuel Cardona	Max Planck Institut, Stuttgart
Ren-Fang Chang	National Bureau of Standards
Richard Chang	Yale University
Noel A. Clark	University of Colorado
Elisha Cohen	Bell Laboratories, Murray Hill
Herman Z. Cummins	City College of City Univ. of New York
Mireille Delaye	Massachusetts Institute of Technology
Ralf Dornhaus	Yale University
Richard A. Ferrell	University of Maryland
Daniel S. Fisher	Bell Laboratories, Murray Hill
D. B. Fitchen	Cornell University
Paul Fleury	Bell Laboratories, Murray Hill
Edward J. Flynn	Bell Laboratories, Murray Hill
J. M. Friedman	Bell laboratories, Murray Hill
S. Geschwind	Bell Laboratories, Murray Hill
Orest J. Glembocki	Brooklyn College, City Univ. of NY
J. Woods Halley	University of Minnesota
B. I. Halperin	Harvard University
John C. Hensel	Bell Laboratories, Murray Hill
Jonathan P. Heritage	Bell Laboratories, Holmdel
R. M. Hochstrasser	University of Pennsylvania
Pierre C. Hohenberg	Bell Laboratories, Murray Hill
David L. Johnson	Northeastern University
John Kirtley	IBM Research Center, Yorktown Heights
Miles V. Klein	University of Illinois
Melvin Lax	City College of City Univ. of New York
T. K. Lee	City College of City Univ. of New York
J. D. Litster	Massachusetts Institute of Technology
Kenneth B. Lyons	Bell Laboratories, Murray Hill
Paul Martin	Harvard University
Samuel L. McCall	Bell Laboratories, Murray Hill
Horia Metiu	University of California, Santa Barbara
D. L. Mills	University of California, Irvine
Dean L. Mitchell	National Science Foundation
Benjamin R. Mollow	University of Massachusetts, Dorchester
C. A. Murray	Bell Laboratories, Murray Hill
William J. O'Sullivan	University of Colorado
John B. Page	Arizona State University
Deva N. Pattanayak	City College of City Univ. of New York
Peter S. Pershan	Harvard University
Stanley J. Pickart	National Science Foundation
A. Pinczuk	Bell Laboratories, Holmdel
Fred H. Pollak	Brooklyn College, City Univ. of NY

R. Romestain   Bell Laboratories, Murray Hill
Charles Schnabolk   Brooklyn College, City Univ. of NY
James F. Scott   University of Colorado
Robert Silberstein   Brooklyn College, City Univ. of NY
J. J. Song   University of Southern California
Harry L. Swinney   University of Texas, Austin
Toyoichi Tanaka   Massaschusetts Institute of Technology
H. R. Trebin   City College of City Univ. of New York
James C. Tsang   IBM Research Center, Yorktown Heights
Narkis Tzoar   City College of City Univ. of New York
Chandra Varma   Bell Laboratories, Murray Hill
Claude Weisbuch   Bell Laboratories, Murray Hill
Andrei N. Weiszmann   College of Staten Island, CUNY
Peter Wolff   Massaschusetts Institute of Technology
D. J. Wolford   IBM. Research Center, Yorktown Heights
John M. Worlock   Bell Laboratories, Holmdel
William Yao   City College of City Univ. of New York
Peter Yu   IBM Research Center, Yorktown Heights
Dirk Zwemer   Bell Laboratories, Murray Hill

SUBJECT INDEX